GENERAL PHYSICS

AN ELEMENTARY TEXT-BOOK FOR COLLEGES

HENRY CREW, Ph. D.

New York

1908

ILLUSTRATED &

PUBLISHED BY

E-KİTAP PROJESİ & CHEAPEST BOOKS

www.facebook.com/EKitapProjesi

www.cheapestboooks.com

Copyright, 2016

By

e-Kitap Projesi

Istanbul

ISBN:

978-625-6004-58-0

In these days of numerous text-books the author who dares still further to increase the number owes to the public, at least, an explanation in which he shall set forth his purpose, however far short of accomplishment he may have fallen.

Accordingly, in this volume, the author's hopes have been: —

(i) To prepare a text-book adapted to the needs of first-year students in physics at Northwestern University. One fear in this connection is that the transfer of a course of lectures from flexible manuscript to rigid type metal may correspond to that stage of development which, in the life history of an animal, is known as *rigor mortis*. This difficulty, it is hoped, may be avoided by varying, from year to year, the illustrative phenomena employed in the lecture demonstrations.

(ii) To keep the treatment elementary, and yet include all the fundamental principles of physics; and at the same time to bind them together with "connective tissue" in such a way as to make clear to the student the essential unity of the subject.

(iii) Not merely, or even mainly, to impart information, but to set before the student a large and compact body of truth obtained by a method which shall remain for him, throughout life, a pattern and norm of clear and correct thinking.

To succeed completely in an undertaking so ambitious as that suggested by the threefold purpose just stated is more than the author dares to hope.

To my friend, Mr. John Mackenzie of Minneapolis, I am indebted for those paragraphs in Chapter III which describe Professor Osborne Reynolds' remarkable theory of gravitation.

The imperfections of the present text would have been many more except for the clever revision of the proof by my colleague, Professor R. R. Tatnall, and by my friend, Professor A. A. Knowlton of the Armour Institute. Special thanks are due Mr. J. H. Spencer of the United States Weather Bureau, to the *Youth's Companion*, and to the De Laval Steam Turbine Company, for the use of cuts. Mr. John A. Fleming of the Carnegie Institution was good enough to revise those paragraphs dealing with the secular variation of the earth's magnetic field. To my sister, Caroline L. Crew, I am indebted for the excision of many awkward English constructions.

H. C.

EVANSTON, ILLINOIS,
April 8, 1908.

TABLE OF CONTENTS

CHAPTER IV

Some Special Properties of Matter

CHAPTER V

Waves

CHAPTER VI

Sound

CHAPTER VII

Theory of Heat

CHAPTER VIII

Magnetism

CHAPTER IX

Electrostatics

CHAPTER X

Electric Currents

CHAPTER XI

Light

CHAPTER XII

Optical Instruments

GENERAL PHYSICS

INTRODUCTORY

PHYSICS, in the present sense of the word, is not an ancient science, but a creation of the last three hundred years. Compared with astronomy and medicine, it is a mere upstart. When, however, one considers the work of the Egyptians in handling tremendous blocks of stone, such as those used in the construction of the pyramids, or the work of the Greeks in devising powerful catapults, or the work of the Romans in distributing water through well-constructed aqueducts, or any one of their numerous clever devices, such as the steelyards, locks, and surgical instruments which have been recovered at Pompeii, it becomes evident that the ancients were familiar with a wide range of physical science. That there were among them keen observers is patent to every one who has considered the Hermes of Praxiteles or the columns of the Parthenon. The skill of the ancients in hunting and the accuracy with which they observed the motions of the planets are proverbial.

Nor were they wanting in the desire for a unitary view of the world about them. This search for unity is evinced in the idea of Heraclitus that igneous vapor (fire) was the one great principle, in the idea of Thales that aqueous vapor was the key to the situation, in the idea of Anaximenes that air was the essential principle, and in the notion of Democritus that the origin of all things was to be sought in the atom. The failure of these various views gave rise to the general opinion that four "elements"— earth, air, fire, and water — constituted an irreducible minimum for the physical universe. The Ptolemaic system of crystalline spheres, cycles, and epicycles is another product of their eagerness for unity. Well, then, in what respect does modern physical science differ from the ancient? Why is the beginning of this science supposed to lie only some three hundred years back of the present?

The answer to this question, like that to most others which begin with "why," is not easy; and all that can be done here is to point out briefly *how* ancient differs from modern science rather than to explain *why* this difference exists. During the fifteenth century there seemed to come over the minds of thinking men a change, marked especially by an increased freedom from the trammels of authority. Such freedom had already, even as early as the fourteenth century, made itself evident in painting, architecture, and music. A new literature had found expression in Dante, Petrarch, and Boccaccio, whose works were now being disseminated by the newly invented art of printing. An increased boldness was shown by navigators, which culminated in the discovery of America. This same century was marked by the birth of Copernicus, with Paracelsus and Leonardo da Vinci for contemporaries. Without shedding any new light upon the "cause" of this changed attitude of mind, it may perhaps be best described as "the spirit of the Renaissance." Previous to this time, those great generalizations which we now call "physical laws" seem to have been rather fortuitous and unfruitful; after this period, the search for law, for the constant elements in the behavior of nature, seems to have been more a matter of deliberate and definite plan, while the results of the search have been tremendously prolific of other important results.

It was upon such a background as this, and into such a changed world as this, that on the 15th of February, 1564, at Pisa, in Italy, the founder of modern physics was born. But, while all science is the creation of man, it is equally true that no science is the creation of any single man. In what sense, therefore, can Galileo be spoken of as the founder of modern physics? Just this: he it was who first discovered the fundamental principles of modern dynamics, the one branch of learning upon which the entire superstructure of the physical sciences has been erected. His formulation of these principles was, as we shall later see, imperfect; but his suggestion that bodies change their motion when, and only when, acted upon by some external force — or, if you please, the substitution of the conception "force" in place of "cause" of motion — has proven to be the most fertile principle, as well as the first fundamental and perfectly general principle, of modern dynamics, a principle which finds its later and more perfect expression in Newton's Three Laws of Motion. Among the ancients, Archimedes (287–212 B.C.) appears to be the one man who approached the modern method in physics. He established a firm foundation for the science of statics — the conditions of equilibrium for bodies at rest; but this is a very special case of dynamics, and the achievements of Archimedes, which early attracted the attention of Galileo, — remarkable and valuable as they impress us to-day, — bore little fruit until the idea of force as the condition for change of motion was introduced by the great Italian. From Galileo's time, on, motions cease to be classified as "natural" and "violent." *All* motions are henceforth "natural," and are proper subjects for special investigation. The faith and skill in experiment exhibited by Galileo set a new pace, one which had possibly never before been reached, and one which has certainly seldom since been surpassed.

The experimental method once established, new facts and new generalizations accumulated with marvelous rapidity; and the subject of natural philosophy, enriched by the labors of Hooke, Boyle, and Newton in England, Descartes and Mersenne in France, Huygens in Holland, Torricelli and other pupils of Galileo in Italy, became so bulky that certain natural cleavage lines began to appear.

Certain phenomena were soon recognized as peculiar to living bodies — bodies which possess the ability to assimilate food, the power of reproduction, sensation, etc., and these came to be studied under the head of natural history and medicine. This entire group, including, as it now does, botany, zoölogy, physiology, anatomy, paleontology (as distinguished from geology), bacteriology, etc., has, by wide consent, been given the name of the **natural sciences**. All matter, other than that of living bodies, is said to be inanimate; and it is with this class of bodies especially that the **physical sciences** have to do.

In the early days, nearly all the phenomena of the physical sciences were studied under the head of **Natural Philosophy**, a term which was employed in contradistinction to **Natural History**, and one which included such diverse subjects as the motions of the solar system, engineering structures, and meteorology. But Physics, or Natural Philosophy, in its narrow and proper sense, does not deal with *all* the phenomena of dead matter. There are, for instance, certain bodies which are so distant from us that they can be studied only by specially designed instruments such as the telescope. Accordingly, the study of the fixed stars, the planets, comets, etc., is pursued under a special department of physical science called **Astronomy.** There is another exceedingly large and important branch of physical science which deals especially with those changes which occur in homogeneous bodies — mainly changes of composition. The systematic study of these changes is carried on under the head of **Chemistry.**

Those properties of inanimate bodies which are not primarily connected with a change in their chemical composition are then the especial subjects of study pursued under the head of **Physics.**

Besides these there are other important, but more specialized, branches of physical science which deal with particular groups of bodies. Thus the comprehensive science which takes up the study of the earth's crust from a physical and chemical standpoint is called **Geology**; while **Meteorology,** a science of great significance and of increasing practical value, deals exclusively with the physical phenomena of the earth's atmosphere.

Built upon the physical and natural sciences, is still another group of studies distinguished from the preceding "pure" sciences mainly by the purpose and attitude of mind on the part of the student. Such are the various branches of engineering, which are largely branches of applied physics; such also is the study of modern medicine, a science which has become immediately dependent upon Biology, Chemistry, Physics, Physiology, and Anatomy.

In attempting to roughly locate Physics among the various sciences, we have so far considered only those groups of phenomena which are associated with matter, and have left to one side the phenomena of mind, also the historical sciences which deal with the purposes and the controlling wills of individuals and nations. To one side also have been left those sciences which deal with the consensus of sane opinion and with the expression of the will of the entire race, sciences which Münsterberg calls "Normative" and which include Logic and Mathematics.

The place of Physics in the family of the sciences is then a very modest one, for it deals only with a particular group of phenomena, associated with inanimate matter, and limited essentially to those changes in which the composition of the various substances involved remains unaltered. Within this limited domain the purpose of the physicist is to discover and to describe the universal and constant modes in the behavior of nature. Very often these results can be given a simple and accurate mathematical formulation. With what splendid success this work has been pushed forward will appear when we come to study in more detail the work of the men who have followed the great Galileo, men such as Newton, Faraday, and Maxwell in England; Fresnel and Ampère in France; Gauss, Weber, Kirchhoff, Helmholtz, and Hertz in Germany; Volta and Melloni in Italy; Franklin, Henry, and Rowland in America.

CHAPTER I

KINEMATICS

Position

1. Kirchhoff (1824–1886), a distinguished German physicist, first clearly pointed out that the aim of the physicist is to describe the motions of bodies completely and in the simplest possible manner. But since a body is a limited portion of matter, which we may consider as made up of a number of small particles, it will be simpler to consider first the motion of a *particle* and then pass to the consideration of *bodies* in motion. However, since motion is merely change of position, it will be convenient first to learn how the position of a particle is defined. The word "particle" is used always to denote a body so small that its dimensions may be neglected in comparison with other distances involved. Thus a baseball may be considered a particle when one is thinking only of its path through the air; but in the pitcher's hands it becomes a body of very appreciable size, a body which he sets into rotation about a definite axis in order to give it any desired "curve."

Position of a Particle in a Plane

2. When geographers describe the position of a town, they assume that the reader already knows the position of some point on the earth, which may be called a point of reference. Such a point of reference for the surface of the earth is the intersection of the meridian through Greenwich with the equator. This fixed, the location of a town is definitely described when two things are known about it; namely, its *latitude* and its *longitude.* Thus the island of St. Helena is in 15° 55′ south latitude and 5° 43′ west longitude.

A similar method is frequently used in physics; *e.g.*, the position of a point P (Fig. 1), with reference to another point O, *in the same plane*, is described when two mutually rectangular straight lines are drawn through the point O, and the perpendicular distances of the point P from these two lines respectively are given. **The point of reference O is generally called the origin. The horizontal line of reference** $\overline{OX}$ (Fig. 1) **is generally called the axis of** X; **and the** vertical line **of** reference $\overline{OY}$ **the axis of** Y. **The longitude** x **of** the **point** P **is called its abscissa**; while **the** latitude y **is known as its ordinate.** Abscissas are positive when drawn to the right from the axis of Y; negative when drawn to the left. Ordinates are positive when drawn upward from the axis of X; negative when drawn downward. These quantities x and y are known as the **rectangular coördinates** of the point P; they are a special kind of Cartesian coördinates, so called after Descartes, who first employed an algebra of this kind.

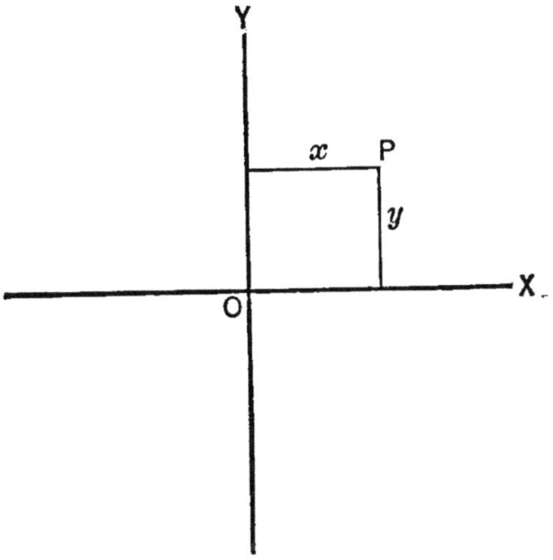

FIG. 1. — Showing one method of locating a point in a plane.

In the laboratory we have frequent occasions to draw curves, and this is generally done by locating points on coördinate paper in the manner just described.

Navigators and surveyors generally employ a different method of locating one point with reference to another in the same plane. **They give the straight-line distance from the first point to the second, then they give the direction of the** line **joining the** points. **This** method **also is frequently used in physics and is known as the method of** polar **coördinates.** The idea will be clear from Fig. 2, where O is the point of reference and $\overline{OX}$ any known direction. There is no doubt about the exact position of a particle at P as soon as we know the distance r between the points O and P, and the angle θ between the lines OP and OX. **The angular distance θ we**

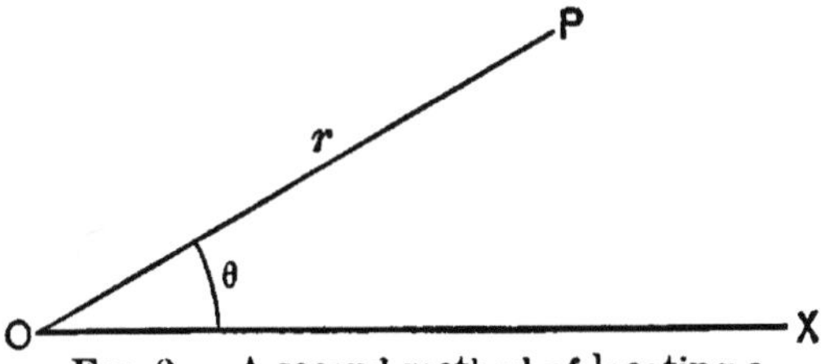

FIG. 2. — A second method of locating a point in a plane.

shall always consider positive when measured in a direction contrary to that in which the hands of a clock move, a direction which is called counter-clockwise. In like manner, θ is considered negative when measured in a clockwise sense. The line $\overline{OX}$ we may call the line of reference.

Position of a Particle in Space

3. If a point on the earth's surface is to be described still more completely, we may choose the sea level as a plane of reference, and give the altitude of the point. So in mathematics and physics one frequently describes a point by locating it first in a certain plane as described in the preceding sections, and then stating the position of this plane with respect to some other plane of reference. It will be observed that, in general, *three specifications are needed to completely describe the position of one point with reference to another;* and when the numerical values of these three coördinates are once known, the position of the point is given without ambiguity. Thus if the rectangular coördinates of a particle in space have the values, —

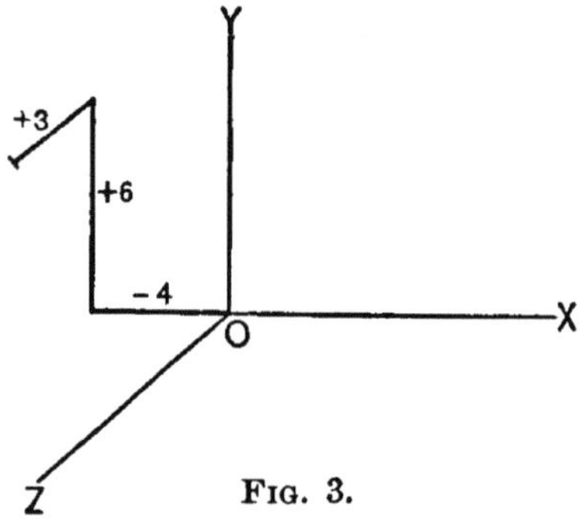

Fig. 3.

$$\begin{cases} x = -4, \\ y = +6, \\ z = +3, \end{cases}$$

one can proceed to locate it as indicated in Fig. 3.

Position Vectors

4. But a third and simpler way to think of the position of a point is to consider it as completely specified by a single straight line running from the point of reference to the point under consideration. If in Fig. 4 we know, not only the length, but also the direction and sense of the straight line $\overline{OP}$, we know all that can possibly be known about the position of the particle at P.

To any quantity such as $\overline{OP}$, possessing direction, sense, and amount, the name "vector" has been given. This class of quantities includes many of the most important studied in physics.

A vector quantity is simply one which *resembles a limited straight line* in three particulars; namely, *in the possession of direction, sense, and numerical magnitude.* A vector which locates one point with reference to another, as in Fig. 4, is called a position vector, and is a type of vector quantities in general, of which we shall meet many examples in later pages. The usefulness of the vector, in representing the position of a point, lies in its clearness and simplicity; for purposes of numerical calculation it has small value.

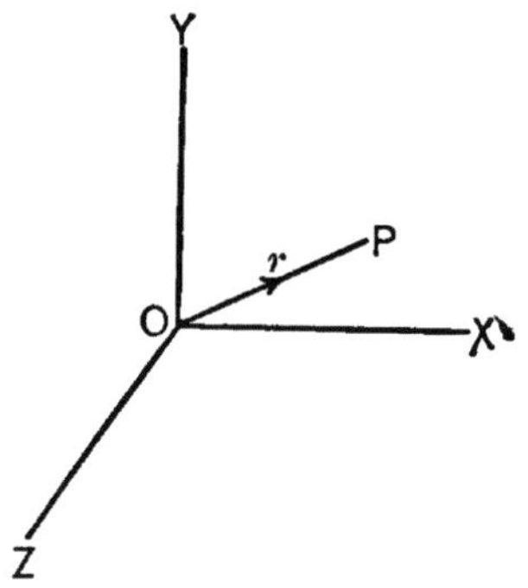

FIG. 4.

DIGRESSION ON SCALAR QUANTITIES

5. The mere length of a straight line is the same in whatever direction the line may lie. The *length* of a straight line is, therefore, *not* a vector quantity. It is independent of direction. **Quantities which do not have direction are called scalars.**

Philadelphia is a city 90 miles from New York. This distance is a scalar quantity. The information contained in this statement tells me only that Philadelphia lies somewhere on a circle of 90 miles radius described with New York as a center. But "90 miles southwest of New York" gives me a vector quantity, which locates the position of Philadelphia without ambiguity. A pound of sugar, three yards of calico, two gallons of water, a volume of anything, any amount of matter, are illustrations of scalars.

SUMMARY

6. Summarizing, then, the position of a particle is nearly always described by one or the other of the three following methods:—

(i) *Rectangular coördinates*, in which the position of a point is defined by giving its distance from each of three mutually perpendicular planes. These distances are called the rectangular coördinates of the particle and are usually denoted by x, y, z. The three mutually perpendicular reference planes are indicated in Fig. 3 by YOZ, ZOX, and XOY.

(ii) *Polar coördinates*, in which the position of a particle is described by its distance r from the origin, and by its latitude and longitude on a sphere whose center is at the origin and whose radius is the distance r just described. These three quantities are called the polar coördinates of the particle and are usually denoted by the letters r, θ, and ϕ respectively.

(iii) *Vectors*, a graphical method in which the position of a particle is located as the terminal point of a single straight line drawn in a definite sense and in a definite direction from the origin.

Position of a Body

7. Having thus far considered the position of a *particle* only, we now take up the position of a *body* which has already been defined as a limited portion of matter. For the present let us deal only with perfectly rigid solids, that is, with bodies whose size and shape are each constant. If we consider a body merely as a system of rigidly connected particles, it is evident that its position is completely defined only when the position of *each* particle in the system is known. But since the body is rigid, this is equivalent to saying that its position is completely determined when we know the positions of any three points in the body, provided of course that these points do not lie in the same straight line. In other words, if a straight line and one point outside this line be located in a body, the position of the entire body is completely defined. It is clear, therefore, that a body may have many different positions in which two points in the body are fixed; for under this condition it may be rotated through any angle about the straight line joining these two points. As we shall see later, it is often quite as necessary to consider the angular as well as the linear position of a body; while for a single particle angular position is a term which has no meaning.

Measurement of Lengths

8. The actual determination of the position of any point in the world about us leads at once to the measurement of a distance, by which is meant simply the comparison of any assigned distance with a given distance which we agree to take as unity.

As later experience will show, comparison is the essential feature of all measurement. Ten yards of cloth is simply a piece of cloth which is ten times as long as a certain piece of wood called a yard stick.

Units of Length. Centimeter. Foot

9. **Almost the entire scientific world has agreed to take as the standard length the distance between two marks on a certain platinum-iridium bar preserved in the International Metric Bureau at Sèvres, near Paris. This distance is called a meter. The one hundredth part of the distance between these two marks is taken as the unit and is called a centimeter.** For many laboratory measurements this smaller unit will be found more convenient than the meter.

It will be observed that this definition is purely arbitrary, a mere matter of convention. The fact that the meter is so chosen as to be nearly equal to one ten-millionth of the distance from the earth's equator to its north pole is interesting; but it does not enter into the definition of the unit of length. As a matter of fact, the length of the earth's quadrant has been found to be more nearly 10,000,880 meters. The student who is not familiar with the metric system will find it interesting to estimate his own height in centimeters, also to guess at the length and width of this page in centimeters; and then test the accuracy of his estimates by actual measurement with a meter stick.

English and American engineers generally employ the yard as the standard of length and the foot as the unit of length.

In Great Britain the yard is, by definition, the distance between two rulings on a certain bronze bar preserved in the Standards Office at London. In the United States the yard is defined as 36/39.37 of the standard meter at Sèvres.

Measurement of Areas. Unit of Area

10. Having settled upon the centimeter as a unit of length, the simplest possible unit of area is the area of one square centimeter. In consideration of this fact, the unit of area almost universally employed in science is the **square centimeter.** Very seldom, indeed, are areas measured by direct com-

parison with a unit of area. More frequently it will be found convenient to measure the *linear* dimensions of the surface in question, and, from these, to *compute* by geometry the area of the surface. Engineers, of course, employ the square foot as the unit of area.

Measurement of Volumes. Unit of Volume

11. Standards of volume, such as the bushel, the quart, the cubic foot, are already familiar. **In scientific work the cubic centimeter is adopted as the unit of** volume, and for the same reasons that led to the adoption of the square centimeter as the unit of area. An acquaintance with the metric system makes the wisdom of this choice at once apparent.

In the case of volumes whose linear dimensions can be measured, *e.g.* a parallelopipedon or a sphere, it is frequently more convenient to compute the volume than to measure it directly.

Angular Position. Unit Angle

12. When two straight lines do not coincide in direction, they are said to have different angular positions. The difference of their directions is the angular displacement of one line with respect to the other. This angular displacement is expressed by surveyors and astronomers in degrees. But navigators use "points" of the compass, or $11\frac{1}{4}°$, as units. In physics and mathematics the unit of angle mostly employed in the description and discussion of phenomena is different from either of these. It is defined as follows: If two straight lines lying in the same plane differ in direction, they will intersect. About the point of intersection as center, and in the plane of these two lines, draw a circle. **These lines are said to include a unit angle when they intercept an arc of this circle equal in length to its radius. See Fig. 5. This unit angle is called the radian.**

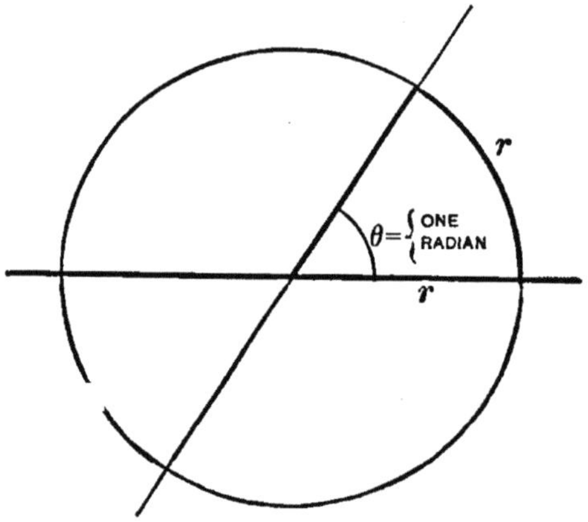

Fig. 5. — Unit angle.

It is evident, since the circumference of a circle is a length equal to 2π times the radius, that in one right angle there will be one quarter of 2π radians, *i.e.* $\frac{\pi}{2}$ radians. Compute, and on the margin of this page record the number of degrees in one radian.

Problems

1. Explain just how it is that any limited straight line is a vector quantity, while the length of such a line is a scalar quantity.

2. What light does the etymology of the word "vector" shed upon its present meaning?

3. How many radians in the angle subtended by a semicircle?

4. The volume of a cube is 921 c.c. Find the area of one face of the cube.

5. Draw a pair of axes at right angles to each other. With reference to these axes, locate the points A, B, and C which have abscissas and ordinates as follows: —

Point	Abscissa	Ordinate
A	$+2$	$+7$
B	$+5$	-10
C	8	-4

6. How many cubic centimeters in the volume of a hemisphere of 8 cm. radius?

7. Draw three mutually rectangular axes of coordinates and locate the two points whose coördinates are $(-2, +3, +4)$ and $(+1, +5, +2)$.

8. Find the length of the vector joining the two points whose positions are given in the preceding problem.

9. Which of the two following points is farther from the origin, $(+5, -9)$ or $(+7, +7)$?

10. The two mutually perpendicular sides of a right-angled triangle are 9 and 20 centimeters, respectively, in length. Find the area of the triangle in square centimeters, and the length of the hypothenuse in centimeters.

11. Compute the number of seconds of arc in one radian.

12. Locate the following position vectors $\begin{Bmatrix} r = 30, \theta = 15° \\ r = 30, \theta = 345° \\ r = 30, \theta = 195° \end{Bmatrix}$.

13. Considered as a unit of volume, what advantage has the cubic centimeter over the cubic inch?

14. Indicate in three dimensions the position of a point whose polar coördinates are $r = 25$, $\theta = 45°$, $\phi = 90°$. See § 6.

15. Locate the two following points by use of polar coördinates and determine which of them is nearer to the line of reference, $(r = 12,\ \theta = 15°)$ and $(r = 6,\ \theta = 30°)$.

CHANGE OF POSITION

I. *Case of a Particle*

13. Since the position of a particle can always be defined by a straight line of a definite length drawn in a definite direction, it is evident that, if the particle moves to a new position, the new line which describes this new position will, in general, differ from the old one in its two characteristic features, viz. in length and in angulăr position. But this line may change in length only; in which case the particle moves along a straight line. It may change in angular position only; in which case we have motion on the surface of a sphere.

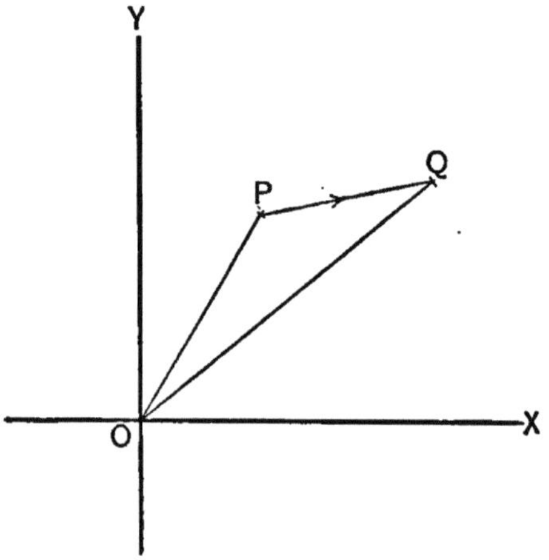

FIG. 6.—Change of position, a quantity having direction.

Suppose now that a particle at P (Fig. 6) has moved to a point Q such that the line $\overline{OP}$ has changed in length and also in angular position. If we take the old position of the particle P as a new point of reference, it is then clear that the *change* of position will be described by the straight line, $\overline{PQ}$, joining the new and old positions.

Evidently, then, we may describe the new position Q of the particle either by giving the two straight lines $\overline{OP}$ and $\overline{PQ}$, or by simply giving the one straight line $\overline{OQ}$. For, by either process, we definitely locate the new position Q.

The upshot of the whole matter is that change of position is described in exactly the same way as is position itself; namely, by stating the length, sense, and direction of a straight line.

Digression on the Addition and Subtraction of Vectors

14. Referring to Fig. 7, let us denote the vector $\overline{OP}$ by p, the vector $\overline{PQ}$ by $\bar{q}$, and the vector $\overline{OQ}$ by r. Then it will be clear that the new position Q of the particle may be located in either one of two different but strictly equivalent methods. By one method, the particle at Q is located as the terminal point of the vector r. By the other method it is the terminal point of two vectors p and $\bar{q}$, placed end to end with their arrows in the same sense. **And since $\bar{r}$ locates the point Q at the same distance and in the same direction from the origin, $\bar{r}$ is said to be the resultant or vector sum of $\bar{p}$ and $\bar{q}$; while p and $\bar{q}$ are said to be the components of r.** In the algebra of vectors this fact is expressed by writing

$$p+q=r, \qquad \text{Eq. 1}$$

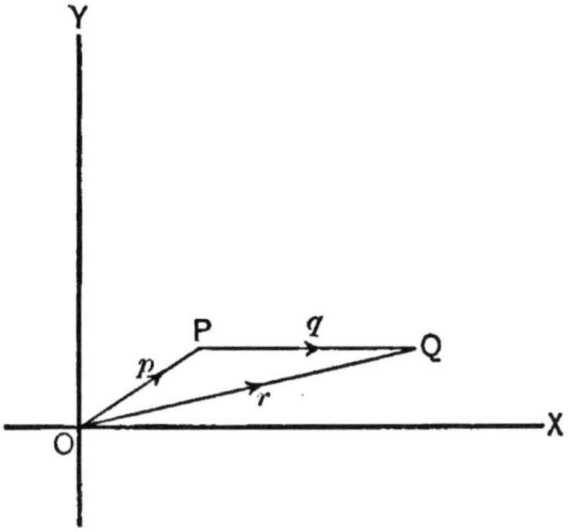

FIG. 7.—Vector addition.

an equation which contains the whole story of vector addition. In ordinary algebra, which deals only with scalar quantities, we should have

$$p+q>r,$$

but this inequation refers merely to lengths and not at all to directions.

The resultant of any number of vectors is obtained by precisely the same process, namely, placing these vectors end to end, in any order, *but all in the same sense*, and then joining the initial point of the first vector to the terminal point of the last vector.

This graphical method gives rise to the vector polygon, as illustrated in Fig. 8, where r is the resultant of the vectors $\bar{p}$, q, and $\bar{s}$. From what precedes it will be seen that vector addition means nothing more than what is implied in the statement that, so far as mere transportation is concerned, it is all the same whether one travels by train directly from Chicago to Pittsburg, or goes first from Chicago to Toledo, thence to Cleveland, and thence to Pittsburg.

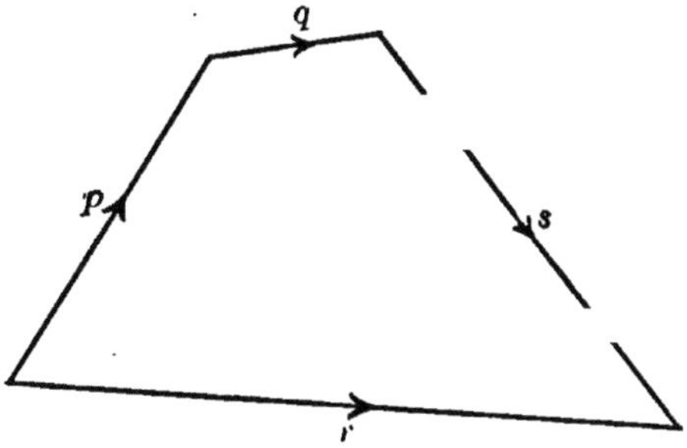

FIG. 8.—Polygon of vectors.

Since the sign of a vector determines its sense, one sees that the subtraction of any given vector from another is precisely the same as adding this vector to the other, having first reversed its sense.

When a particle receives a displacement which is represented by any given vector, say $\overline{OP}$, Fig. 9, the question often arises as to what the amount of the displacement is in some other direction, say $\overline{OH}$.

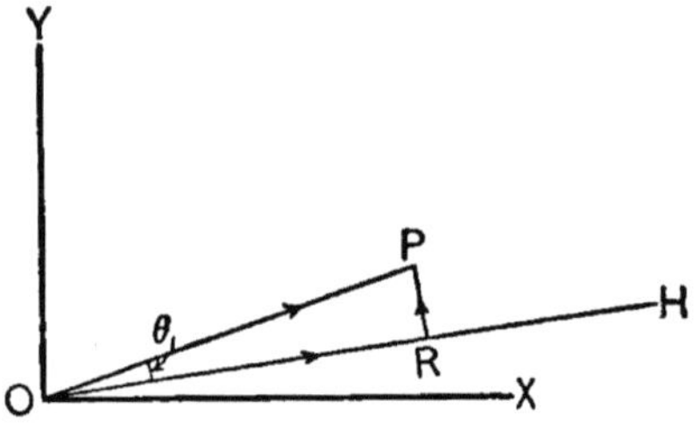

FIG. 9.—Illustrating the resolution of a vector.

This is found by letting fall from the terminal point P a perpendicular upon the line $\overline{OH}$. If R denote the foot of this perpendicular, we may consider the vector $\overline{OP}$ as broken up into two components $\overline{OR}$ and $\overline{RP}$, of which the former lies entirely along $\overline{OH}$; the latter has no component along $\overline{OH}$. From the definition of a cosine it is evident that

$$OR = OP \cos \theta. \qquad \text{Eq. 2}$$

II. *Case of a Rigid Body*

15. *As* has been observed (§ 7), the position of a rigid body is fixed when any three points in the body, not lying in the same straight line, are fixed. Accordingly, we see that the positions of bodies change in several *apparently* different ways.

(1) If only two points are fixed in the body, as, for instance, a wheel and axle held between two pivot points, AA' (Fig. 10), then the body is free to rotate about the line joining these two points. **The motion of a rigid body about a fixed axis is called a pure rotation.** A bicycle wheel lifted from the ground and spun is an instance.

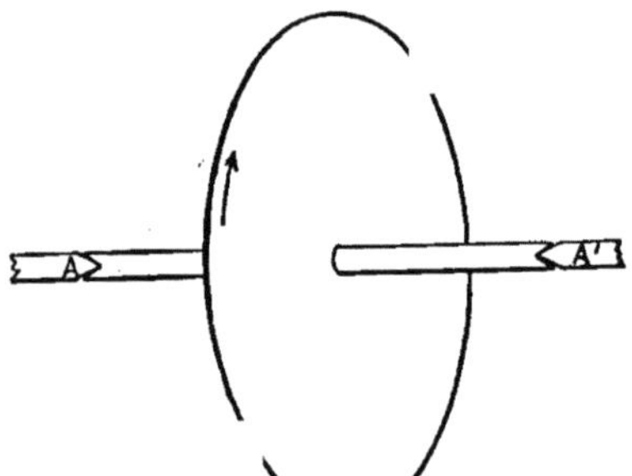

FIG. 10.—A case of pure rotation.

(2) It may happen that only one point is fixed in the body, as in the case of a walking-stick with its lower end on the ground. Here the body is free to rotate about *many* different lines as axes. It may, indeed, rotate about any straight line passing through the lower end of the stick.

(3) It may be that no point in the body is fixed, but that the body is compelled to move in such a way that *each point of the body describes a path equal and similar to the path of every other point.* Such a motion is called a **pure translation.** It

occurs in the piston rod of a steam engine. A sled sliding to and fro in its own tracks also represents the motion.

(4) Lastly, a body may be rotated *and* translated simultaneously. Such a case is represented in the wheel of a bicycle while being ridden.

The student who pursues this subject farther will find that a rigid body may be changed from any one position to any other position by simply translating it along a certain straight line and rotating it through a definite angle about a definite axis. Consequently the first three cases just considered are special cases of the fourth. The fact then is that every motion of any rigid body turns out to be either a pure translation or a pure rotation or a combination of these two; in other words, a rigid body may suffer a linear displacement or an angular displacement, or a combination of the two.

As we have already seen, linear displacements are represented by vectors; so also are angular displacements. For imagine the rough outline in Fig. 11 to represent any rigid body having one point O fixed. Take this point as origin. Imagine the body to be turned through any angle θ about an axis $\overline{OP}$. Such an angular displacement is what we have just called a pure rotation. It requires for its description three specifications, namely, (i) the direction of the axis of rotation, (ii) the amount of rotation, *i.e.* the angle through which the body has turned, and (iii) the sense of rotation. In short, this angular displacement requires a vector to represent it.*

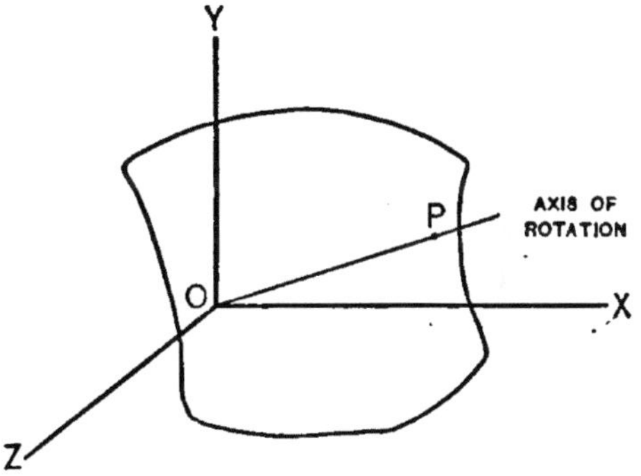

FIG. 11. — Angular displacement, a vector.

Thus, if a grindstone be rotated through 90°, its angular displacement would be represented by a vector drawn in the direction of the axis of the grindstone, and of a length $\frac{\pi}{2}$, the sense being such that if one looks in the direction of the arrow

* The student's attention is here especially directed to the fact that in adding vectors which represent angular displacement, the resultant is not independent of the order in which the vectors are added, except when the angular displacements are infinitesimal.

on the vector, the stone will appear to rotate in a clockwise direction. **Such a relation between the direction of the vector and the direction of the rotation is** exactly that **which exists between the direction of advance and the direction of rotation in an ordinary woodscrew.** It **is hence called the right-handed screw relation.** It is important to observe that the rule for the addition of vectors applies to angular displacements only when these are very small.

In the later parts of physics, we shall meet many quantities occurring in pairs similar to these two analogous displacements, linear and angular, which we have just been considering.

Degrees of Freedom

I. *Case of a Particle*

16. It is interesting to observe that when a particle is free to change its position by moving in any direction whatever, it can move in three, and only three, directions such that its motion along any one of these directions has no component along either of the other two. And it is evident that these three directions are mutually perpendicular; for if they were inclined at any angle other than 90°, say θ, a displacement r along one axis would have a component $r \cos \theta$ along the other. These three directions are represented, therefore, by any three rectangular axes. In view of these facts, a particle is said to have three **degrees of freedom.** In other words, a particle has three possibilities in the way of translational motion. It cannot rotate because, by definition, it has no appreciable dimensions.

II. *Case of a Rigid Body*

17. The case is quite different with an extended body. Consider a coin, a metal cylinder, or any rigid body whatever; if it is perfectly free, it can be translated parallel to any of the three rectangular axes. Like the particle, it has three degrees of translational freedom. Now imagine this body supported on a single pivot, in the same manner as a spinning top. Here one point is fixed so that the body has no freedom of translation; yet it is clearly possible for the body to rotate about any one of three mutually perpendicular axes passing through this fixed point. A rotation about any one of these axes will have no component of rotation about either of the other axes. A rigid body not constrained in any way is said, therefore, to have three degrees of freedom of rotation, making in all six degrees of freedom for any unhampered body, namely, three of translation and three of rotation.

The student will find it both interesting and important to discover, in any particular case of constrained motion, just how many degrees of freedom a body has ; for the advanced student will find later that the whole behavior of a body may be described by use of as few coördinates as the body has degrees of freedom.

Interesting Special Case. Uniplanar Motion

18. When the motion is in a single plane, the rigid body can be transferred from any one position to any other position (in that plane, of course) by a pure rotation alone.

To prove this theorem, consider any two positions of the body as represented by AB and $A'B'$ in Fig. 12. Bisect AA' and also BB'. Draw perpendiculars through these points of bisection. Denote by O the point where these perpendiculars intersect. Then the axis about which the body AB must be rotated to bring it into the position $A'B'$ is perpendicular to the plane, and passes through the point O. For evidently the triangle AOB may be rotated about this axis until it is exactly superposed upon the congruent triangle $A'OB'$.

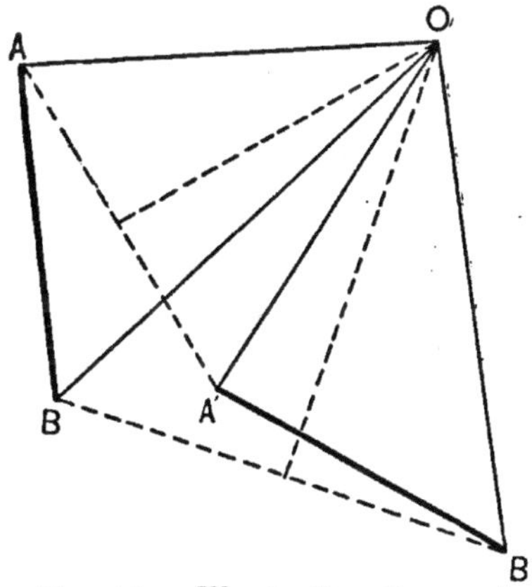

FIG. 12. — Illustrating the most general uniplanar motion.

Problems

1. How many degrees of freedom has a particle which is compelled to move in a plane?

2. A marble free to roll on a floor, but not free to leave the floor, has how many degrees of freedom ?

3. Design a rigid body having one degree of freedom of translation and one of rotation.

4. How many degrees of freedom has one of the balls on the governor of a steam engine?

5. Most motions that are employed in actual machinery are constrained, *i.e.* they have less than six degrees of freedom. What, in your estimation, is the special case which occurs most frequently in machines?

6. In the following motions of rigid bodies distinguish clearly which are cases of pure translation, which are cases of pure rotation, and which are cases involving both translation and rotation: —

i. The motion of a door on its hinges.
ii. The motion of a carriage wheel with reference to its axle.
iii. The motion of a carriage wheel with reference to the road on which it travels.
iv. The motion of a rocking-chair as one rocks in it.
v. The motion of the eyepiece as one focuses a pair of opera glasses.
vi. The motion of the blade, with reference to the handle, when one opens a pocketknife.

7. What vector will describe the final position of a man who walks 10 miles N.W., then turns and walks 10 miles S.W.?

8. Two boats sail from the same port; at the end of 5 hours one has traveled 50 mi. due E., the other $\sqrt{5000}$ mi. S.E. What vector will then represent their relative positions? *Ans.* 50 miles due S.

9. The coordinates of two points are $(+2, +3)$ and $(-1, +4)$. Find the vector which joins this pair of points.

10. A particle moves in a circle starting from a given point which we may call the origin. Find the vector which describes the position of the particle when $\frac{2}{3}$ of the circumference has been traversed.

11. Plot the following four points and show that they are the corners of a parallelogram: $(+3, +2)$; $(+4, +1)$; $(+2, +5)$; $(+3, +4)$.

12. What are the rectangular components of the vector which joins the following two points: $(+7, +3)$; $(+10, +5)$?

Rate of Change of Position. Velocity

19. It is to be carefully observed that *up to this point* we have been concerned not at all with the history of the body between the time of its leaving its old position and the time of its reaching its new position. We do not care how long a body may have been in changing its position, nor does it matter through what "cause" it may have changed its position. We have been concerned only with a method for describing, most simply and most completely, the change in position of the body. When, however, it is asked *how* this change occurred, we are led to the consideration of *Time* and *Velocity*.

In the world about us we find motion to be a state of matter which is quite as natural — perhaps even more natural — than rest. Displacement may be considered as the result of motion. But very often we shall be concerned rather with the rate of motion. Accordingly, we must now consider the method of describing a rate of motion, *i.e.* a velocity.

When a man buys a railroad ticket from New Orleans to Boston, he pays for "displacement," and, in general, the price of a railroad ticket varies directly as the amount of displacement. But one's choice of a train depends largely upon its speed, or rate of displacement. The difference in price between a good watch and a poor one depends principally upon the rate at which the angular position of the hands change. If this rate is very uniform, the watch is well made; if the hands do not move at a fairly constant rate, the watch is poorly made.

While etymologically the same, the words "rate" and "ratio" now have different meanings; the former is a special case of the latter. For while "ratio" denotes the quotient of *any* one quantity by another, "rate" is employed in physics to mean only the ratio of some quantity to a *time*. Rate, unless otherwise specified, is always an exact synonym of **time rate**; time always appears in the *denominator* of any rate. The rate of interest is the amount of money one pays for the use of a dollar, divided by the number of years he has used the dollar. The death rate is the total number of deaths among a thousand people during any period, divided by the number of years in that period.

But before we can measure time rates we must decide upon a standard and a unit of time. These are described in the following paragraph.

Time. Unit of Time

20. The mean (i.e. the average) time occupied by the sun in one apparent revolution about the earth is known as the mean solar **day.** This interval is universally employed as the standard of time. **The** $\frac{1}{86400}$ part **of a mean solar day is called a second and is used by the scientific world as the unit of time.**

Intervals of time can be compared with marvelous accuracy by means of watches, clocks, or chronographs.

Definition of Speed

21. More than a century ago it was proved by Lavoisier, the great French chemist, that all matter has a continuous existence; from which fact it follows that, as a particle moves from one point to another, it must occupy every intermediate position along the route. This route is then a continuous geometrical curve. Accordingly the **path** of a particle is defined as the continuous locus of all the successive positions which it occupies. We are now able to define clearly what is meant by **speed**, namely, **the rate at which a particle moves along its path.** Direction of motion, it will be observed, does not enter into this definition. A man who, during fifteen minutes, walks a mile along a winding cowpath travels with precisely the same average speed as another man who walks a mile in fifteen minutes along a straight city pavement.

Let us denote by p the distance which a particle has moved along any path during the time t. Then for purposes of computation, the average speed of the particle is expressed as follows : —

$$\text{Average Speed} = S = \frac{p}{t}. \qquad \text{Eq. 3}$$

But if the speed of the particle is varying from point to point, as is usually the case in nature, then denote by p_1 the distance traversed by the particle at the end of the time t_1; and by p_2 the distance traversed by the end of the time t_2. Now if t_1 and t_2 be so chosen that t_1 is very little earlier than t; and t_2 a very little later than t, the speed at any instant t will be the limit approached by the ratio $\frac{p_2 - p_1}{t_2 - t_1}$. Accordingly one writes

$$\text{Speed at instant } t = S_t = \left[\frac{p_2 - p_1}{t_2 - t_1}\right]_{t_2 \doteq t_1} \qquad \text{Eq. 4}$$

where the subscript $t_2 \doteq t_1$ indicates that t_2 approaches indefinitely near to t_1.

The necessity for some expression of this kind will appear when one asks such a question as, "What is the speed of a baseball thrown vertically into the air?" The question has no meaning unless one specifies some particular instant of time; for the ball has every speed from the initial speed with which it left the hand of the thrower to zero, which is its speed at the top of its path.

Speed is, therefore, really, a *limit*. The conception of a limit is something which each student should bring with him from a study of elementary geometry ; or, not having acquired it there, he should now master it once for all, since the idea is one which is essential even in elementary physics.

Unit of Speed

22. Having adopted the centimeter as the unit of length and the second as the unit of time, we have no trouble in settling upon the proper scientific unit of speed; namely, a speed of one centimeter per second. The engineer, however, generally employs the foot per second for his unit of speed.

Problems

1. Find the average speed, in miles per hour, of a steamer which makes the passage from New York to Liverpool, 3032 mi., in 6 da.

2. A 24-hour train from New York to Chicago, 926 mi., must make what average speed?

3. A shot travels a distance of 4 kilometers in 10.5 sec. Find its average speed in meters per second.

4. A speed of 1 ft. per second is equivalent to a speed of how many centimeters per second?

5. A gun is fired at a distance of 1 kilometer; the report is heard 3.25 sec. later. From this observation, compute the speed of sound in air.

6. Light travels from the sun to the earth in about 8 min. The distance of the sun from the earth is about 95 million mi. Find the approximate speed of light in miles per second. What is the value of this speed in meters per second?

7. One man walks at the rate of 3 mi. per hour; another walks at the rate of 7 km. per hour. Which one travels faster?

8. Assuming that the circumference of the earth at the equator is 25,000 mi., find the speed of a particle on the equator rotating once in 24 hours. Express this speed in miles per second, also in meters per second.

Definition of Velocity

23. It is evident from the preceding examples that speed alone is not sufficient to determine the rate at which a particle is changing *position*. Consider, for instance, a particle which moves through a small distance from P_2 to P_3 along the curved path indicated in Fig. 13, where P_2 is located by the vector r_2 and P_3 by the vector r_3. In order to describe the rate of change of *position*, it is necessary to know not only how far the particle moved in passing from P_2 to P_3, but also in what direction the line joining P_2 and P_3 lies.

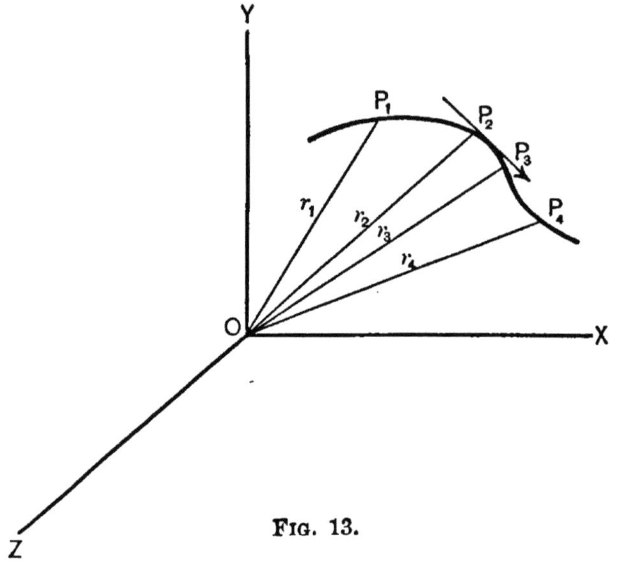

FIG. 13.

It is clear, of course, that when a particle moves, it always moves at any particular instant, in some one direction, *and* in some one sense. To completely describe the motion at any instant it is necessary, therefore, to know not only the speed but also the direction and the sense of the motion at that instant. This can be most conveniently done by use of **a vector called "velocity" which has the same direction and sense as that of the motion and which has a length equal to the speed of the motion.** Such a vector completely describes the velocity of the particle at the particular instant under consideration. Velocity is, therefore, a vector quantity, and speed a scalar. To illustrate: "A boat leaves her pier and steams in a straight line at a speed of 10 miles an hour for 3 hours." From this information, all that we know of the boat's position is that she now lies somewhere on a circle of 30 miles' radius, having the wharf for center. If, however, her velocity (instead of her speed) had been given by saying that "she steamed for 3 hours at a 10-mile rate in a northeasterly direction," her present position would be accurately known. **Velocity** is, accordingly, defined as the **time rate of change of position** of the particle. Or, in terms of algebra, if we denote velocities by V,

$$V = \left[\frac{\overline{P_1P_2}}{t_2 - t_1}\right]_{t_1 \doteq t_2} \qquad \text{Eq. 5}$$

Note carefully the difference between this equation, which defines velocity, and Eq. 4, which defines speed; the numerator in (5) is a vector, while the numerator in (4) is a scalar. From this follows the

Distinction between Speed and Velocity

24. Speed represents merely the rapidity of motion, while velocity represents rapidity, direction, and sense. Speed is defined by a single number, while velocity demands three specifications; namely, amount, direction, and sense. Imagine two trains traveling along the same piece of double track, the one going north 30 miles an hour, the other going south 30 miles an hour. Both trains have the same speed and the same direction; but their velocities are as different as they can be; for they are directly opposite in sign, *i.e.* in sense.

Velocity of a Body

25. We leave now the discussion of a particle and pass to the consideration of an extended body, *i.e.* of a body whose dimensions cannot be neglected in comparison with other quantities involved. And since we have found (§ 15) that any change in the position of a body can be defined by not less than two vectors, a translation and a rotation, it follows that the *rate* at which a body changes position will require for its description two velocities.

If the motion of the body is one of translation only, it is the same as that of a car along a straight piece of railroad; the linear velocity is the same for every particle in the car; and this velocity is completely described when the speed of any one point on the car and the direction of the track are given.

But suppose the car is placed on a turntable (Fig. 14), and the table then made to rotate; particles near the end of the car will move rapidly, those near the center will move slowly. In general, the speed of a particle will vary as the distance from the axis of the turntable. How shall we describe a motion which is apparently so complicated? This is very easily done when we recall that practically the only quantity changing is the angular position of the car, and that this change is the same for every part of the car.

Imagine this car to have turned through an angle of θ radians.

Consider next any *horizontal* line drawn in the car, the back of a seat, or the edge of a window sill ; each has been turned through the same angle θ. Indeed, the new position of the car is definitely fixed, when the motion is one of rotation only,

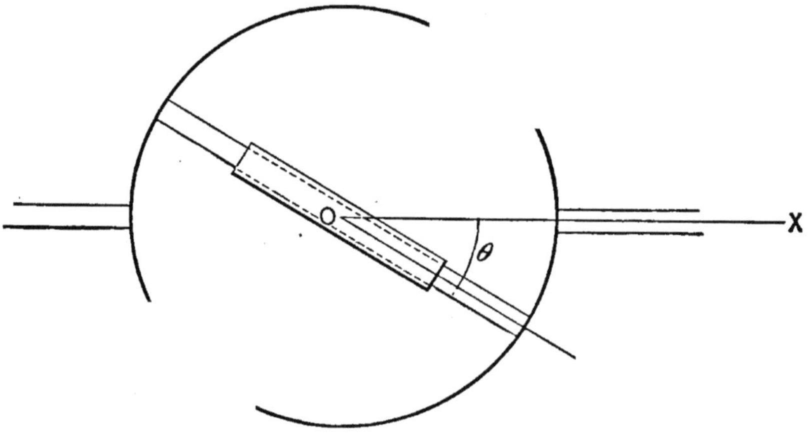

FIG. 14.

by saying that *the axis of rotation is vertical, the amount of rotation,* θ. The rate at which the angle θ varies with time is called the **angular speed** of the car (regardless of the axis about which the rotation takes place).

ANGULAR SPEED

26. *And*, in general, if θ_1 be the angular position of any line in a body at a time t_1, and θ_2 its angular position at time t_2, then its average angular speed ω is defined and completely described by the following equation:

$$\omega = \frac{\theta_2 - \theta_1}{t_2 - t_1}$$ Defining equation for *average* angular speed. Eq. 6

Unit angular speed is evidently that with which a body must rotate in order to describe one radian per second.

The engineer generally employs a unit of angular speed which is only about one tenth as large as this, namely, *one revolution per minute;* for which he uses the symbol "R.P.M."

If, instead of the average angular speed, we wish to measure the actual angular speed at any particular instant, we must then take the average speed over an indefinitely small interval of time, including this particular instant. This operation is indicated in the following notation :

$$\omega = \left[\frac{\theta_2 - \theta_1}{t_2 - t_1}\right]_{t_1 = t_2}$$ Defining equation for angular speed *at any instant.* Eq. 7

ANGULAR VELOCITY

27. The numerical value of ω (in Eq. 7) together with the direction and sense of the axis about which the rotation occurs are the three specifications which describe any angular velocity. It is evident, then, that one specification is sufficient to define angular speed, but three specifications are required to define angular velocity.

The graphical representation of angular velocity is quite as simple as that of linear velocity; for we have only to lay off along the axis of rotation as many units of length (centimeters) as there are radians per second in the angular speed.

With this convention, the straight line $\overline{OX}$ (Fig. 15) completely represents an angular velocity; for its length measures the angular speed and its direction is the direction of the axis of rotation, while the arrow indicates the sense of rotation.

An angular velocity is said to be positive when, on looking along the positive direction of the axis OX, the body appears to rotate in the same sense that a right-handed screw appears to rotate when one looks along the direction of its advance. When the rotation is in the opposite sense, it is called negative.

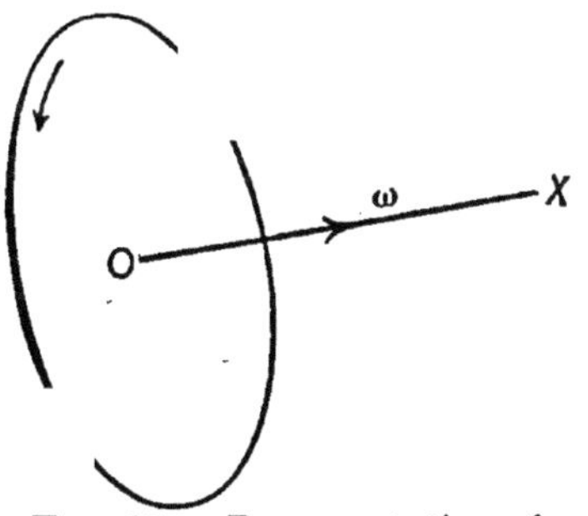

FIG. 15. — Representation of angular velocity by a straight line.

The rate at which the angular velocity of the earth is changing is fairly well known; for its axis changes direction at a constant rate and describes a complete circle in the sky once in every 25,000 years. No variation in the angular *speed* of the earth has, however, certainly been detected; although it is well known that the friction of the tides must be slowing up the earth to some slight, though as yet unmeasurable, extent. This phenomenon is beautifully illustrated by spinning any ordinary top. From Eqs. 4 and 6 it is clear that angular speed is a quantity which is strictly analogous to linear speed; while Eqs. 5 and 7 show that linear and angular velocities are likewise strictly analogous.

Relation between Linear and Angular Speed

28. Consider any point P (Fig. 16) in a rigid body rotating about any fixed axis, a wheel on its axle, for instance. Let the perpendicular distance of this point from the axis of rotation be r. If the angular speed of the body is ω, it will rotate once in $\frac{2\pi}{\omega}$ seconds.

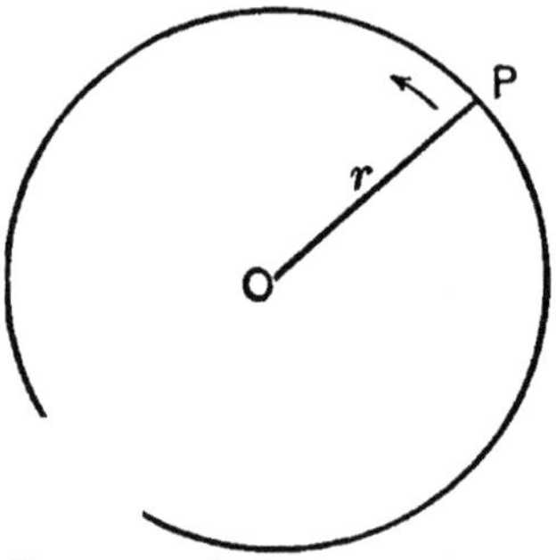

Fig. 16.—Relation between linear and angular speed.

But while the body performs one rotation, the particle at P has moved over a distance of $2\pi r$.

Its average *linear* speed is, therefore,

$$\frac{\text{Distance}}{\text{Time}} = \frac{2\pi r}{\left(\frac{2\pi}{\omega}\right)} = r\omega = S.$$

In the case of a fixed axis, the angular speed of the body, and the linear speed of any particle on it, are, therefore, related by the following equation: —

$$S = r\omega. \qquad \text{Eq. 8}$$

Problems

1. A boy attaches a small stone to the end of a string which is 80 centimeters long, and then swings it about his head at the rate of 3 turns per second. What is the linear speed of the stone in centimeters per second?

2. The belt of a driving pulley travels at a speed of 15 m. per second. The pulley rotates 10 times per second. Assuming that the belt does not slip on the face of the pulley, compute the diameter of the pulley.

Ans. $\frac{75}{\pi}$ cm.

3. In the case of a watch, compute the angular speeds of the second hand and the hour hand, respectively.

Ans. $\frac{\pi}{30}$ and $\frac{\pi}{21600}$.

4. What is the angular speed of a bicycle wheel making 30 revolutions per minute? *Ans.* π radians per second.

5. The front wheel of a bicycle is 28 in. in diameter. Find the angular speed of the wheel about its axle when it is ridden at the rate of 12 mi. per hour. *Ans.* 15.09 radians per second.

6. How may the ordinary peg top be used to illustrate the case of a body having a constant angular speed, but at the same instant a variable angular velocity?

7. Show that the fly wheel of an engine, when the steam is cut off, illustrates an angular velocity which is uniform in direction, but variable in speed.

8. What is the aim of the clockmaker, to produce an instrument which will give constant angular speed or constant angular velocity?

Change of Velocity. Addition and Subtraction of Velocities

29. Up to this point we have considered a particle which is urged only by one single velocity. It often happens, however, that a particle has impressed upon it two or more different velocities at the same time. For instance, if a man be sitting in a moving street car, his motion with respect to the earth is directly along the rails. If, however, he change his seat to one directly opposite, on the other side of the car, his motion with respect to the earth, as he walks across the car, is not along the rails, nor across the rails, but in a diagonal direction, $\overline{AC}$, as indicated in Fig. 17. $\overline{AB}$ is a straight line which indicates the

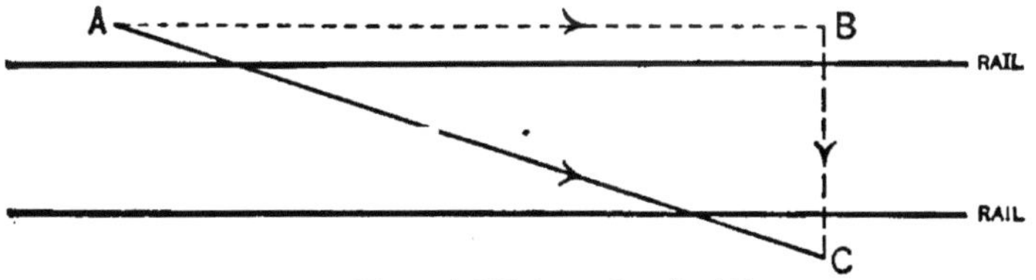

FIG. 17. — Addition of velocities.

velocity of the car; $\overline{BC}$ is a straight line which represents the velocity of the man with respect to the car. These vectors are added as are all other vectors, and $\overline{AC}$ is said to be the *resultant velocity* of the *components* $\overline{AB}$ and $\overline{BC}$. $\overline{AC}$ is, therefore, the straight line which represents the man's motion with respect to the earth.

Since velocity is a vector quantity, it is evident that we may at once apply to it the general rules for vector addition and subtraction. For these see § 14 above. Thus having given a particle whose speed is v and whose direction is defined by θ, its velocity will be the vector $\overline{OP}$, in Fig. 18. Let us suppose the motion to take place in the $\overline{XY}$ plane. Then if we desire the velocity of the particle in a direction parallel to the axis of X, it is only necessary to project $\overline{OP}$ on the axis of $\overline{OX}$. We thus obtain $\overline{OH}$ as the required velocity. In like manner $\overline{HP}$ will be the velocity parallel to the axis of Y; or in terms of algebra

$$\left.\begin{aligned}\overline{OH} &= \overline{OP}\cos\theta,\\ \overline{HP} &= \overline{OP}\sin\theta,\\ \overline{OH}^2+\overline{HP}^2 &= \overline{OP}^2.\end{aligned}\right\} \qquad \text{Eq. 9}$$

In like manner, if it is desired to find the component of the velocity $\overline{OP}$, resolved in any other direction, say, along the line $\overline{OR}$, which makes an angle ϕ with $\overline{OP}$, we have

$$\overline{OR} = \overline{OP}\cos\phi. \qquad \text{Eq. 10}$$

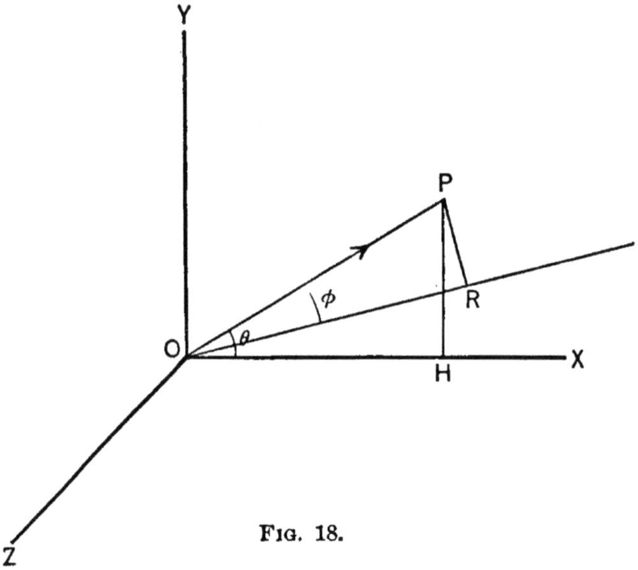

FIG. 18.

This same result may be expressed graphically by saying that the component of a velocity in any given direction is the projection of the vector which represents that velocity upon a straight line drawn in the given direction. Angular velocities are compounded and resolved in precisely the same manner as linear velocities.

ALL MOTION RELATIVE

30. In the case above considered, where a man changes his seat from one side of a car to the other, the question might have been asked: What is his motion *with respect to the car?* The answer would then have been that his velocity with respect to the car is completely represented by the straight line $\overline{BC}$, Fig. 17.

Whether, then, the man's velocity is represented by $\overline{AC}$ or $\overline{BC}$, depends entirely upon whether we take the earth or the car as our reference point.

And so it is with all velocities. Each one is a velocity with reference to some point which for the time being we consider as fixed. If a man be seated in a railway car which is traveling at the rate of 30 miles an hour, we may say that the man is at rest or in motion, according as we take for our reference point some part of the car or some part of the earth.

Problems

1. Represent by a diagram the velocity of a boat which is being rowed east, at the rate of 4 mi. an hour, while the current is carrying it south at the rate of 1 mi. an hour.

2. A wheelman is riding north at the rate of 8 mi. an hour against a head wind which is retarding him at the rate of 2 mi. an hour. What speed would he make if there were no wind?

3. What speed would this same wheelman make if he turned around and rode with the wind?

4. A canal boat is towed at the rate of 3 mi. an hour. How fast must a man walk its deck in order that he may remain at rest with respect to the towpath?

5. A wheel rider travels east at the rate of 6 mi. per hour while the wind is blowing from the north with a speed of 6 mi. per hour. From what direction will the wind appear to strike the rider?

6. A boat is rowed on the river so that its speed in still water would be 6 mi. an hour. Suppose the river flows at the rate of 4 mi. an hour. Make a diagram showing the direction in which the boat must head in order that its motion with reference to the bottom of the river may be at right angles to the current.

7. A wheel having a diameter of 28 inches revolves once in two seconds. Find the distance traversed in one hour by a particle on the circumference.

8. Two trains of the same length are running with the same speed on parallel tracks, but in opposite directions. Their combined length is 800 ft., and they pass each other in 6 sec. What is the velocity of the trains relative to the track? From Shearer's *Notes and Questions in Physics*, No. 49.

9. An express train runs due north, from Chicago to Milwaukee, at the rate of 50 mi. an hour, during a west wind which is blowing at the rate of 25 mi. an hour. Find the direction of the smoke which leaves the locomotive.

10. Find the resultant of the following three velocities, 10, 6, and $10\sqrt{2}$, which make the angles $+45°$, $0°$, and $-30°$ respectively with the horizontal.
Ans. 24.2 in a horizontal direction.

11. Two velocities each of 16 cm. per second include an angle of 120° between them. Find the resultant.

Rate of Change of Velocity. Acceleration

31. Having now considered three fundamental ideas (position, change of position, and velocity), we proceed to the study of a fourth fundamental quantity; namely, rate of change of velocity.

If all the motions we encounter in nature had velocities which were uniform both in direction and in speed, the vocabulary of physics would be even smaller than it now is. Although the behavior of a good clock or the rotation of the earth on its axis are marvelously close approximations to constant angular speeds, it is an interesting fact that no instance of perfectly constant speed has yet been discovered in nature or invented by man. The motion of a cutting tool on a shaper is fairly constant so far as direction is concerned; but its speed is by no means constant.

Definition of Linear Acceleration

32. If the velocity of a particle is not uniform, it must be because either the direction of the motion, or the speed of the motion, is changing, or both. For these are the only two continuously variable elements in velocity. When either or both of these changes occur, *i.e.* when the velocity of the particle is changing, the motion is said to be accelerated (*ad* + *celero*).

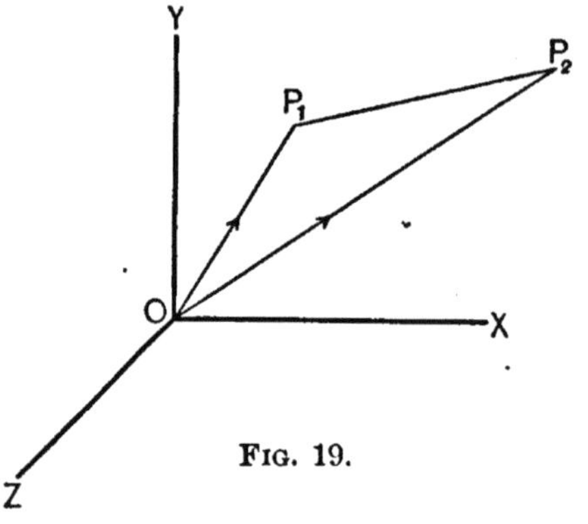

Fig. 19.

If the vector $\overline{OP_1}$ (Fig. 19) represents the velocity of the particle at a time t_1, and the vector $\overline{OP_2}$ represents it at a time t_2, we know from the properties of vectors that the total change in velocity during the interval $t_2 - t_1$ is represented by the vector $\overline{P_1P_2}$. In other words, the straight line joining P_1 and P_2 is as many centimeters long as there are units of speed in the *increment* of velocity $\overline{P_1P_2}$. Not only so, but the direction of this straight line is the same as that of the added velocity.

To the rate, $\frac{\overline{P_1P_2}}{t_2 - t_1}$, has been given the name **acceleration.** Let us denote this acceleration by a, the change in velocity by v, then

$$a = \frac{\overline{P_1P_2}}{t_2 - t_1} = \frac{v}{t_2 - t_1}. \quad \text{Defining equation for } \textit{average} \text{ acceleration.} \qquad \text{Eq. 11}$$

Acceleration, then, is simply the rate at which the particle is gaining or losing velocity. This rate is not always constant; consequently, the average acceleration will depend in such a case upon how long an interval, $t_2 - t_1$, is covered by the average. In order to get the actual acceleration at any particular point of time t, we must take the average over an indefinitely small interval of time, including t.

$$a = \left[\frac{\overline{P_1P_2}}{t_2 - t_1}\right]_{t_2 \doteq t_1} = \left[\frac{v}{t_2 - t_1}\right]_{t_2 \doteq t_1} \quad \text{Defining equation for acceleration } \textit{at any instant t.} \qquad \text{Eq. 12}$$

Acceleration is accordingly the limit approached by the ratio between an increment of velocity and the corresponding increment in time. If, therefore, a particle is moving with uniform velocity, its acceleration is zero. Whether the particle is moving fast or slowly makes no difference. So long as its velocity is not changing, its acceleration is zero.

Unit of Acceleration

33. From this "defining equation," our unit of acceleration is obtained exactly as we obtained our unit of velocity, viz., we make each of the terms in the right-hand member unity, then a must be unity. We thus arrive at the following definition: **Unit acceleration is an acceleration in which unit change of velocity is produced in unit of time.**

Since, in pure science, the centimeter per second is the unit of velocity, the natural unit of acceleration is the *centimeter per second per second*, often written cm./sec.^{2}. Thus a projectile fired from a gun gains during each second a velocity of approximately 981 centimeters per second toward the center of the earth. Accordingly the acceleration of a body falling freely under gravity is said to be 981 cm./sec.^{2}. In engineers' units this quantity would read approximately 32.2 ft./sec.^{2}.

Angular Acceleration

34. Angular velocity, like linear, depends for its value upon two variables, namely, the direction of the axis of rotation, and rate of rotation. If, in any spinning body, either of these two variables changes, the body is said to have an angular acceleration. Thus the shaft of a turbine steamer may be rotated with a constant angular speed, but if the steamer is changing her course, the angular velocity of the shaft is changing, and the shaft is said to have an angular acceleration. If, at another time, the boat be moving in a perfectly straight line and the speed of the turbine be varying, the motion of the shaft is also said to have an angular acceleration.

In general, the velocity of a spinning body varies from instant to instant; as, for example, the fly wheel of a stationary engine; so that if we wish to obtain the angular velocity at a particular instant t, we must choose t_2 and t_1 in such a way that t_2 shall be a time a very little later than t, and t_1 a time a very little earlier than t. Then, if we denote by ω_1 the angular velocity at the instant t_1, and by ω_2 the angular velocity at the instant t_2, the angular acceleration at any instant t will be the limit approached by A as $t_2 - t_1$ approaches zero.

$$A = \left[\frac{\omega_2 - \omega_1}{t_2 - t_1}\right]_{t_2 \doteq t_1} \qquad \text{Eq. 13}$$

Defining equation for angular acceleration.

Since the numerator in the right-hand member is a vector quantity, while the denominator is a scalar, it follows that the angular acceleration A is also a vector quantity, and is, therefore, compounded and resolved as are other vector quantities.

Two Important Special Cases

35. Having now considered in some detail the three fundamental conceptions of kinematics, namely position, velocity, and acceleration, we proceed to the study of two highly interesting special cases.

Case I. *Motion of a Particle along a Straight Line: Direction of Velocity Constant; Speed Variable*

36. If, in Fig. 20, the velocity of a particle be changed from $\overline{OP_1}$ to $\overline{OP_2}$ (the points O, P_1, P_2 all lying on the same straight line), then the acceleration results from a change of speed only. For the velocity of the particle is all the while along the straight line $\overline{OP_2}$. Here the acceleration is in the same direction as the motion of the particle.

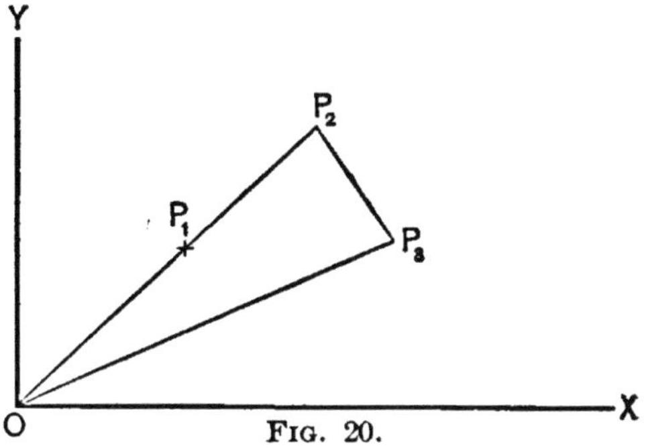

FIG. 20.

Let us imagine that the speed of a particle changes from OP_1 to OP_2 in any interval of time t, and that, throughout the motion, *the acceleration is constant.* Let us call the acceleration a; then the particle will gain a units of speed during each second. The *total gain of speed* during any time t is therefore given by the following expression: —

$$s = at. \qquad \text{Eq. 14}$$

If, now, we happen to know the speed of the particle at the beginning of the interval and call it S_0, then the actual speed at the end of the interval will be the original speed plus the speed gained, viz.: —

$$S = S_0 + s,$$

or

$$S = S_0 + at. \qquad \text{Eq. 15}$$

This equation is very useful; for it not only describes the variable speed S at each instant of time, but it also enables us to compute the value of any one of the four quantities involved as soon as we know the other three.

In case a particle loses speed, then we have a negative value for s, and hence a negative value for a in the equation, $a = \frac{s}{t}$. For in these problems, t is always positive.

Not infrequently we shall want to know something more about the motion of the particle than its mere speed. We often want to know what the position of the particle is after a certain interval during which its motion has been accelerated.

If the acceleration a is a constant, then the gain in speed as time goes on will be proportional to the time,

$$s = at.$$

And the *mean* speed of the particle from the beginning to end of the interval is the value of the speed at the middle of this interval, viz. the speed represented by the line *LM*, in Fig. 21, but this speed is, by definition, the mean of S_0 and $S_0 + at$, which is $S_0 + \frac{at}{2}$. Call this average speed, S. We may now take this average value of the speed and substitute it for S in Eq. 3, § 21, thus obtaining

$$p = St = \left(S_0 + \frac{at}{2}\right)t,$$

or

$$p = S_0t + \tfrac{1}{2}\,at^2. \qquad \text{Eq. 16}$$

where p is the length of path (or simply the distance) covered by the moving particle.

For the special case which we are considering, namely, motion in a straight line, this equation, together with Eq. 15, $S = S_0 + at$, tells the whole story. One gives us the speed of the particle when we know the acceleration and the duration of its action; the other gives us the change in the position of the particle.

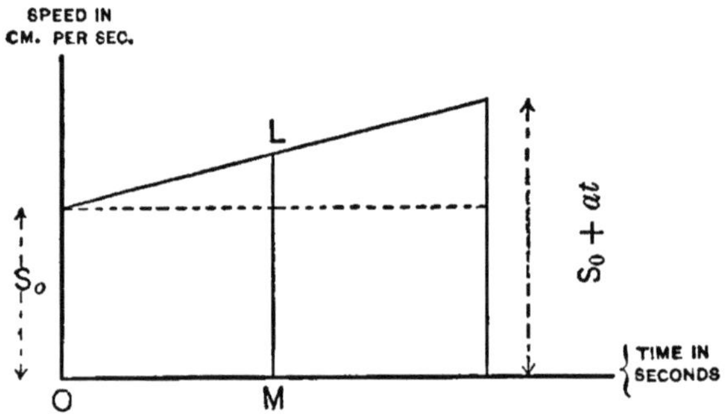

FIG. 21.

If between these two equations we eliminate t, we get another useful expression giving us the speed of a particle in terms of its acceleration and the distance traversed, namely

$$S^2 = S_0^2 + 2\,ap. \qquad \text{Eq. 17}$$

To solve any problem, then, under this special case which we have just been considering (direction of velocity constant: speed variable), one has merely to observe which of the above five quantities (S, S_0, a, t, and p) are given, which required, and lastly which of these three equations contains the desired and the known quantities. He has then one equation to solve for the unknown quantity.

Problems

1. A bicycle and rider start from rest at the top of a hill. At the end of 16 seconds the foot of the hill is reached, with a speed of 640 cm. per second. Find the average acceleration of the wheel during the descent of the hill. Here S, S_0, and t are given. *Ans.* $a = 40$.

2. Having computed the acceleration, use this and the speed at the foot of the hill to compute the length of the hill. *Ans.* $x = 51.2$ m.

3. A train running at 36 km. per hour is stopped by a sudden application of the brake. What acceleration must be produced by the brakes in order to stop the train in 8 sec.? Here, again, S, S_0, and t are given. *Ans.* $a = -125$.

4. When steam is turned on again, a constant acceleration of 50 units is produced. How long will it be before the train has again acquired its original speed of 36 km. per hour? *Ans.* 20 sec.

5. Over what distance will this train have traveled while coming to rest in the third problem? *Ans.* 40 m.

6. How far will the train travel while again acquiring its original speed in the fourth problem? *Ans.* 100 m.

7. A football player running north at the rate of 8 m. per second reverses his velocity and in an interval of 2 sec. is again running, this time south, with a speed of 8 m. per second. Find his mean acceleration during the interval. *Ans.* $a = -800$.

8. A body is simultaneously urged to move with speeds of 40, 15, and 20 respectively. Can it remain at rest?

9. A train which is uniformly accelerated starts from rest, and at the end of 3 sec. has acquired a speed with which it would travel through 1 km. in the next 5 min. Find the acceleration.

10. Two bodies, whose respective velocities are accelerated in every second by 30 and 50 cm. per second, begin to move toward each other at the same instant. At first they are 2 km. apart, and have no initial velocity. After how many seconds will they meet?

11. A body tends to move with equal speeds in two directions inclined to each other at 120°. Find its path and resultant velocity.

12. One man A is standing at a point 1500 ft. east of another man B. At the same instant both men start to walk east, B at the rate of 4 mi. an hour, and A at the rate of 3 mi. an hour. How long will be required for B to overtake A?

13. A particle has an initial speed of 20 cm. a second, and receives an acceleration of 8 cm. per second every second. What distance will it travel in 15 sec.?

14. A wind is blowing from a point intermediate between north and east. The northerly component of its velocity is 10 mi. an hour, and the easterly component is 36 mi. an hour. Find the whole velocity.

15. Write in the margin of your book the special form which Eqs. 15, 16, and 17 assume when the particle starts from rest.

CASE II. *Uniform Motion in a Circle: Direction of Velocity Variable; Speed Constant*

37. If, however, in Fig. 22, the velocity of a particle be changed from $\overline{OP}_2$ to $\overline{OP}_3$, where the vector $\overline{OP}_3$ is constantly equal in length to $\overline{OP}_2$, *there is no change in speed; but there is a change in the direction of the velocity. The motion is, therefore, accelerated.*

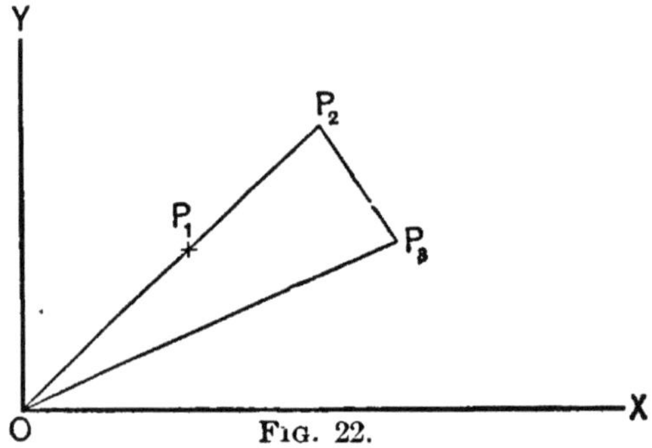

FIG. 22.

Such a motion takes place when a particle moves in a circle at a uniform speed. It is approximated in the following instances: when a boy whirls about his head a small mass attached to the end of a string; a small particle of gravel sticking to the wheel of a bicycle in motion; the earth in its revolution about the sun; the motion of the moon in its orbit about the earth; each particle which goes to make up the fly wheel of an engine.

In each of these cases, let us consider the center of the circular path O (Fig. 23) as a fixed point of reference. And let us draw a straight line from O through the point where the particle lies when we first begin to observe its motion. Call this line $\overline{OX}$.

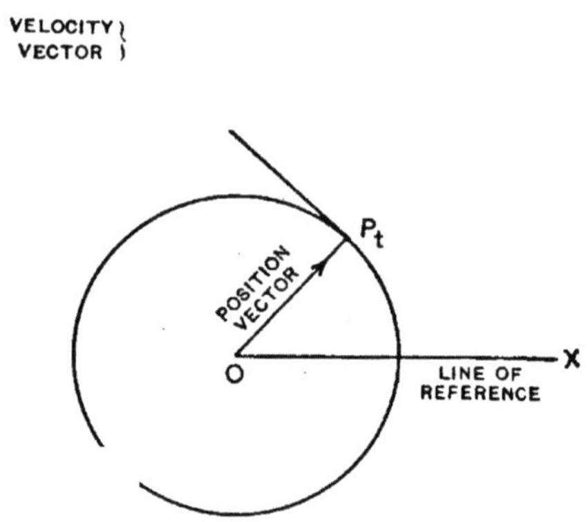

FIG. 23.

For a clear understanding of this motion, it is next essential that the student distinguish carefully between the position, the path, and the velocity of the moving particle.

The position of a particle P_t at any instant t is completely given when we know the particular radius which joins the particle and the center O at that instant.

By the path of a particle is meant (§ 21) the line which is made up of the successive positions of the particle.

The velocity of the particle at any instant is evidently in a direction which is tangent to the circle at the point where the particle is at that particular instant.

Imagine the particle at any time to be at the point P_t in Fig. 23. The direction of the velocity will there be perpendicular to the radius $\overline{OP_t}$. And the same is true at any other instant, for in circular motion the straight line which represents the velocity and the radius which represents the position of the particle are always at right angles to each other.

If we wish to represent the velocity completely, we must choose for the constant length of this tangent line as many centimeters as there are units in the constant speed of the particle.

In Fig. 24 let P_1, P_2, P_3, etc., represent positions of a particle P at small intervals of time t. Imagine the particle to be moving with uniform speed in a circle. $\overline{OV_1}$, $\overline{OV_2}$, $\overline{OV_3}$ (drawn at right angles to $\overline{OP_1}$, $\overline{OP_2}$, $\overline{OP_3}$, respectively) may then represent the *velocities* of the particle in its positions P_1, P_2, P_3, respectively.

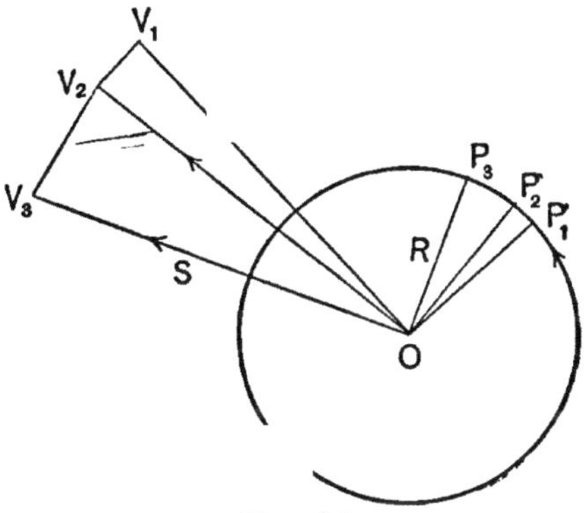

FIG. 24.

We can now determine at once the acceleration of the particle. For, if the angular speed of the radius $\overline{OP_1}$ be ω, so that the speed of the particle P is $S = \omega R$, then the angular speed of $\overline{OV_1}$ is also ω, because it keeps always just 90° ahead of $\overline{OP_1}$; and the point V_1 will describe a circle of radius S, with a speed ωS. Now the total change in the velocity of the particle P — remember there is no change in speed — during the time t is completely represented by the line $\overline{V_1V_2}$. For $\overline{V_1V_2}$ is the velocity which must be added to the velocity $\overline{OV_1}$ to give the velocity $\overline{OV_2}$. And hence, by definition,

$$\text{Acceleration of the particle} = \frac{\overline{V_1V_2}}{t}.$$

But $\dfrac{\overline{V_1V_2}}{t}$ measures the speed of the point V, and is hence equal to ωS, in a direction at right angles to $\overline{OV}$.

Hence the acceleration of a particle moving with speed S in a circle is completely described by saying that the direction of the acceleration is parallel to $\overline{PO}$, and the amount of the acceleration is equal to ωS. Or we may say the **acceleration is always toward the center and is numerically equal to ωS.**

And since (Eq. 8, § 28) $S = \omega R$, we may also write

$$\left.\begin{array}{r}\text{Acceleration of a particle}\\ P\text{ moving in a circle}\end{array}\right\} = \omega S = \omega^2 R = \frac{S^2}{R}. \quad \text{Eq. 18}$$

If we denote the time of one revolution of a particle in a circle —an interval known as the **period** of the motion—by T, then we have $T = \frac{2\pi}{\omega}$; and hence substituting for ω, in Eq. 18,

$$\text{Acceleration towards center} = \frac{2\pi S}{T} = \frac{2\pi R\omega}{T}.$$

We have now a complete description of the motion of a particle which moves with uniform speed in a circle. The student should be able to put into words the meaning of each of these different expressions. For example, the last form, $\frac{S^2}{R}$, says that "the acceleration of a particle moving with uniform speed in a circle is constant, and is numerically equal to the linear speed of the particle squared and divided by the radius of the path."

General Case of Linear Acceleration

38. We have now considered, in some detail, two special cases of acceleration, in one of which the speed alone varies, in the other of which the direction alone varies.

The importance which attaches to these two cases arises partly from the fact that the acceleration of any particle whatever, moving at any given instant in any path, however tortuous, may always be resolved into two components, one along the path, and one normal to the path; and these two components belong respectively to the two special cases which we have just studied. We thus have the general result:—

$$\left.\begin{array}{c}\text{Total acceleration of}\\ \text{any particle}\end{array}\right\} = \left\{\begin{array}{l}\text{Vector sum of the acceleration along}\\ \text{the path and the acceleration normal}\\ \text{to the path.}\end{array}\right.$$

The advanced student will discover in Vector Analysis a much more elegant method of treating this subject.

Addition and Subtraction of Accelerations

Since acceleration is a vector quantity, it follows at once that accelerations are added and subtracted in the same manner as velocities and other vectors. See § 14.

Definition of Angular Acceleration

39. Consider any spinning body, such as the propeller shaft in a steamer, a top, the earth, the fly wheel of an engine, or a grindstone. It is at once clear that the angular velocity with which any of these rigid bodies is rotating may change in either one of two ways, as already intimated in § 34, namely, —

(i) The angular speed of the body about its axis may change (as, for instance, that of a fly wheel when the engine is just starting or stopping), and

(ii) The direction of the axis of rotation may change (as, for instance, that of the propeller shaft, when the steamer is turning around without slowing up her engines). A body which is changing in either one of these two respects is said to have an **angular acceleration**. This quantity is a vector and a strict analogue of linear acceleration. Denoting angular velocities by ω_1 and ω_2, and angular acceleration by A, we have already (§ 34) obtained the defining equation

$$A = \left[\frac{\omega_2 - \omega_1}{t_2 - t_1}\right]_{t_1 \doteq t_2} \qquad \text{Eq. 13}$$

Unit of Angular Acceleration

40. When the rate at which a body is rotating changes by one radian per second during each second, the value of its angular acceleration is unity.

Two Important Special Cases

41. Corresponding to the two special cases of linear acceleration, we have here also two special cases.

Case I. *Direction of Axis of Spin Constant; Rate of Spin Variable*

42. Here the whole story is told by three simple equations which are identical in form with Eqs. 15, 16, and 17, § 36. Denoting angles by θ, angular speeds by ω, angular acceleration by A, and time by t, one has

$$\omega = \omega_0 + At, \qquad \text{Eq. 15}'$$

$$\theta = \omega_0 t + \tfrac{1}{2} At^2, \qquad \text{Eq. 16}'$$

$$\omega^2 = \omega_0^2 + 2A\theta. \qquad \text{Eq. 17}'$$

CASE II. *Rate of Spin Constant; Direction of Axis of Spin Variable*

43. Here again the analogue is complete; the axis about which the acceleration occurs is at right angles to the axis of spin; and the amount of acceleration is given by an expression which is identical in form with Eq. 18, § 37, namely,

$$A = \left\{\begin{matrix}\text{Acceleration about}\\ \text{axis normal to axis}\\ \text{of spin}\end{matrix}\right\} = \Omega\omega, \qquad \text{Eq. } 18'$$

where, as before, ω is the angular speed with which the body is spinning, while Ω is the angular speed with which the axis of spin is changing direction

This case is well illustrated by the ordinary gyroscope; also by the motion of the earth, a body in which the constant rate of spin ω is 2π radians a day, and in which the axis of spin has its direction changed at the rate Ω by the pull of the moon upon the equatorial belt of the earth. Observe that just as in Eq. 18 the three quantities a, S, and ω are vectors, each at right angles to the other two, so here in Eq. 18′, A, Ω, and ω are three mutually rectangular vectors.

The student will find it an interesting exercise to prove that angular and linear accelerations at any point on a rotating rigid body are connected by the following equation,

$$a = rA, \qquad \text{Eq. } 8'$$

an equation which is strictly analogous to Eq. 8, p. 28.

This subject is one whose detailed discussion at this point would lead us too far afield; but it is hoped that every student will at least get a fair grasp of the exact analogy between translation and rotation in Kinematics, and thus reduce the complexity of the subject by at least one half.

ALGEBRAIC SUMMARY OF KINEMATICS

Linear	**Angular**
Fundamental Quantities	
Length and time; p and t.	Angle and time; θ and t.

Linear	**Angular**

Position

Defined by a vector whose length measures a distance.	Defined by a vector whose length measures an angle.

Velocity

$V = \left[\dfrac{\overline{P_1P_2}}{t_2 - t_1}\right]_{t_1 \doteq t_2}$	$\omega = \left[\dfrac{\theta_2 - \theta_1}{t_2 - t_1}\right]_{t_2 \doteq t_1}$

$$V = r\omega.$$

Acceleration

$a = \left[\dfrac{\overline{V_1V_2}}{t_2 - t_1}\right]_{t_2 \doteq t_1}$	$A = \left[\dfrac{\omega_2 - \omega_1}{t_2 - t_1}\right]_{t_1 \doteq t_2}$

$$a = rA.$$

Special Case I. Direction Constant: Speed Variable

$S = S_0 + at,$	$\omega = \omega_0 + At,$
$p = S_0t + \frac{1}{2}at^2,$	$\theta = \omega_0t + \frac{1}{2}At^2,$
$S^2 = S_0^2 + 2ap.$	$\omega^2 = \omega_0^2 + 2A\theta.$

Special Case II. Direction Variable: Speed Constant

$a = S\omega.$	$A = \Omega\omega.$

Problems

1. Twenty seconds after the current is turned on an electric motor, the armature, starting from rest, has acquired an angular speed of 6000 radians per second. Find the average angular acceleration.

2. The fly wheel of an engine is subject to a negative angular acceleration of 4 radians per second per second. If the wheel is making 220 R.P.M. when steam is shut off, how long will it take the fly wheel to come to rest?

3. A particle is accelerated vertically upward with an acceleration of 3, and horizontally eastward with an acceleration of 4. Find the total acceleration of the particle.

4. A particle on the surface of a fly wheel of 120 cm. radius, rotating with an angular speed of 4 radians per second, is accelerated toward the center at what rate?

5. How fast must a bicycle wheel be made to revolve in order that a small particle of mud attached to the tire may experience an acceleration of 12 cm./sec.^{2}? It is assumed that the diameter of the wheel is 70 cm.

6. What data would you need in order to compute the acceleration of the moon toward the earth as it revolves about the earth in an orbit which is practically a circle?

7. A grindstone is set in motion with an angular speed of 3 radians per second. When left to itself it rotates through an angle of 60 radians before stopping. Find the angular acceleration.

8. The shaft of a hoisting windlass is set into motion (starting from rest), under an angular acceleration of 200 radians per second per second. What angle will it have turned through by the end of 3 seconds?

References

MAXWELL. *Matter and Motion* (Soc. Promotion Christian Knowledge), pp. 128. Price 40 cents.

MACH. *Science of Mechanics*, trans. by McCormack (Open Court Pub. Co.), pp. 534. Price $2.50. Excellent chapter on Galileo's achievements, pp. 128–155.

LODGE. *Elementary Mechanics* (Chambers, London), pp. 208. Price $1.00.

JEANS. *Theoretical Mechanics* (Ginn & Co., Boston), pp. 364, Chapter I.

Ency. Brit., art. "Weights and Measures."

SLATE. *Principles of Mechanics* (Macmillan), pp. 299, Chapters I and II.

DUFF. *Elementary Experimental Mechanics* (Macmillan), pp. 267, Chapters I–IV.

FRANKLIN AND MCNUTT. *Elements of Mechanics* (Macmillan), pp. 283.

CHAPTER II

SIMPLE HARMONIC MOTION

CASE I. LINEAR MOTION

44. There is another motion with which all students, even beginners, are already familiar in a general way, but which few beginners are able to describe in a manner intelligible to others or useful to themselves. This motion is typified by that of a particle P (Fig. 25) supported by a vertical spiral spring, and vibrating freely in a vertical straight line. The motion of any point on a guitar string which has just been plucked and let go furnishes another illustration of this motion. If a body be suspended by a rubber band and then slightly displaced up or down and suddenly released, its motion also will be that which we are about to study.

FIG. 25.

In each of these cases observe that, when the body vibrates freely,—

(*a*) The motion is one of translation; because the particles of the body describe equal and similar paths.

(*b*) The path is a limited straight line.

(*c*) The particle moves most rapidly at the middle of its path, and comes to rest at each end of its path.

(*d*) The motion gradually dies down, unless kept up by some "outside" means.

Such would be a popular and qualitative description of the motion.

CASE II. ANGULAR MOTION

45. In the case of solid bodies having a fixed axis of rotation, we frequently meet a similar motion, and one with which, again, all students are familiar in a general way.

The pendulum of a clock is approximately a rigid body whose angular position is constantly changing. It typifies the motion we are about to describe. The motion of the balance wheel in a watch movement also illustrates the motion under consideration. If a slender lath be clamped at one end, while the free end is pulled aside and suddenly let go, this free end will vibrate with practically the same motion as that of a pendulum left to itself.

Observe that, in each case, when the body vibrates freely, —

(*a*) Its motion is an angular one.

(*b*) The amount of swing (angular displacement) never exceeds, in any one case, a certain limited angle.

(*c*) The angular speed is most rapid at the middle, and ceases altogether at the end of the swing.

(*d*) The vibration gradually dies down, unless kept up by some means external to the system we are studying, as, for instance, by the weight of the clock or the coiled spring of the watch.

These two motions, the linear and the angular, are each close approximations to what is called **Simple Harmonic Motion** ; and are often indicated by the letters S.H.M.

The student should note here, once for all, that a mere name of a physical quantity or phenomenon, such as the above, while exceedingly convenient, does not define it. We now proceed to the exact

Definition of Linear Simple Harmonic Motion

46. Imagine a point P (Fig. 26) to move with uniform speed in the circumference of a circle. Imagine any diameter drawn in the circle. And from the point P, in each of its successive positions, imagine a perpendicular let fall upon this diameter. The motion of the foot of this perpendicular, H, to and fro along the diameter, is a simple harmonic motion. It is evident that the speed of the point H near either end of the diameter will be much less than at the center O. In this respect the motion of H is like that of a pendulum bob.

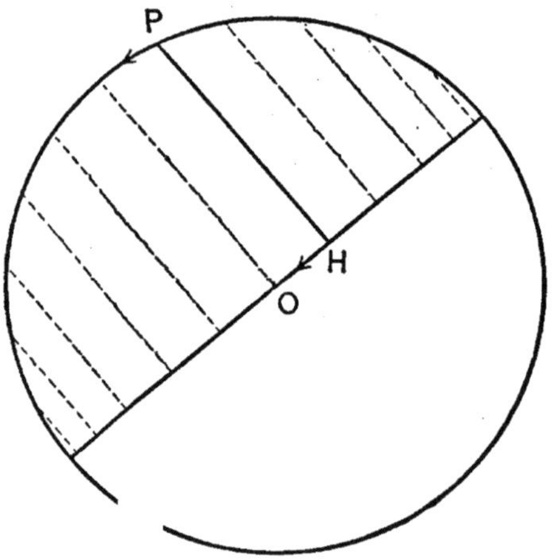

Fig. 26. — Simple Harmonic Motion.

The same definition may be put more elegantly, as follows: **Simple harmonic motion is the projection of uniform circular motion upon a diameter of the circle.** We shall presently see that this motion can be accurately described (defined) in a manner even simpler than this.

The student frequently confounds simple harmonic with circular motion, and should, therefore, at this point, carefully distinguish between these two motions. The *actual* motion of one of Jupiter's moons is very nearly circular, while its *apparent* motion is practically simple harmonic. As these moons revolve about Jupiter, we see them move alternately to the right and to the left, but we do not observe their motion toward us or away from us. Hence their motion is apparently simple harmonic. For S.H.M. is merely uniform circular motion seen edgewise.

Circle of Reference

47. The circle in which we have imagined the point P (Fig. 26) **to move with uniform speed is called the circle of reference.** We shall also employ the center of this circle O as our point of reference

Amplitude of S.H.M.

48. The radius of the circle of reference is the maximum distance to which H can recede from O in either direction. This maximum distance, which will be denoted by A, is called the **amplitude** of the S.H.M. **Accordingly the amplitude of S.H.M. is defined as the radius of the circle of reference.**

Period of S.H.M.

49. Let us denote by S the uniform speed of the point P in the circle of reference. In one complete revolution of the circle this point will travel a distance $2\pi A$. The time required for this round trip will, therefore, be $\frac{2\pi A}{S}$ seconds. This interval of time, generally denoted by T, is a constant, and is called the **period** of the S.H.M. **Accordingly the Period is defined as the time occupied in one round trip of the actual moving point H to and fro across the diameter of the circle.** The reciprocal of the period T is called the **frequency**, and is generally denoted by n.

PHASE OF S.H.M.

50. The angular position of the radius vector OP at any instant t is measured by the angle θ (Fig. 27) which this radius makes with the line of reference $\overline{OX}$. This angle θ, expressed in radians, is called the **phase** of the S.H.M. at the time t. **Phase of S.H.M. is, therefore, defined as the angle between the line of reference and the radius** vector **drawn to the point moving in the** circle **of** reference.

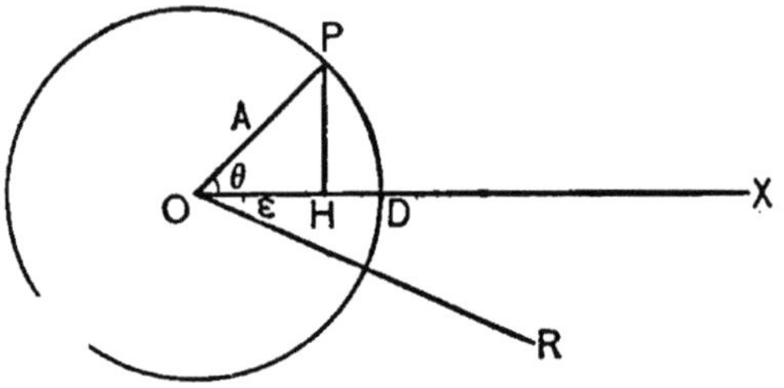

FIG. 27.

Imagine the point to start from D, at the time $t = 0$, and move about the circle of reference in a counter-clockwise sense, the phase will increase from 0 to 2π; but if the motion is clockwise, *i.e.* negative, the phase will decrease from 0 to -2π. Evidently the phase is continually changing as time goes on. It is, therefore, a variable quantity; and when the angle and the time are each measured from the line of reference $\overline{OX}$, it is very easy to express the angle θ in terms of the time t by saying —

$$\theta = \omega t, \qquad \text{Eq. 6}$$

where ω is the angular speed of the radius A.

EPOCH

51. But let us now suppose that the line of reference from which the angle θ is measured is *not* OX, but some line OR, such as indicated in Fig. 27, while the line of reference for times remains the same. Let ϵ denote the angle between the reference line for angles and the reference line for times; *i.e.* let the angle $R\widehat{O}X$ be indicated by ϵ. Then the proper expression for the phase will be, not ωt, but $\omega t + \epsilon$. **This angle ϵ, which is the value of the phase at the time $t = 0$, is called the epoch of the S.H.M.**

DISPLACEMENT OF S.H.M.

52. *The point which is actually moving with S.H.M. is represented by H in Fig.* 27. **The distance of this moving particle,**

at any instant, from the center of the circle of reference is called its Displacement. We shall denote this displacement OH by x. Then in the right-angled triangle OPH, Fig. 27, it is evident from the definition of a cosine that

$$\cos\theta = \frac{x}{A},$$

or

$$x = A\cos\theta.$$

But by the definition of constant angular speed, we have

$$\theta = \omega t,$$

and hence

$$x = A\cos\omega t. \qquad \text{Eq. 21}$$

Or if angles be measured from some line OR, Fig. 27, instead of OX, we have the most general expression for S.H.M., namely,

$$x = A\cos(\omega t + \epsilon). \qquad \text{Eq. 22}$$

This equation tells the whole story concerning the displacement. Every possible value which x can assume is easily computed as soon as we know the corresponding value of the variable t and of the constants A, ω, and ϵ.

To completely describe any S.H.M. we must, then, tell the following four things about it: —

(1) The **amplitude** A, *i.e.* the radius of the circle of reference.

(2) The **period** T, or, what is the same thing, ω; for

$$\omega = \frac{2\pi}{T}.$$

(3) The **phase** at any time t, namely, $\omega t + \epsilon$.

(4) The **direction of motion** of the point P in the circle of reference, *i.e.* the sign of ω must be given.

These having been given, nothing can be added that will make the description more definite.

Linear Acceleration of S.H.M.

53. In our previous study of circular motion, we found that whenever a particle moves in a circle with uniform speed, it is always accelerated toward the center. The amount of this acceleration we found (§ 37) to be $\omega^2 A$, where A is the radius of the circle and ω is its angular speed.

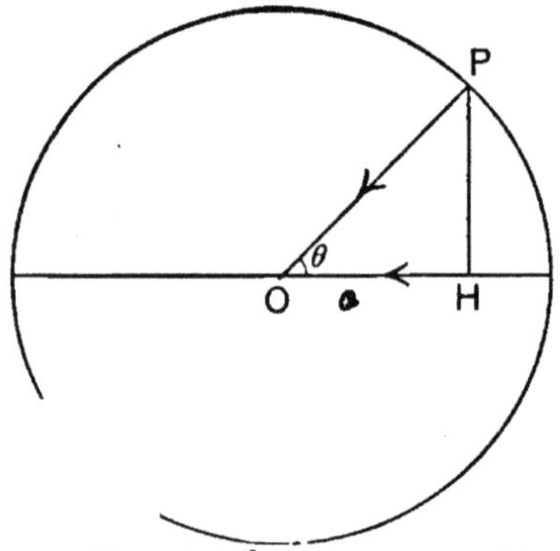

FIG. 28. — Acceleration of S.H.M.

In other words, the acceleration in uniform circular motion is represented by a vector of length $\omega^2 \cdot PO$ drawn to O in the direction PO.

Hence the acceleration in S.H.M., which is the projection of uniform circular motion, is represented by the projection of this vector, that is, by a vector of length $\omega^2 \cdot HO$ drawn to O in the direction HO. See Eq. 2, § 14.

Let us denote by a this acceleration of S.H.M., then

$$a = \omega^2 PO \cos\theta = \omega^2 PO\left(\frac{HO}{PO}\right) = \omega^2 A\left(\frac{-x}{A}\right) = -\omega^2 x. \qquad \text{Eq. 23}$$

This equation shows us, when once we know the *displacement* of any S.H.M., how to obtain the corresponding *acceleration.* This is the fundamental equation of S.H.M., and may perhaps be written in the following form :

$$\frac{\text{Acceleration}}{\text{Displacement}} = \text{A negative constant.}$$

In other words, the acceleration is proportional to the displacement and opposite in sense.

The relation between displacement, velocity, and acceleration will be clear from Fig. 29.

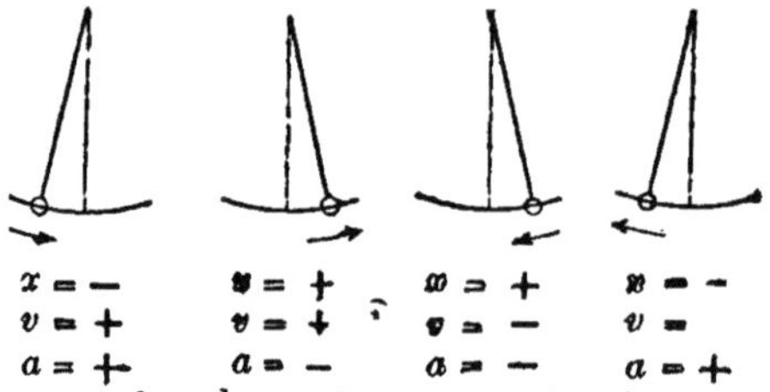

a and *x* always have opposite signs.

FIG. 29.

This equation (23) also gives us an exceedingly useful expression for the period of a S.H.M. For, since

$$T = \frac{2\pi}{\omega} \text{ and } \omega^2 = -\frac{a}{x}, \text{ we have } T = 2\pi\sqrt{\frac{x}{-a}}. \qquad \text{Eq. 24}$$

On account of damping such as that due to frictional resistance, the amplitude of all *actual* vibratory motions, except those in which energy is supplied from outside the vibrating system, continually diminishes. Hence we do not meet any motions in nature which are accurately simple harmonic; but in thousands of cases the approximation is so close that in treating these cases we may, with impunity, use the laws of S.H.M.

Angular Simple Harmonic Motion

As we have already seen (§ 45), bodies rotating about a fixed axis may have an angular motion which is analogous to the linear motion which we have just been studying. A most excellent illustration is to be found in the motion of the balance wheel of a watch.

The angular displacement θ of such a body is described in terms of three constants and the time exactly as in the case of linear S.H.M. Thus

$$\theta = \Theta \cos(\omega t + \epsilon), \qquad \text{Eq. } 22'$$

where the constants are

$\Theta =$ maximum angular displacement, or amplitude,

$\omega = \frac{2\pi}{T}$, where $T =$ period of oscillation,

$\epsilon =$ phase at time $t = 0$,

while the variables are θ and t.

Just as in the case of linear S.H.M. it was shown (Eq. 24) that the period depends only upon the ratio of displacement to acceleration, so here it may be shown in the same way that

$$T = 2\pi\sqrt{\frac{\text{Angular displacement}}{-(\text{Angular acceleration})}} = 2\pi\sqrt{\frac{\theta}{-A}}. \qquad \text{Eq. } 24'$$

In other words, the criterion of angular S.H.M. is that the angular acceleration shall be proportional to the angular displacement, but opposite in sense.

The Harmonic Curve, or the Composition of Uniform Rectilinear Motion with a S.H.M. in a Direction at Right Angles

54. If one walks along the smooth sand of the lake shore or the ocean beach, dragging his walking stick behind him and at the same time vibrating it to and fro sidewise, he will trace upon the sand a wave-shaped line which is of considerable importance in Physics and which is known as the Harmonic Curve.

It is the resultant of two motions which are at right angles to each other and which are described by the following equations:

(i) $y = A \cos \omega t$, Simple Harmonic Motion.

and

(ii) $x = Vt$. Uniform Rectilinear Motion.

Eliminating t between these two equations of motion, we have the equation of the path, namely,

$$y = A \cos \left(\frac{\omega}{V}\right) x. \qquad \text{Eq. of Harmonic Curve 25}$$

Such a motion is evidently periodic, the same value of y recurring whenever $\frac{\omega x}{V}$ changes by 2π, as illustrated in Fig. 30.

FIG. 30. — Path resulting from S.H.M. and uniform rectilinear motion.

This curve is also known in mathematics as the cosine- or sine-curve.

Problems

1. Show, either graphically or algebraically, that when the particle H (Fig. 28) is passing the center and has its maximum speed, the acceleration is zero.

2. Show that when the particle H (Fig. 28) reaches the end of its swing and for an instant stands still, the acceleration is maximum.

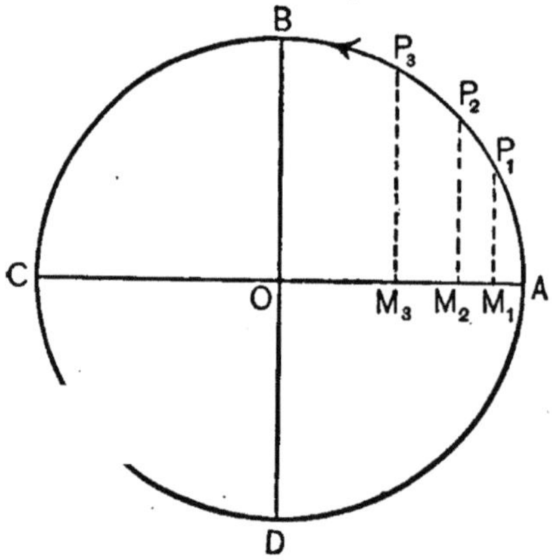

FIG. 31.

3. The displacement in a certain S.H.M. at a given instant is -32. At the same instant the acceleration is $+\frac{1}{2}$. Find the period of the S.H.M.

Ans. 16π.

4. The period of a S.H.M. is 6.28 seconds, and the acceleration at a given instant is -4. What is the displacement at the same instant?

5. In any S.H.M. when is the displacement a maximum? When is it a minimum?

6. At what part of its path does a particle in S.H.M. move most rapidly?

7. A man walks at a uniform rate in a circular track $ABCD$, Fig. 31. Another man starts from A at the same time and walks along the diameter

AC so that the line joining them is perpendicular to AC. What kind of motion will the second man have? Where will he walk the fastest? The first goes clear around in 20 minutes. What is his angular speed? What is the periodic time of the second man? Shearer's *Questions* 407.

8. (i) Draw a curve with time as x, and distance of M (Fig. 31) from O as y.

(ii) Draw the corresponding time-velocity curve.

(iii) Draw the corresponding time-acceleration curve.

Shearer's *Questions* 410.

CHAPTER III

SOME GENERAL PROPERTIES OF MATTER

55. Up to this point we have been considering some of the motions of particles or bodies, and have confined our attention entirely to the motion, not at all to the moving object. We next proceed to study some of those properties which all pieces of matter have in common; properties possessed alike by bodies at rest and by bodies in motion, and by bodies of every possible chemical composition. Properties of this kind are called **general properties.**

We shall consider four of these, viz. **Inertia, Gravitation, Capacity for Energy,** and **Elasticity.** Afterwards we shall investigate some of the **special properties** of inanimate matter, such as hardness, magnetic quality, transparency, color, and we shall then see how these special properties are employed to classify the different kinds of bodies.

Matter is something with which we are familiar *in a general way* from our earliest years; on the other hand, investigators in physical science have spent centuries in studying the various peculiarities of matter, and have not yet succeeded in defining it in terms of anything simpler. Much has recently been discovered to commend the view that matter will ultimately be found to consist of electricity in motion; but at present this idea is merely a suggestive hypothesis.

I. Inertia

56. One of the commonest experiences of life is that of "sizing up" a body by pushing, pulling, or "hefting" it.

When one sees a barrel being rolled along the street, he can always tell whether it is empty or is filled with something that has considerable weight. The empty barrel rebounds to a greater extent as it passes over a small stone, and is more easily turned aside from a straight course.

Even a wagon drawn by a team behaves so differently when

empty and when loaded that an observer does not need to look into the box of the wagon to see whether it contains a load or not.

Imagine three balls, each six inches in diameter, each painted black, one a rubber football, one a wooden ball from a bowling alley, one an iron cannon ball. Imagine them all started rolling with the same speed over a moderately smooth sidewalk. Any boy or girl can tell instantly, by observing the behavior of the three balls, which one is filled with air, which one is made of wood, and which one of iron.

Everybody knows that the empty barrel is more easily set in motion than the full one, and that it is more easily stopped than the full one. Any one can tell, by kicking the barrel, whether it is full or empty. In like manner, any one can go up to the three black balls, each of the same size and color, each at rest on the ground, and can distinguish, by a push with the foot, the iron ball from the wooden ball, the wooden ball from the football. But just how is one able to arrive so quickly at the correct conclusion? The answer is that every one knows from his previous experience with matter these two facts: —

(i) That it requires an *effort* to put matter into motion when the matter is already at rest; that it requires an effort to stop matter after it has once been set in motion.

(ii) That, with any given body, the *amount* of effort is greater in proportion as the change of motion is greater; and that, for any given change of motion, the *amount* of effort is greater in proportion as the amount of matter in the body is greater.

The first of these facts is described by saying that **all matter possesses inertia.** Now inertia is simply the Latin word for laziness; but this laziness of matter differs in one essential respect from that exhibited by ourselves — on certain more or less rare occasions — in that matter hesitates quite as much to stop when once in motion as to start when once at rest. Does your own experience indicate that liquids and gases, as well as solids, possess inertia?

In all the experience of the human race no exception has been found to the statement that all matter possesses inertia. How, then, shall we explain the fact that a marble started rolling on a level blanket so soon stops? Here the blanket forms a slight hill in front of the marble, and hence exerts an effort, so to speak, against the marble. But suppose the marble put into motion on a smooth level table ; here the marble rolls much farther before it stops, but the very slight roughness of the table is sufficient to bring the marble to rest. The effort which a sheet of level plate glass can oppose to the motion of the marble is slighter still, yet sufficient to stop the marble by and by. The smoother the road and the freer the bearings are from friction, the longer a bicycle will run when coasting on level ground.

Suppose now that a bicycle wheel is entirely free from friction, not only the friction of the bearings, but the friction of the air. How long will it continue to spin when the frame is held above the ground and the wheel set in motion ? **All our experience goes to show that, under these ideal conditions, the wheel would never stop.** The earth in its rotation on its axis comes very nearly being such a body ; the earth is, indeed, so perfectly free from any effort to stop it, that it keeps on moving, year after year, with essentially the same speed.

How nearly constant this speed of rotation is may be judged from the estimate of Sir George Darwin (*Ency. Brit.*, Art. "Tides," Sec. 48), that the effect of the tides in retarding the motion of the earth is perhaps 0.44 second in a century.

In like manner, if a body has a motion of pure translation, (§ 15), the natural and logical inference is that the body will keep on moving indefinitely *if no effort is exerted against it.* This inference was first made by Galileo on the ground of the following experiment.

A body is allowed to roll down the inclined plane *AF*, of fixed height *AB* (Fig. 32). The body on reaching the foot of the plane *AF* with a constant and definite velocity, is allowed to roll up a second incline of variable slope. Galileo observed that as the slope of the second plane diminished it required a longer and longer time for the ball to reach an elevation equal to *AB*. From this he argued that if friction were eliminated and the ball were allowed to roll along a perfectly smooth and horizontal plane *FG*, the time required would be infinite. In other words, there would be no retar-

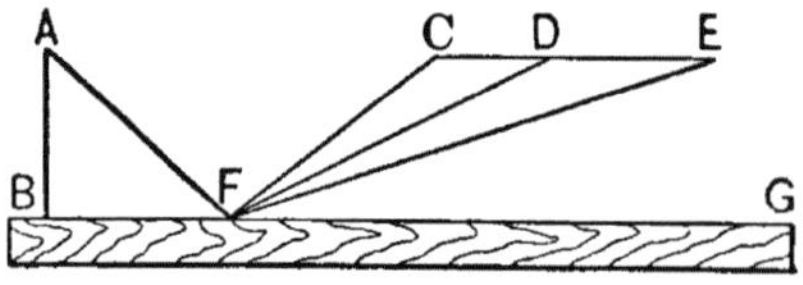

FIG. 32. — Galileo's experimental evidence for the principle of inertia.

dation along such a plane, and hence the motion would be uniform in direction and speed.

57. **But as a matter of fact in all cases of actual motion we find that there is some outside influence at work which interferes with the uniformity of the motion.** Thus, in the case of a bullet fired vertically upwards, we find that the bullet does not continue to move upwards forever, but soon returns to the earth. In the case of a kite, on the other hand, the combined effort of the string and the weight of the kite are not sufficient to bring the kite down. An equally peculiar case is that of an empty bottle corked and then immersed under water. The bottle at once rises to the surface.

In order to explain such various motions in a really simple and comprehensive way it will be necessary first to define two new terms, viz., *Mass* and *Momentum*.

COMPARISON OF MASSES

58. In the preceding paragraphs it has been pointed out that we all know something about estimating the amount of matter in a body by the difficulty of starting or stopping it.

It is customary **in the physical sciences to employ** the word "mass" **to denote the inertia or, if you prefer, amount of** matter **in a body**; and we shall hereafter use the word "mass" to mean simply the inertia or amount of matter in a body, nothing more, nothing less.

The next problem which arises is to find some quantitative method for estimating the mass of a body. The solution may probably be best obtained by considering some simple experiment such as the following: Let two small iron wagons, or a pair of roller skates, be placed on a smooth horizontal table and attached to each other by means of a stretched rubber band *AB*,

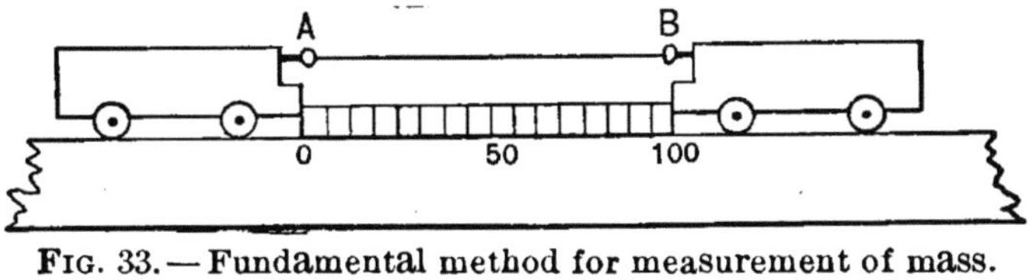

FIG. 33. — Fundamental method for measurement of mass.

as indicated in Fig. 33. Let them be held apart at a distance of 100 centimeters by means of a meter stick. If now the wagons are made to run with very little friction, we may assume, when the meter stick is quickly removed and the wagons approach each other, —

(i) That the only external influence at work is that of the contracting rubber band, and

(ii) That this rubber band pulls equally hard upon each of the wagons.

These are, at this stage, pure assumptions; but they are not violent assumptions; and they are amply justified by the experience of every one.

From the moment when the wagons are released until they collide they will acquire an increasing — and hence accelerated — motion. Let us indicate by M_A and M_B the respective masses of the wagons and their loads, and by a and b the respective accelerations which the wagons have at any instant. Then by definition, *i.e.* by convention of the scientific world, the ratio of the masses M_A and M_B is equal to the ratio $\frac{b}{a}$, a fact more easily expressed by the following equation: —

$$\frac{M_A}{M_B}=\frac{b}{a}. \qquad \text{Eq. 26}$$

Defining equation for ratio of masses.

But since the time during which the stretched rubber acts is the same for each wagon, we may, in view of Eq. 16, substitute for the ratio $\frac{b}{a}$, the ratio of the distances which the respective wagons traverse before collision. Let us observe, then, the point on the meter stick at which the collision occurs, and denote by p_A and p_B the distance from this point to the respective ends of the meter. We shall then have

$$\frac{a}{b}=\frac{p_A}{p_B},$$

and hence

$$\frac{M_A}{M_B}=\frac{p_B}{p_A}. \qquad \text{Eq. 27}$$

Another equation describing the ratio of two masses.

From Eq. 15 it follows that we might have obtained the ratio of these two masses from the velocities of the respective wagons just before collision. For since the accelerations a and b are in the same ratio as the velocities acquired during any time t, we should have

$$\frac{M_A}{M_B}=\frac{V_B}{V_A},$$

where V_A and V_B are the velocities (measured with respect to the table) with which the wagons collide.

A balance for measuring the velocities of collision, and thus comparing masses on this principle, has been devised by the English physicist, Hicks. For description, see Duff's *Mechanics*, p. 80.

UNIT OF MASS

59. Having once learned how to compare (*i.e.* measure) masses, we have only to select a certain standard of mass, in order to be able to determine the mass of any body whatever.

There is a piece of matter preserved in the International Bureau of Weights and Measures at Sèvres, near Paris, a piece of platinum-iridium, weighing about two pounds, which the scientific world has arbitrarily agreed to call the standard kilogram, *i.e.* 1000 grams. **The gram, which is therefore one thousandth part of this standard of mass, is taken as the unit of mass.**

When hereafter we shall speak of the mass of a body as "25 grams," we shall mean that the body has 25 times as much matter in it as there is in one gram; that the effort required to start it moving with a given speed is 25 times that required to start one gram moving with the same speed; or, what happens to be the same thing, that it weighs 25 times as much as one gram.

DIGRESSION ON FUNDAMENTAL AND DERIVED UNITS

60. The standard of mass, and also the standards of length and time, have been defined in a purely arbitrary way. See §§ 9 and 20. They do not depend upon any other standards for their value. **A unit which is chosen in this arbitrary manner, without reference to other units, is called a fundamental unit. In modern physics there are practically only three fundamental units, viz. the unit of mass, the unit of length, and the unit of time.**

All other units are on a different basis, as will be seen by considering one of them, say the unit of speed. The whole scientific world uses as unit speed, that speed with which a particle will traverse unit distance in unit time. In the metric system we say unit speed is one centimeter a second. The unit of speed thus depends upon the units of time and length. It is, therefore, said to be a **derived unit**.

Derived units are defined as those which depend for their value upon the fundamental units. In like manner, the unit of acceleration, the unit of area, and the unit of volume are all derived units. So are all the units employed in this text-book, except the three fundamental ones just mentioned. Units which are based upon the centimeter, gram, and second are spoken of as "C.G.S. units."

The engineer uses a different set of fundamental units, namely, *length, force*, and *time*. We have not yet studied the subject of force; but it will be allowable here to anticipate, merely to say that the engineer's unit of force is the weight of a certain mass of metal preserved in the Standards Office at Westminster, London, and called the **Standard Pound**.

Mass in engineering practice is, therefore, a derived unit; and the size of this unit is about 32.2 times the mass of the standard pound. It is not often used; but when used is frequently called the slug.

Dimensions of Units

61. The powers to which the three fundamental units enter into the various derived units are called the "dimensions" of the derived units. Thus, if we denote the units of length, mass, and time by L, M, and T, respectively, we may write the following table at once from the defining equations for the quantities there listed: —

Quantity	Dimensions
Linear speed	L^1T^{-1}
Linear acceleration	L^1T^{-2}
Angular displacement	0
Angular speed	T^{-1}
Angular acceleration	T^{-2}

The idea of dimensions is useful principally in detecting any lack of homogeneity in the equations of physics. Thus take Eq. 16, or any other equation in this book, and it will be found that, in each term of the equation, the fundamental units enter to the same power. Thus if, in attempting to write from memory the expression for p (Eq. 16), one were to put down

$$p = S_0 t^2 + \frac{1}{2} at^2$$

he could at once detect the error; for each of the terms on the right-hand side must be of the same dimensious as p, namely, L^1; while as the expression now stands the term $S_0 t^2$ has the dimensions $L^1 T^1$.

Dimensional equations are generally written in square brackets, thus, $[V] = [L^1 T^{-1}]$.

Digression on Density

62. From the earliest times it must have been observed that there are certain bodies which float on water, while others sink, more or less rapidly, and also that whether or not a body floats depends not at all upon its size. Thus it is well known that lead, volume for volume, is a heavier substance than wood. But note carefully that this statement does *not* imply that any particular piece of lead has a greater mass than some particular piece of wood.

To describe the manner in which mass and volume are related the term "density" has been introduced and is defined as follows: **The average density of any body is simply the ratio of its mass to its volume.***

If we denote the mass of a body by M, its volume by V, and its density by D, then

$$D = \frac{M}{V}. \quad \text{Defining equation for average density.} \qquad \text{Eq. 28}$$

The gram was intended to be, and is very nearly, equal to the mass of one cubic centimeter of water at a temperature of 4° C. The *mass* of any body of water is, therefore, numerically equal to its *volume*, provided we use the centimeter and the gram as units.

In terms of Eq. 28, $M = V$ in the case of water; and hence the density of water is very approximately unity. More accurately, —

1 cubic centimeter of water at 4° C = 0.999980 gram.
Congrès international de Physique I. 96 (1900).

* Or, as Newton puts it, in the first two lines of the *Principia*, "*The quantity of matter is the measure of the same, arising from its density and bulk conjointly.*"

Digression on Moment of Mass and Center of Mass

63. Let us consider any rigid body of mass M as made up of a large number of particles. Call the masses of these particles m_1, m_2, m_3, etc.; their positions in the body may be stated by means of rectangular coördinates $(x_1y_1z_1)$, $(x_2y_2z_2)$, $(x_3y_3z_3)$, etc., respectively. Next form the sums indicated in the three following equations: —

$$\left.\begin{aligned} m_1x_1 + m_2x_2 + m_3x_3 + \text{etc.} &= \Sigma(mx) \\ m_1y_1 + m_2y_2 + m_3y_3 + \text{etc.} &= \Sigma(my) \\ m_1z_1 + m_2z_2 + m_3z_3 + \text{etc.} &= \Sigma(mz) \end{aligned}\right\} \qquad \text{Eq. 28 A}$$

Each of these products is called a "moment of mass." Those in the first equation are the moments of mass with respect to the YZ-plane; those in the second equation are the moments of mass with respect to the XZ-plane; while the moments of mass, for each particle, with respect to the XY-plane are given in the third equation.

Let us now choose a point whose distance x from the YZ-plane is such that

$$\bar{x} = \frac{\Sigma(mx)}{M}, \qquad \text{Eq. 28 B}$$

and whose distance, y, from the XZ-plane is such that

$$\bar{y} = \frac{\Sigma(my)}{M}.$$

In like manner, make the distance from the XY-plane such that

$$\bar{z} = \frac{\Sigma(mz)}{M}.$$

The point thus determined by these three rectangular coördinates $(\bar{x}, y, z,)$ is called the **center of** mass of the body.

If the origin of coördinates be chosen at the center of mass, it follows that $x = y = z = 0$; and hence

$$\Sigma(mx) = \Sigma(my) = \Sigma(mz) = 0. \qquad \text{Eq. 28 C}$$

It follows, therefore, from Eq. 28 C that the **center of mass of any body is the point about which the moment of mass is** zero. If we pass a plane through the center of mass, it does not follow that there are as many particles of unit mass on one side the plane as the other; but it *does* follow that the moment of mass of the particles on one side is the same as the moment of those on the other. A briefer and clearer way of viewing this subject is perhaps the following. Let each particle m be completely located with reference to any fixed origin by means of a position vector $\bar{r}$. Then the position vector of the center of mass R is by definition expressed in the following equation: —

$$R = \frac{\Sigma(m\bar{r})}{M}. \qquad \text{Eq. 28 D}$$

As will be seen later, in the discussion of gravitation, the center of mass is practically identical with the center of gravity.

DEFINITION OF LINEAR MOMENTUM

64. **The product obtained by multiplying the mass of a body by** its linear velocity is **one so** frequently **employed in physics that a** special **name has been given to** it, viz. **Linear Mo**mentum.

Since velocity is a vector quantity and mass a scalar quantity, it is evident that their product, momentum, is a vector quantity, and that its direction is the same as that of the velocity factor. The product of the mass and linear *speed* of a body gives the merely numerical part, the scalar factor, of momentum.

The momentum of a bicycle ridden north at the rate of six miles an hour is by no means the same as that of the same wheel ridden east at six miles an hour. The mere numerical values of these two momenta are the same; the difference of momenta in these two cases is keenly appreciated when we consider the difficulty involved in avoiding a collision with another wheel going south. For instance, in a crowded street, when the wheels are proceeding from north to south, one after another, the danger to another wheel moving east or west six miles an hour is enormously greater than the danger to the same wheel going north or south six miles an hour.

Let m denote the mass of a particle and $\bar{v}$ its velocity, then

Linear momentum of a particle $= mv$. **Defining eq. for linear momentum.** Eq. 29

The reasonableness of this definition will be seen when we recall that the mass of a body is one of the factors which measure the difficulty of starting or stopping the body. But, as every one knows, the difficulty of starting or stopping a body in translation depends also upon how fast you start it going, or upon how fast it is moving when you attempt to stop it. It depends, as Newton said (Definition 2, *Principia*), upon the "quantity of motion" in it; or, as we now say, upon its **linear momentum.**

Since, then, the difficulty of stopping a body in translation depends both upon its mass and upon its velocity, we say,

$$\text{Linear momentum} = \text{Mass} \times \text{Velocity}.$$

Rotational Inertia and Angular Momentum

65. Two children who are seesawing, one at each end of a board, know very well that the seesawing goes on more smoothly and with greater uniformity when they are out near the ends of the board than when they are in near the middle. But when they are near the middle, the board is much more quickly started or stopped by a third party taking hold of the end. The mass of the board and the children is the same in each case; only the distribution of the mass is different.

The children and the board constitute a typical rotating system. The axis of rotation is the line of contact of the board and the log over which the board is balanced. The point to which attention is here directed is that **the resistance which a body offers to being set in rotation, i.e. its rotational inertia, is not measured by its mass merely, but depends also upon the distribution of its mass,** *i.e.* upon the position of the mass.

Imagine two fly wheels to have equal masses. In the first case, we shall suppose the mass to be distributed mostly near the axis of rotation; in the second case, the mass is placed far out, mostly near the rim of the wheel. Which wheel is more easily set into rotation?

The farther away a particle is from the axis of rotation, the greater its leverage. Consequently it is found, when we attempt to impress a *linear* speed v upon any particle rotating about a fixed axis, that the effectiveness of the particle in resisting us is measured by the product of its linear momentum mv and its distance r from the axis of rotation. Thus the ability of a particle to keep a fly wheel in rotation is measured by the distance of the particle from the shaft, as well as by its linear momentum.

The product $mv \cdot r$ is **called** the **angular momentum** of the particle m.

$$\text{Angular momentum} = mv \cdot r.$$

But since $v = r\omega$, we may write

$$\text{Angular momentum} = mr^2\omega \qquad \text{Defining eq. for angular momentum.} \qquad \text{Eq. 30}$$

And just as we define

$$\text{Linear momentum} = \text{Linear inertia} \times \text{Linear velocity,}$$

so we may define

Angular momentum = Rotational inertia × Angular velocity.

Employing this analogy, which we shall find amply justified as we advance, it follows that the rotational inertia of a particle is measured by mr^2, where m is the mass of the particle and r its distance from the axis of rotation. This product which measures the rotational inertia of any particle is generally, though with little propriety, known as the **moment of inertia**, and is usually denoted by I. Observe that the rotational inertia of a particle is not invariable, as is its linear inertia, but depends upon the position of the axis of rotation with reference to the rotating mass.

Passing from a particle to a rigid body, it is found by experiment that the moment of inertia of a body is the sum of the moments of inertia of all the particles of the body, a result which may be most briefly expressed by the following equation:

$$\text{Rotational inertia} = I = \Sigma(mr^2) \qquad \text{Eq. 31}$$

$$\text{Angular momentum} = I\omega \qquad \text{Eq. 32}$$

Problems

1. How is it that, in a fly wheel weighing one ton, the effect of the wheel in steadying the motion of the engine is very much increased by placing most of the mass out near the rim of the wheel?

2. If an extra pound is to be added to a bicycle wheel, where will it prove to be the greater hindrance to starting, when placed in the tire or in the hub? Why?

3. Why is it that in the balance wheel of a watch most of the material is placed in the rim of the wheel?

4. An iron hoop, circular in shape, and very thin, weighs 200 g. Its radius is 15 cm. Find its rotational inertia about an axis which passes through the center of the hoop and is perpendicular to the plane of the circle.

5. The angular momentum of a certain fly wheel is 25,000,000 C.G.S. units. Its angular speed is 20 radians per second. Find its moment of inertia.

66. At the beginning of this chapter, we cited some familiar instances of the behavior of matter in motion. We then digressed for a moment to increase our vocabulary by the definition of four very important physical quantities; namely,

(1) Mass (Linear Inertia).
(2) Moment of Inertia (Rotational Inertia).
(3) Linear Momentum.
(4) Angular Momentum.

Returning now to the consideration of inertia, we shall treat first the behavior of a body which is left to itself.

Case I. No External Influence at Work

Newton's First Law of Motion

67. As we all know, it is not an easy matter "to leave a body to itself"; but Galileo and Newton have correctly inferred the behavior of a body when left to itself, and have described this behavior in the following simple way:—

If a body is in translation under no external influence, its linear momentum remains constant. This fact is generally called Newton's *First Law of Motion*, and is described algebraically by the following equation:

$$mv = \text{constant.} \qquad \text{Eq. 33}$$

In like manner if a body is in rotation under no external influences, its angular momentum remains constant. Or, more briefly,

$$I\omega = \text{constant.} \qquad \text{Eq. 34}$$

Experiment shows that these two statements are true for bodies at rest as well as for bodies in motion, and that they are true for a body of any size, shape, or composition. But they are true only under the conditions stated; namely, that the body, or system of bodies, is affected by no outside action. Before one can say whether or not a body is acted upon by outside influences, he must be very careful to define just what the body is which he is studying. This done, the distinction between "internal" and "external" influences is easy.

If, therefore, a heavy fly wheel, set in rotation and "left to itself," gradually diminishes in angular speed, we may fairly expect to find the wheel is *not* really left to itself and is *not* entirely free from external influences. In such an expectation no one has yet been disappointed.

If a skater coasting on level ice finds himself moving more and more slowly, he suspects resistance to his motion in the air, and in the frictional resistance between the ice and the skates.

If a baseball, thrown vertically into the air, changes its velocity (as it does, first diminishing, then stopping, and then increasing in speed), we may rest satisfied that an outside influence is at work. In this case it is popularly called "gravity."

Since, then, bodies left to themselves move with uniform velocity, we may say that motion is a state of matter quite as natural as rest. For bodies in motion and left alone keep in motion; while bodies at rest and left alone remain at rest.

Case II. An External Influence at Work

Force and Torque; Newton's Second Law

68. In the preceding pages we have frequently been compelled to speak of the motion of a body as influenced by "resistance," by "external action," by "some outside influence," or by "some external effort." These terms have all been used to express the same idea; but such loose, vague, and varying names are not in keeping with the spirit of modern physics. We accordingly take the first opportunity to define exactly what is meant by these various external influences which change the momentum of a body. For the common feature of all these actions which affect a body from without is that they do not permit the body to move with unchanged momentum; they continually alter the velocity of the body, and hence change its momentum.

It is now nearly three hundred years since men began to observe that changes of momentum are brought about always by the interaction of one body upon another; that the train of circumstances which precedes any particular change of motion is, in general, complicated and far-reaching. The "cause of motion," in other words, usually presents insurmountable difficulties, and leads to metaphysical obscurities. Accordingly, physical science, at a very early stage of its history, cut loose entirely from all considerations of the "cause of motion," *i.e.* of these external influences which we speak of as producing change of momentum, and adopted in its stead the ideas of "force" and "torque." **All these external influences, however complicated and whatever their nature, whether magnetic, capillary, or gravitational, are all grouped under two heads. They are called Forces if they change the linear momentum of the body, and Torques (or Moments of Force) if they change the angular momentum of the body.**

Thus, at a single stroke, physical science rids itself of all hazy metaphysical considerations and confines its attention to observable and measurable quantities, such as masses and velocities. This achievement is due largely to Galileo and Newton. In the future, therefore, when we speak of bodies "acted upon by" forces, we shall understand that these forces are merely names which cover our ignorance of the real causes of motion; but we must not forget that these forces stand for definite physical quantities which can be accurately measured and can be used to predict future results with a precision which is almost incredible.

Newton, looking for the shortest and clearest description of what had been observed to happen where masses, either in motion or at rest, are acted upon by forces, arrived at the following statement, which is generally known as Newton's *Second Law of Motion:* —

The change in the linear momentum of a body is proportional to the force acting upon the body; and the direction of the change is the same as the direction of the force.

A similar law describes the behavior of a body in which there occurs a change of *angular* momentum.

The change in the angular momentum of a body is proportional to the torque (or moment of force) acting upon the body; and the axis and direction of the change are the same as the axis and direction of the moment of force.

DEFINITION OF FORCE

69. Force is defined as that which changes the linear momentum of a body; and it is measured by the time rate at which the linear momentum is changed.

This idea we owe to Galileo (1564–1642). In popular usage force has an endless variety of meanings ; in physics it is never used except in this one sense. The student should observe carefully that in this definition nothing whatever is said as to the origin or source of the force which produces the change of momentum. The source of the external influence may be electric, magnetic, or gravitational ; it matters not. If the momentum is changed, some external force is acting on the body ; and the measure of this force is the ratio between the change of momentum and the time occupied in this change. This is the whole story about force.

It may be expressed in the shorthand of algebra as follows : Let F denote the force acting upon a body of mass m, let v_1 be its linear velocity at the time t_1, and v_2 its linear velocity at the time t_2; then the

Average Force

acting upon the body during the interval $t_2 - t_1$ is given by the following equation : —

$$F = \frac{mv_2 - mv_1}{t_2 - t_1} = m\left(\frac{v_2 - v_1}{t_2 - t_1}\right). \quad \text{Defining equation for average force.} \qquad \text{Eq. 35}$$

If the interval $t_2 - t_1$ be very small indeed, the force at any instant t lying in the interval $t_2 - t_1$ is given as follows : —

$$F = m\left[\frac{v_2 - v_1}{t_2 - t_1}\right]_{t_1 \doteq t_2}$$

Since we have already (Eq. 12) agreed to call the rate of change of velocity by the name acceleration, we may also write this as follows : —

$$F = ma. \quad \text{Defining equation for force.} \qquad \text{Eq. 36}$$

and define **force as the product of mass and acceleration,** a definition which is exactly equivalent to that given above.

Of the two factors in this defining equation, one, m, is a scalar, and the other, a, is a vector quantity. **Hence force** itself is **a vector quantity**, and can, therefore, be represented in direction, sense, and amount by a straight line.

Definition of Torque

70. In a precisely analogous manner, **torque is the name given to the time rate at which the angular momentum is changed.** Let L denote the torque acting upon a rigid body whose rotational inertia (moment of inertia) is I and which is capable of rotation about some given axis; let ω be the amount by which its angular velocity is changed during the time t; then the

Average Torque

acting upon the body during the interval t is given by the following equation: —

$$L = \frac{I\omega}{t} = I \times \text{rate of change of angular velocity,}$$

or, Eq. 13,

$$L = IA, \qquad \text{Eq. 37}$$

where A denotes the angular acceleration of the rigid body about the given axis.

Relation between Force and Torque

71. As a matter of fact many bodies are set into rotation by means of "a single force" applied to the body at what is practically a single point. Thus, one starts a grindstone or a coffee mill by means of a push of the handle; the beam of a balance is set into rotation by means of a force applied at the knife-edge.

The effect of such a force acting on any such body is in general to change not only its motion of translation, but also its motion of rotation. The amount of this change in rotational speed will obviously depend upon two things: (1) the magnitude of the force applied, and (2) the distance of the line in which it acts from the axis of rotation.

Thus, if one wishes to shut a door and can exert only a limited force, he knows that this force will be much more effective when applied near the edge of the door than when applied near the hinge line.

In like manner, when one wishes to turn a heavy wheel, he

always takes hold near the rim and not near the hub. He does this in order to make the perpendicular distance between the force and the axis as great as possible.

Hence, the effect of a force in producing rotation about any axis can be measured by the following product, viz. Fr, where F is the force acting, and r is the perpendicular distance from the axis to the direction of the force. See Fig. 35.

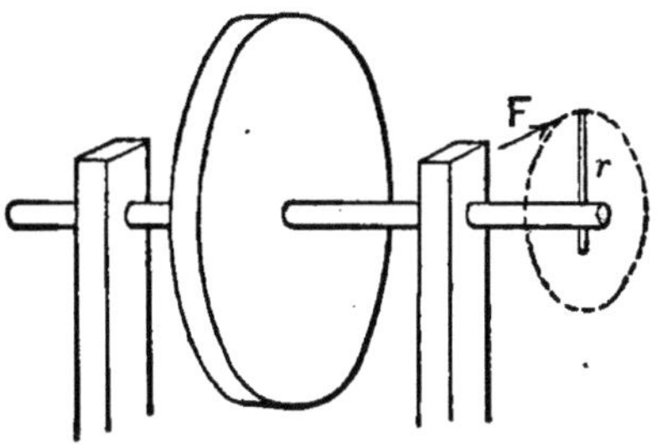

FIG. 35.—Moment of force on a particle, at the end of an arm r.

Let us denote the moment of force by L, then

$$L = rF, \qquad \text{Eq. 38}$$

This simple relation between force and moment of force, $L = rF$, should be thoroughly mastered by every student; for it is the connecting link between force and torque, and is generally considered as the defining equation for torque. Outside of engineering circles torque is generally called "moment of force," because "moment" is a Latin word which means importance; and since the product Fr measures the importance of the force F in producing rotation, it is appropriately called moment of force. The word is used here then in the same sense in which Shakespeare employs it when he speaks of "enterprises of great pith and moment."

ADDITION OF FORCES

72. Since forces are directed quantities, they are added in the same manner as are all directed quantities; that is, in the same manner as displacements, velocities, momenta, accelerations, viz. by placing end to end the lines which represent the single forces. **The line joining the extremities of the broken line so formed represents the resultant force, or simply the resultant.**

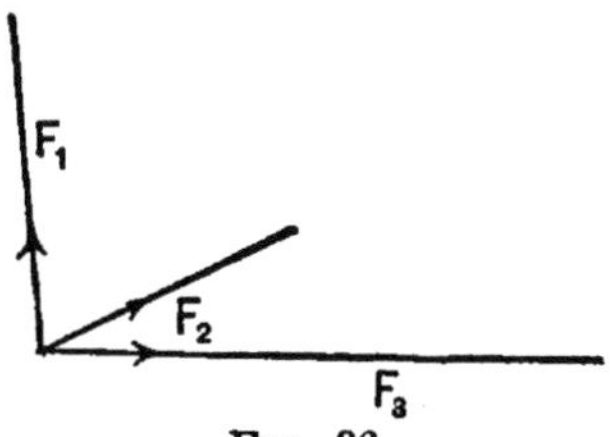

FIG. 36.

Each force produces its own effect quite independently of the others. This is one of the essential features, although merely a tacit implication, of Newton's Second Law.

Imagine the three vectors F_1, F_2, F_3, indicated in Fig. 36, to represent in direction and in amount three different forces acting at any point O. The line $\overline{OA}$ in Fig. 37 will then represent a force which alone will change the momentum of a body at the same rate as will the three forces F_1, F_2, F_3, combined. This is all that is meant by a resultant.

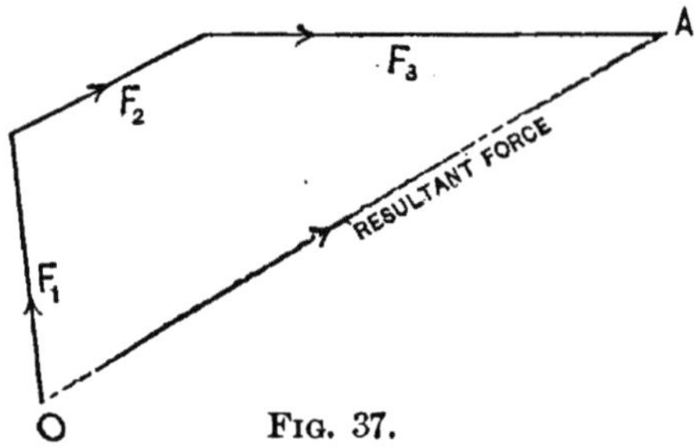

FIG. 37.

Imagine the sense of the force F_3 to be reversed; the line $\overline{OA}$, in Fig. 38, will then be the resultant. The numerical values of the forces are the same in Fig. 38 as in Fig. 37; so are their directions; but the resultants in these two cases are very different. Accordingly, the student should bear in mind that the sense of a force is, in general, quite as important a factor as its amount, or its direction.

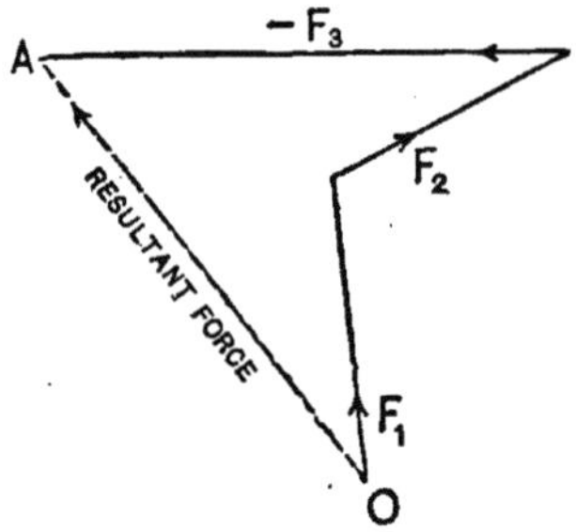

FIG. 38.

73. Very often **the resultant of a given system of forces is zero; in this case the body acted upon is said to be in** equilibrium.

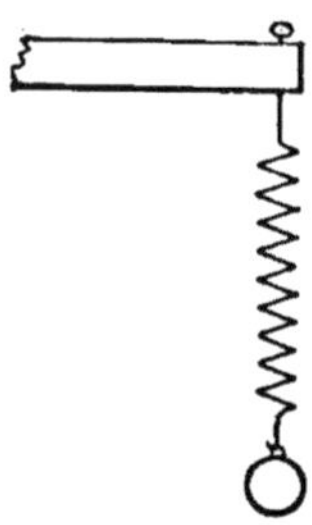
FIG. 39.

That large branch of Mechanics which deals with forces which are in equilibrium is called Statics. Thus, when a metal ball is suspended by a spiral spring as shown in Fig. 39, the weight of the ball, acting downwards, is exactly balanced by the pull of the spring upwards; and either of these forces is to be measured by the acceleration it would produce in the ball if the other force were absent.

UNIT OF FORCE: THE DYNE. THE POUND

74. Since the force acting upon any body is the product of the acceleration which it produces in this body multiplied by the mass of the body, we have no liberty of choice as to a unit of force. For, provided we wish to be at all consistent, we must maintain our previously employed units of acceleration and mass. But these we have already defined in §§ 33 and 59. Hence the unit **of** force which is **called a "dyne" is that force which** will **produce** unit **acceleration in a mass of one** gram.

Or, what is the same thing, unit force is that force which is capable of changing the momentum of a body at the rate of one unit per second.

Or, what is the same thing still, we may define the unit of force as that force which will change the velocity of a mass of one gram at the rate of one unit (that is, one centimeter per second) per second. Note that "dyne" is simply the Anglicized form of the Greek word for force. The student will be interested in proving that the dimensions of force are MLT^{-2}.

The engineer employs as a unit of force the weight of a one-pound mass at London; in other words, his unit of force is that force which the standard pound exerts upon the bottom of the box which contains it at the Standards Office in London.

In order to distinguish between the mass of a pound and the force of a pound, it is well to write the former "*lbs.*" and the latter "*pounds.*"

The unit of torque has no special name; its value is evidently (Eq. 38) the torque of one dyne acting through an arm of one centimeter in a direction perpendicular both to the arm and to the axis. The engineer, starting from the same equation, uses the pound-foot as his unit of torque. (This unit is to be carefully distinguished from the *foot-pound*, which the engineer uses as a unit of work, and in which the distance and the force are measured in the *same* direction instead of at right angles as here.)

Statics

Case of a Particle

Let us consider first the case of a single particle acted upon by any number of forces all lying in the same plane. If these forces produce no acceleration of the particle, they are said to be in equilibrium. From what has preceded it will be evident that the condition for equilibrium here is as follows: —

Polygon of forces. *Represent all the forces by limited straight lines. Starting with any one of these forces, lay them off end to end. If the polygon thus formed is closed (that is, if the terminal point of the last vector coincides with the starting point of the first vector), the forces are in equilibrium.*

If there are only three forces acting upon the particle then they must form a triangle when placed end to end. This is, of course, a special case of the preceding. Another way of expressing this result, due to Lami, and often called "Lami's theorem," is to say that three forces are in equilibrium when each is proportional to the sine of the angle between the other two. For most purposes the graphical method of determining equilibrium is convenient; but when numerical computations are concerned, the algebraic method is often better.

Algebraic Condition for Equilibrium

Let us denote by R_1, R_2, R_3, etc. the various forces which act upon the particle; and by θ_1, θ_2, θ_3, etc. the respective angles which these forces make with the axis of X. Then if X is the sum of all the components along the direction of the X-axis, and Y the sum of all the components along the Y-axis, we shall have as the condition of equilibrium,—

$$X = R_1 \cos\theta_1 + R_2 \cos\theta_2 + R_3 \cos\theta_3 + \text{etc.} = \Sigma(R\cos\theta) = 0.$$
$$Y = R_1 \sin\theta_1 + R_2 \sin\theta_2 + R_3 \sin\theta_3 + \text{etc.} = \Sigma(R\sin\theta) = 0.$$

For if there is no force acting along either of two perpendicular axes, then it must follow that the resultant $\sqrt{X^2 + Y^2}$ is zero and the particle is in equilibrium.

What form will the condition of equilibrium take when the particle is free to move not only in two dimensions, but in three?

Case of a Body

We shall here consider only the case of a body which is free to rotate about a single fixed axis, such, for instance, as a grindstone or the fly wheel of a stationary engine. In seeking for the condition of equilibrium of such a body we have to remember that the question is now one of torques, and that the perpendicular distance from the axis of rotation to the line of the force is a factor which is quite as important as the amount of the force.

Since the resultant torque L is to be zero, we may express the condition of equilibrium as follows:—

$$L = F_1 p_1 + F_2 p_2 + F_3 p_3 + \text{etc.} = \Sigma(Fp) = 0 \qquad \text{Eq. 39}$$

where F_1, F_2, F_3, etc. are the forces and p_1, p_2, p_3, etc. are the perpendiculars let fall from the axis to the lines of action of the respective forces.

Thus in the case of a windlass where the force F acting through the arm a raises a weight W acting through an arm b, we have, as the condition of equilibrium

$$Fa - Wb = 0.$$

Observe that in the case of torques it is quite as important to give to each moment of force its correct sign as it is to assign the proper sense to forces.

THE COUPLE

One of the most important combinations of forces occurring in nature is that in which the resultant of the linear forces, ΣF, is zero, while the torque is not zero. The typical case here is that of two forces which are equal, parallel, and oppositely directed, but which do not lie in the same straight line. Call the two forces F_1 and F_2; then since $F_1 + F_2 = 0$, it follows that they can produce no change of linear velocity in the body upon which they act. But, consider their torque with respect to any point lying, either between or outside the lines of the two forces, and it will be seen that they exert a torque whose value is $F_1 r$, where r is the least distance between the two forces. A couple will therefore produce angular acceleration. Hence the two equal forces F_1 and F_2 cannot be said to be in complete equilibrium unless they act along the same straight line.

Problems

1. An iron rod 4 ft. long weighs 10 lb., and has suspended from one end a weight of 4 lb., from the other end a weight of 8 lb. At what point must the rod be supported in order to be in equilibrium?

2. A pendulum bob weighing 2 lb. is pushed aside by the hand until the suspension string makes an angle of 30° with the vertical. What horizontal force must the hand exert in order to do this.

3. An iron bar carrying a weight of 3 lb. at one end balances when supported at a point 4 ft. from that end. If the bar weighs 2 lb. per foot length, what is the total length of the bar?

4. A man can just close a door against a strong wind by using a horizontal force of 50 lb. at a distance of 2 ft. from the hinge line. What force will be required at a distance of $2\frac{1}{2}$ ft.

5. Find the center of mass of a uniform right-angled triangle?

6. Prove that when a particle is in equilibrium under the action of three forces, the resultant moment of these three forces about any line is zero.

7. Two parallel forces, of which one is 5 lb., have a resultant of 6 lb., acting at a distance of 2 ft. from the larger force. Find the distance between the parallel forces.

8. A square sheet of metal 8 cm. on the side has a square of 2 cm. on the side cut out of one of its corners. Find the center of mass of the remaining figure.

9. Find the center of mass of four masses, of 3, 6, 8, 4 lb., respectively, which are placed in a straight line with the following distances separating them, 6 in., 2 ft., and 3 ft., respectively.

10. Three men are carrying a log. The rear end is carried by one man; the forward end is supported over a stick each end of which is carried by one of the other two men. Where must this stick be placed under the log to divide the load equally between the three men?

11. What couple will two equal parallel and opposite forces of 3 dynes produce when acting at opposite ends of an arm 0.1 mm. long?

12. A string 80 in. long is stretched by a force of 5 lb. What force must be applied at the middle of the string to depress it 1 in.? Neglect the weight of the string.

13. An I-beam is supported by two pillars, one at each end, the load on the respective pillars being 100 and 200 lb. When the pillars are shifted so that each stands 1 ft. from the end of the beam, the loads are 90 and 210 lb. respectively. Find the length of the beam.

14. Prove that when the arms of a balance are unequal, the true weight of a body may be found by observing its apparent weight, first in one scale pan, then in the other, and afterwards taking the square root of the product of their apparent weights.

Two Interesting Special Cases

I. *Centripetal and Centrifugal Force*

75. We have found that, in general, the force acting upon any body is to be measured by the product of the mass of the body and its acceleration; and when a particle moves in a circle with constant speed, we have found (§ 37) that it is always accelerated toward the center with an acceleration which is measured by the square of its linear speed divided by the radius of the circle.

$$a = \frac{v^2}{r} = \omega^2 r. \qquad \text{Eq. 18}$$

If now we wish to know what force F is required to keep this particle moving in a circular path, we have only to multiply the mass of the particle m by its acceleration. Thus,

$$F = m\frac{v^2}{r} = m\omega^2 r = mv \cdot \omega. \qquad \text{Eq. 40}$$

This expression gives the force which one must exert upon a string in order to change the direction of the linear momentum mv at the rate ω.

This force is sometimes called **centripetal** force because it measures the pull of the string towards the center; it is sometimes called **centrifugal** force because it measures the pull of the stone outward. As we shall see later (§ 77) both names are appropriate; for **each** represents **one of the two points of view from which** every force **may be considered.** These laws of uniform circular motion were discovered by Huygens (1625–1695), the great Dutch physicist and astronomer.

II. *The Gyroscope; Precession*

76. The rotational analogue of this centrifugal force is one of the most interesting phenomena of mechanics and astronomy.

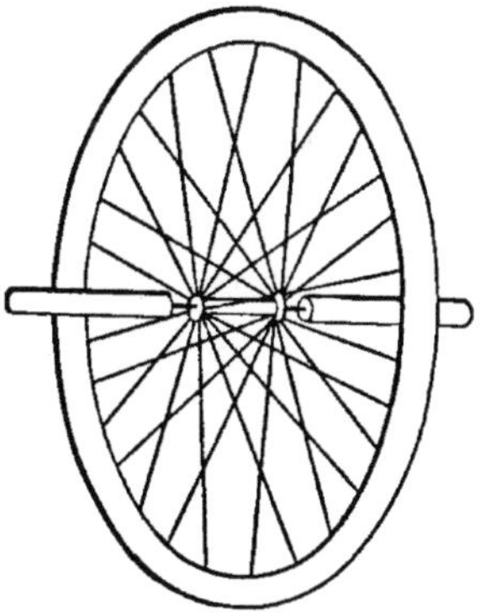

FIG. 40. — Bicycle wheel detached so as to easily show precession.

Corresponding to the linear momentum of a rigid body is the angular momentum of a rigid body, such, for instance, as that of a detached bicycle wheel, shown in Fig. 40. Such a bicycle wheel held in one hand can be set spinning rapidly with an angular velocity which may be denoted by Ω. The angular momentum of the wheel will, by analogy, then be $I\Omega$. And the torque L required to change the direction of this angular momentum at the constant rate ω, will, by the same analogy, be $I\Omega\omega$; or in terms of algebra, —

$$L = I\Omega \cdot \omega. \qquad \text{Eq. 41}$$

If the spinning wheel is supported by both handles with its axle horizontal, there is no torque acting upon it, and hence no tendency for the axle to change direction; but if the wheel be set spinning about a horizontal axis and held lightly by one handle, then the weight of the wheel exerts a torque about another horizontal axis perpendicular to the axis of spin, and the torque L makes the axis of spin, therefore, rotate about a vertical axis. This rotation of the axis of spin is called **precession.** The advanced student will find this phenomenon beautifully exhibited by the earth, upon the equatorial belt of which the moon exerts a torque L.

The simplest illustration is the ordinary peg top; but some of the various forms of gyroscope, usually found on sale at the expositions and toy stores, are more convenient.

Problems

1. A mass of 500 g. is acted upon by a force of 3500 dynes. Find the acceleration produced in the mass. Starting from rest, under this acceleration, what speed will the mass acquire in 8 sec.?

Ans. $a = 7$, $v = 56$.

2. A force of 20 dynes is able to produce an acceleration of 2 in a certain body. What is the mass of the body? *Ans.* $m = 10$ g.

3. What force would be required to change the momentum of a body from 25 to 150 in 12.5 sec.? *Ans.* $F = 10$ dynes.

4. A boy swings about his head, in a circle of 1 m. radius, a bullet whose mass is 20 g. What is the pull on the string when the bullet is making 1 revolution per second? *Ans.* $8000\,\pi^2$ dynes.

5. A locomotive whose mass is 100,000 kg. is rounding a curve of 400 m. radius at a speed of 10 m. per second. What is the force which the locomotive exerts upon the outer rail? This problem is similar to the preceding; the only difference is that here the flange of the wheel, or the outer rail, holds the locomotive in its circular path, while in the preceding the string holds the bullet in its circular path.

Ans. 2500 million dynes.

6. Why does a bicycle rider "lean in" as he rounds a corner?

7. How does a cyclist "right" himself by turning his front wheel toward the side to which he is falling? The centrifugal force, *i.e.* the righting force, acting upon the wheel and rider is measured by $\frac{mv^2}{r}$. This may be increased by increasing v, or by diminishing r.

8. An engine fly wheel is spinning about a horizontal axis with an angular momentum of 50,000,000 C.G.S. units. How great a torque will be required to rotate the axle of the flv wheel (still spinning) about a vertical axis at the rate of $\frac{1}{2}$ radian per second? About what axis must this torque be exerted?

On Systems of Bodies; Newton's Third Law of Motion

77. If we take acceleration as the test for the presence of force, questions such as the following are likely to arise: A railroad train is running on a level track at the uniform rate of 24 miles an hour, there is no change in the speed or in the direction of the velocity; accordingly the acceleration of the train is zero. This conclusion is plain, simple, unavoidable. But how can this be, when the engine is exerting a force of millions of dynes on the train?

The answer is simply this: The pull of the engine is only one among many forces acting upon the train. In these forces we must include the friction of the rails, the friction of the bearings, and the friction of the wind. These forces are all opposing the pull of the engine. If now the train be running *uniformly* at the rate of 24 miles an hour, we conclude that these opposing forces are exactly equal and opposite to the pull of the engine; and hence that the total force acting upon the train is zero. If steam be cut off the engine, the opposing forces accelerate the motion of the train in such a way as to diminish its speed. When the train comes to rest, the opposing forces become zero; and hence the total forces are again zero and the acceleration is again zero. When the engineer lets on steam, the forward pull of the engine is greater than the backward pull of the opposing frictional forces. Hence the motion of the train is accelerated, this time in such a way that its speed is increased. This acceleration will continue until the frictional forces are exactly equal and opposite to the pull of the engine. At this point the acceleration is again zero.

From this illustration the student will observe that great care is necessary to include in any system under discussion all the various parts of that system. The track and the air retarding the train are essentially parts of the system.

When the forces which aid any uniform motion are added to those which oppose the motion, the sum is always zero. But even when the motion is *not* uniform, it is found that the mass which is being accelerated exhibits a force of reaction just opposite in sense and equal in amount to the resultant of all the other forces acting upon the body. This is an observed fact of nature which Newton was the first to accurately describe. **This he did by saying that action is always equal and opposite to reaction; or the mutual actions of any two bodies are always equal and oppositely directed.**

This statement is known as Newton's *Third Law of Motion.* This force of reaction in accelerated bodies is beautifully illustrated by the following experiment: —

With a piece of light string support an iron ball, from two to four inches in diameter, as indicated in Fig. 41. To the lower side of the ball attach a piece of the same kind of thread.

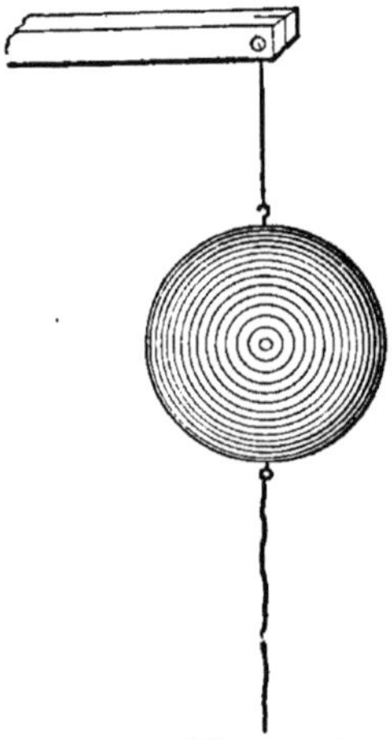

FIG. 41. — Illustrating reaction which is measured by mass × acceleration.

Taking hold of this lower thread, one may by pulling break either thread he likes. If he pulls slowly, the lower thread will be affected only by the pull of the hand, while the upper thread will be stretched both by the pull of the hand *and* by the weight of the iron ball. In this case, of course, the upper string will break first.

If, on the other hand, one pulls the lower string quickly, it will as before feel the pull of the hand alone. But the upper string will be stretched practically by the weight of the ball only; for in this case the ball will be so rapidly accelerated that the reaction of the ball (its mass × its acceleration) will be sufficient to break the lower string before the ball has moved through a distance sufficient to appreciably stretch the upper string.

The student will find it an excellent habit to look at every force from these two points of view.

When a body, or system of bodies, is not accelerated, we may consider it is acted upon by two or more forces whose sum is always zero, or by two or more moments of forces whose sum is always zero.

When a body *is* accelerated we may consider the force of reaction as one of the forces acting upon the body; and from this point of view, which is due to D'Alembert, *all forces are balanced forces.*

Relations of Newton's Laws to Each Other

78. The first law describes the behavior of a body when there is no external force acting upon it; or if you prefer, it states the conditions under which there are no external forces. The second law describes the behavior of bodies when external forces are acting. It is clear, therefore, that the first law is merely a special case of the second law, telling us what happens when the external forces have a particular value, namely, zero.

To clearly grasp the relation of the third law to the other two it is helpful to recur to the idea of Maxwell that when two bodies interact there is a "dynamical transaction" taking place between them. "Just as in commercial affairs the same transaction between two parties is called Buying when we consider one party, Selling when we consider the other, and Trade when we take both parties into consideration" (*Matter and Motion*, Art. 39).

The third law tells us then what happens when we include in our system all the bodies whose action we are considering. Under these conditions, that is, under no external forces, the third law says that the forces remaining are equal. The "forces remaining" may be called *internal* forces.

Put in other words, the first law furnishes one, but not the only, criterion for the presence (or absence, if you like) of force. The second law tells us how to measure force, an idea due to Galileo. The third law expresses the fact that forces always occur in pairs which are equal and opposite.

Problems

1. A bicycle rider exerts a force of 10 million dynes on a pedal. Suppose the force to be at right angles to the pedal crank, and imagine the crank to be 18 centimeters long, what moment of force does the rider exert? *Ans.* $L = 180 \times 10^6$.

2. At one end of a board which is 20 ft. long is placed a stone weighing 40 lb. At the other end of the board is placed a stone weighing 90 lb. Neglecting the weight of the board, find the point at which it must be supported in order that these two stones may just balance each other. *Ans.* $6\frac{2}{13}$ ft. from end.

3. When a bicycle is being ridden through a wet or slushy street, how is it that the water collects in a little ridge at the middle of the tire?

4. In opening my pocket-knife I catch the blade at a point 8 cm. from the pin about which the blade turns; and, at this point, I exert a force of 500,000 dynes. What moment of force is required to open the knife?

5. A mass of 20 g. is revolving uniformly once in 5 sec. in a circle of 30 cm. radius. Find the centrifugal force.

6. A skater weighing 70 kg. describes a circle of 30 metres radius with a speed of 6 m. per second. Find the horizontal force tending to throw him over. Why does he incline himself toward the center of the circle?

7. How many times a minute must a mass of 3 g. revolve in a horizontal plane at the end of a string 50 cm. long in order that the force on the string may equal 1000 dynes?

8. An angler, fishing with rod and line, hooks a fish. Will the fish appear to pull harder if the rod is a long one or a short one? Why?

9. Taking a period of the moon's revolution around the earth as 28 days and its distance from the center of the earth as 240,000 mi., calculate its acceleration.

II. Second General Property of Matter: Gravitation

79. Children at a very early age learn to look toward the floor for things that leave their hands. This indicates what is probably their first acquaintance with the great fact that the earth attracts all bodies to itself. This single fact was practically all that men knew of gravitation until the sixteenth century, when Galileo, by his researches along this line, instituted the era of modern physics.

Before the time of Galileo (1564–1643), the ideas of men concerning gravitation were simply chaotic. They had observed that some bodies, such as stones, fall very rapidly; that others, such as feathers and thin paper, fall very slowly; that heavy bodies fall with a constantly increasing speed; that lighter bodies, such as small raindrops, or feathers, fall with a practically uniform speed; that in some media certain bodies do not fall at all, as, for instance, wood in water, or iron in mercury. Galileo was the man who first harmonized these apparently contradictory facts, and described them all in a very simple way.

Up to this period nothing was known as to how far a body would fall in a given time, or as to what speed it would acquire in a given time, or as to what speed it would attain in falling a given distance.

The difficulty in getting at these facts, *i.e.* the difficulty in discovering *how* a body falls, lay partly in the circumstance that the ancients had no accurate method for measuring small intervals of time, partly in the circumstance that the speed of falling bodies, after a minute interval of time, is liable to be very rapid, and partly in the fact that the predecessors of Galileo had little or no idea of obtaining knowledge by experiment.

For measuring small intervals of time, Galileo employed an ingenious water clock, consisting of a large vessel of water with a small opening in the bottom. Just before the beginning of the observation, this opening was closed by the finger. At the beginning of the observation the finger was released and the water allowed to flow until the end of the observation, when the opening was again closed. The water flowing out was collected in a cup and weighed, the weight of the water being proportional to the duration of the observation. To prevent the attraction of the earth from giving the falling body too high a speed, he allowed the body to roll down a smooth inclined plane of small slope.

But first he satisfied himself that all kinds of matter would fall the same distance in the same time. One ball might be made of copper, another of iron; the time of descent was the same in each case. The resistance of the air was practically eliminated when bodies of the same size and nearly the same density were let fall. But, in many cases, large bodies and small bodies occupy *practically* the same time in falling from the same height. On the other hand, as every one knows, a bunch of cotton does *not* fall as rapidly, in air, as does a bullet of lead.

But Galileo's conclusions were that, in all such cases, the difference in speed was due to the resistance of the air. He regretted that he had no means of producing a vacuum, so that he might appeal to a direct experiment; accordingly he employed a series of fluids of different densities, such as mercury, water, air, and found that as the density of the fluid diminished the bodies descending in these fluids fell more nearly at the same rate, *i.e.* the same distance in the same time. **He argued, therefore, that, in the limit (i.e. in a vacuum), where the density is zero, all bodies, when starting from rest, whatever their size, shape, or composition, fall through equal distances in equal times.**

How thoroughly Galileo was justified in this conclusion was proved later, about the year 1650, when the air pump was invented by Otto von Guericke. One of the first experiments to be tried in the newly discovered vacuum was to place a feather and a coin in the same tube, and see whether they would fall at the same rate in a region devoid of air. The result was such as to show that Galileo was perfectly correct in thinking that the earth impresses the same velocity upon all falling bodies, if allowance is made for the resistance of the air.

80. Galileo was now ready to begin his experiment on the inclined plane, to which we have referred above. This is very conveniently repeated in the following form (see Fig. 42): From one side of a room to another stretch a brass or steel wire of one or two millimeters diameter. The wire should have a slope of about one foot in eight or nine, and should be stretched very taut.

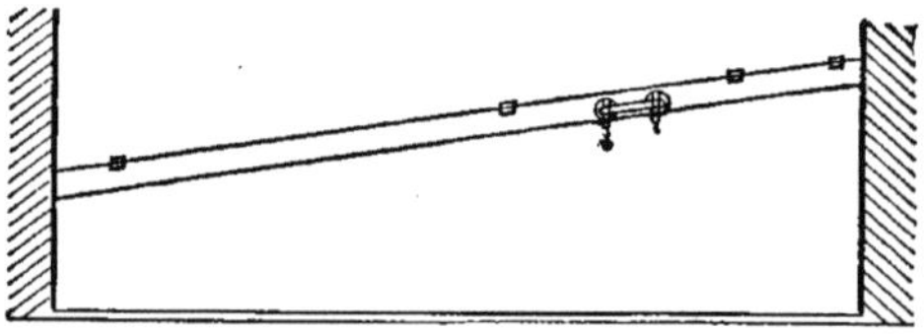

FIG. 42. — Galileo's experiment.

A small carriage to run on this wire track can be easily made by soldering to a small brass bar the straps of two small and very light wooden pulleys (see Fig. 43). The straps of the pulleys must be sawed through on the other side, so that they may slip over the wire. A small piece of metal, say a brass sphere, hooked on to the lower part of the carriage, will keep the whole upright on the wire track.

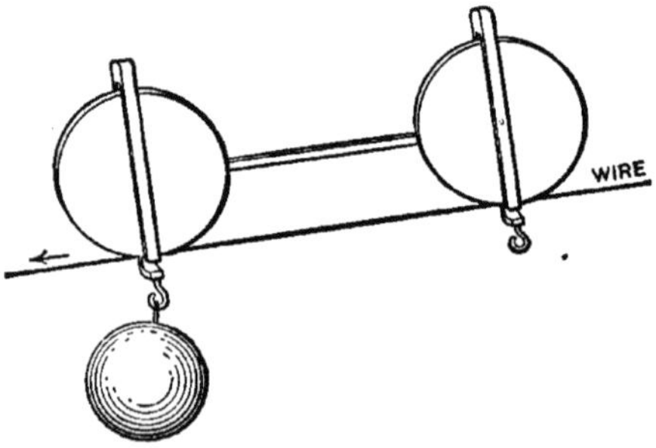

FIG. 43. — Carriage for Galileo's experiment.

A metronome, a clock, or a telegraph sounder should beat seconds in the room. Parallel to the stretched wire, and about six inches above it, should be stretched a string, on which some small indices of cardboard are strung, as shown in Fig. 42.

On the beat of the seconds pendulum the carriage is released at the upper end of the inclined track. The position of the carriage at the end of one second is marked by sliding a cardboard index along the string to the proper place. A second observer, or a very soft buffer, to catch the carriage at the lower end of the track, is convenient. Another trial enables one to mark the position of the carriage at the end of the second second, a third observation gives the position of the carriage at the end of the third second, and so on. Having thus marked off, on the inclined plane, the positions of the falling body at the ends of successive seconds, one may tabulate his results as they are given in the first two columns of the following table. From these the acceleration along the plane is easily computed by means of our general equation for acceleration in terms of distance and time, Eq. 16. These results are given in column 3 and are, it will be noticed, nearly constant, thus indicating that *the acceleration along the plane does not vary, although the velocity does vary.*

Time since Carriage was Released t	Total Distance passed over by Carriage in Time t	Acceleration a along Inclined Plane: computed from Formula $x = \frac{1}{2}at^2$	Acceleration in a Vertical Direction: computed from Formula $g = 12.7\,a$
1 sec.	30 cm.	60	762
2 sec.	118 cm.	59	749
3 sec.	273 cm.	61	775
4 sec.	478 cm.	60	762
5 sec.	726 cm.	58	737

Observer — E. W. Gray.

Another way of stating the result which Galileo obtained by making a large number of observations of this kind, is to say that the constancy (approximate, of course) of the figures in the third column means that the distance traversed by a body sliding freely down an inclined plane varies directly as the square of the time, a result which is graphically expressed in Fig. 44.

We must next recall that the acceleration a which we have been measuring is only one component of the total acceleration, namely, that component which acts along the inclined plane. Knowing this component, BD in Fig. 45, and knowing that the total acceleration of gravity, which we shall hereafter denote by g, is vertical, we can easily find the value of g. For the triangles BCD and AOX are similar; and hence

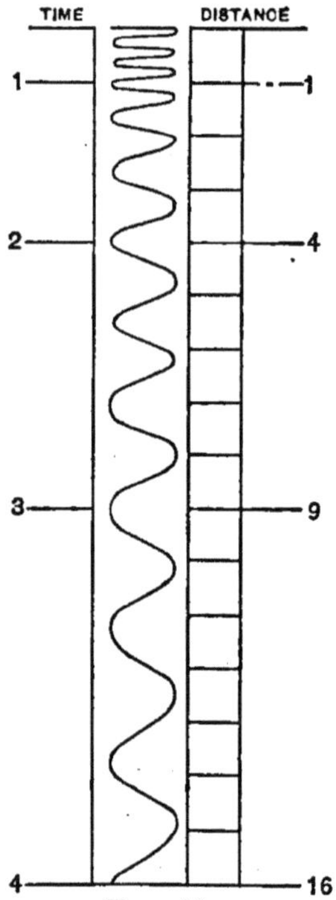

FIG. 44.

$$\frac{g}{a} = \frac{AX}{AO} = 12.7 \left\{ \begin{array}{l} \text{in the experiment} \\ \text{recorded above.} \end{array} \right.$$

The values computed from a in this manner are given in the fourth column of the table above.

In this manner Galileo inferred that this particular acceleration, namely, the acceleration produced, upon a freely falling body, by the earth, is also a constant. This is a capital discovery; a fact with whose consequences we are so familiar that we are perhaps a trifle blind as to its importance.

As every one knows, gravity accelerates a body most rapidly along the vertical direction, and does not accelerate a body at all in a horizontal direction. Indeed, all that we mean by a "horizontal" direction is one in which the acceleration of gravity has no component.

This device is, therefore, competent not only to show that the acceleration of gravity is a constant, but also to give us the numerical value of this constant. Since, however, no allow-

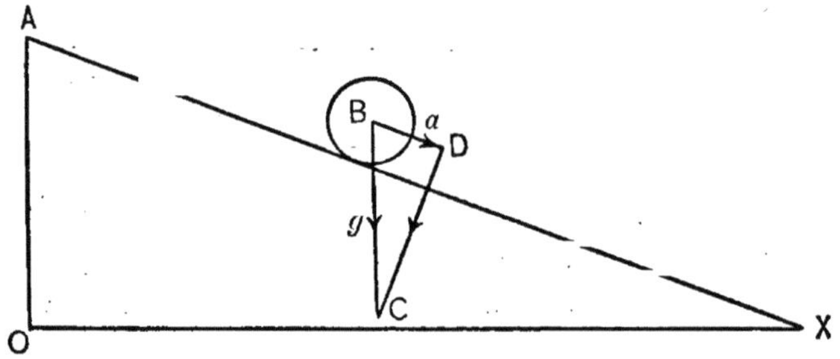

FIG. 45.

ance has here been made for friction, or for the sag of the wire, or for the fact that a part of the body is rolling instead of sliding, the results are about 25 per cent too small.

How nearly constant the acceleration of gravity is at various parts of the earth's surface will be shown by the following table : —

PLACE	OBSERVER	OBSERVED VALUE OF g IN $\frac{m}{\text{sec.}^2}$
London	Kater	9,81282
St. Helena	Foster-Lütke	78736
New York	Foster	81613
Cape Horn	Foster	81676
Guam	Lütke-Freycinet	78599
Paris	Defforges	81012
Biskra	Defforges	79701
Chicago	Defforges	80375
Salt Lake City	Defforges	80050
Denver	Defforges	79983
Mount Hamilton	Defforges	79916

— *Congrès international de Physique,* Paris, 1900, Vol. 3, p. 368.

81. Having found that the acceleration of gravity is constant, Galileo at once made the following inference; namely, that the speed at the end of any interval of time is proportional to the first power of the time. This consequence, we have already seen (Eq. 15, p. 43), follows in *all* cases of constant acceleration. The inference may be tabulated as follows :—

Relation between Time of Fall and Speed Acquired

TIME	SPEED
1	g
2	$2g$
3	$3g$
4	$4g$
5	$5g$
6	$6g$
etc.	

By definition, g is the number by which the speed of a freely falling body is changed in each second. Consequently, in t seconds, the total *change* in speed will be gt; or, if the body starts from rest, the results in the preceding table are expressed algebraically as follows :—

$$S = gt.$$

If the falling body has, at time $t = 0$, a speed of S_0, our expression for the *actual speed* at any instant t becomes

$$S = S_0 + gt.$$

SUMMARY OF LAWS OF FREELY FALLING BODIES

82. These are exactly the laws of a uniformly accelerated particle which we have studied in Chapter I, only here the numerical value of the acceleration g is practically always 981 centimeter/second^2 units.

Relation between Speed and Time

$$S = S_0 + gt. \qquad \text{Eq. 15}$$

Relation between Distance and Time

$$x = S_0 t + \tfrac{1}{2} gt^2. \qquad \text{Eq. 16}$$

Relation between Distance and Speed

$$S^2 = S_0^2 + 2gx. \qquad \text{Eq. 17}$$

83. But when a body is *not* falling freely, one obtains its acceleration by the general method of first finding the resultant of all the forces acting upon it and then dividing this resultant by the mass of the body.

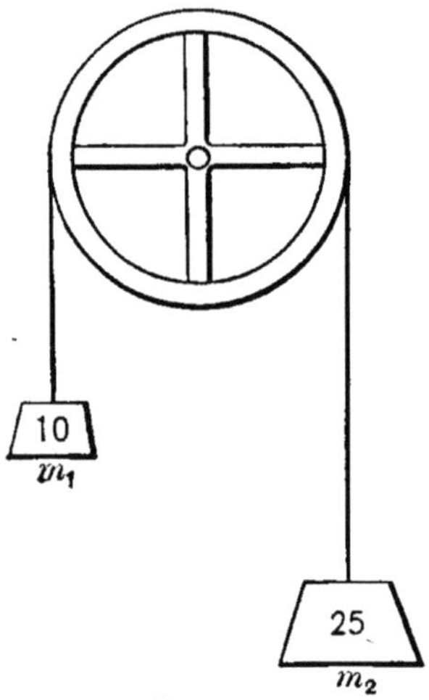

FIG. 46.

Thus if two masses m_1 and m_2 be strung over a light frictionless pulley by means of a thin string, as shown in Fig. 46, the mass which falls is not a "freely falling body." The same may be said of a body which rolls or slides down an inclined plane. Referring to Fig. 46, which represents the essential feature of Atwood's machine (a classical instrument devised for the purpose of diluting gravity, so to speak), we proceed to write the equations of motion for the two bodies m_1 and m_2 there represented.

The data of the problem are, say, the masses of the bodies and the acceleration of gravity g. It is required to find the acceleration a of the system and the force T with which the string is stretched.

Consider first the body m_1; all the forces acting upon it are—

(i) the tension T of the string acting upward,

(ii) its weight $m_1 g$ acting downward; and

(iii) the reaction of the mass m_1 against its upward acceleration, which amounts to $m_1 a$, acting downward.

These three are the only forces acting upon m_1, and they all act along the same straight line; hence its equation of motion is —

$$T - m_1 g - m_1 a = 0. \qquad \text{Eq. 42}$$

Considering now the mass m_2, the forces acting upon it are —

(i) the tension T, acting upward — the tension being the same in every part of the string,

(ii) the weight $m_2 g$ acting downward, and

(iii) the reaction of the mass m_2 acting upward against its downward acceleration a; the acceleration for each of the two bodies being the same in amount but opposite in sense.

Since these are the only forces acting upon m_2, and since they all act in the same straight line, the equation of motion for m_2 is

$$T - m_2 g + m_2 a = 0. \qquad \text{Eq. 43}$$

These two equations, since they contain but two unknown quantities, are competent to give us the values of each of them. Eliminating between them first T and then a, we obtain —

$$a = \frac{m_2 - m_1}{m_2 + m_1} \cdot g, \qquad \text{Eq. 44}$$

and

$$T = \frac{2\, m_1 m_2}{m_1 + m_2} \cdot g. \qquad \text{Eq. 45}$$

A method identical with this gives one the solution of practically any problem of bodies falling over frictionless inclined planes or pulleys, or combinations of these two. A little later we shall learn how to take friction into account.

84. On the other hand a body may be falling freely and also have an initial velocity in a horizontal direction, as, for instance, a bullet fired from a gun which is held horizontal. If we take the muzzle of the gun as the origin of coördinates, the equations of motion for the bullet are

$$x = Vt, \qquad \text{Eq. 5}$$

and

$$y = -\tfrac{1}{2} g t^2. \qquad \text{Eq. 16}$$

Between these two equations one may eliminate t and thus obtain the path of the bullet, namely,

$$y = -\frac{g}{2V^2} x^2, \qquad \text{Eq. 46}$$

an equation which the advanced student will recognize as that of a parabola.

If the initial velocity is not horizontal then one must substitute for V the horizontal component of the initial velocity; likewise the vertical component of the initial velocity will enter into the equation for the vertical motion. But these are matters which, like the resistance of the air, must be left for later study.

Problems on Gravitation

In considering the following questions, the resistance of the air is to be neglected and the acceleration of gravity assumed to be 981, using the centimeter and second as units.

1. A stone let fall from a window 20 m. high will strike the ground with what speed? *Ans.* 1980 cm. per second.

2. A boy throws a baseball to a height of 39 m. With what speed does the ball leave his hands? With what speed will it return to his hands?
Ans. 27.66 m. per second.

3. A tuning fork makes 250 vibrations a second. How many vibrations will it make while it falls from rest a distance of 10 cm.? How many while falling 20 cm.? In the latter case, compare the number of vibrations performed in the first and the second decimeters of the fall.
Ans. 35.7 vibrations; 50.4 vibrations.

4. A gun is held horizontal at a distance of 300 cm. above the ground. How long after the gun is fired until the bullet reaches the ground? During this interval, how far will it travel in a horizontal direction if it leaves the muzzle of the gun with a speed of 500 m. a second?
Ans. 0.78 sec.; 390 m.

5. A block is allowed to slide down a smooth inclined plane of 30° slope. What speed would it acquire at the end of 4 sec. if its motion were not opposed by friction? *Ans.* 1962 cm. per second.

6. Determine the acceleration and the tension in each of the following diagrams:— *Shearer* 157.

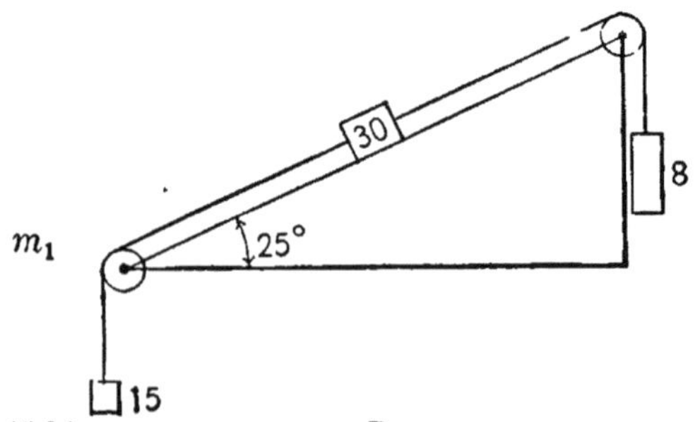

A FIG. 46 *bis* *B*

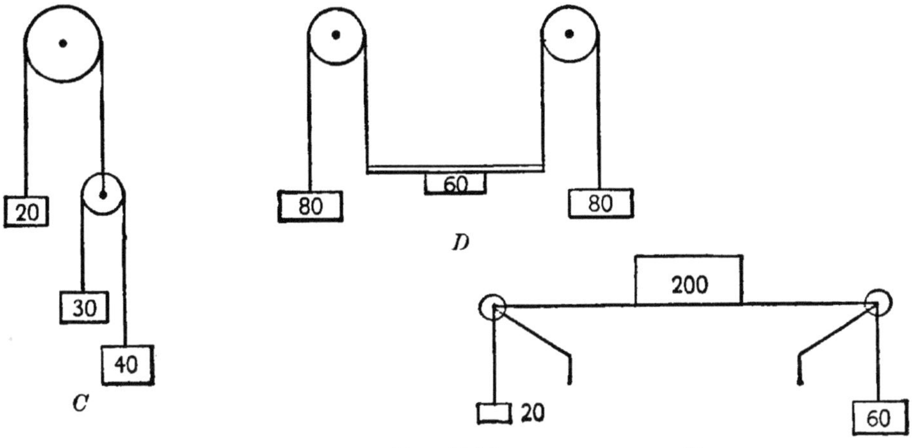

FIG. 46 *bis* E

7. A man who weighs 150 lb. steps into an elevator just as it is starting up with an acceleration of 200 C.G.S. units. By how much will his weight appear to increase?

8. Starting from rest, an automobile travels 400 ft. during the first ten seconds. Suppose the engine to exert a constant pull, what must be the inclination of a hill in order that the machine will ascend it at constant speed?

9. Two weights whose sum is 4 lb. are strung over a massless and frictionless pulley; the difference between the two weights is 2 oz. Find the acceleration which gravity will produce in the system.

10. Find the time required for a body to slide down an inclined plane whose length is 32 ft. and which makes an angle of 30° with the horizontal.

11. An ore bucket descending a mine shaft breaks the cable at an instant when the bucket is moving with a speed 20 ft. per second and is at a height of 240 ft. from the bottom of the mine. How long after the break will the bucket strike the bottom?

THE PENDULUM

85. The first *accurate* information concerning the *amount of the earth's acceleration* was obtained by Huygens (1629–1695), the inventor of the pendulum clock.

Galileo had already rediscovered what had previously been known and what is now familiar to most boys, that a pendulum takes practically the same time to vibrate, whether the angle through which it swings be large or small. **In other words, he showed that the period of a pendulum does not depend upon its amplitude.** Mersenne, in 1644, had determined the length of a pendulum which would beat seconds. But Huygens went a step farther, and showed how this period *does* depend upon the acceleration of gravity, and how a pendulum may be used to measure this acceleration with accuracy.

Determination of the Period of the Simple Pendulum

86. Definition. **A heavy particle suspended from a fixed point by means of a thread whose mass is negligible is called a simple pendulum.**

When such a pendulum is pulled aside through an angle θ (Fig. 47), the particle is urged back toward its position of equilibrium with an acceleration which may be easily computed as follows: —

The simple pendulum is represented by the line OB, the heavy particle by B, its position of equilibrium by D.

Let the vertical line $\overline{BC}$ represent the acceleration of gravity. This may be replaced by two components: one, $\overline{AC}$ acting along the direction of the string, balanced by the pull of the string toward the support O, and hence not affecting the particle at B, the other component $\overline{BA}$ acting practically at right angles to the string and urging the particle back into its position of equilibrium. And since the triangles ABC and OBD are similar, this latter acceleration a bears to g the same ratio that $\overline{BA}$ bears to $\overline{BC}$, and hence

$$\frac{a}{g} = \frac{\overline{BA}}{\overline{BC}} = -\sin\theta.$$

The acceleration represented by $\overline{BA}$ we call negative because it is toward the left.

But when the angle θ is small, we may replace $\sin\theta$ by θ without appreciable error. Hence $a = -g\theta$, where θ is, as usual in physics, measured in radians.

Let l denote the constant length of the pendulum, $l\theta$ will then be the linear displacement of the particle from its position of rest. This displacement we call positive because it is toward the right.

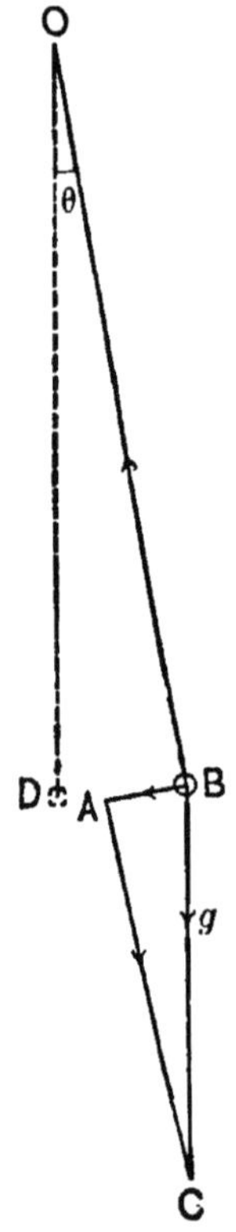

Fig. 47.

Accordingly, we have

$$a = -g\theta = \text{acceleration,}$$

and

$$x = l\theta = \text{displacement.}$$

Eliminating θ, we have

$$a = -\left(\frac{g}{l}\right)x,$$

where $a \propto -x$; for g and l are constants.

That is, the acceleration varies as the displacement, and is opposite in direction; since when the particle is displaced to the right the acceleration is to the left, and *vice versa.*

This motion therefore satisfies the criterion for simple harmonic motion. And hence we know that the particle B moves to and fro with a S.H.M. in a path which, for small amplitudes, is practically straight.

Since the motion of the pendulum is simple harmonic, we may apply to it at once the principles of S.H.M. and write for the period T

$$T = 2\pi\sqrt{\frac{\text{linear displacement}}{-\text{linear acceleration}}} \qquad \text{Eq. 23}$$

$$= 2\pi\sqrt{\frac{l\theta}{g\sin\theta}}$$

$$= 2\pi\sqrt{\frac{l}{g}}, \text{ approximately.} \qquad \text{Eq. 24}$$

This simple expression enables one to compute the value of g. For T and l are easily measured; and then we have for the computation of g the following laboratory equation: —

$$g = 4\pi^2\frac{l}{T^2},$$

where it is to be noted that T is the time of one "complete vibration," *i.e.* of a swing "to *and* fro," not "to *or* fro."

Digression on Center of Gravity

87. One would naturally next proceed to a discussion of the actual physical pendulum, such as is employed in clocks and in the laboratory. But before this can be intelligently taken up, we must add to our vocabulary another term, namely, **center of gravity**.

The mass of any particle multiplied by the acceleration of gravity measures the force with which this particle is drawn towards the earth, — a force which is ordinarily called the weight of the body.

For any body of moderate size it is found by experiment that the direction in which the weight of any particle acts is the same for all parts of the body; in other words the vector g does not vary from one part of the body to another. The total weight of a body is, of course, the resultant of the individual weights of its individual particles.

We have found (§ 14) a general method for the addition of all vectors. This method when applied to any number of forces gives us the amount and the direction of the resultant; but there is one very important thing which this rule does *not* tell us, namely, the point at which the resultant is applied.

Let us, for instance, string two heavy masses M_1 and M_2 (Fig. 48) upon a light rod. Let W_1 be the weight of one mass and W_2 the weight of the other. We then know, by the rule for addition of vectors, that the total weight of the system R is $W_1 + W_2$. But through what point of the system this total weight R acts we have yet to discover.

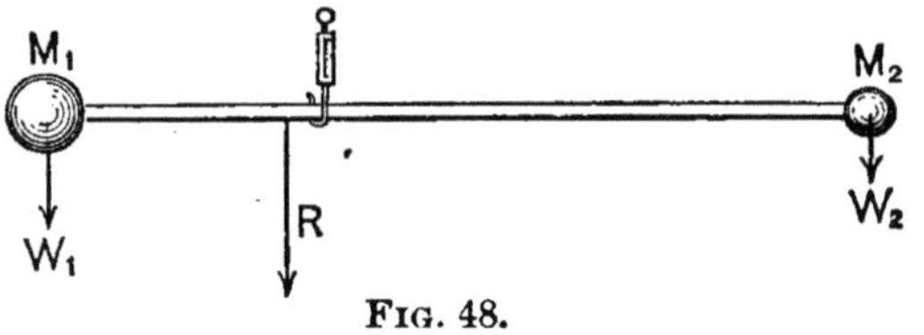

FIG. 48.

From whatever point we may hang the system, the weight of it, as read off on the spring balance, will be the same. But there is only one point on the bar at which we can place the balance and have the system remain horizontal. That is, there is only one point of support about which the weights will offer no moment of force tending to rotate the bar. When once we have placed the hook of the balance at this point of zero moment, we have discovered the point through which the resultant R acts. The weight is not zero, but the arm of the resultant weight is then zero; or, what is more strictly true, the sum of all the moments of force tending to rotate the system about this point of support is zero. This point is known as the "center of gravity" of the system. **The center of gravity is accordingly defined as the point through which the total weight of the body — considered as a single vertical force — acts.** That the center of gravity is identical with the center of mass, when we assume the weights of the particles in the body as parallel forces, may be proved as follows: Take Eq. 28 B § 63 and multiply each side by g. One then has

$$Mg\,\bar{x} = \Sigma(mgx),$$

and two similar equations for $\bar{y}$ and $\bar{z}$. Now Mg is the total weight of the body; and mgx is the moment of force of the particle m. These equations, therefore, mean that the sum of all the moments of force tending to rotate the body are equivalent to the torque of the total weight of the body acting at the center of mass $(\bar{x}, \bar{y}, \bar{z})$. But such a point is, by definition, also the center of gravity.

Center of Gravity by Experiment

88. From this definition it is evident that if a body be supported through this point it will be in equilibrium; *i.e.* it will not rotate, because there is no resultant moment of force acting upon it, and it will not be translated because the point at which it is supported is fixed. This leads to an easy method for determining, *by experiment*, the position of the center of mass in a body.

If a body is straight and slender, as in the case of a wire or a walking stick, we may locate the center of mass at once by balancing the body over a knife-edge, thus discovering the point about which the moment of force is zero.

But if the body be a lamina, *i.e.* a thin flat plate, we may infer from symmetry that the center of mass lies somewhere in the plane of the body. And by balancing the body about a knife-edge we may determine in this plane a straight line $\overline{AB}$ (Fig. 49), which contains the center of mass. If, now, we choose another position for the knife-edge, say $\overline{A'B'}$, and balance the body on it again, we shall locate the center of mass on another straight line.

Since now we know that the center of mass lies on each of these two straight lines, we know that it must lie at their intersection, *C*. We shall hereafter frequently denote center of mass by the letters *c.m.*

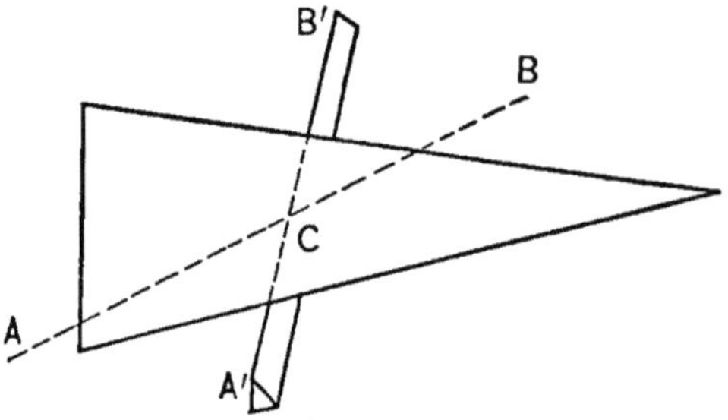

FIG. 49. — Determination of center of mass by experiment.

In the case of an extended body, *i.e.* one all three of whose dimensions are considerable, we should need to balance the body about three different lines; for in this case, each balance would determine one vertical *plane* in which the *c.m.* lies; and three such planes would be required to locate, by their intersection, the *c.m.* of the extended body.

In many cases of this kind the *c.m.* can be located by symmetry, as in the case of a sphere or a parallelopipedon or a cylinder; but only where the density of the body is uniform or symmetrical. In the case of "loaded" dice, for instance, the center of mass would *not* coincide with the center of figure.

It is a general principle of dynamics, first formulated by Galileo, that the center of mass always seeks the lowest possible position in space. This we now recognize as a special case of a still more general principle, — that the potential energy of a body tends to become a minimum.

But however this principle is stated, it suggests still another slightly different method for getting the *c.m.* of a body extended in all three dimensions; namely, instead of balancing on a knife-edge, suspend the body from some one *point* by a thread; the *c.m.* will seek its lowest position, and will, therefore, lie in the straight line determined by the thread.

If now the body be suspended from another point, a second line will be determined in which also the *c.m.* lies. The intersection of these two lines must occur, else the body would have *two* centers of mass. This intersection must therefore be the *c.m.* In this way two observations, instead of three, suffice to locate the *c.m.* of a three-dimensional body.

From this digression, we return now to the

Determination of the Period of the Physical Pendulum

89. Imagine any rigid body which may be perfectly irregular in shape, or more or less regular as shown in Fig. 50. Let it be suspended at the point S so that it swings freely in a vertical plane. Such a body is called a physical or compound pendulum. Denote the *c.m.* by G and let SG' indicate the position in which the line SG comes to rest. Denote the distance of G from S by a and the deflection of the pendulum at any instant by θ. If the mass of this rotating body is M, the only force acting upon it is its weight Mg acting through the *c.m.*, G. This force exerts a torque $Mg \cdot GG'$, equal to $Mg \cdot a\theta$, tending to restore the pendulum to its position of rest. And since this torque is proportional to the displacement and opposite in sense, the motion will be simple harmonic. We may, therefore, at once write the period, by analogy with Eq. 24,

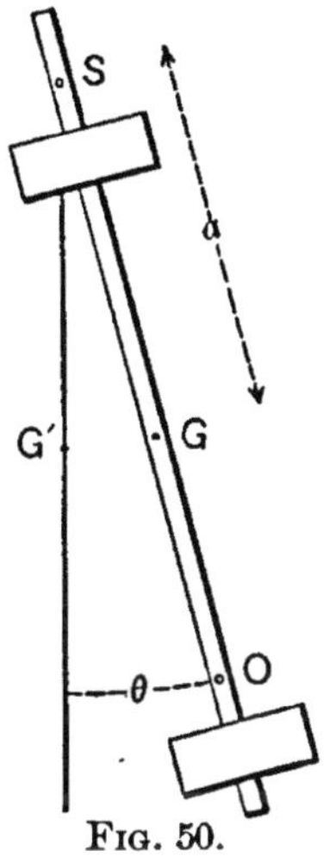

Fig. 50.

$$T = 2\pi\sqrt{\frac{\text{Angular displacement}}{-\text{Angular acceleration}}} = 2\pi\sqrt{\frac{\theta}{-A}}.$$

But, Eq. 37, § 70, $$A = \frac{\text{Moment of force}}{\text{Moment of inertia}} = \frac{Mga\theta}{I}$$

Hence $$T = 2\pi\sqrt{\frac{I}{Mg \cdot a}} \quad \text{Period of physical pendulum.} \qquad \text{Eq. 47}$$

If a simple pendulum of length l is to have the same period, it will be necessary that

$$l = \frac{I}{Ma}. \qquad \text{Eq. 48}$$

This value of l is known as the equivalent length of the physical pendulum.

Reversible Pendulum

90. The advanced student will find in the reversible pendulum a beautiful device by which the length of the equivalent simple pendulum can be directly measured in the laboratory. Here one can only say that if the pendulum vibrates with equal periods about two points S and O (Fig. 50), which include the center of gravity G in the straight line joining them, then the distance SO is the length of equivalent simple pendulum.

Problems

1. Shortly after the invention of the pendulum clock, one was sent from Paris, where it kept correct time, to Cayenne, in South America. In the latter place the clock lost $2\frac{1}{2}$ minutes daily. The length of the pendulum was adjusted so that the clock kept correct time in Cayenne; and in this condition the instrument was sent back to Paris, where it now *gained* $2\frac{1}{2}$ minutes daily. Explain this phenomenon.

2. It is well known that metals expand on heating. What will be the effects of the cold weather in winter and the hot weather of summer upon the rate of a clock having a brass pendulum?

3. A clock having a pendulum 60 cm. long keeps correct time. The pendulum is lengthened to 60.5 cm. How many seconds will the clock lose each day? *Ans.* 362 sec.

4. The period of a given pendulum at Denver is exactly 1 sec. The period of the same pendulum at Chicago is 0.9995 sec. The acceleration of gravity at Denver is 979 C.G.S. units. What is its value at Chicago? *Ans.* 979.97 C.G.S. units.

5. There are 454 g. in 1 lb. What is the force, in dynes, with which the earth attracts a mass of 1 lb., at a place where $g = 980$ C.G.S. units? *Ans.* 444920 dynes.

6. Find the weight of 7 kg. in dynes; $g = 960$. *Ans.* 6,720,000 dynes.

7. A spring balance is made and graduated at a town where $g = 983$. At another town, where $g = 979$, the spring balance is used in a meat market. Who gains by this error, the man who buys the meat or the man who sells? *Ans.* The buyer.

8. From the basket of a balloon hangs a spring balance. This balance carries a mass of 100 g. The balloon ascends with acceleration of 220 C.G.S. units. What will be the apparent weight of the 100 g. during the ascent? *Ans.* 122.4 g.

9. A ball thrown vertically upward is caught by the thrower 5 sec. later. Neglect the resistance of the air, and compute the height to which the ball was thrown. *Ans.* 30.66 m.

10. The radius of the earth is 636,000,000 cm. At what linear speed must a body move along the surface of the earth to lose all apparent weight? *Ans.* Eq. 18, $\frac{v^2}{636 \times 10^6} = 981$, $v = 7899$ m. per second.

11. In which of the following cases does the *c.m.* lie inside, and which outside the body: a teacup, a straight copper wire, the same wire bent into the arc of a circle, a hoop, a silver half-dollar, a baseball, a football?

12. Locate, by geometry, the *c.m.* of a triangular sheet of brass. If a corner of this triangle be clipped off along a line parallel to the opposite side, the *c.m.* will be displaced along what line?

13. A meter scale (the mass of which is negligible) is divided into decimeters, and at the end of the first and eighth decimeters are attached masses of 1 and 9 k. respectively. Where is the *c.m.* of the system?

Ans. At the end of 73rd cm. on the scale.

UNIVERSAL GRAVITATION

91. Hitherto we have been considering a special case of gravitation, namely, the acceleration which the earth produces upon small bodies near its surface.

Newton's discovery of the law which governs the general case of gravitation must always be reckoned one of the most brilliant achievements of the human mind. The circumstances which led up to this discovery are something like the following: —

Copernicus (1473–1543) had suggested, on grounds of simplicity and symmetry, that the sun and not the earth was the center of what we now call the solar system; but neither he nor any of his successors up to the time of Newton had shown what it was that kept the planets in motion about the sun.

Kepler (1571–1620) in the meanwhile had succeeded in describing, with a high degree of accuracy, the motion of the planets about the sun. His description is as follows: —

(i) Each planet moves in a nearly circular ellipse, with the sun in one focus,

(ii) at such a speed that the line from the center of the sun to the center of the planet sweeps out equal areas in equal times,

(iii) while the square of the year, measured in any unit of time, bears to the cube of the planet's distance from the sun a ratio which is constant for each member of the solar system.

Next came Newton (1642–1727), who, building upon the dynamics of Galileo, argued from Kepler's second statement that *no* force from behind is necessary to drive the planet along its orbit, but only a deflecting force acting toward the sun, a force which is continually balancing the centrifugal force of the planet. His next step was to infer from Kepler's first statement that the value of this deflecting force varies inversely as the square of the distance which separates the planet from the sun. While from Kepler's third statement he inferred that the force of attraction between the planet and the sun is proportional to the mass of the planet, the constant of proportionality being the same for each of the planets. Following this, he concluded that, unless there be something more peculiar than mere size in the mass of the sun, this attraction must be proportional also to the mass of the sun.

Before summarizing these results of Newton, let us recall the fact that the dimensions of each of these planets are so small compared with their distances from the sun that they may be treated as particles. Newton showed by mathematical computation that if we suppose the sun to attract each of the planets with a force which varies directly as the mass of the planet, and inversely as the square of the distance between the planet and the sun, this supposition will explain, in the most satisfactory manner, all the motions of the planets. Going one step farther, he generalizes and supposes that *any* two particles, or masses which may be considered as particles, m and m', attract one another with a force depending only on the product of the masses mm', and the square of their distance r. Expressing this in the language of algebra,

$$F = G\frac{mm'}{r^2} \text{ dynes,} \qquad \text{Eq. 49}$$

where G is a numerical constant found to be 6.48×10^{-8} very approximately.

The clever methods by which Cavendish, Boys, Poynting, and others have determined the value of this "constant of gravitation" G is a matter which must be postponed for advanced study.

Newton also proved that this supposition would explain the circular path of the moon. And by an elegant demonstration, he showed that the reason that ordinary bodies at the surface of the earth are heavy is that the earth attracts them just as it does the moon. The whole earth acts as if its mass were concentrated at its centre, so that bodies at the surface of the earth act as if attracted by a particle whose mass is that of the earth, and whose distance is 4000 miles. What we call the weight of a body thus becomes a special case of the forces described in Eq. 49, which is known as Newton's Law of Universal Gravitation. **Let the student bear in mind, however, that this law simply describes the behavior of a mass m in the presence of another mass m', and, as Newton himself clearly pointed out, does not explain gravitation even in the slightest degree.** Why bodies attract one another is as much a mystery as ever.* How bodies attract one another Newton discovered and described in the most perfect manner.

The Balance

92. The balance is known to most people as an instrument for comparing the weights of two different bodies, *i.e.* for comparing two different forces. The grocer when he sells us a pound of coffee deals out an amount such as the earth will attract with a force which English-speaking people have agreed to call "the weight of one pound."

* In this connection the granular pressure theory advanced by Osborne Reynolds, Professor of Engineering, Owens College, Manchester, should be mentioned. It purports to explain gravitation as being due to the push or pressure of a universal granular medium which fills all space, is under enormous pressure, and has a very high density. There is no attractive force inherent in masses of matter themselves. In spaces where no matter exists the grains of the medium, which are almost infinitely small, are arranged in what Reynolds calls "normal piling," such as that of spherical cannon balls. Where matter exists the grains are in "abnormal piling," there being a smaller number of grains per unit volume in the latter than in the former. Matter indicates a certain deficiency of grains in the medium, a "singular surface of misfit," a crack or gap in the otherwise uniform medium, and dilatations and strains, due to the pressure, are set up in the granular medium between such "negative inequalities" or "singular surfaces of misfit." This causes curvature in the normal arrangement of the medium between the "negative inequalities" or atoms and the normally piled grains of the medium, and the pressure of the medium under such conditions pushes the "negative inequalities" or atoms together according to the law of universal gravitation.

As Reynolds states, this theory leads to an inversion of ideas hitherto conceived or preconceived as to matter and mass, inasmuch as it maintains that matter is absence of mass. The theory also purports to explain cohesion, and to give a key to the explanation of practically all other physical phenomena upon purely dynamical principles.

The weight of one pound is simply the pull of the earth* upon a certain piece of platinum preserved in the Standards Office at London.

The essential features of the ordinary balance are : —

1. A rigid beam carrying three knife-edges, one near each end, and a third, as nearly as possible, midway between these two.

2. Two scale pans, supported, one on each knife-edge, at the end.

3. The beam when carrying the knife-edges and the pans must be so adjusted as to come to rest in a definite position. usually horizontal. This is accomplished by so placing the knife-edges and so distributing the mass in the beam that the *c.m.* of the whole system lies just *below* the central knife-edge. Why would not neutral equilibrium answer quite as well as stable ?

4. The whole system must be sensitive to a small moment of force. This is accomplished by making the distance from the central knife-edge to the *c.m.* very small. How does this secure the desired result ?

5. A pointer, or index of some kind, generally fastened to the beam, for the purpose of indicating the position of equilibrium.

Principle of the Balance

93. In **physics** the **balance is employed to compare two masses by comparing their** weights. This is done as follows : The position of equilibrium is observed before any mass has been added to either pan, that is, when the pans are empty. The body whose mass is required, having been placed in one pan, standard masses are then placed in the other pan until the moments of force which their weights exert on the beam are just equal. Since these two masses tend to rotate the beam in opposite senses with equal moments, the total moment will be zero. This equality and opposition of moments is indicated by the pointer coming back to the same position of equilibrium which it had when the pans were empty. **Thus the immediate purpose of**

* Strictly speaking, "the weight of a pound" is the pull of the earth on the platinum mass diminished by the centrifugal force due to the earth's rotation. The weight of a body is, therefore, about one third of one per cent less than the pull of the earth upon it.

the balance is simply to tell us when we have equal moments of force.

Now if, in addition, the arms of the balance (that is, the respective distances between central and end knife-edges) are equal, we know that the *weights* of the two bodies in their respective pans are equal.

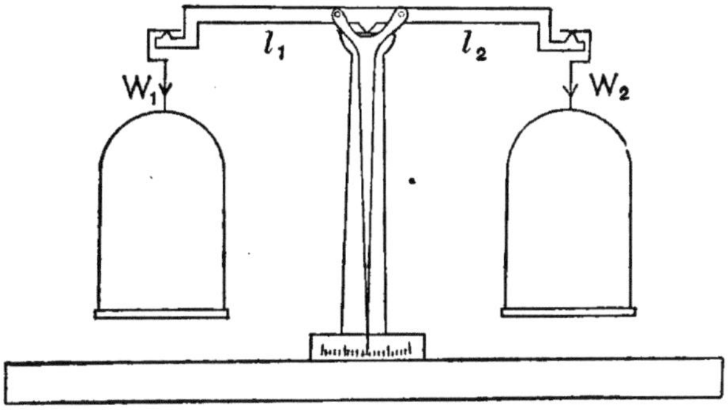

FIG. 51.

Let l_1 and l_2 (Fig. 51) indicate the length of the arms, W_1 and W_2 the weights of the two masses M_1 and M_2. Then, by proper adjustment of the masses, we make

$$W_1 l_1 = - W_2 l_2,$$

so that the total moment L is zero.

$$L = W_1 l_1 + W_2 l_2 = 0.$$

If, now, $$l_1 = - l_2,$$

then $$W_1 = W_2,$$

or $$M_1 g_1 = M_2 g_2.$$

But Newton and Galileo have shown that, at any one point on the earth's surface, the acceleration of gravity is the same for all bodies. Expressed in a more useful form, *i.e.* in an algebraic form, this would read, —

$$g_1 = g_2,$$

and hence $$M_1 = M_2.$$

So that really the equality of masses is an inference from the equality of moments. And this inference would not be allowable if the earth acted as a magnet, attracting different bodies with a force depending upon the kind of material as well as the mass of material in them.

The earth, indeed, is a great magnet, directing the compass needle, as we all know; but, as we shall see later, it does not attract the needle as a whole. It simply *rotates* the needle into one position, but does not *translate* it; since the lines of magnetic force at any one point on the earth's surface are practically parallel.

Aside from the Tower of Pisa experiments by Galileo, and from the consistency of the results obtained by the balance, Newton furnished conclusive evidence by the use of a hollow pendulum that gravity acts equally upon all substances without regard to their chemical composition. For a description of this experiment see the *Principia*, Book III, Prop. VI.

Problems

1. At the surface of the sun the acceleration of a falling body is 26 times as great as at the surface of the earth. A man who weighs 150 lb. at the earth would weigh how much at the sun?

2. The rotation period of Jupiter is 10 hr.; its radius is 11 times that of the earth. Compute what fraction of its weight a body would lose in being carried from the north pole to the equator.

3. The earth is nearly 3 times as far away from the sun as the planet Venus. If the earth were transferred to the orbit of Venus and made to move in it, what length of year would be necessary in order that the centrifugal force should just balance the sun's attraction. Assume the orbit circular. See Eq. 40, § 75.

4. Prove that the "dimensions" of the "gravitational constant" are $M^{-1}L^3T^{-2}$.

5. If a hole were drilled through the earth along any diameter, the attraction of the earth on any body placed in this tube would vary directly as the distance of the body from the center of the earth. Neglecting friction, what kind of motion would be executed by a body dropped into this tube at the surface of the earth?

6. In the preceding problem prove that a body would fall through the earth, so to speak, in approximately 42½ minutes.

7. A fly is inclosed in a corked bottle. The bottle and fly are weighed with great accuracy (i) when the fly is standing still on the bottom of the bottle and (ii) while the fly is flying about inside the bottle but without touching the glass. How will these two weights compare?

8. Sound travels at the constant rate of 1100 ft. a second. At what height above the surface of the earth must a body be let fall so as to reach the earth at the same instant as the sound of a pistol shot fired at the same place and time from which the falling body starts?

III. Matter as a Vehicle of Energy

94. Work is something of which every one has, or ought to have, a fairly definite idea. The farmer knows that it requires twice as much work to plough a two-acre field as to plough a one-acre field. For while the horses exert no more force on the plough in the first case than in the second, the total length of furrow is twice as great in the first case as in the second. In like manner, we all know that it requires twice as much work to carry 2000 bricks to the top of a house as to carry 1000 bricks to the same height. For while the distance is the same in each case, the weight of 2000 bricks is twice that of 1000 bricks.

In general, the quantity of work done depends upon two factors, and upon two factors only: (1) the force exerted, and (2) the distance through which this force is exerted.

Definition of Work

95. Consequently, the term **work** is employed in physics to denote one definite quantity, viz. **the product of a force multiplied by the distance through which it is exerted, both distance and force being measured in the same direction.**

A mass *B* (Fig. 52) is attracted by the earth with a force of 1000 dynes. What work will be required to roll this mass up an inclined plane whose length $\overline{OA}$ is 24 centimeters? If work were defined simply as "the product of force by distance" the answer to the problem would be $1000 \times 24 = 24{,}000$ units of work. But using the proper definition, we observe that since here the force, *i.e.* the weight of the body *B*, is vertical, we must measure the distance also in a vertical direction.

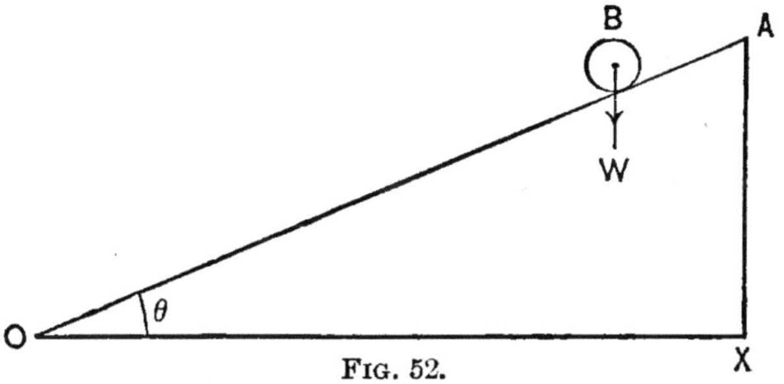

Fig. 52.

Evidently, therefore, we must know the height $\overline{AX}$ before we can solve the problem at all. In other words, we must know how steep the inclined plane is before we can estimate the work required to carry the body *B* through the distance $\overline{OA}$.

Suppose the height $\overline{AX}$ to be 12 centimeters. The work in this particular case will then amount to 1000×12 units of work.

So, in general, to obtain the work done we have merely to multiply the force exerted by the distance resolved along the direction of the force; or, what amounts to the same thing, multiply the distance by the force resolved along the direction of the displacement.

Let us denote the work by W, the force by F, and the displacement by l. Then the defining equation for work may be written as follows: —

$$W = Fl\cos\theta = Fx, \qquad \text{Eq. 50}$$

where θ is the angle between the directions of the two vectors, F and l; and x is the component of l in the direction of F. This algebraic definition is of course identical with the one given above in words.

96. Rotational Analogue. The student will find it an interesting exercise to prove that, in like manner, the work done by a torque L acting about any given axis is measured by the product of the torque, and the angular displacement θ about this same axis. Thus

$$W = L\theta, \qquad \text{Eq. 51}$$

an expression which is convenient in computing the amount of work done in winding a watch, turning a dynamo, etc.

Units of Work: The Erg and Joule

97. Since now the unit of force is the dyne, and the unit of distance the centimeter, **the only consistent unit of work is that amount of work which is done when a force of one dyne is exerted through a distance of one** centimeter.

This unit is of such frequent occurrence in physics that it has been given a special name. It is called an **Erg**, which is simply the Greek form of our English word "work."

A larger, and for many purposes, therefore, more convenient, unit is the **Joule**, which is defined as 10^7 ergs. This unit will be found especially convenient when we come to consider the work done by an electric current.

The engineer's unit of work is, of course, the work done in exerting a force of 1 pound through a distance of 1 foot. It is called the **foot-pound.**

Problems

1. How many ergs of work will be done in exerting a force of 6 dynes through a distance of 18 cm.? *Ans.* 108 ergs.

2. What work will be required to hoist 10 kg. of water against an acceleration of 981 units from a well whose depth is 12 m.?

Ans. 11,772 million ergs.

3. Suppose forces to be measured in "pounds weight" and distances in feet; how many foot-pounds of work are required to carry 100 lb. of brick to the top of a building 42 ft. high? *Ans.* 4200 foot-pounds.

4. If the weight of one kilogram be taken as the unit of force and the meter as the unit of length, what will be the unit of work?

Ans. One kilogrammeter.

5. If 2640 foot-pounds of work are required to propel a bicycle for a distance of one mile, what is the average force required to propel the wheel?

6. A magnet attracts a piece of iron with a force of 8000 dynes. How much work will be performed by the magnet in moving the iron through a distance of 0.15 mm.?

7. A force of one million dynes is required to operate a "double action" bicycle pump. If the length of stroke is 12 cm., how much work will be required to make 50 complete strokes of the pump?

Energy

98. We shall see later that a body may be able to do work in consequence of

1. Its position.
2. Its speed (angular or linear).
3. Its temperature.
4. Its electrical condition.
5. Its chemical composition.

Etc.

The power of doing work is called energy, and the system is said to possess energy when it can do work. Since the work which a system can do is the measure of its energy, **the same unit is employed for energy as for work, viz., the erg.**

If we were to attempt to study all the forms of energy which a system may possess, we should find ourselves at once in the midst of the entire subject of physics. Accordingly, we shall here consider only the energy which a mechanical system may have by virtue of

1. The relative position of its parts; and
2. The relative speed of its parts.

Every one is familiar with many instances of these two forms of energy.

The boy who starts at the top of a hill on a bicycle knows that his energy of position is such that, without pedaling, it will carry him to the bottom of the hill and some distance beyond. If, coming in the opposite direction, he approaches the foot of the hill with sufficient speed, the energy due to his motion may be ample to carry him to the top without any pedaling. In this case the system consists of the bicyclist and the earth.

An uncoiled spring will not run a watch; but if, using a key, we change the relative position of its parts, by coiling it up, it will run the watch for the next twenty-four hours. When one drives a nail, he gives the hammer a high speed. The energy due to this speed does the work of forcing the nail into the wood.

In the language of physics, the energy which a body possesses in virtue of its position is called Potential Energy; while the energy which a body possesses in virtue of its speed is called Kinetic Energy.

The energy of the Niagara River just at the top of the falls is nearly all potential, but just before reaching the bottom of the falls it is nearly all kinetic. The energy of the iron block —the ram—used in a pile driver is all potential at the top of the stroke, but all kinetic just before striking the pile at the bottom of the stroke.

The question now arises as to how much work a body can do in virtue of its position, and how much work a body can do in virtue of its speed. In other words, what are the proper measures for potential and kinetic energy respectively? Before answering these two questions we must digress for a moment to consider the

Dissipation of Energy

99. It is well known to every one that a ball, or a bicycle, which is allowed to run down one hill from a certain height h (Fig. 53) will not, with the speed thus acquired, ascend an adjoining hill to the same height. The ball, or the wheel, always stops short of its original elevation. The

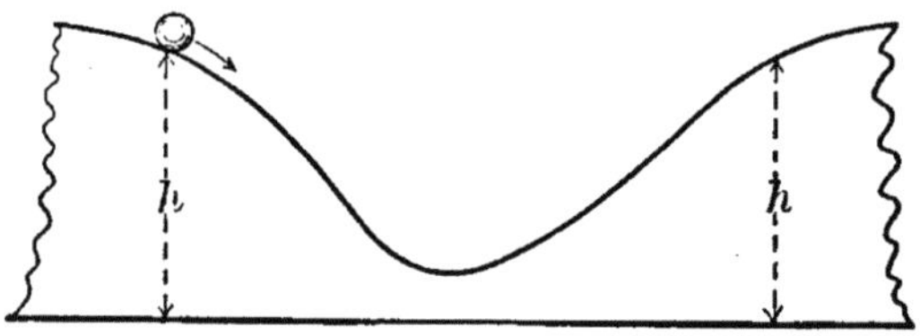

Fig. 53.

explanation of this is something which does not now concern us. The important fact is that the kinetic energy of the wheel at the bottom of the hill is not quite sufficient to do the work of carrying the wheel to its original height.

The same is observed to be true of a pendulum as it swings from its position of rest at *A* to its position of rest at *C* (Fig. 54). The point *C* is always a little lower than the point *A*, *i.e.* the kinetic energy of the pendulum at *B* is not quite sufficient to do the work of carrying the bob to its original height.

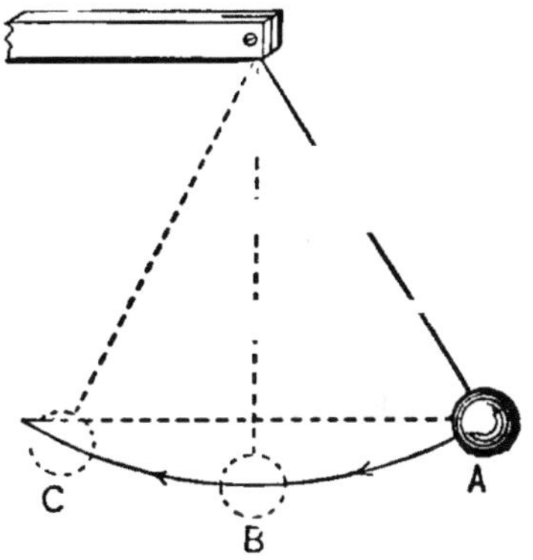

FIG. 54.

In every case of this kind it has been observed that we never get out of a machine all the work which we put into it. Some of the mechanical energy of a system is always wasted when we attempt to use that energy by transforming it, or transferring it.

As we shall learn later, this unavoidable waste of energy is brought about by friction; a portion of the mechanical energy is transformed into heat energy, where it is of less use to us than before. This tendency of energy to assume a more and more useless form is known as the principle of the Dissipation of Energy.

Conservation of Energy

100. But the most important discovery along this line, perhaps the most important discovery of the nineteenth century, is the following: that *when allowance has been made for unavoidable wastes*, the sum of the kinetic and potential energies of a body or system of bodies never changes, unless through some external influence.

"Waste of energy" is a very common expression; but, as we shall learn by and by, this never means destruction of energy. There is no evidence that the slightest bit of energy has ever been annihilated. Energy is, in this respect, like matter. No human being has ever succeeded in creating or destroying any amount of either one. When, therefore, we speak of allowing for "unavoidable wastes," we refer to that energy which has been unavoidably changed into heat or some other form of energy which, for our purpose, is useless.

The law of the Conservation of Energy expresses the fact that the sum total of the energy in any isolated system remains the same. This energy can change from one form to another, it tends constantly to become less and less available; but, so far as is known, it has never changed in quantity.

That this grand principle was more or less clearly suspected and grasped by Galileo, Stevinus, Leibnitz, Huygens, and Newton there can be little doubt; but it is also very certain that the clear and definite establishment of the principle is due to Helmholtz, Joule, and Kelvin, the last named of whom is still (1907) living. If a single date he asked, perhaps the year 1847, in which Helmholtz's *Memoir on the Conservation of Energy* appeared, will mark more nearly than any other the rise of the doctrine of the conservation of energy. And yet the French Academy of Sciences many years before (1775) had announced that it would not receive any further communications upon the subject of perpetual motion; which practically affirms their belief that energy cannot, by any human power, be created.

The evidence for and the full import of this principle the student can hardly expect to comprehend until he is more or less familiar with the subjects of heat, electricity, and chemistry.

Measure of Kinetic Energy of Translation

101. By the application of this principle to many special cases, the proper measure of the kinetic energy of a body in translation has been found to be as follows: —

If a force F acts upon a mass m initially at rest, for a time t, it will produce in it a speed of translation s, such that (§ 69)

$$F = ma = m \cdot s/t.$$

If the force F remains constant, then the acceleration will remain constant, and the mass m will travel, during this time, a distance x, such that (§ 36)

$$x = \tfrac{1}{2}\, at^2 = \tfrac{1}{2}\,(s/t)\, t^2 = \tfrac{1}{2}\, st.$$

But the work done by a constant force F exerted through a distance x is Fx, and therefore

$$Fx = \frac{1}{2} ms^2. \qquad \text{Eq. 52}$$

Now this work has all gone toward putting the mass m in motion, since, by supposition, there has been no waste from friction or other causes. We see, therefore, that the right-hand member of this equation, half the product of the mass by the square of the speed, is the proper measure of the work put into the body. Relying now upon the conservation of energy, it is seen that this product also measures the amount of work which we shall be able to get out of the body, *i.e.* measures its kinetic energy of translation. Let us denote this energy by K.E., then

$$\text{K.E.} = \frac{1}{2} ms^2 \qquad \text{Eq. 53}$$

Analogue. Kinetic Energy of Rotation

102. Consider any particle m of a rigid body which is rotating about an axis O (Fig. 55) with a constant angular speed ω. The linear speed of the particle m will be $r\omega$, where r is the distance of the particle from the axis of rotation, Eq. 8. Hence according to Eq. 53 the kinetic energy of the particle m will be $\frac{1}{2} mr^2\omega^2$. Obviously we may consider the entire rigid body as made up of such particles as m; let us call them m_1, m_2, m_3, etc., and suppose them situated at distances r_1, r_2, r_3, etc. Then the total kinetic energy of the body will be the sum

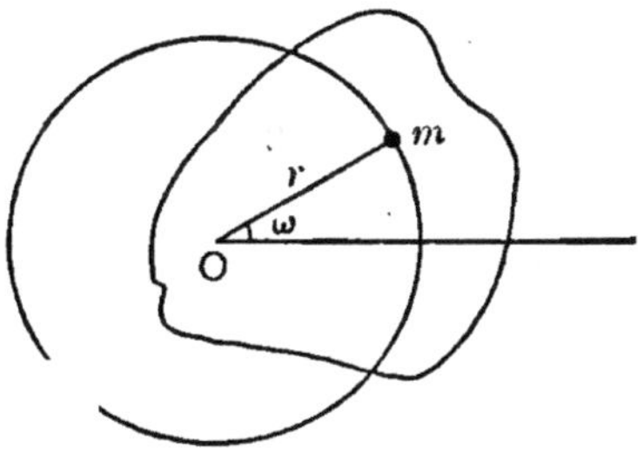

FIG. 55.

$$\text{K.E. of body} = \frac{1}{2} m_1 r_1^2 \omega_1^2 + \frac{1}{2} m_2 r_2^2 \omega_2^2 + \frac{1}{2} m_3 r_3^2 \omega_3^2 + \text{etc.}$$

$$-\frac{1}{2} \Sigma(mr^2\omega^2).$$

But since, by hypothesis, the body is rigid, every part of it is rotating with the same angular velocity, a fact which may be expressed by writing

$$\omega_1 = \omega_2 = \omega_3 = \text{etc.} = \omega$$

Hence $\text{K.E. of body} = \frac{1}{2}(m_1 r_1^2 + m_2 r_2^2 + m_3 r_3^2)\omega^2 = \frac{1}{2}\omega^2\Sigma(mr^2)$.

Let us now denote the quantity in the parenthesis, which for any given body is a constant, by I. We may then write

$$\text{K.E. of body} = \tfrac{1}{2} I\omega^2, \qquad \text{Eq. 54}$$

which is strictly analogous to Eq. 53.

The quantity I, whose defining equation becomes

$$I = m_1 r_1^2 + m_2 r_2^2 + m_3 r_3^2 + \text{etc.} = \Sigma(mr^2), \qquad \text{Eq. 55}$$

is, of course, the rotational inertia already described in § 65, which the student should now reread. The principal object of a heavy fly wheel on an engine is to supply energy of this type, which shall keep the machinery going while the piston rod is in the neighborhood of the "dead point," where it exerts a relatively small torque.

Measure of Potential Energy

103. We turn next to a system whose parts have a definite relative position, and ask how much work can be gotten out of it. For instance a coiled watch spring, a stretched rubber band, a pond of water at a definite height above a waterwheel, a stone raised to a certain height above the ground. How shall we measure the potential energy of each of these systems?

Experiment has answered this question in a clear and simple manner. **The amount of work done in bringing the system into its present position is, barring frictional losses, exactly equal to the potential energy now in the system.**

The weight of the water in the pond F multiplied by the average height x through which it was pumped, is at once the work required, Fx, to give the water its present potential energy and is also the measure of its present potential energy; or, denoting potential energy by P.E., we have

$$\text{P.E.} = Fx.$$

It must not be forgotten that here allowance is to be made for the energy wasted in friction in the pump.

The student will observe that in like manner the energy of the coiled watch spring is simply the average moment of force which one exerts upon the key L multiplied by the angle through which he turns the key, say θ. Or if the torque L

be spent entirely in producing angular velocity, and if the body, say a fly wheel, start from rest,

$$L\theta = \tfrac{1}{2} I\omega^2,$$

the analogue of Eq. 52.

As a single, simple, concrete illustration of the manner in which the law of the conservation of energy is used quantitatively, imagine a ball let fall from a second-story window whose height above the ground is h. The total energy of the ball just before it is allowed to drop, taken with reference to the ground, is mgh, where m is the mass of the ball and g the acceleration of gravity. Therefore, according to the law, mgh is the energy at any intermediate height, say x, where the velocity of the falling ball is v. Hence

$$\tfrac{1}{2} mv^2 + mgx = mgh = \text{constant}$$

an equation which may at once be used for computing the velocity of the ball at any given height x, the value of h being given as one of the data of the problem.

Definition and Conditions of Equilibrium

104. At this point it becomes necessary to digress slightly from the main thread of our discussion to consider a certain class of critical conditions which are most clearly described in terms of potential energy. **If the speed of a body is zero, the body is said to be at rest; but if the acceleration is zero, the body is said to be in static equilibrium.** Thus a pendulum bob at the end of its swing is in rest, but it is not in equilibrium ; while the same bob at the instant of passing through its lowest point is in motion, but it is also in static equilibrium. Rest is not the criterion of equilibrium. The condition of zero acceleration is the absence of force and torque, which may be expressed by writing

$$\Sigma(F) = \Sigma(L) = 0.$$

If we confine our attention to systems in which friction is not the determining factor, then it is a matter of daily experience that bodies may be in equilibrium in a variety of ways. A lead pencil may, with a little care, be made to stand upon an end which has not been sharpened, and it is said to be in equilibrium; but a breath of wind or the slightest blow is sufficient to "start" it, and when once started, it falls over, "of itself," we say. If the pencil be a round one, it will rest on its side in *any* position we may give it, so long as it lies on a horizontal plane. Here, again, it is said to be in equilibrium; but if it be displaced slightly, it will still be in equilibrium in its new position; for there is no force or moment of force acting upon it in any position so long as it lies on its side.

Let us again consider such a body as a brick lying on any one of its sides. We all know that if the brick be slightly lifted at one end, *i.e.* if the brick be slightly rotated about one edge, it will not go on and "of itself" rotate still farther; but, if released, will return to its original position of equilibrium.

These three cases typify *all* the cases of equilibrium met with in nature. Their essential features are the following: —

1. *Unstable Equilibrium.* When a body is supported in such a way that its center of mass is lowered by a slight displacement, its equilibrium is said to be **unstable**; for the slightest displacement will introduce a moment of force tending to rotate the body still farther. For illustration, imagine a cone inverted and balanced on its apex. The *c.m.* when the cone is ever so little inclined moves *down* a spherical surface, and hence tends to move farther, of itself.

2. *Neutral Equilibrium.* If a body is supported in such a way that its *c.m.* remains at the same height above the point or line of support, its stability is said to be **neutral**; for when disturbed no moment of force will be introduced to displace it farther or to restore it to its original position. Such a case is represented by a right cone lying on its side in a level plane. The *c.m.* lies on a level, and has, therefore, no tendency to move in any direction.

3. *Stable Equilibrium.* When a body is supported in such a way that the slightest displacement will raise its *c.m.*, the body is said to be in stable equilibrium; for the slightest displacement will introduce a moment of force tending to restore the body to its original position. Such a case is found in a right cone standing on its base. Any small motion of such a cone starts the *c.m.* uphill, so to speak; and when left to itself, the *c.m.* tends to roll back, downhill.

It should be carefully observed that in each of the three different cases of equilibrium the *c.m.* of the displaced body moves in such a way that the potential energy becomes less. If, therefore, in the given position the potential energy of the body is a maximum, the equilibrium is unstable; if the potential energy is a minimum, the equilibrium is stable: but, if the potential energy does not vary when the body is displaced, the equilibrium is neutral.

SUMMARY OF THE PRINCIPAL DISCOVERIES CONCERNING ENERGY

105. 1. Energy is known only in connection with matter. We know of no matter devoid of energy, and know of no form of energy which is not associated with matter.

2. If no energy be given to or taken away from a system, the sum of its kinetic and potential energies remains constant.

3. When a system is left at rest, its potential energy tends to assume the form of kinetic energy. This great law of nature may be described in another way by saying that there is a universal tendency for the potential energy of a system at rest to decrease.

POWER

106. One important feature of any mechanical system is the rate at which it can do work. Two steam engines may be similar in design and yet so different in dimensions that one can do twice as much work as the other in any given time, both using steam from the same boiler.

In everyday life, power is a word which is employed with considerable variety of meaning; but in physics it is always used to denote the time rate of doing work. Accordingly the defining equation becomes

$$\text{Power} = P = \left[\frac{W_2 - W_1}{t_2 - t_1}\right], \qquad \text{Eq. 56}$$

where, of course, W denotes work and t, time.

The fundamental idea involved in power is rapidity of working. The reason it is more fatiguing to ride a wheel uphill than to ride it along a level stretch at the same rate is that, in the former case, the rate of working is higher. The main difference between a 10-horse power motor and a 5-horse power one is not that the one has done or is doing twice as much work as the other, but that the one *can* do work twice as rapidly as the other.

Units of Power

107. Three units of power are in common use,

(i) The **erg per second**, which, has no special name, and is inconveniently small.

(ii) The **joule per second**, which, in honor of James Watt, the inventor of the steam engine, is called a watt, and

(iii) The horse power, introduced by James Watt; defined as **33,000 foot-pounds** per minute, and generally indicated by the symbol HP.

$$1 \text{ HP} = 746 \text{ watts.}$$

108. Having now completed a brief survey of the five fundamental conceptions of mechanics; namely, *inertia*, *momentum*, *force*, *energy*, and *power*, we pass to some simple concrete problems in energy which, it is hoped, will make the foregoing principles the intellectual property of each student.

Applications of the Principles of Energy to Mechanics

I. *The Lever*

109. The lever is a machine which, in various forms, is familiar to every lad. It has a variety of objects. Sometimes it is employed to increase the force which one is able to apply at some particular point, as in raising a heavy stone or cracking a nut. Sometimes it is employed to multiply the motion of a point, as in the index on the dial of a steam gauge; sometimes, as in sugar tongs, pliers, and tweezers, to reach otherwise inaccessible places.

The principle of the lever was first correctly enunciated by the Greek scholar Archimedes (287–212 B.C.). The essential feature of all levers is a more or less rigid bar rotating about the edge of some body as an axis. This edge is called the fulcrum of the lever.

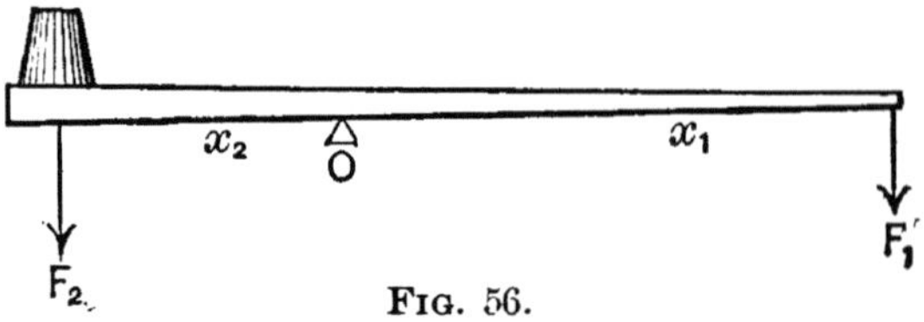

Fig. 56.

Since hardly anv energy is here wasted in friction, one obtains by use of a lever practically as much work as he puts into it.

Let F_1 (in either Fig. 56 or Fig. 57) be the force exerted on the end of the lever whose length is x_1, and F_2 the force exerted upon the arm of length x_2. Since these forces are each supposed to be at right angles to the arm, the distance through which the force F_1 acts is $x_1\theta$, and the distance through which F_2 acts is $x_2\theta$, where θ is the angle through which the lever is rotated. The work put into the lever is therefore $F_1x_1\theta$. And the work gotten out of the lever $F_2x_2\theta$ must, on the principle of conservation of energy be the same, or

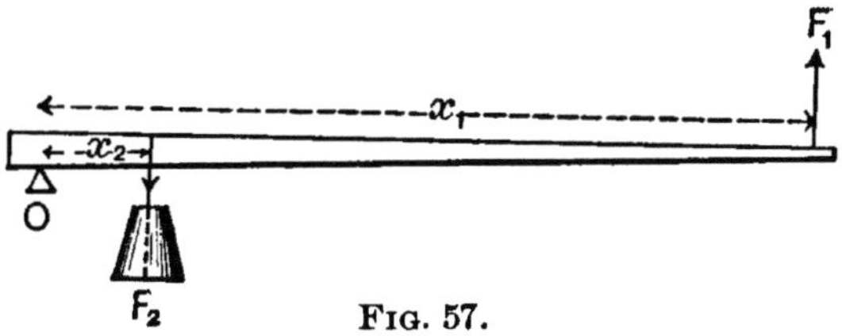

FIG. 57.

$$F_1x_1\theta = F_2x_2\theta,$$

and hence $$F_2/F_1 = x_1/x_2.$$

The physical meaning of this equation is that the force which one can obtain by use of a lever is to the force which he applies to the lever inversely as the lengths of the arms.

The ratio F_2/F_1 is sometimes called the **mechanical advantage** of the lever.

Great care must always be observed in estimating the length of the arms of a lever. The fulcrum at O is the axis of rotation. The distance from the axis to the point where the force is applied is the one arm x_1; the distance from the axis to the point where the opposing force is applied is the other arm x_2. These definitions are illustrated in each of Figs. 56 and 57.

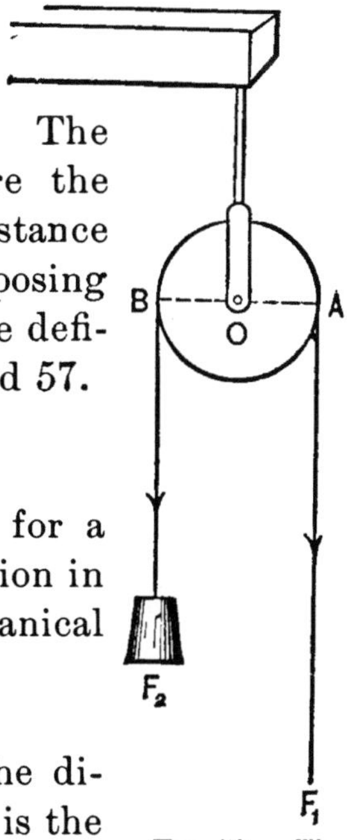

FIG. 58. — The pulley a special case of the lever.

II. *The Pulley*

110. The pulley is simply a lever used for a particular purpose, viz. to change the direction in which a force is applied and to gain mechanical advantage.

Fixed Pulley

111. The bar constituting the lever is the diameter of the pulley $\overline{AB}$ (see Fig. 58). O is the axle (axis) about which the bar is continually rotating. And hence O is the fulcrum.

Since here the arms $\overline{AO}$ and $\overline{BO}$ of the lever are equal, it is evident that $F_1 = F_2$. **So that the only advantage offered by the fixed pulley is a change of direction;** a pull downward can be transformed into a pull upward.

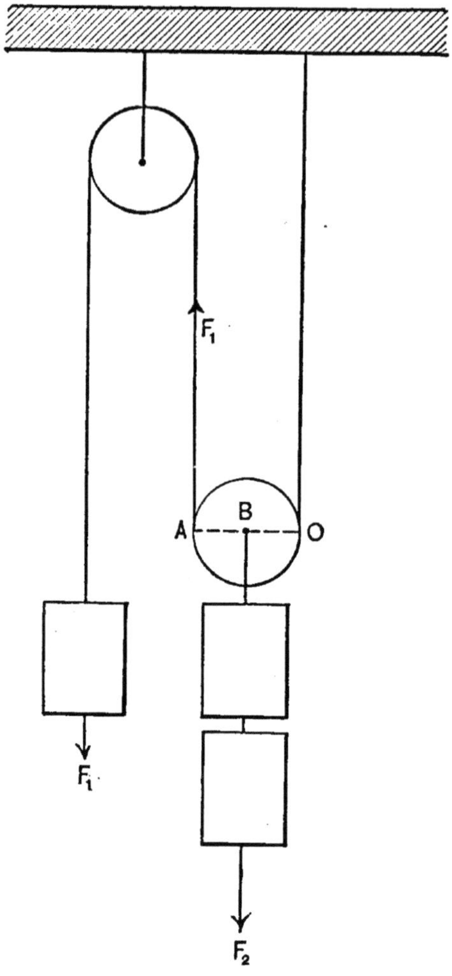

FIG. 59.

Movable Pulley

112. When a pulley is arranged as in Fig. 59, it is evident that *at any particular instant* the point O may be considered as fixed, and as the fulcrum about which the lever rotates.* The force at A acts, therefore, through an arm twice as long as that of the force at B.

Now it is found by experiment that when any point P on the rope is lifted through a distance x_1, the point B on the pulley is lifted through a distance x_2, which is only half as great; in symbols,

$$x_2 = x_1/2.$$

Since the work done by the force F_1 is equal to the energy stored up, we have

$$F_1 x_1 = F_2 x_2 = F_2 x_1/2.$$

$$F_2/F_1 = 2,$$

or the mechanical advantage of a movable pulley is 2.

Block and Tackle

113. In the case of the block and tackle (Fig. 60) we secure a mechanical advantage of 2 for each movable pulley added. So

* A line drawn through the point O, perpendicular to the plane of the pulley block, is called the instantaneous axis of rotation. Where is the instantaneous axis of rotation of a carriage wheel rolling on the ground?

that with a machine such as that figured one is able to exert a force six times as great as without its aid. For the force F_1 is exerted downward through a distance six times as great as that through which the force F_2 is exerted upward.

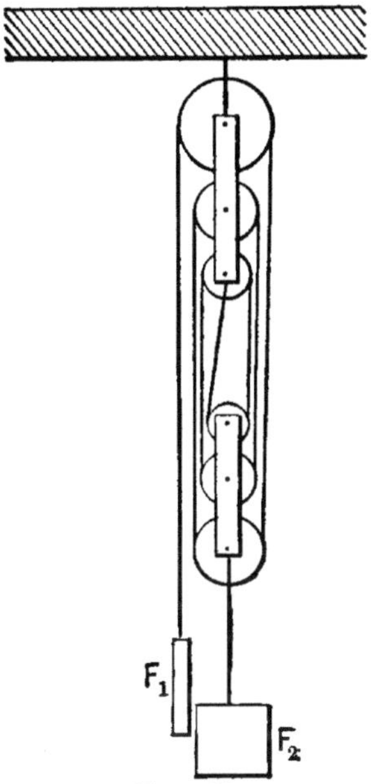

FIG. 60.

And, in general, to obtain the mechanical advantage of such a machine one has only to observe the distance through which the force F_1 is exerted in order to exert the force F_2 through the distance x_2. Call this observed distance x_1.

Then it matters not what the mechanism of the block and tackle is, *provided only that no energy is lost inside*, the mechanical advantage may always be computed as follows:—

The work put into the machine is F_1x_1, the work obtained is F_2x_2,

and $$F_2/F_1 = x_1/x_2.$$

Therefore x_1/x_2, an easily observed ratio, is the required mechanical advantage. In the case represented in Fig. 60, evidently

$$x_1 = 6\,x_2.$$

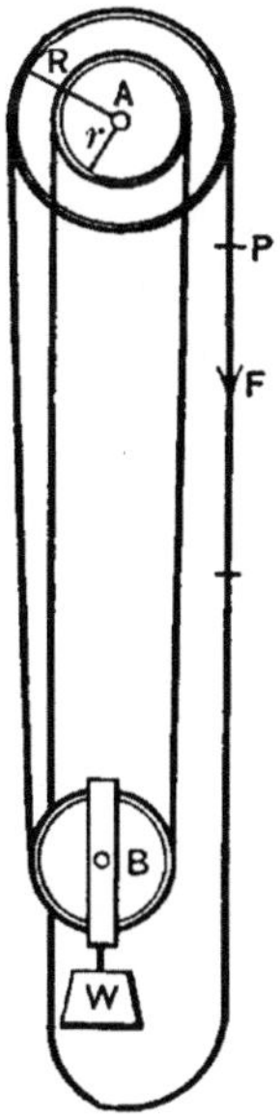

FIG. 61.

The Differential Pulley

114. An interesting form of pulley and one very widely used in large machine shops is that known as the **differential** pulley and shown in Fig. 61. It consists of two parallel fixed pulleys, cast in a single piece A, and one movable pulley B, to which the weight W is attached.

Of the two fixed pulleys one has a radius a little larger than that of the other. Let us denote their radii by R and r. Then if the force F be applied to the free side of the chain coming from the large pulley, and if any link, say P, of this chain be displaced through one circumference of the larger fixed pulley, the work done by the force F will be $F \cdot 2\,\pi R$.

During this same time the work done on the weight W (which is lifted by one chain and lowered by the other) is

$$\tfrac{1}{2}W(2\,\pi R - 2\,\pi r).$$

Hence, neglecting friction and employing the law of the conservation of energy, we have

$$F \cdot 2\,\pi R = W\pi(R - r),$$

or

$$\frac{W}{F} = \frac{2R}{R - r} = \begin{cases}\text{mechanical advantage}\\ \text{of differential pulley.}\end{cases}$$

III. *The Inclined Plane*

115. The inclined plane is a device frequently employed to avoid the direct lifting of heavy weights. Barrels are frequently loaded into wagons by use of ***skids***, which are simply inclined planes. Goods are often taken aboard steamers by inclined gangways. What advantage does such a device offer? The energy required to raise a mass to a certain height above the ground is quite independent of the path by which it has been raised, *provided friction be negligible.*

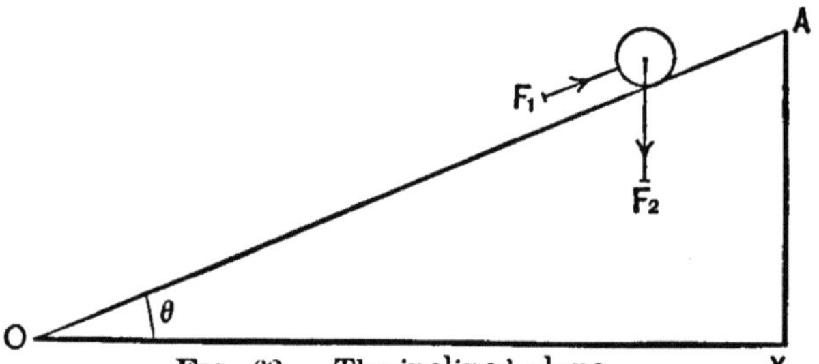

FIG. 62. — The inclined plane.

Let F_1 be the force required to push the body along the incline $\overline{OA}$ (Fig. 62); and F_2 the weight of the body. Let the length of the slope $\overline{OA}$ be x_1 and the vertical height $\overline{XA}$ be x_2.

Then the work put into the mass by the force F_1 is F_1x_1, and the potential energy of the mass at the top of the plane is F_2x_2. But by the law of the conservation of energy these two are equal.

$$\therefore F_2x_2 = F_1x_1,$$

or,

$$F_2/F_1 = x_1/x_2 = 1/\sin\theta,$$

where θ is the angle AOX which measures the steepness of the inclined plane. **The mechanical advantage of the inclined plane is, therefore, $1/\sin\theta$; that is, the mechanical advantage varies inversely as the sine of the angle of slope.** This statement is sometimes called the law of the inclined plane.

In road building the ratio of the rise to the length of the road is called its "grade"; but the ratio of the rise to the length of base is called its "pitch." The numerical value of the grade is evidently the sine of the angle of slope.

Problems

1. Show that, in the case of a windlass, such as is used for hoisting stone to the top of a building, the mechanical advantage of the windlass is R/r, where R is the radius of the crank operated by the man, and r is the radius of the axle on which the rope is wound. The windlass should be considered as a special case of the lever.

2. A lever is 14 cm. long. Where must the fulcrum be placed in order that a weight of 8 g. at one end may just balance a weight of 16 g. at the other end? *Ans.* $4\frac{2}{3}$ cm. from end.

3. A meter stick is suspended at a point 40 cm. from one end. A kilogram placed at one end of the long arm will be balanced by what weight placed at the end of the short arm? *Ans.* 1500 g.

4. When a weight is prevented from sliding down an inclined plane by a force acting along the plane, show that the ratio of this force to the normal pressure on the plane is the same as the ratio of the height to the base of the plane.

5. What is the greatest weight which a man weighing 150 lb. can lift by means of one movable and one fixed pulley? The man is not supposed to be anchored to the ground in any way. *Ans.* 300 pounds.

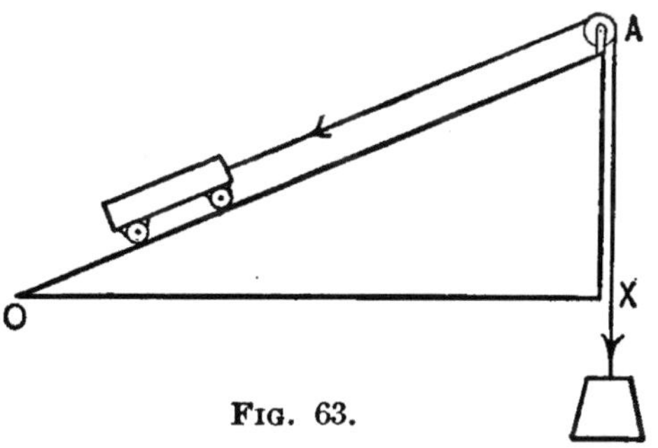

FIG. 63.

6. A man weighing 150 lb. is lowered into a well by means of a windlass, the arm of which is 30 in. long and the axle of which is 8 in. in diameter. Find the force required to let him down with uniform velocity.

7. A screw, the pitch of which is 5 mm., is turned by means of a lever which is 120 cm. long. What force will be required to raise 2000 kg.?

8. A weight of 300 lb. is raised 3 ft. by means of a block and tackle, the block of which has three sheaves. Determine the force required and the space through which it has acted.

9. The pedal crank of a bicycle is 6½ inches long. The rider exerts a force of 100 lb. on the pedal. Find the force exerted on the chain when a 6-inch sprocket is used.

10. A body is displaced against a force which varies directly as the amount of displacement. The initial value of the force is 2 lb., the final value 10 lb., and the displacement 6 in. Find the work done.

11. A belt which drives a pulley travels at the rate of 12 ft. per second. The pull of the belt is 50 lb., the diameter of the pulley is 3 ft. Find the torque exerted upon the pulley and the power which the belt is transmitting.

Work of Friction

116. When two surfaces are brought into contact, a number of interesting phenomena are observed. There is, for instance, a certain amount of adhesion, which becomes very great in the case of two highly polished surfaces, such for instance, as two pieces of optically plane plate glass. This adhesion is very marked in the case of objects covered with gold leaf, in the case of graphite and paper when an ordinary pencil is used, in the case of silver deposited on the back of a mirror, etc.

But there is another very different, and, so far as every-day life is concerned, more important, phenomenon exhibited when two solid surfaces are brought together, namely, one of these surfaces cannot be moved over the other, even at uniform speed, without the exertion of force. **The resistance which either one of the bodies offers to this motion is called the force of friction and is said to be due to friction.**

Frictional forces, like all other forces, are vector quantities, and are to be compounded and resolved as are other forces.

In machinery, friction is at times a most useful feature, at other times a most wasteful one. Thus, for instance, it is by friction that belts are able to drive pulleys; but on the other hand, it is through friction that a very large portion of the energy given to an ordinary machine is wasted. And it is on this ground that a discussion of friction finds place under the head of energy.

The origin of friction doubtless lies in the interlocking of the small hills and valleys, slight rugosities, that remain on even the most highly polished surfaces. As might be expected, therefore, friction diminishes as polish increases. Note that just the opposite is true of adhesion.

Sliding Friction

117. The principal experimental facts regarding the behavior of one solid body sliding over another may be summarized as follows: —

The frictional resistance between solid bodies is (i) *proportional to the force with which the two bodies are pressed together;* (ii) *is independent of the area of the surfaces in contact; and* (iii) *is, within wide limits, independent of the relative speed of the bodies.* The first of these facts may be put into useful quantitative form as follows: Let N be the normal force pressing the two bodies together, and F the force of friction; then for any two given materials

$$F = fN, \qquad \text{Eq. 57}$$

where f is a proportionality constant which is known as the **coefficient of friction** for the two materials in question.

The following table will give some idea of the values which f assumes in practice: —

Materials	Coefficient of Friction
Brass on cast iron	0.19
Wrought iron on cast iron	0.20
Wrought iron on wrought iron	0.14
Iron on ice (skates)	0.16 to 0.032
Oak on oak, fibers parallel	0.48
Oak on oak, fibers crossed	0.32
Leather on oak	0.27 to 0.38
Leather on metals, dry	0.56
Leather on metals, wet	0.36
Leather on metals, oily	0.15

A helpful point of view from which to regard the coefficient of friction is the following: If a body B be placed on an inclined plane as indicated in Fig. 64, its weight may be resolved into two rectangular components, one parallel to the plane, the other perpendicular to the plane. Let the slope of the plane be variable; and let ϕ be the angle which it makes with the horizontal at any instant: $mg \sin \phi$ will then be the component urging the body down the plane; while $mg \cos \phi$ will be the normal component.

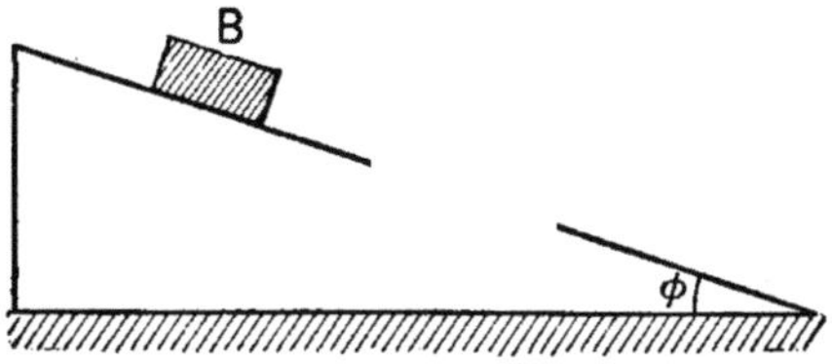

Fig. 64.

Let us now suppose the angle ϕ to increase until the body B is just on the verge of slipping, or has just begun to slip, down the plane. At this point, the friction, which acts up the plane, reaches its maximum value, and Eq. 57 becomes

$$F = mg \cdot \sin \phi = f \cdot mg \cos = fN,$$

and hence

$$f = \tan \phi. \qquad \text{Eq. 58}$$

The angle ϕ thus defined is known as the **angle of repose.**

The rate at which energy is absorbed by friction when one body is drawn a distance s over another during an interval t is then

$$\text{Power} = P = \frac{\text{Work}}{\text{Time}} = \frac{Fs}{t} = \frac{fNs}{t} = fNv \qquad \text{Eq. 59}$$

where N is the load and v the speed.

Rolling Friction

118. Why are ball bearings, instead of ordinary bearings, used in a bicycle? The reply is, in order to substitute rolling friction for sliding friction, since rolling friction is very much less than sliding friction. But the question now arises, why should there be *any* friction in, say, a smooth iron car wheel rolling over a level smooth iron track? The reply is that the iron track is slightly and temporarily indented so that the

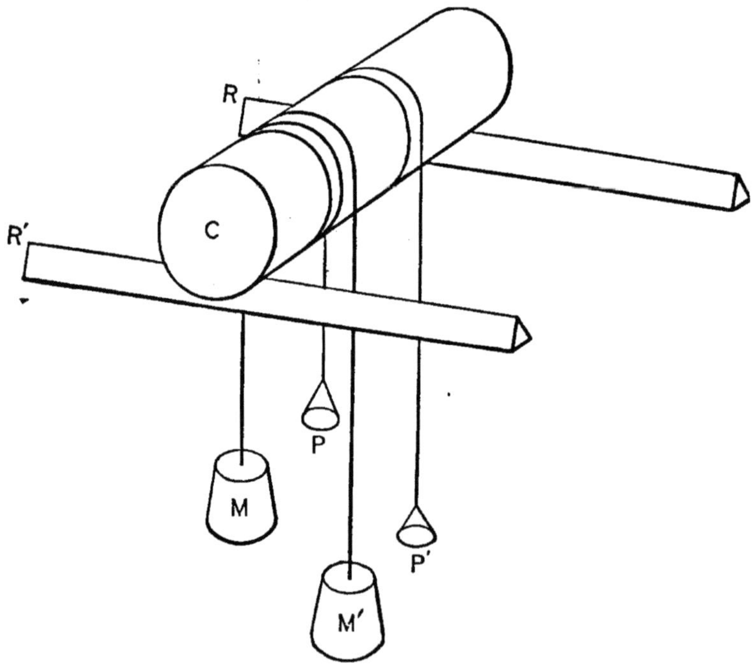

FIG. 65.

wheel has really to run uphill all the time. Every one who rides a bicycle for any distance over a level sandy road soon becomes familiar with this phenomenon in an exaggerated form and keenly realizes the uphill feature of the problem.

Regarding methods for observing this rolling friction quantitatively, perhaps the following is the simplest. See Fig. 65.

Let a metal cylinder C roll upon a pair of rails R and R'. A pair of equal masses M may be strung over the cylinder by means of string to furnish the load. A pair of equal scale pans P and P' supported by a thread, which is wrapped about the cylinder several times in the manner indicated, can receive small weights just sufficient to roll the cylinder in one direction or the other. Two forces now act upon the cylinder,

(i) the load made up of the weight of the cylinder and the suspended masses M; call this force N: and

(ii) the weight W placed in the one pan or the other. If the cylinder is to remain in equilibrium as regards translation, the reaction of the rails Y must just balance these two forces,

or $$N + W - Y = 0.$$

In like manner, if the cylinder is to remain in equilibrium as regards rotation, the moments of all the forces acting upon it must add up to zero. We know the points through which N and W act; but as yet we do not know the point through which the reaction Y acts. Let O, Fig. 66, denote the lowest point of the cylinder, and x the distance between the lines of action of N and Y, respectively. Then taking moments of force about the point O as an axis of rotation, one has

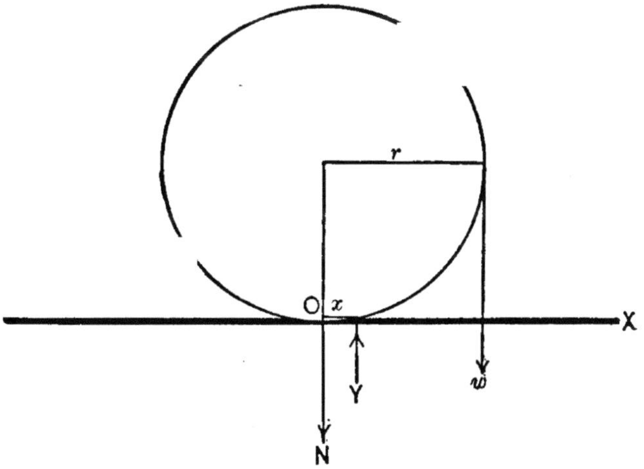

FIG. 66.

$$N \cdot 0 - Y \cdot x + W \cdot r = 0,$$

or, solving for x,

$$x = \frac{Wr}{Y}, \qquad \text{Eq. 60}$$

an equation which could not be true unless the cylinder were in contact with the rail at some finite distance x from the lowest point on the cylinder, which in turn shows that the cylinder is constantly rolling uphill. The energy used up in rolling friction is therefore spent in distorting the material of the rail — a fact which will be clearer after we have considered the subject of elasticity, to which we now proceed. A much more extended study of friction is essential for all students looking forward to engineering.

Problems

1. Is rolling friction the rotational analogue of sliding friction in the same sense that torque is of force?

2. A fly wheel whose moment of inertia is 20 million C.G.S. units is rotating with an angular speed of 240 R.P.M. A friction brake brings this wheel to rest in 20 sec. Find the torque exerted by the brake.

3. A brick lying on an oak plank begins to slide down when one end of the plank is raised so that it is 4 ft. higher than the other. If the coefficient of friction of oak and brick is 0.64, what is the length of the plank?

4. What is the nature of the friction between the axle and the hub of a wagon, rolling or sliding?

5. How does oil act to diminish friction?

6. What is the total force of friction on a train which is pulled at a constant speed of 20 mi. an hour by an engine exerting 50 horse power?

7. What force of friction will be required to bring a mass of 4 gr. to rest in a space of 500 cm., the initial horizontal velocity of the body being 100 cm. per second?

8. The coefficient of friction of iron on ice is, say 0.16. A boy is able to pull 100 pounds in a horizontal direction. What is the greatest load he can move by the use of a sled shod with iron runners?

9. It requires a force of 10 tons to move a given freight train. If the coefficient of friction between the driving wheels and the rails is 0.14, what is the least weight which the locomotive may have in order to start the train?

10. What is the angle of repose for one oak plank lying on another, the fibers being parallel? See Table, § 117.

11. What power will be absorbed in pulling a 400-lb. cake of ice over a level wooden platform at the rate of 5 ft. per second? Take coefficient of friction as 0.20.

12. A horse pulls a sleigh at the rate of 8 miles per hour. The sleigh with its occupants weighs 500 lb. If the coefficient of friction is 0.16, what power must the horse exert?

13. Referring to Problem 6, § 84, solve the case represented in Fig. *A*, when the coefficient of friction between m_2 and the inclined plane is 0.33.

14. Referring to Problem 6, § 84, solve the case represented in Fig. *B*, when the coefficient of friction between the mass "30" and the inclined plane is 0.20.

15. A locomotive can pull a train of 500 tons weight along a level piece of track. What weight can it pull up a grade of 1 in 60, the coefficient of friction being the same in each case?

IV. *Elasticity*

119. In the preceding pages we have had frequent occasion to deal with a "rigid" body, by which we have meant a **perfectly rigid body**; *i.e.* one the relative position of whose parts admits of no change whatever. Experience tells us that no such bodies are met with in nature; although we have many close approximations. A strip of hard and brittle window glass may be bent with ease. Cold steel can be rolled or drawn into various shapes. While *for all ordinary operations* the metal parts of a pair of shears, the frame of a bicycle, the beam of a balance, etc., are rigid bodies, there are many cases in which we are forced to recognize the fact that each of these bodies changes form or volume. **Bodies that can be changed in either size or shape are said to be elastic.**

In the case of a collision, it may become painfully evident that one's bicycle frame is not a rigid body. In case one is weighing a large load, the balance beam is liable to bend slightly and thus lower the *c.m.* of the balance, and hence diminish the sensibility of the balance.

A large variety of the phenomena of elasticity are familiar to us all; one need only mention the pneumatic tire, the rubber band, the bending of an oar, the "straining" of a ship, the vibrations of a steel bridge.

120. To bring some degree of order out of this chaos of experience is the object of this chapter. And this is not very difficult. For, indeed, careful examination shows that, after all, there are only two kinds of changes which may occur in the configuration of elastic bodies; viz.

1. Change of size, *i.e.* Compression or Dilatation. } Strains.
2. Change of shape, *i.e.* Distortion. }

When a body is changed in either size or shape, it is said to be strained. The size of a body may be strained, and the shape of a body may be strained; and these are the only strains that are known. Since all known bodies are subject to one or both of these changes, elasticity is said to be a general property of matter.

Distinction between Solids and Fluids

121. But while all bodies may be changed as indicated, they behave very differently *during* the change.

If we attempt to alter the *shape* of a piece of glass tubing by slightly bending it, we shall find that this will require a moment of force; that, if we remove the force, the rod regains its original shape. In like manner, we should find that force would be required to change the *size* of the glass rod, and that this force would need to be maintained as long as the changed size is maintained.

Now imagine a second case in which the glass tube is full of mercury or coal gas while it is being bent: the mercury will offer no permanent resistance whatever to a change of shape; it conforms perfectly, without requiring any moment of force, to the shape of the containing vessel. The mercury has, in fact, no elasticity of *shape*. The same is true of the coal gas. But if one attempts to *compress* the mercury or the coal gas into a smaller volume, he finds this operation requires force; and that this force must be maintained so long as the diminished volume is maintained. Mercury and coal gas have, therefore, elasticity of *size;* but they have no *shape* of their own.

All bodies in nature fall into one or other of the two cases we have just considered; and this fact leads to the following definitions: —

1. Bodies which exhibit elasticity both of shape and of size are called **solids**.

2. Bodies which possess elasticity of size, but not of shape, are called **fluids**.

As is well known to every one, this distinction between fluids and solids is not a distinction between different substances, but between different bodies. For the same substance may, under different circumstances, assume each of these two states — one needs only to recall the case of ice and water.

Distinction between Gases and Liquids

122. While mercury and air are alike in possessing only elasticity of size, **they differ very much in the manner in which they exhibit this elasticity of size.**

No one ever heard of a closed vessel partly full of air; a closed vessel of *any* size containing *any* amount of air is always full of it; and, at ordinary temperatures, the air is distributed uniformly throughout the entire vessel. Contrast this with mercury or water, which may or may not completely fill the vessel.

Definitions: —

1. "There are certain fluids any portion of which, however small, is capable of expanding indefinitely so as to fill any vessel, however large. These are called **gases**.

2. "There are certain other fluids a small portion of which, when placed in a large vessel, does not expand so as to fill the vessel uniformly, but remains in a collected mass at the bottom, even when the pressure is removed. These fluids are called **liquids**." — MAXWELL, Art. "Constitution of Bodies," *Ency. Brit.*

123. We might summarize the preceding classification of bodies as follows: —

- Bodies
  - Rigid — not known in nature.
  - Elastic
    - Solids.
    - Fluids
      - Gases.
      - Liquids.

QUANTITATIVE CONSIDERATION OF ELASTICITY

Coefficients of Elasticity

124. A builder who is putting up a roof or a bridge always finds himself between Scylla and Charybdis. The bridge must be strong enough to sustain the heaviest load easily; and, at the same time, it must contain as little useless material as possible. In order to secure this latter result, the engineer must have definite and quantitative information concerning the elastic properties of the steel to be used in the bridge; in particular, he must know what elongation (stretch) will occur in a specimen of this steel, having a definite length and a definite area of cross section, when it is given a definite load to carry

In like manner, the engineer who designs a cotton mill must know what moment of force a steel shaft will safely transmit before he can determine the proper size of shafting to be used in the mill. The designer of a steam engine must likewise be definitely informed concerning the elastic properties of steam. A certain amount of steam has been admitted to the cylinder of an engine and the port closed; what pressure will it exert when the piston has reached the end of the stroke? Answers to such questions as these have been obtained by the following method: —

General Definition of Stress and Strain

125. When two or more forces are applied to a body in such a way that their resultant is zero, there is no acceleration and no evidence, therefore, from the motion of the body, that such individual forces are acting. Such forces are said to be in equilibrium. Their existence is shown by the strains which they produce in a body.

"A stress is an equilibrating application of force to a body.

"A **strain** is any definite alteration of form or dimension of a solid." — KELVIN.

Elasticity is measured by the resistance which a body offers to change either of size or of shape, *i.e.*

$$\text{Elasticity} = \frac{\text{Stress}}{\text{Strain}}. \qquad \text{Eq. 61}$$

Elasticity of Size. I. Length

126. When a force is applied to a body in such a way as to alter its length, either by compressing it or by stretching it, the force is necessarily distributed over the cross section of the body.

The longitudinal stress at any point in a body is defined as the ratio of the force to area at that point. Longitudinal stress, then, is not a force, and is not measured in dynes; but it is measured in dynes per square centimeter.

A longitudinal stress always changes both the length and the cross section of the body to which it is applied. But in most practical cases the change in cross section is small; and we consider only the alteration in length. The total change in length depends not only upon the stress, but also upon the length of the body. Hang a kilogram on the end of a wire one meter long: the elongation will be a certain amount. If, now, the same kilogram is hung from the end of a wire of the same kind two meters long, the elongation will be twice as great. But the wire is strained at each point in its length by the same amount in each case. Consequently **longitudinal strain** is defined as **the ratio of the total elongation to the length.**

127. In the case of a wire, let

$L =$ length of wire;
$a =$ area of cross section of wire;
$F =$ stretching force applied to wire;
$e =$ total elongation produced by F.

The longitudinal stress $= F/a$ dynes per square centimeter and the longitudinal strain $= e/L$ centimeters per centimeter.

Now the longitudinal elasticity of any material is measured by the ratio of the longitudinal stress to the longitudinal strain. And hence for this ratio, which is generally known as **Young's* modulus**, we have in the case of a wire

$$\textbf{Young's modulus} = M = \frac{\textbf{Longitudinal stress}}{\textbf{Longitudinal strain}}$$

or

$$M = \frac{\left(\frac{F}{a}\right)}{\left(\frac{e}{L}\right)} = \frac{FL}{ea}. \qquad \text{Eq. 62}$$

From this equation it is evident that, in order to determine M by experiment, we must measure the four quantities in the right-hand member of the equation. As a stretching force it is often most convenient to use the weight of a mass m; viz. mg. If we employ a round wire of diameter d, its area will be $\pi d^2/4$. Accordingly Eq. 62 may be written

$$M = \frac{4\,mgL}{\pi e d^2},$$

* The life of Thomas Young, a London physician, forms an interesting chapter in the history of physical science. See his biography in *Ency. Brit.*

in which all the quantities on the right are directly measurable. Such an equation we have called a **laboratory equation**.

128. M has been measured for a large number of substances, and has been found to be constant, for any one substance, within certain limits. These limits vary from substance to substance. For steel and glass these limits are not very wide; for rubber these limits are exceedingly wide. And hence the popular impression that rubber is highly elastic. As a matter of fact, glass or iron is vastly more elastic than rubber; for to produce a given strain in glass much greater stress is required than to produce the same strain in rubber. One can test this by allowing two small balls, one of rubber and one of glass, to fall from equal heights. The flattening of the rubber when they rebound to equal heights is much greater. (Try this on oiled glass or marble plate.)

129. In such an equation as that which defines Young's modulus many important physical facts have been summarized. The student should early acquire the habit of interpreting the physical meaning of each equation which he employs.

Solving Eq. 62 for e, we have

$$e = Fl/aM,$$

from which we may write at once: —

The elongation

- $\propto$ directly as the stretching force (Hooke's Law).
- $\propto$ directly as the length.
- $\propto$ inversely as the area of cross section of the wire.
- $\propto$ inversely as the longitudinal elasticity; *i.e.* the elongation depends also upon the material of which the wire is made.

Each of these predictions is easily tested in the laboratory.

Hooke's Law

The fact that, for a rod of given length, cross section, and material, the elongation varies as the stretching force is one of the earliest important discoveries in the subject of elasticity.

It was made by Robert Hooke in 1660 and published by him in 1676 under the form of the following anagram: —

c e i i i n o s s s t t u u

In 1678 he rearranged these letters and translated them as follows: "*Ut tensio sic vis;* that is, the Power of any spring is in the same proportion with the Tension thereof." Observe that at this early date he uses *power* in the sense in which we now employ the word *force;* while his *tension* is equivalent to our *extension.*

The best evidence for the truth of Hooke's law is the fact first pointed out by Stokes, that all elastic solids can be made to vibrate isochronously; for equal periods of vibration, at all amplitudes, means that the force acting upon the body is proportional to the displacement.

The extreme convenience of this modulus can here be merely indicated by saying that, if an engineer wishes an iron rod of length L to carry a certain load, which we may call F, without producing an elongation greater than a certain amount e, our equation tells him at once just how large a cross section a the rod must have; viz. —

$$a = FL/eM.$$

As indicating the approximate numerical values of the modulus, the following will serve: —

For Bessemer steel, M	$= 22 \times 10^{11} \frac{\text{dynes}}{\text{cm}^2.}$	$= 31 \times 10^6 \frac{\text{pounds}}{(\text{inch})^2}$
Cast iron	$- 12 \times 10^{11}$ "	$- 17 \times 10^6$ "
Brass wire	$- 10 \times 10^{11}$ "	$- 14 \times 10^6$ "
Copper wire	$- 12 \times 10^{11}$ "	$- 17 \times 10^6$ "

Elasticity of Size: II. Volume

130. Let us denote the volume of a body by V. Imagine the volume to be subjected to pressure so that it is diminished by an amount v.

The ratio of the diminution to the total volume is called the voluminal strain. Suppose further that this change of volume has been produced by applying a force F to a surface of area a; as, for instance, when air is compressed in an ordinary bicycle pump. **The ratio of the total force to the area over which it is applied is the voluminal stress; and is what we call pressure.**

In accordance with our general equation (§ 125), the elasticity of volume for any material is measured by the ratio of the voluminal stress to the voluminal strain. This ratio is generally known as the **bulk modulus**, and is commonly denoted by k.

$$\textbf{Bulk modulus} = k = \frac{\textbf{Voluminal stress}}{\textbf{Voluminal strain}},$$

or

$$k = \frac{\left(\frac{F}{a}\right)}{\left(\frac{v}{V}\right)} = \frac{FV}{av} \qquad \text{Eq. 63}$$

The reciprocal of the bulk modulus is what is known as the **compressibility** of a substance.

Approximate values of certain bulk moduli are given in the following table: —

Substance	Bulk Modulus	
	In grams per sq. cm.	In pounds per sq. in.
Water at 20° C.	23.2×10^6	0.33×10^6
Brass	1090×10^6	15.48×10^6
Copper . . .	1202×10^6	17.10×10^6
Steel	1518×10^6	21.61×10^6
Glass	405×10^6	5.76×10^6

Smithsonian Tables, Nos. 93, 94.

Elasticity of Shape

131. In a strictly analogous manner is defined the elasticity of shape, more frequently called the rigidity modulus.

$$\textbf{Rigidity modulus} = n = \frac{\textbf{Shearing stress}}{\textbf{Shearing strain}}.$$

But this subject is somewhat intricate, and must be left for later study. Before leaving it, however, we merely remark that the rigidity modulus of any material measures its strength in resisting a shear, and hence its fitness for mill shafting, for propeller shafts of steamships, etc. Without attempting to define shearing stress or shearing strain, we can only say that when a rod or shaft is twisted, a shearing stress is exerted upon it; and this produces a shearing strain, *i.e.* one surface is displaced over another, just as would happen if the rod were cut in two by a pair of shears.

For a shaft, or wire, of circular cross section, the experimental facts are summarized in the following equation, where, if the shaft be clamped at one end, θ is the amount of twist, in radians, which will be produced by a torque L, applied at the free end: —

$$\theta = \frac{2l}{\pi n r^4} L \qquad \text{Eq. 64}$$

In other words, the amount of twist (angular displacement) varies

(i) directly as the torque L which produces the twist (Hooke's Law).

(ii) directly as the length of shaft l.

(iii) inversely as the rigidity modulus n.

(iv) inversely as the fourth power of the radius of the shaft r.

Eq. 64, together with the following table, will enable the reader to solve many practical problems concerning mill shafting, etc.: —

SUBSTANCE	RIGIDITY MODULUS	
	In grains per sq. cm.	In pounds per sq. in.
Brass	$320 - 410 \times 10^6$	$4.6 - 5.8 \times 10^6$
Copper . . .	$393 - 473 \times 10^6$	$5.6 - 6.7 \times 10^6$
Steel . . .	$746 - 829 \times 10^6$	$10.6 - 11.8 \times 10^6$
Glass . . .	$235 - 380 \times 10^6$	$3.3 - 3.9 \times 10^6$
Wood	$7 - 12 \times 10^6$	$0.1 - 0.17 \times 10^6$

Smithsonian Tables, No. 78.

Torsion Pendulum

132. If one suspends a metal cylinder or bar, by means of a wire as shown in Fig. 67, it will oscillate in a horizontal plane with a period T. If I denote the moment of inertia of the disk, L the torque at any instant, as defined above, and A the angular acceleration at this same instant, then since the acceleration is always in a sense opposite to that of the torque which produces the twist,

$$L = IA. \qquad \text{Eq. 37}$$

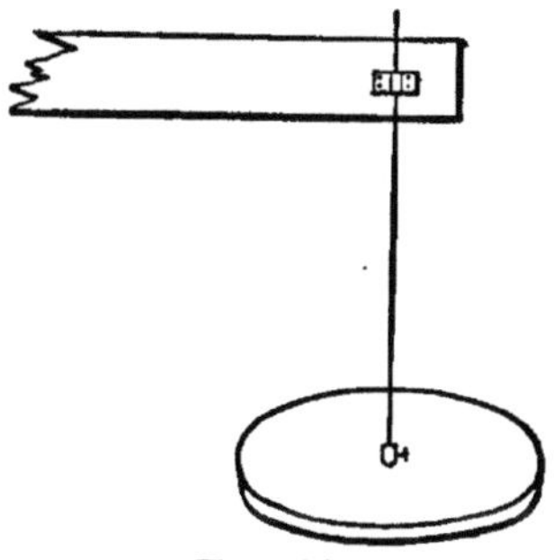

FIG. 67.

Eliminating L between this and Eq. 64, one has

$$A = -\frac{\pi n r^4}{2lI}\theta = - \text{constant} \times \theta.$$

Hence the motion is simple harmonic and has the following period: —

$$T = 2\pi\sqrt{\frac{2lI}{\pi n r^4}}, \qquad \text{Eq. 65}$$

which is a convenient **laboratory equation** for measuring n, the modulus of rigidity for the material from which the wire is manufactured.

Problems on Elasticity

1. A wire whose original length was 64.4 cm. has been stretched so that it is now 70.0 cm. long. Find the longitudinal strain of the wire.

Ans. .0869.

2. A wire has a radius of 2 mm. and carries a load of 40 kg. Find the longitudinal stress on every cross section of the wire. Express your result in C.G.S. units.

Ans. $\frac{981000000}{\pi}$.

3. A certain metal rod has an area of cross section equal to 3 mm². The length of the rod is 150 cm. Young's modulus for this rod is 12×10^{11} C.G.S. How much will the rod be elongated by stretching it with a force of 8,000,000 dynes?

Ans. $\frac{1}{30}$ cm.

4. I wish to support a weight of 10 kg. by use of a wire 2 m. long. Young's modulus for the wire is 8×10^{11} C.G.S. What cross section must the wire have in order that it may not stretch more than 1 mm. when the load is attached.

Ans. 0.0245 cm².

5. A very slight stress applied to a rubber band will produce a very great strain. Would you, therefore, say that rubber is a highly elastic or highly inelastic substance? What gives the popular impression that rubber is very elastic?

6. A steel rod requires a large stretching force to produce even a small elongation. Is it, therefore, very elastic or very inelastic?

7. Under ordinary conditions it requires a much larger pressure to compress a gallon of water by a certain amount than it does to compress a gallon of air by the same amount. Which, would you say, has the greater elasticity of volume?

8. Two wires suspended from a beam have each the same length and carry the same load. One wire is of brass, the other of Bessemer steel; the former has a diameter of 3 mm.; find what diameter the latter should have in order that the load will produce the same elongation in each wire.

9. What are the "dimensions" of longitudinal stress? Of longitudinal strain? Of the bulk modulus?

10. A piece of mill shafting 1 in. in diameter can just withstand a certain torque L without injury. How much stronger will be a piece of $1\frac{1}{2}$-inch shafting from the same material?

11. Two torsion pendulums are alike in all respects except that the first is made of wire which is $\frac{1}{64}$ in. in diameter and 6 in. in length, while the second is made of wire $\frac{1}{32}$ in. in diameter and is 8 ft. long. Compare their periods of oscillation.

Algebraic Summary of Dynamics

Linear	Angular

Fundamental Quantities

Linear	Angular
Length, inertia, and time.	Angle, rotational inertia, and time.
x, m, t	θ, I, t

$$I = r^2 m.$$

Velocity

Linear	Angular
$v = \dfrac{x}{t}$	$\omega = \dfrac{\theta}{t}$

$$v = r\omega.$$

Acceleration

Linear	Angular
$a = \dfrac{v}{t}$	$A = \dfrac{\omega}{t}$

$$a = rA.$$

Simple Harmonic Motion

Linear	Angular
$T = 2\pi\sqrt{\dfrac{(-x)}{a}}$	$T = 2\pi\sqrt{\dfrac{(-\theta)}{A}}$

Momentum

Linear	Angular
mv	$I\omega$

$$I\omega = r \cdot mv.$$

Force and Torque

Linear	Angular
$F = ma$	$L = IA$

$$L = rF.$$

Algebraic Summary of Dynamics — *Continued*

Angular	Linear
Centrifugal Force and Precessional Couple	
$mv\omega$	$I\Omega\omega$
Energy	
K.E. $= \frac{1}{2} mv^2$ P.E. $= Fx$	K.E. $= \frac{1}{2} I\omega^2$ P.E. $= L\theta$
Power	
Fv	$L\omega$

CHAPTER IV

SOME SPECIAL PROPERTIES OF MATTER

133. In the preceding pages we have considered four properties of matter which are perfectly general. It is an experimental fact that matter possessing any one of these, possesses all the others. This by no means exhausts the list of general properties, but it probably includes those which are most important as parts of the intellectual equipment of a liberally educated man.

Properties of Liquids

134. We shall first consider some of the special properties of liquids, and from the three following points of view: —

I. Behavior of Liquids at Rest and under Gravity; Hydrostatics.

II. Behavior of Liquids in Motion under Gravity; Hydraulics. Wave motion in liquids will, however, be treated in the following chapter, along with waves in strings.

III. Behavior of Liquids at Rest and freed from Gravity; Surface Tension.

I. Hydrostatics

135. There are few people who are not familiar with the fact that the water in a teapot stands at the same height in the spout as in the pot. The danger of a leak in the hull of a boat arises from the fact that the boat will continue to "fill" until the water stands at the same level inside and out. Mercury in a U-tube stands at the same level in each arm, provided the diameter of the tube is uniform or large. If, in any of these cases, the level is disturbed for an instant, it is quickly restored when the force producing the disturbance is removed.

This behavior of liquids has been summarized and described in a very brief and simple manner. But we must first digress to add another word to our vocabulary; namely, *pressure.*

Definition of Pressure

136. In physics the term "pressure" is employed always to denote the ratio between the force applied to any surface and the area of that surface. Let us denote pressures by P, forces by F, areas by A, then

$$P=\frac{F}{A}$$ Defining equation for *average* pressure over area A.

$$P=\left[\frac{F}{A}\right]_{A\doteq 0}$$ Defining equation for pressure *at any point*. Eq. 66

From this definition the **dyne per square centimeter** follows as the natural **unit of pressure.** But no unit can be considered thoroughly practical until it is expressed in some more concrete form. Besides, this unit is very small, and hence the International Congress on Physics which met at Paris in 1900 adopted a larger unit, namely, **one million dynes per square centimeter,** and called it the **barye** — (from the Greek *βάρος*, *weight*).

As we shall find, a little later (Eq. 67), this definition of the *barye* is equivalent to assuming a standard value for the acceleration of gravity. For since the accepted value for the density of mercury is 13.5950, it follows from $ghD=10^6$ that $g=980.692$. Normal atmospheric pressure also becomes exactly $\frac{76}{75}$ baryes.

The engineer's unit is sometimes the pound per square foot, more generally the pound per square inch. Thus steam gauges are practically always graduated in pounds per square inch.

The definition of a fluid (§ 121) as a body which is devoid of elastitity of shape (rigidity) is equivalent to saying that a fiuid cannot sustain any tangential stress.

137. From this definition follows at once the experimental facts described in the first two of the following propositions.

Proposition I. **The stress (pressure) across any surface drawn in a liquid at rest is normal to that surface.**

Proposition II. **At any point in a liquid at rest the stress (pressure) is the same in all directions.** Pascal's Theorem.

138. The two preceding propositions would be equally true for liquids which were not acted upon by gravity. In the case of all actual liquids, we have to consider among the forces which act upon them, not only the pressure due to the containing vessel, as indicated in Fig. 68, but also the pressure due to the weight of the liquid itself.

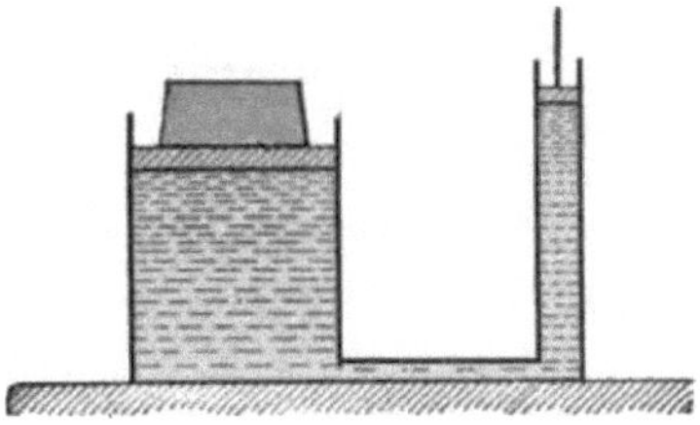
FIG. 68.

This fact is familiar to all boys who have observed when in swimming that they can dive to very moderate depths without being troubled by water getting into their ears; but that, if they exceed a depth of eight or ten feet, water almost invariably gets into their ears. Builders know that when a tank is put into the attic of a house, the walls of the tank must be made strong, *not* in proportion to the *volume* of water which it is to hold, but in proportion to the *depth* of water in the tank. These and all similar phenomena are described as follows:

Proposition III. **The pressure at any point in a liquid at rest under gravity is proportional to the depth.**

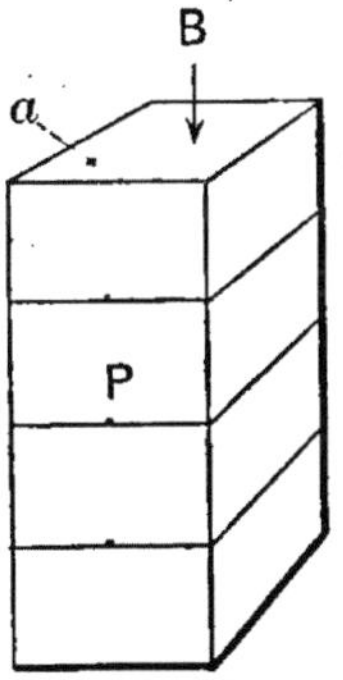

FIG. 69.

The facts are summarized more compactly in the symbols of algebra as follows: The pressure at any point in the liquid will be the total weight of the liquid supported by any small area about that point, divided by the area. To this must be added the pressure of the air upon the free surface of the liquid. Let

P = pressure at any point in liquid (see Fig. 69);
B = pressure of earth's atmosphere on free surface of liquid;
W = weight of liquid supported by area a.

Then $$P = W/a + B.$$

But from the definition of density (Eq. 28)

$$W = ahD \cdot g,$$

where h = vertical distance of area a from free surface;
D = density of fluid, here supposed to be uniform;
g = acceleration of gravity.

Hence $$P = hD \cdot g + B. \qquad \text{Eq. 67}$$

If we now neglect all pressure of the atmosphere, and consider that due to the liquid alone, we may write

$$P \propto h,$$

which is *Proposition III* stated above.

The important idea here is that in any given liquid the difference of pressure, $P_2 - P_1$, between any two points depends only upon their difference of level, $h_2 - h_1$. Thus

$$P_2 - P_1 = Dg(h_2 - h_1).$$

The fact that the pressure is independent of the form of the containing vessel is generally known as the "Hydrostatic Paradox."

139. In *Proposition II* only the variation of pressure with direction is considered. But consider a vessel filled with water in one case and mercury in another. At any given point in the vessel how will the pressure change when the liquids are changed?

This question also is definitely answered by Eq. 67, since it shows

$$P \propto D.$$

We have then as a corollary to *Proposition III* the following theorem easily verified by experiments on liquids in a U-tube.

Proposition IV. **Other things remaining the same, the pressure at any given point in a liquid varies directly as the density of the liquid.**

140. Another important experimental fact, also an inference from Eq. 67, is the following:

Proposition V. **The pressure is the same at all points in a horizontal plane of a liquid at rest under gravity.**

For if the pressure were not the same at all points, we should have present a force tending to drive a particle from one point to another on a horizontal surface. And this we do not find in nature. When a spherical pebble sinks in quiet water, there is no lateral pressure thrusting it either to one side or the other.

A simple experiment illustrating this fact is that sketched in Fig. 70, where the height (above the outside level) of the liquid in the inverted glass bulb indicates a variation of pressure as the lower end of the tube is raised or lowered, but shows no variation when the lower end of the tube is moved about in any horizontal plane.

Among those surfaces over which the pressure is uniform may be reckoned the top of the liquid, the bounding surface, more generally called the **free surface.**

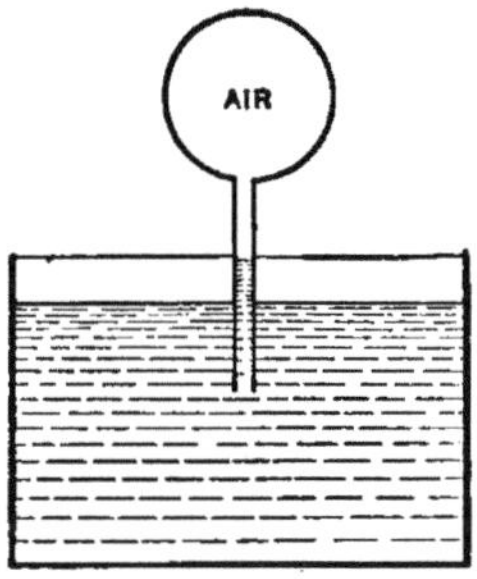

FIG. 70.

141. If we assume the converse of *Proposition V*, which is also amply verified by experiment, then it follows as a corollary that **the free surface of a liquid is horizontal.** But all that is meant by a horizontal surface is one which stands perpendicular to the direction of the force acting upon the liquid. By experiments such as that outlined below, this particular case has been generalized into the following theorem:

Proposition VI. **Any free liquid surface is, at each point, perpendicular to the resultant of all the forces acting upon the liquid particle at that point.**

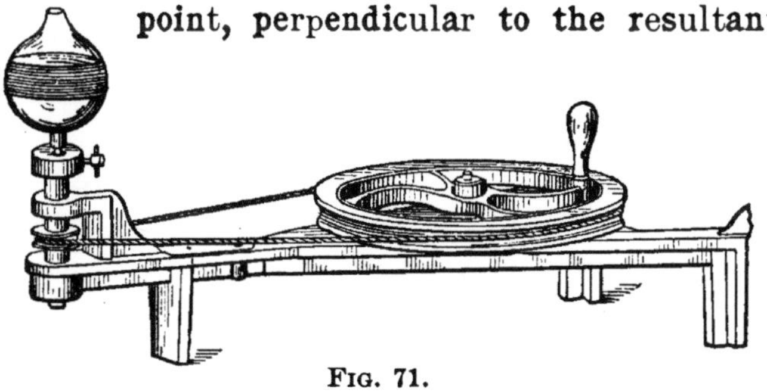

FIG. 71.

To illustrate, attach a vessel of colored liquid to a whirling table as shown in Fig. 71. The curved form which the free surface assumes when the vessel is rotated is at every point normal to the force F resulting from the weight mg of a liquid particle m, and from the centrifugal force $mr\omega^2$, acting upon the same particle. The vertical line in Fig. 72 is the axis of rotation of the whirling table; the curve is a cross section of the free surface. The resultant force F, or the normal to the liquid surface,

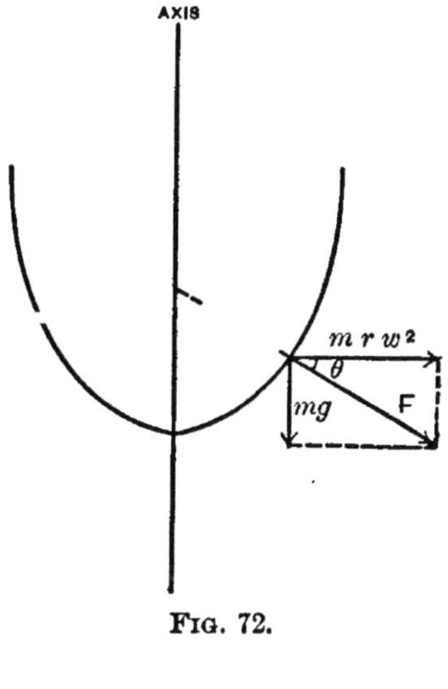

FIG. 72.

therefore makes an angle θ with the horizontal whose tangent is

$$\tan \theta = \frac{g}{r\omega^2}.$$

Free surfaces are, therefore, not always horizontal.

The Principle of Archimedes

142. The modern marine engineer, in designing a boat, computes the weight of all the material which enters into its construction. Often the iron used in the hull is actually weighed. Knowing the dimensions of the boat, the designer is then able to say, generally within less than an inch, and before the boat is launched, just how deep she will lie in the water. The simple manner in which this prediction is made will be evident from the following: —

Whether a body floats or sinks when placed in a liquid, one fact is patent to everybody, — the body apparently loses weight. But just *how much* weight a body loses under these circumstances was first determined by Archimedes (287–212 B.C.), the founder of our present system of hydrostatics.

It is evident that, since the pressure increases with the depth, the bottom of an immersed body will be subjected to a pressure upward which is greater than the pressure downward to which the top of the body is subjected. The pressure of the water on a body will, therefore, always exert a buoyant force.

FIG. 73.

Archimedes saw very clearly, by a simple deduction from experiment, that if from any vessel of liquid (Fig. 73) we imagine a portion of the liquid to be removed, the position of the remainder being unchanged, the upward force exerted by the pressure on the surface so exposed would just be able to support the weight of the liquid removed. Not only so, but it is evident that this upward force acts through the center of mass of the displaced fluid; since, when the fluid is replaced, there is no moment of force tending to rotate it. This point is called the **center of buoyancy.**

By such reasoning he established the following theorem, which is known as the Principle of Archimedes.

Proposition VII. **The resultant force of a fluid on a body immersed in it acts vertically upward through the center of mass of the displaced fluid; and is equal to the weight of the displaced fluid.**

Manifold applications of this principle will be met in the laboratory.

Stability of Floating Bodies

143. The equilibrum of any body — and hence also of a floating body — is secured only when the following conditions are satisfied (see § 104): —

(i) The sum of all the forces (of translation) acting upon it is zero.

$$\Sigma F = 0.$$

(ii) The sum of all the moments of force acting upon it is zero.

$$\Sigma L = 0.$$

When a body which will float is placed in the water, it will therefore sink to such a depth that the water displaced is equal to its own weight. The first of the two conditions is then satisfied. But when the body is first put into the water its center of gravity G will not, in general, lie in the same vertical line with the center of buoyancy B. The result is a couple, as indicated in Fig. 74, which tends to rotate the body until B and G *do* lie in the same vertical line. Then both of the above conditions for equilibrium will be satisfied.

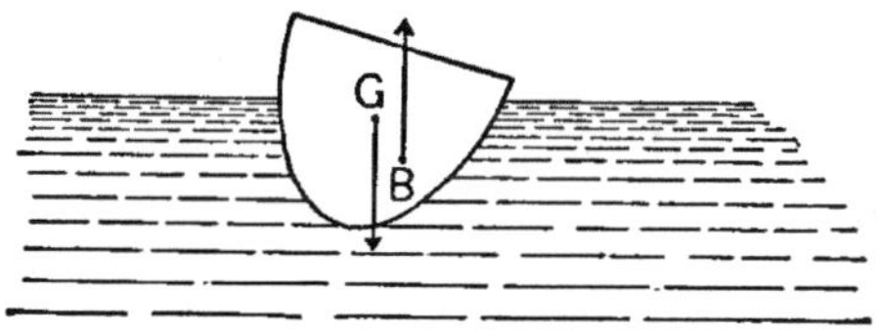

FIG. 74.

But the owner of the boat now imposes a third condition. He demands not only that B and G shall lie in the same vertical line, but also that the boat shall be right side up. How this condition is satisfied by the builder of the boat may be most easily explained in terms of a point which is called the "metacenter" and which is defined as follows: —

When a body "rolls" in the water, as indicated in Fig. 75, the center of buoyancy moves in the body, and thus describes a curve BB', Fig. 75 (*c*). **The center of curvature of this curve**

described by the center of buoyancy in its motion through the body is known as the metacenter. The metacenter may, therefore, be considered as the point in a floating body at which the upward thrust of the water is applied. **The condition for stability in a vessel is, therefore, that its metacenter shall be higher than its center of gravity.** For then if a boat heels over, there will be called into play a moment of force tending to right it; but if the metacenter lies *below* the center of gravity, then when a boat rolls ever so little there will be a torque tending to roll it still farther.

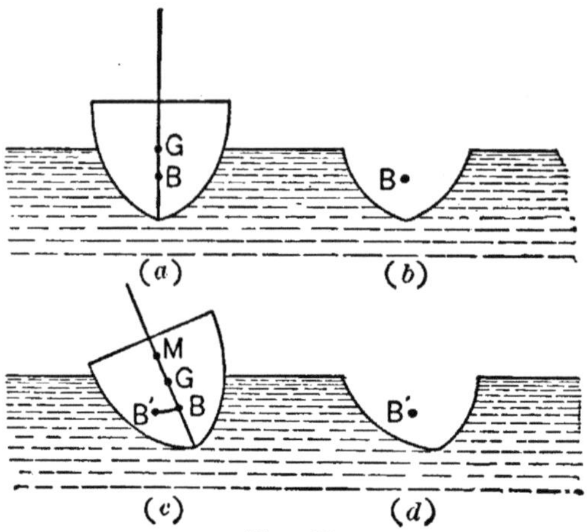

Fig. 75.

The weight of a vessel being constant, it is clear that stability varies directly as the height of the metacenter above the center of gravity. Things are generally made fast and cargoes stowed in such a way that they will not shift; hence the center of gravity may be considered as a fixed point.

In sailing vessels — even more than with steamers — stability is a consideration of prime importance. In the battleship, stability is more difficult to secure, since the armor belt must be placed high, and also because the guns are more effective when placed high — both circumstances tending to elevate the center of gravity. The capsizing of the British gunboat *Captain* in the Bay of Biscay with great loss of life in 1870 was due to insufficient metacentric height.

II. Hydraulics

144. In the problems of hydrostatics, we have been considering only cases in which at every point of the liquid the forces were in equilibrium, because the pressures were equal. But now let us consider what happens when two bottles are connected as in Fig. 76. At C there is a stopcock which retains the water at a height h in the vessel A. So long as the stopcock is closed, the pressure on any element of the liquid in the small tube is the same on the right as on the left; but when the cock is opened, the pressure on the right is only that of the atmosphere, while that on the left is considerably greater. The result is that the water flows from the left to the right until it stands at the same level in each bottle.

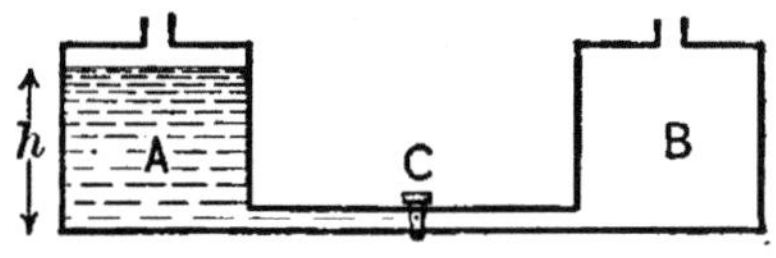

FIG. 76.

How does this happen? One reply — and a perfectly valid one — is that this is merely a special case of Newton's Second Law. But in order to look at the matter from another point of view, let us compute the potential energy of a mass of water m :—

1. When it is all confined in the bottle A.
2. When equally divided between the bottles A and B.

Imagine the bottles each of the same diameter and of uniform diameter throughout.

Let h denote the height of the liquid in A before the stopcock C is opened. If, now, we neglect the very small amount of liquid in the connecting tube, the center of mass of the liquid when it is all in A will lie at a height $h/2$; and the potential energy of the water with reference to the bottom of the bottle is therefore $mg \cdot h/2$.

When, however, the stopcock C has been opened, the water is shared equally by the two bottles. The center of mass in each bottle stands at a height $h/4$; the potential energy of the water in each bottle is, therefore $(mg/2)(h4/)$; and the total potential energy of the water now is $2\,(mg/2)\;h/4$ or $mgh/4$ instead of $mgh/2$ as in the preceding case. While neither the *mass* nor the *weight* nor the *volume* of the liquid has changed, we observe that the potential energy has diminished by 50 per cent. The motion of the liquid — the fact that "water seeks its level" — is, therefore, simply a special case of the general dynamical principle that the potential energy of a system when left to itself tends to become as small as possible. See § 105.

What becomes of the 50 per cent of energy which, in this particular case, ceases to be potential?

The laws of fluid friction — very different from those for solids — must be postponed for later study.

SPEED OF EFFLUX. TORRICELLI'S THEOREM

145. When a liquid is flowing through a pipe, such, for example, as that which connects the two bottles in Fig. 76, one needs frequently to know not only to what height the water will rise, but also with what speed the liquid will flow under given conditions. Owing to the difficulty of handling fluid friction, a general answer to this question has never been obtained.

But there are many cases in which friction becomes negligible; these were first solved by Daniel Bernoulli (1700–1782), the solution being described in a theorem which bears his name. We can here consider but a single special case, which was solved experimentally in 1643 by Torricelli, the lifelong friend and pupil of Galileo, two centuries before the law of the conservation of energy had been established.

Imagine a tank of water, in the side of which is placed an aperture A (Fig. 77) at a distance h below the upper surface. With what speed will the water flow from the aperture? Neglecting friction, it is permissible to apply the law of the conservation of energy to the system.

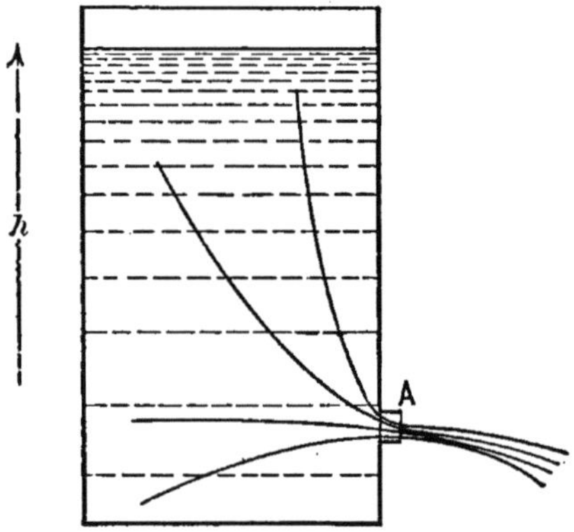

FIG. 77.

Let m denote the mass and v the speed of liquid which passes the aperture during any small interval of time, such that the upper level of the liquid will not be appreciably altered. The kinetic energy of this mass as it leaves the aperture is $\frac{1}{2}mv^2$. This energy has been gained at the expense of work done by gravity. The change in potential energy is measured by the amount of work done in raising the mass m from the nozzle to the upper surface of the liquid against gravity. This is mgh. Since the increase of kinetic energy must just equal the diminution in potential, we have

$$\frac{1}{2}mv^2 = mgh,$$

or

$$v = \sqrt{2gh}. \qquad \text{Eq. 68}$$

This is the theorem of Torricelli, which may also be derived as a special case of the elegant theorem of Bernoulli given below.

146. It is to be observed that the quantity of liquid which passes through an orifice in a given time cannot be computed from this speed and the area of the aperture, the reason being that the stream lines which crowd together at the opening interfere with one another and thus contract the jet. Accordingly, in order to compute the output of liquid in any time t, one must determine a, the area of the smallest cross section of the jet, called the *vena contracta*.

Bernouilli's Theorem

147. In general the motion at any point in a fluid is varying from instant to instant; but there are many cases of great practical importance in which the velocity of the fluid varies from point to point, but remains constant at any given fixed point.

A motion of this kind which may vary with position, but which does *not* vary with time, is said to be **steady.** The path which any particle of fluid describes in steady motion is called a **stream line.** From the definition, it is evident that one stream line never intersects another; so that we may consider a tube of the fluid bounded by these stream lines as an isolated pipe through which the fluid is flowing without friction. This is sometimes called a **tube of flow.** Five tubes of flow are represented in Fig. 77 *bis*.

Let us now consider the still more special case in which the fluid is incompressible. Water is, for all practical purposes, such a fluid; for an increase of pressure amounting to one atmosphere will alter its volume by only one part in 20,000. If the pipe be large so that friction is negligible, we may apply to any given mass of this incompressible fluid, under steady motion, the principle of the conservation of energy; in other words, we may equate to a constant the total energy of a given mass of the fluid as it moves along through its tube of flow.

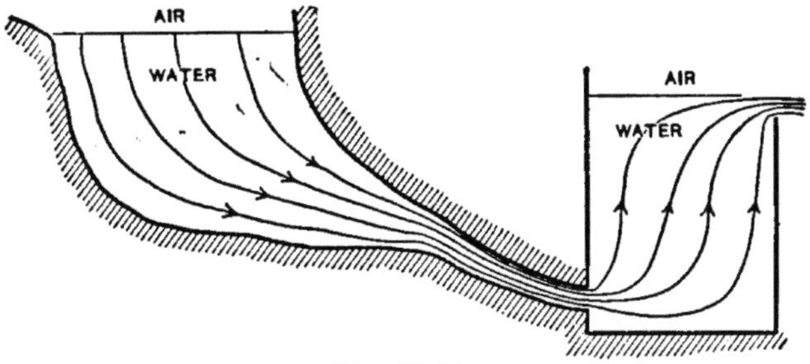

FIG. 77 *bis*.

Let us consider the mass of unit volume which is measured by the density and indicated by ρ. Denote the velocity of the fluid by v; then the *kinetic* energy of unit volume is $\frac{1}{2}\rho v^2$. The *potential* energy of unit volume due to *position* is clearly ρgh, where h is its elevation above the horizontal plane of reference. In addition to this, the unit volume has potential energy due to the fact that it is under pressure. This last statement will be clear when one considers that the work required to force a volume, v, of incompressible fluid into a tank under constant pressure, p, is pv. The area of the piston multiplied by p is the force; the product of the force by the length of stroke of the piston is pv, and is also the work done.

Equating the total energy per unit volume to a constant, we have then for any particular tube of flow,

$$\frac{1}{2}\rho v^2 + \rho gh + p = H = constant, \qquad \text{Eq. } 68\tfrac{1}{2}$$

which is the theorem of Bernouilli (1738). Upon this fundamental equation the entire science of hydraulics is built.

Variation of Pressure with Speed

147 bis. We can consider here only a single illustration of this theorem — a very special case; namely, that in which the fluid flows through a horizontal pipe, a case in which, by proper choice of axes, we may write $h = 0$. Bernouilli's Theorem then becomes

$$\frac{1}{2}\rho v^2 + p = constant, \qquad \text{Eq. } 69$$

which means that **when particles flowing along a stream line reach a portion of the tube where the speed is greater, they at the same time reach a point where the pressure is less, and vice versa;** for otherwise the right-hand member of the equation could not remain constant.

Accordingly if a fluid be forced through a pipe with a constriction in it, such as shown in Fig. 78, the speed at b will be greater than either at a or at c. Now it has been shown experimentally by attaching upright tubes at these three points that the pressure is least where the speed is greatest, namely, right at the constriction. This is the experimental fact which is described by Eq. 69; and is

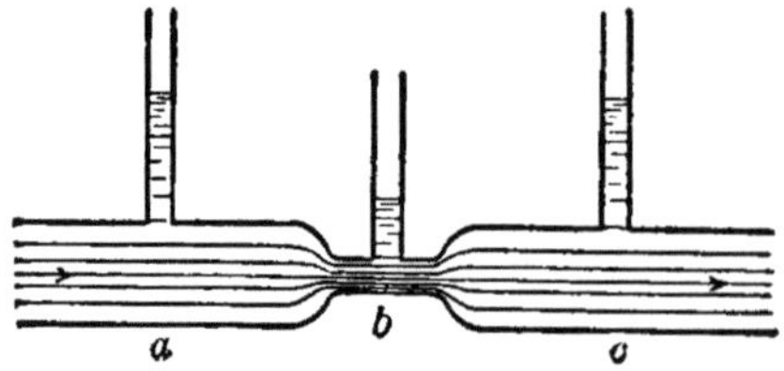

FIG. 78.

exactly what one would expect when he remembers that it is the larger pressure at *a* which imparts the extra speed at *b*. This particular case was first described by the Italian Venturi in 1797 and is the principle used in the "Venturi water meter," invented by the American engineer, Clemens Herschel, in 1887.

148. The various other applications of this principle which are encountered in everyday life are numerous and interesting. Following is a short list, each of which is to be explained in terms of Eq. 69.

(i) The ordinary atomizer; an air jet produced by squeezing the bulb lowers the pressure sufficiently to raise the liquid from the bottle to the nozzle.

(ii) The Bunsen filter pump or aspirator.

(iii) The forced draught in a locomotive furnished by exhaust steam.

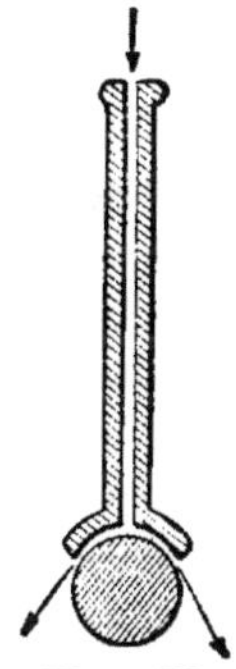

FIG. 79.

(iv) The ball nozzle, shown in Fig. 79, where a jet of fluid is blown so rapidly past the ball that the atmospheric pressure on the lower side of the ball holds it up snug against the socket.

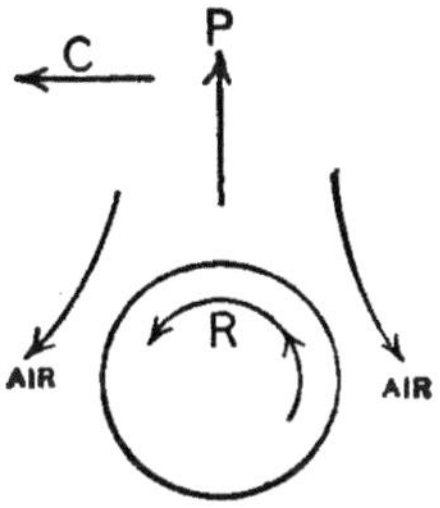

FIG. 80.

(v) The curved baseball, where the passage of the air over the ball is greatly facilitated on the side at which the direction of rotation is opposite to the direction in which the ball has been thrown — the left-hand side in Fig. 80; while the obstruction to passage is greater on the side that is rotating in the same direction in which the ball is being translated. Hence, if a ball leaves the pitcher's hands in the direction indicated by the arrow P, and rotates in the direction shown by the arrow R, it will curve toward the side indicated by the arrow C.

Problems

1. A solid cube 12 cm. on each edge is made of a material whose density is 13.5. What is its weight in a vacuum? Find its apparent weight in water. Density of water = 1. *Ans.* 23,328 g. 21,600 g.

2. A body whose volume is 100 $cm.^3$ has a density of 1.4. Find its apparent weight in alcohol whose density is 0.8. *Ans.* 60 g.

3. A piece of wood whose density is 0.6 floats on water. The volume of the wood is 40 $cm.^3$ What is the volume of the water displaced? *Ans.* 24 $cm.^3$

4. What variation in pressure would be produced by descending to a depth of 76 cm. of mercury? Density of mercury is 13.546 at 20° C.

5. Find the speed of efflux from the bottom of a standpipe which is filled with water to a height of 30 ft.

6. A wire 120 cm. long weighs 40 g. in air; when immersed in water, it weighs 30 g. Find the mean diameter of the wire.

7. If a submarine boat weighs 50 tons and displaces 3000 cu. ft. when immersed, how much water will it have to take in to sink?

Duff, *Mechanics*, p. 258.

8. A rectangular vessel partly filled with water is given an acceleration of 50 cm./sec.^{2} in a horizontal direction. Find the inclination of the free surface of the liquid. *Shearer*, 461 a.

9. How much mercury will be required to fill a cylindrical tube 5 mm. in internal radius and 100 cm. long?

10. What is the specific gravity of a substance such that 3 cu. in. of it weigh a pound?

11. What is the pressure at a point 100 m. below the surface of the sea, the density of the sea water being 1.024?

12. A piece of quartz whose density is 2.7 weighs 50 g. in vacuo. What will be its apparent weight in water?

13. Show that when a cubical box is filled with water the total force exerted on the sides and bottom of the box is three times the weight of the water.

14. A town on the prairie is supplied with water from a standpipe 110 ft. high. What pressure will be available on the first floor of a house in that town?

15. From one end of a balance beam is suspended in water a 50-gram brass mass. From the other end of the beam hangs a piece of quartz, also suspended in water. Given the density of brass as 7.0, the density of quartz 2.7, and the fact that their two weights are in equilibrium, find the weight of the quartz.

16. A silver coin having a density of 9 is dropped into a lake 20 ft. deep. How long will it take the coin to reach the bottom, neglecting friction?

III. Surface Tension. Capillarity

149. When two liquids are brought together, either they mix and diffuse the one into the other, or they do not mix, and consequently exhibit what are known as the phenomena of surface tension. In these phenomena, sometimes called capillary, the very laws of hydrostatics which we have just been studying *appear* to be contradicted.

Arrange two U-tubes (as in Fig. 81), the smaller arm having an inner diameter of from one to two millimeters.

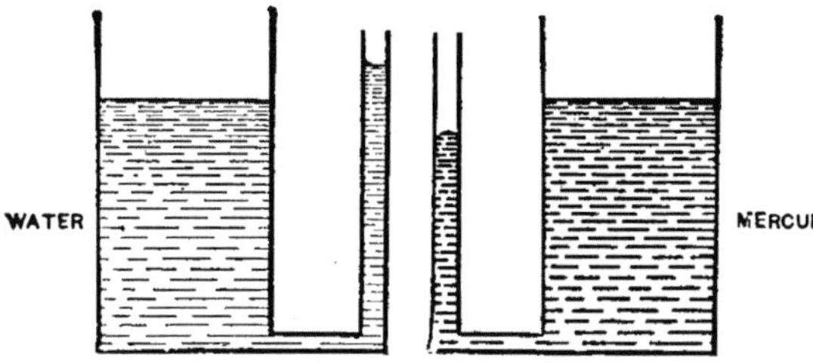

Fig. 81.

Partly fill one U-tube with mercury and the other with water: the mercury in the small arm does not rise so high as in the large; while with the water the reverse is true. Either of these phenomena is sufficient to show that, in communicating vessels, liquids do *not always* stand at the same height. But, as noted above (§ 135), this is only the fact when the communicating vessels have the same diameter, or when each is large.

Any one who has ever placed a small drop of mercury on a plate of glass knows that its center of mass stands at some distance, perhaps a millimeter, above the surface of the glass (see Fig. 82). This phenomenon, *on its face*, would appear to be an exception to the general dynamical principle that the potential energy of a system tends always to a minimum. Why does not the drop flatten out? Why does not its center of mass assume a lower position? Our task now is to show that these contradictions are only apparent, and that the phenomena of the liquids in the U-tube and the phenomenon of the drop of mercury are in perfect accord with general principles already known. **In the discovery of the harmony between any one truth and all other truths lies the essential feature of an "explanation.**

Fig. 82. — Mercury on glass.

150. No one can try the following qualitative experiments without being impressed with the fact that liquids **in air** behave very much **as if** they were contained in a sac or thin membrane, continually holding them in position.

1. A glass rod dipped into a tumbler of water picks up a drop and carries it on the end of the rod **as if** a little rubber bag were tied about the end of the rod and filled with water.

2. A falling raindrop nearly always assumes a spherical shape **as if** it were inclosed in a skin of some sort, and this

skin were continually compressing it into the smallest possible volume.

3. In a camel's-hair brush, when dry, the individual hairs stand apart to some extent, giving it a rather bushy appearance. Dip the camel's-hair brush into a tumbler of water; the same bushy appearance remains. Remove the wet brush from the water to the air; the film of adhering water at once binds the individual hairs together as a band does a sheaf of oats. The liquid **in air** acts **as if** it were covered by a film always tending to contract.

Fig. 83. — A drop of oil freed from gravity.

4. A piece of clean, fine wire gauze may be bent into the shape of a small open box. Water may be poured into this box to a considerable depth without its running through the meshes.

5. One may make a mixture[1] of alcohol and water that will have very nearly the same density as olive oil — a combination known as Plateau's Mixture. A drop of oil placed in this liquid, as indicated in Fig. 83, will not mix with it, nor will it be affected by gravity. How do you explain the perfectly spherical shape which the oil assumes?

6. One of the simplest methods of observing this phenomenon of liquid surfaces is the following, due to Van der Mensbrugge: —

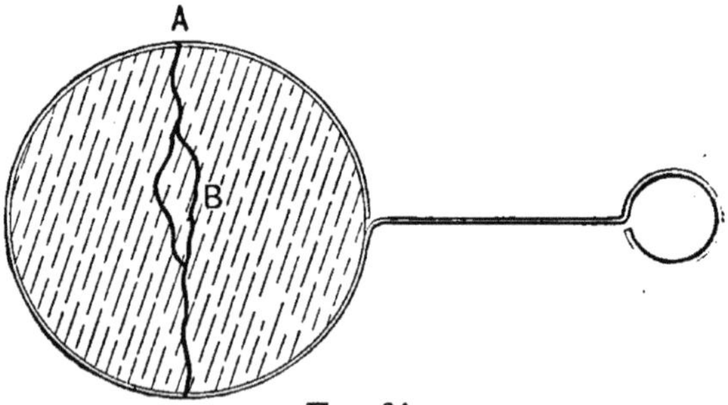

Fig. 84.

Make a strong solution of castile soap. Bend a piece of aluminum wire (about No. 20 or No. 25) into the shape shown in Fig. 84, making the diameter of the ring about two inches. At some point *A* of this ring tie a loop of thread *B*. If, now, the ring be dipped into the soap solution, a film will spread over it. This film has two

[1] About 60 per cent of alcohol (density 0.806) and 40 per cent of water will give a mixture which is nearly right for ordinary olive oil (density 0.917).

surfaces inclosing a thin layer of the soap solution between them. The thread floats on the film; now with the point of a lead pencil puncture the film at some point within the loop of thread *B*.

Each surface of the *film* between the wire and the thread now *tends to become as small as possible*, as is indicated by the fact that the area within the loop tends to become as large as possible, *i.e.* the loop assumes a circular form, as shown in Fig. 85.

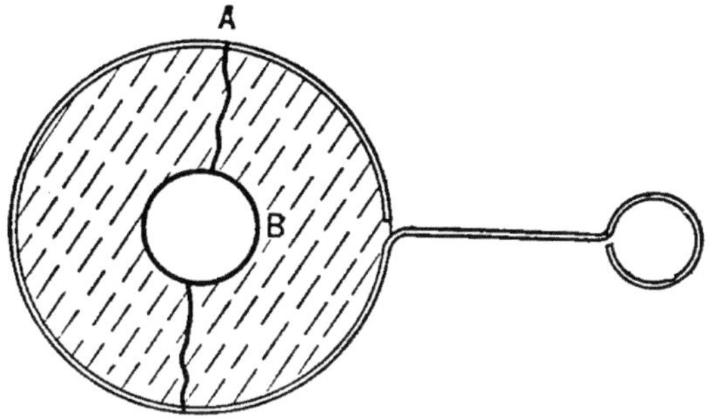

FIG. 85. — Soap film assuming smallest area.

151. Such evidence as this proves that if we enlarge a liquid surface we shall have to do work upon it just as truly as if we stretch a rubber band or inflate a bicycle tire. On the other hand, if a surface contracts, it exerts a force in so doing, and thus does work for us.

A wire is easily bent into the shape indicated in Fig. 86. Dip this wire into a soap solution to cover it with a film. If, now, a small straight wire, *CD*, is laid across the frame, the film on the *B* side may be broken out with the finger. The two surfaces on the *A* side will exert a certain definite force and draw the bar *CD* quickly to the end of the frame. The total force thus exerted upon the bar, divided by twice the width of the film, will give the force per centimeter which a single liquid surface exerts upon the cross bar. This quantity is naturally measured in dynes per centimeter, and is called the **surface tension** of the liquid. Surface tension is denoted by *T*.

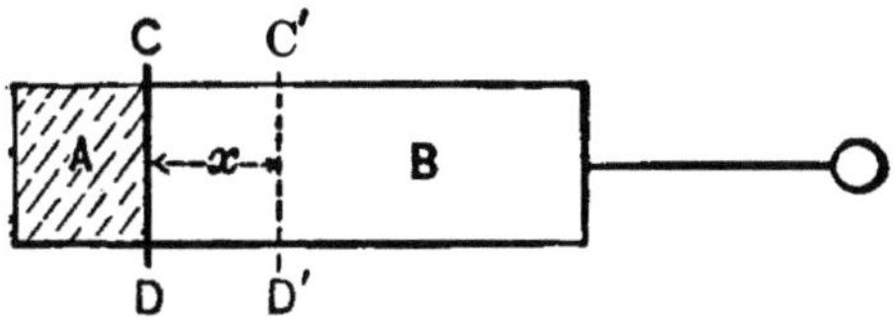

FIG. 86. — Work done by a contracting film.

Suppose the bar $\overline{CD}$ is displaced a distance x, as indicated in the figure. The *force* which we must use to do this is $2\,T \cdot \overline{CD}$ dynes. The *work* which one must perform is $(2\,T \cdot \overline{CD})x$ ergs. The increase in area is $2\,x \cdot \overline{CD}$ square centimeters. For the film has two surfaces.

Consequently, the amount of energy *in unit area* is

$$2x \cdot T \cdot \overline{CD}/2x \cdot \overline{CD} = T.$$

Accordingly we may define the surface tension of any liquid as the amount of potential energy in unit area of its surface, a definition which is strictly equivalent to the one given above.

And now we see that, when a drop of mercury gathers itself up into a little round globule, lifting its center of gravity above the plate on which it rests, it does so in order to make its surface as small as possible, and thus reduce its potential energy to a minimum. So far, therefore, from violating our previous experience, this behavior of liquid surfaces is quite in accord with it.

152. Before examining more closely into the nature of this energy which we have found resident in the surface of a liquid, it will be necessary to show that liquid particles attract their immediate neighbors just as solid particles do. A crayon of chalk is more or less difficult to break, because across any section of it the particles on one side are attracting the particles on the opposite side with considerable force. That this same phenomenon happens in liquids may be conveniently demonstrated by a tube partially filled with water and then freed from air and sealed — what is ordinarily called a "water hammer." The liquid in a tube of this kind may, without difficulty, be made to assume the position indicated in Fig. 87, an experimental fact which it is difficult to explain except by assuming that the glass walls of the tube attract the adjacent particles of water, and these particles in turn attract the neighboring particles of water; and so on, throughout the entire mass of liquid. These forces — which are not at all understood — are sufficiently great to sustain the excess of weight of one column over the other.

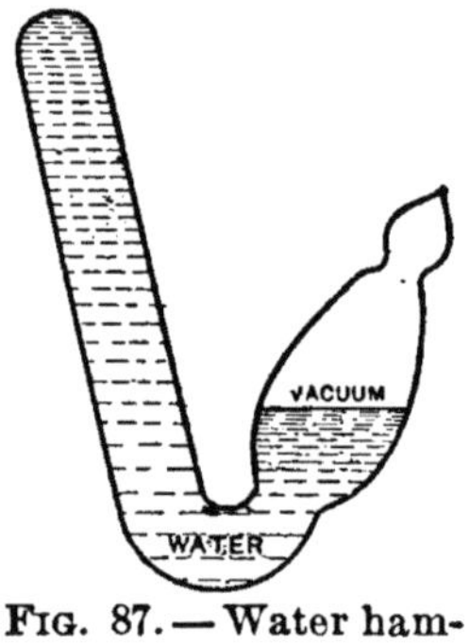

FIG. 87. — Water hammer, illustrating cohesion in liquids.

153. Another fact which is very certain is that there is nothing of the nature of an actual membrane stretched over the free surface of a liquid; for actual ordinary membranes (i) are not perfectly extensible and contractile, and (ii) do not, in general, show the same tensile force in all directions; nor (iii) is this tensile force independent of the thickness of the film.

How, then, is this apparent membrane and this surface energy to be explained? The following answer, for the sake of its admirable clearness and comprehensiveness, is taken from Professor W. F. Magie's *Lectures on Physics*, Sec. 67, (Princeton, 1904):—

"The general laws of hydrostatics depend upon the principle that a liquid subject to the attraction of gravity will be in equilibrium only when its configuration is such that the action of gravity on it introduces no shearing stresses. Now gravity is not the only force which acts on the liquid. Its parts also exert forces of cohesion on each other, and true equilibrium will not be reached until the liquid assumes such a position that these forces of cohesion so act, together with the weights of the parts of the liquid, as to introduce no shearing stresses. We can explain all the phenomena which are treated under capillarity by taking these forces of cohesion into account. It is not necessary for us to know, and in fact we do not know, the way in which the cohesion depends upon the masses of the interacting parts and the distances between them. It is necessary, however, to assume this much, that the force of cohesion exerted by a small part or element of the body acts only on those elements which are in its immediate neighborhood. That is, we assume, as the law of the force of cohesion between elements, that the force between contiguous elements is very great, and diminishes very rapidly when they are separated, so as to become imperceptible, even when the distance between them is still very small. This general law of cohesive forces is illustrated by the behavior of an iron bar when we break it. It requires a very great force to break it; but after it is broken, even though the two surfaces at the break are fitted together again with the utmost nicety, the two parts can be separated with no perceptible effort.

"Because of the short distance within which an element of the liquid acts on its neighbors, those elements which lie below the surface by a depth equal to this distance, which we call the range of action, are in equilibrium under the action of their neighboring elements. It is only those elements which lie in or very near to the surface which are attracted unequally in different directions. These unequal attractions, acting on the elements of the surface film, will produce a peculiar condition in it. This will be the same for all parts of the surface, owing to the minuteness of the range of action, so long as the radius of curvature of the surface is not very small, that is, is not of the same order of magnitude as the range of action. Thomas Young (1773–1829) suggested that the special action of the cohesive forces in the surface film may be represented by supposing the film to be under tension, similar in general to that in a stretched membrane. This tension should be the same, in the case of any given liquid, for all parts of its surface. It is called the surface tension, and its numerical value, when determined for the surface of separation between any two bodies, is a characteristic number for those bodies. The position of the liquid column in the capillary tube, or any of the other phenomena ascribed to capillary action, are, on this view, due to equilibrium between the weights of the parts of the liquid and the forces due to the tensions acting in curved portions of the surface film.

"Another useful concept for the study of capillary phenomena was introduced by Gauss (1777–1855). It is now called surface energy. It is plain from the consideration of the action of the cohesive forces, that when the surface of a liquid mass is enlarged, the potential energy of the liquid is increased on that account. For the surface can only be enlarged by the passage of elements of the liquid from the interior mass into the surface film. Each of the elements in the film is drawn inward toward the interior by a force of cohesion, and hence negative work is done on all the elements which pass out from the interior into the surface film. The negative work thus done is equivalent to an increase in the potential energy of the liquid. The surface, therefore, possesses an energy peculiar to itself, proportional to the extent of surface and characteristic of the bodies separated by the surface. For a given surface its numerical value is the same as that of the surface tension for the same surface.

"By the aid of this concept of surface energy all the phenomena of capillarity may be explained as illustrations of the general principle that the potential energy of a system of bodies at rest tends to become the least possible."

Liquid Jets

154. When a continuous stream of liquid flows from a tap or garden hose and breaks up into small drops, the phenomena occur in such rapid succession that the eye cannot follow them with any certainty. In order to reduce the rapidity with which these events occur Plateau arranged a large drop of olive oil between two rings — one capable of moving up and down above the other — as shown in Fig. 88 — the whole immersed in a mixture of alcohol and water of the same density as oil.

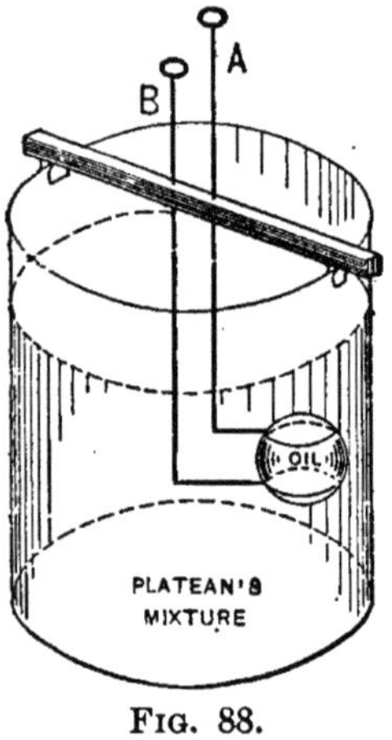

Fig. 88.

By sliding the rod A up, the globule of oil elongates itself in the same way that it would if it were falling freely under gravity from a nozzle. It is found that the column of oil becomes unstable when its length exceeds its circumference, and that the surface tension proceeds first to form a sort of waist and then to pinch the column in two, leaving between the two parts a small drop such as that pictured in Fig. 89. The first three of these figures represent the behavior of the drop of oil in Plateau's mixture; the right-hand figure refers to the jet of water.

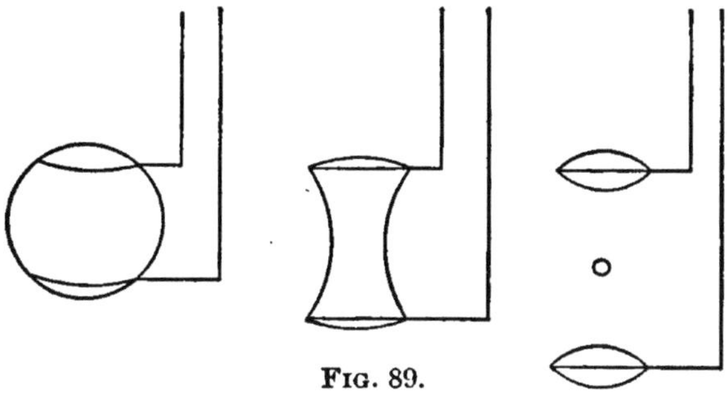

FIG. 89.

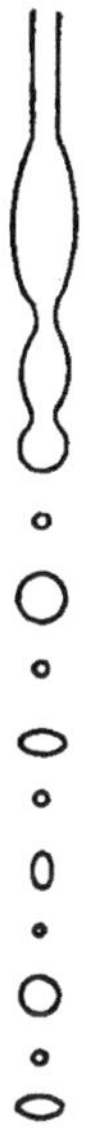

Instantaneous photographs of jets of water, falling freely, show that they are pinched off by surface tension in precisely this manner.

CAPILLARY PHENOMENA

155. When a solid is partially immersed in a liquid, either one of two things may happen; namely, the particles of the solid may attract those of the liquid more strongly than the particles of liquid attract each other, in which case the liquid is said to "wet" the solid; or the particles of the liquid may cohere with a greater force than that with which the particles of the solid attracts them, in which case the liquid "does not wet" the solid.

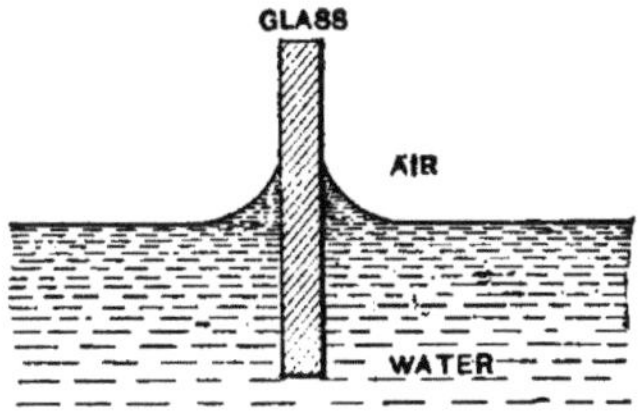

FIG. 90.—A solid which is wet by the liquid.

In the former case the liquid appears to run up the side of the solid, as shown in Fig. 90; while in the latter case the liquid is depressed in the neighborhood of the solid, as indicated in Fig. 91.

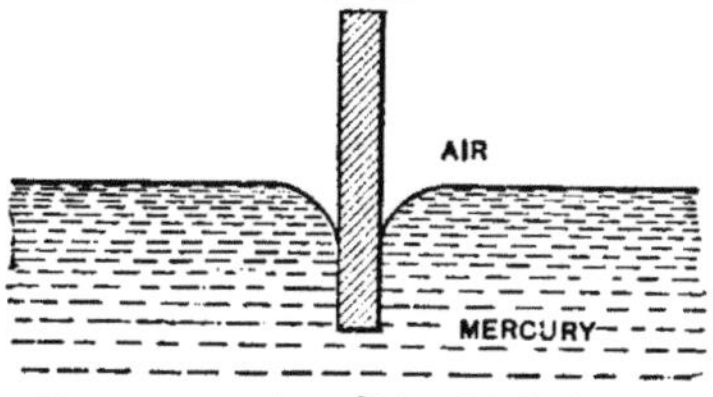

FIG. 91.—A solid which is not wet by the liquid.

The angle at which the liquid surface meets the immersed portion of the solid is perfectly definite and is known as the angle of contact.

156. If the immersed solid be in the form of a fine glass tube as shown in horizontal and vertical section in Fig. 92, it is observed that water will rise to a much greater height on the inside than on the outside.

Let T denote the surface tension of water, r the radius of the tube, a the angle of contact, D the density of water; then h, the excess of height, may be computed as follows: —

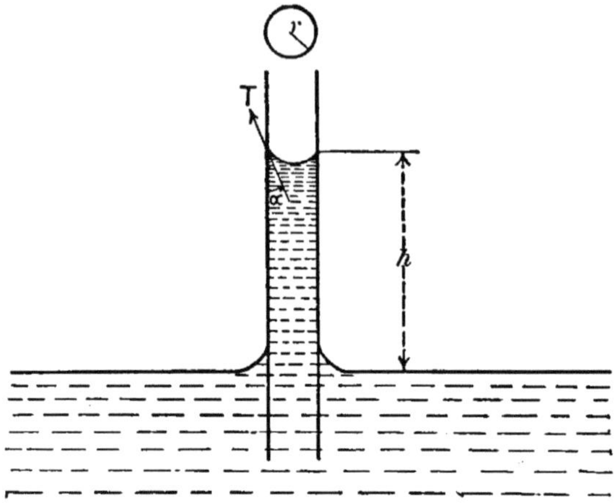

FIG. 92.

The upward component of the force per unit length acting at each point where the liquid surface touches the inside wall of the tube is $T \cos a$. Hence the total force along the direction of the axis of the tube is $2\pi r T \cos a$. The total weight of the column of liquid raised above the level of the outside liquid is $\pi r^2 hDg$. Since these two are the only forces acting upon the liquid column, their sum must be zero. Hence we have

$$2\,T \cos a - rhDg = 0, \qquad \text{Eq. 70}$$

a single equation which serves to determine h, and which shows that, other things remaining the same, $h \propto \frac{1}{r}$, a fact which is known as **Jurin's Law**.

Attraction and Repulsion of Floating Bodies

157. Another curious consequence of the elevation and depression of liquids on the surface of immersed solids is the attraction of floating bodies such as may be observed about the edge of any quiet lake, where all the little twigs that have fallen into the water will be found attached either to each other or to the shore as if tied up to the wharf.

A more convenient method of observing this phenomenon is to take two English walnut shells, and float them in a finger bowl. It will be found that one of these shells will tow the other about with perfect ease, the only connection between the two being the liquid film which rises between them.

This state of affairs is diagrammatically represented in Fig. 93, and may be completely explained by reference (§ 140) to the following general principle; namely, *Over any portion of the free surface of a liquid which is plane, and at all points in the liquid which are at this same level, the pressure is constant and is equal to that of the atmosphere. As one ascends in the liquid from this level the pressure diminishes; as one descends in the liquid, the pressure increases.*

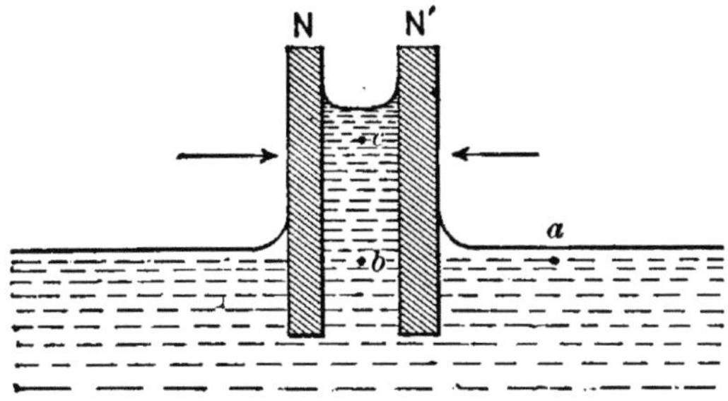

FIG. 93. — Attraction of two bodies, each wet by the liquid.

Thus in Fig. 93, the pressure at *b* is the same as that at *a*; and the pressure at each of these points is that of the outside atmosphere. As one ascends to the point *c* between the two floating nut shells, the pressure diminishes, while outside the liquid the pressure remains that of the atmosphere. The result is a force in the direction from the greater to the less pressure, which holds the two bodies together, so that if one shall be pulled about in the finger bowl it will tow the other behind it.

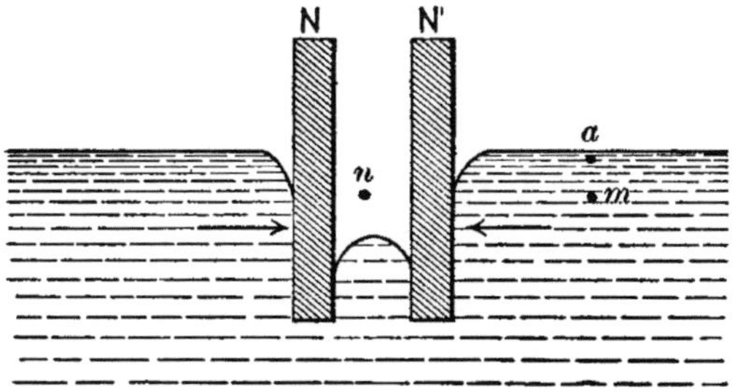

FIG. 94. — Attraction of two bodies, neither of which is wet by the liquid.

158. Oddly enough the same thing happens when *neither* of the two floating bodies is wet by the liquid, as for instance two pieces of paraffine or two pieces of beeswax, a case which is diagrammed in Fig. 94. Here the pressure at *a* is that of the atmosphere; the pressure at *m* is considerably greater; while at *n*, on the same level with *m*, but in the open air, the pressure is only that of the atmosphere.

The result is, therefore, two forces in the direction of the arrows tending to hold the two bodies together.

159. Only one other possibility remains, namely, one of the bodies may be wet and the other one not. When two such bodies float near each other, it will be observed that, in the region *between* them (Fig. 95), the liquid does not rise quite so high against the wetted one as it does on the *outside* of the wetted one. Note also that in the region between the bodies the liquid rises a little higher against the body which is not wet than it does on the outside of this same body.

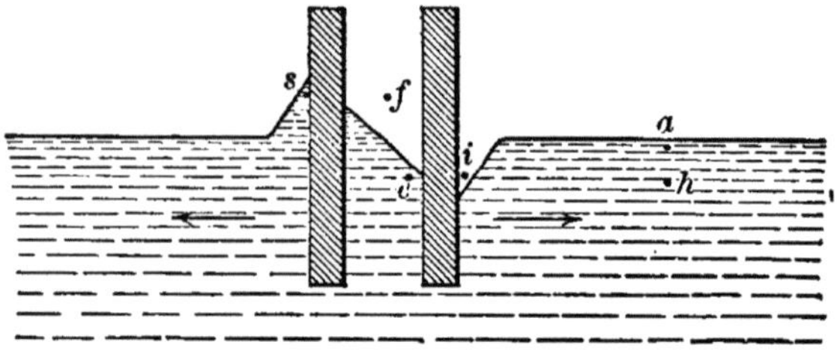

FIG. 95. — Repulsion of two bodies, one of which is wet, the other not.

Following the ordinary laws of dynamics, as in the preceding cases, the pressure at *a*, at *i*, and at *f* is that of the atmosphere; pressure at *h*, and at *e*, which is on the same level, is greater than that of the atmosphere. Hence there is a force from *e* towards *i* as indicated by the arrow.

The pressure at *s*, which is higher in the liquid than *a*, is less than that of the atmosphere, resulting in a force from *f* towards *s*, as shown by the arrow. Accordingly the two bodies appear to repel each other.

Many floating bodies, by being slightly elevated or depressed in the water, may be made to act either as bodies which are wet or not wet by the liquid. It is often convenient thus to illustrate all three cases by the use of only two different bodies.

160. There is a host of other beautiful phenomena which fall under the head of capillarity, such as those of thin soap films, the variation of surface tension with impurities, electric field, etc.; but these must be left for outside reading and observation. A single remark, however, concerning those exquisite soap-bubble experiments of childhood.

Almost any substance when dissolved in water diminishes the surface tension of the water. Dissolved soap has this effect. Whence then arises the peculiar efficiency of a soap solution for producing good soap bubbles? The reply is that the soap enormously increases the viscosity — temporary rigidity, so to speak — of the liquid and thus prevents the bubbles from so quickly breaking.

Problems on Capillarity

1. Two mercury globules, of radii 2 mm. and 1 mm., respectively, are brought into contact. They unite and form one globule. Compare the

surface of this globule with the sum of the surfaces of the two globules before union. Compare the sum of the surface energies before union with that after union.

2. How is it that you can round the end of a glass rod by holding it in a Bunsen flame?

3. Chemists often use a solid glass rod, instead of a funnel, in pouring a liquid from one vessel to another. Explain this.

4. Find the pressure, due to surface tension, inside a spherical drop of pure water whose radius is 2 mm. The value of the surface tension for pure water in air is 75 dynes per centimeter.

5. How high will pure water rise in a glass tube whose internal diameter is $\frac{1}{20}$ mm.? Assume the angle of contact to be zero.

6. Find the pressure in a spherical soap bubble which is 8 cm. in diameter, assuming the surface tension as 30 dynes per centimeter.

7. Explain by a diagram on the blackboard just how it happens that a jet of water flowing from a tap breaks into drops at a short distance below the nozzle.

8. An inverted U-tube supports two soap bubbles, one from each arm of the tube. The diameter of one bubble is 3 cm., the diameter of the other 1 cm. If the tube be left to itself, which bubble will increase in size, and why?

9. How does surface tension determine the size of the drop of liquid which may be held on the end of a glass rod?

10. When a soap bubble breaks, it is observed that the water is thrown violently in many directions. Explain.

11. A test tube is filled with dry sand at the temperature of 20° C. If this sand be moistened by pouring on a little water at 20° C., is there any reason for thinking that the temperature of the mixture will be different from 20° C.?

Properties of Gases

161. Most of the theorems which we have just been studying under the subject of Hydrostatics are quite as true for fluids in general as for liquids; all of them, in fact, except those which refer to the "free surface" of a liquid. For this "free surface" is precisely the thing which we do not find in gases. A gas might, indeed, be defined as a body which has no free surface.

A sailing vessel under way, a tree bending to a breeze, a bank of sand raised by the wind, is each an evidence that this air, upon which we are so dependent for life, is something which can hand over momentum to other bodies. Any one who has ever pedaled a wheel against a head wind knows that air exerts pressure when in motion, and that air, therefore, probably has inertia. The boy who has blown up a paper bag for the purpose of bursting it between his hands knows at least that the inclosed air is a body.

The lad who has lifted a brick with a piece of wet leather and a string (Fig. 96) knows that he must keep the edges of the leather down snug on the brick; and that, when he does so, the earth's atmosphere, without difficulty, keeps the brick up snug against the leather, except in the middle. Here the leather has been lifted, and has in turn lifted the whole column of the earth's atmosphere, leaving, on top of the brick, a space in which there is practically no air.

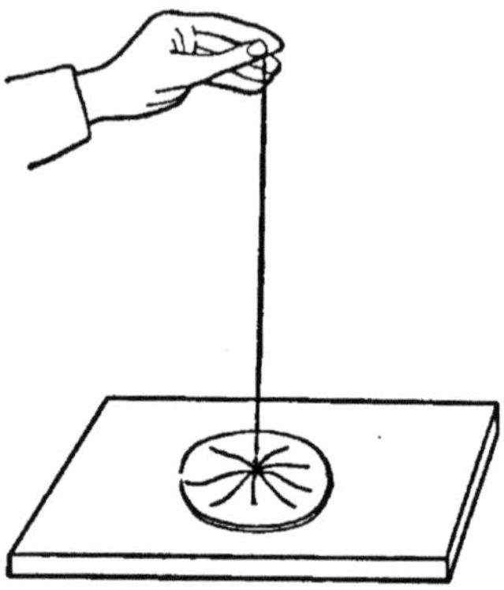
FIG. 96. — The "sucker."

The ancients were, perhaps, essentially as familiar with these facts as we are; at any rate, they were familiar with several kinds of pumps dependent upon these principles. Of the connection between these various facts it appears that they had little or no conception. But since the essential feature of scientific knowledge consists not only in knowing the facts, but also in knowing the connection between them, we proceed to the consideration of some of the properties of gases and of their connection one with another.

162. Before the time of Galileo nearly all the phenomena of pumps were explained by saying that "nature abhors a vacuum." And this *horror vacui* is a fairly good explanation; it will, in any event, explain a large part of the more familiar phenomena, such as those of air pumps and barometers. In addition to this "horror of a vacuum," it was proved by Galileo that air had weight. This he accomplished by weighing a bottle into which he had pumped two or three times its volume of air, and afterward, having allowed the excess of air to escape, again weighing the

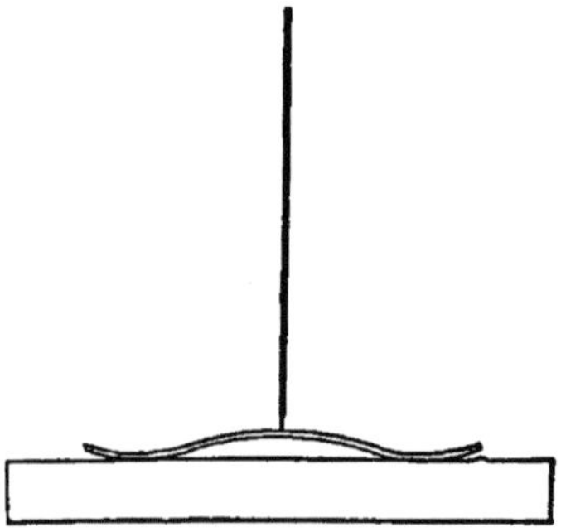
FIG. 97. — Section of sucker shown in Fig. 96.

bottle. See Ostwald's *Klassiker der Exakten Wissenschaften*, No. 11, p. 71, for Galileo's own account of this experiment.

These were two capital discoveries: (1) that nature abhors a vacuum, and (2) that air has weight.

163. Torricelli (1608–1647) showed, by a justly celebrated experiment, the connection between these two facts; namely, nature abhors a vacuum *because* the air has weight. Torricelli's experiment was as follows: Taking a tube more than 76 centimeters long and closed at one end (Fig. 98), he filled it with mercury, and placing his finger over the open end, inverted the tube in a dish of mercury. The column of mercury fell a short distance, but remained standing in the tube approximately at the height of 76 centimeters above the surface of the mercury in the dish.

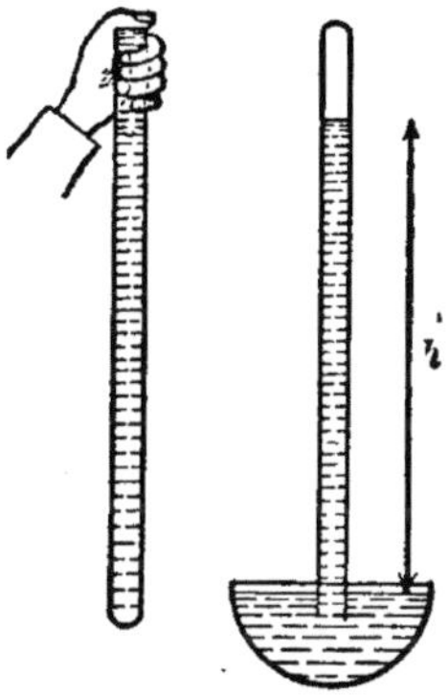

FIG. 98. — Torricelli's experiment.

Torricelli ascribed the support of this column of mercury to the pressure of the atmosphere, and explained the pressure in turn as due to the weight of the atmosphere. He showed, indeed, that the weight of the earth's atmosphere is very approximately that of an ocean of mercury covering the earth to the depth of 76 centimeters.

But if this be the fact of the case, Torricelli saw that the height of mercury in the inverted tube should diminish as one ascends in the earth's atmosphere. This test was shortly made by Pascal, 1648, who carried the inverted tube — such a tube and dish is called a mercurial barometer — to the top of a mountain in France and found that the mercury fell some seven or eight centimeters during the ascent.

The evidence for thinking that we reside at the bottom of an ocean of air, exerting its pressure equally in all directions, was completed by Otto von Guericke, who devised, in 1650, a pump exactly like a single-acting air pump, only the valves were not automatic. This enabled him to remove the greater part of the air from a glass vessel attached to the pump. In the nearly vacant space thus formed he found that the pressure was much less than that of the outer air, and on opening the valve, air would immediately rush in until the pressure within became equal to that without.

Scientists have agreed to adopt 76 centimeters as the standard height of the barometer at sea level; and the pressure required to support a column of mercury of this height is known as the normal atmospheric pressure.

From the observed height of the barometer at any time, we can readily calculate the pressure of the atmosphere. For in Fig. 98 the pressure B over the free surface of the mercury in the basin is that of the atmosphere; while the pressure at the same level inside the tube is that due to a column of mercury whose height is h; and since these two pressures are equal we have, Eq. 67,

$$B = Dgh \text{ dynes per sq. cm.,} \qquad \text{Eq. 71}$$

where D at 0°C is 13.596, $g = 980.69$, and h at the surface of the earth never very far from 76 centimeters. The pressure which the atmosphere exerts upon a barometer is mainly dependent upon two factors; namely, (1) the height of the atmosphere, and (2) the average density of the atmosphere. Anything which changes either one of these changes the "reading of the barometer," *i.e.* the difference in level between the two horizontal surfaces of the mercury in the barometer.

Water vapor when under the same pressure as air has a density which is less than that of air. If, then, there be much water vapor in any portion of the atmosphere, its density will be diminished, consequently its pressure will be diminished, and the mercury column which it supports will be shorter. The barometer is said to "fall." But the same thing happens when the height of the atmosphere is lowered or its pressure is changed by a cyclonic motion. The barometer is, therefore, *not* an instrument for telling whether or not it is about to rain, but for measuring the pressure of the earth's atmosphere. Nevertheless, barometric observations are of great aid to the Weather Bureau in predicting atmospheric changes.

Composition of the Earth's Atmosphere

164. More than a century elapsed between the time of Torricelli and the discovery of the principal elements in the air which we breathe. These were first shown by Lavoisier (1743–1794), one of the founders of modern chemistry, to be nitrogen and oxygen, in the approximate proportion four of nitrogen to one of oxygen. In addition to these there is always present a variable quantity of water vapor, $\frac{3}{100}$ per cent of carbonic acid gas, together with traces of the recently discovered gases, argon, neon, crypton, xenon, and helium.

When we remember that this ocean of air in which we live is invisible, odorless, tasteless, apparently devoid of weight, and almost intangible, it is little wonder that the discovery of its properties came rather late in the history of science. What men have already learned concerning gases is little short of marvelous. And he who thinks that physics is a subject which has to deal *only* with things that can be seen and felt and heard needs only to read the story of *argon* and the chain of facts from which was drawn the inference that this element is present in the earth's atmosphere. In physics, as elsewhere, the discovery of the appropriate idea is at once the important and the difficult matter.

Circulation of the Earth's Atmosphere

165. Few motions have more significance for the human race than those of our atmosphere. Of these there are many; but the trade winds and cyclones are of especial importance and interest. . It is assumed that the student has already made the acquaintance of trade winds in his study of physical geography Accordingly we pass at once to the explanation of cyclones, a name which has been given to those great circular rotating storms which sweep over both our northern and southern hemispheres, and which are responsible for most of what we call "our weather." But since these immense whirls, ranging from 500 to 1000 miles in diameter, acquire their peculiar twist entirely from the earth's rotation, it will be necessary first to consider just how the motion of any body over the surface of the earth is affected by the daily motion of our planet.

Deflection to the Right

166. We are living on the surface of a spinning top whose angular speed is $\frac{2\pi}{86400}$ radians per second. The angular velocity of this sphere, ω, may be represented by the vector $\overline{OP}$ in Fig. 99, where O is the center of the earth and P the north pole. Let E represent a point on the equator and C any locality in latitude λ. Then the vector ω may be resolved along the direction OC. The value of this component $\overline{OH}$ is $\omega \sin \lambda$. The effect of the rotation of the earth, then, at any point C is precisely the same as if the point C were situated at the pole P and the earth had an angular speed of $\omega \sin \lambda$ instead of ω. Let us consider, therefore, how the motion of, say, a bullet would be affected by the earth's rotation (i) when the gun is directed along any meridian at a target situated exactly at the north pole. Imagine the distance of the target from the gun to be 400 yards. The speed of the gun toward the east will then be $400 \left(\frac{2\pi}{86400}\right)$ yards per second, or approximately 1 inch per second. If the muzzle peed is 1800 feet (600 yards) per second it will require $\frac{2}{3}$ of a second for the shot to reach the target. But during this $\frac{2}{3}$ of a second the bullet has drifted some $\frac{2}{3}$ of an inch towards the east; and, hence, if the gun is accurately aimed at the target, the shot will strike $\frac{2}{3}$ of an inch *to the right* of the bull's eye. (ii) Next suppose the muzzle of the gun placed exactly at the north pole and the target placed on a parallel of latitude 400 yards away. The bullet as it leaves the gun now has no easterly velocity impressed upon it; but the target *has*. During the $\frac{2}{3}$ of a second that the projectile is in the air, the target has moved $\frac{2}{3}$ of an inch to the left, the result being that the bullet strikes *to the right* of the bull's eye as before. (iii) Next let us suppose that both gun and target are situated 400 yards apart on the same parallel of latitude. Let M (Fig. 100) denote the muzzle of the rifle and T_1 the position of the target at the instant when the shot is fired. With the north pole P as center, draw a circle through M and T_1. It is

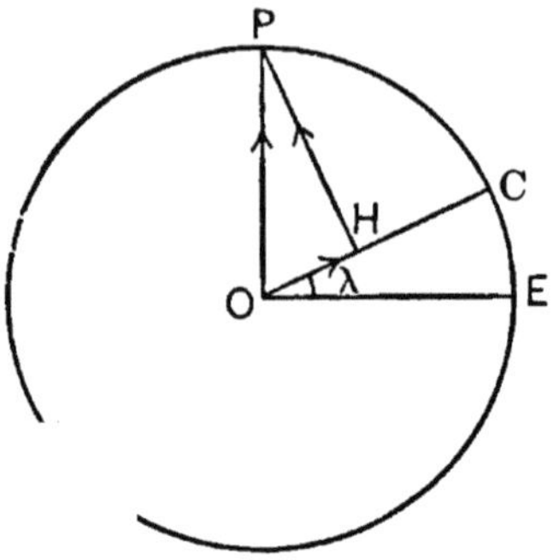

FIG. 99.

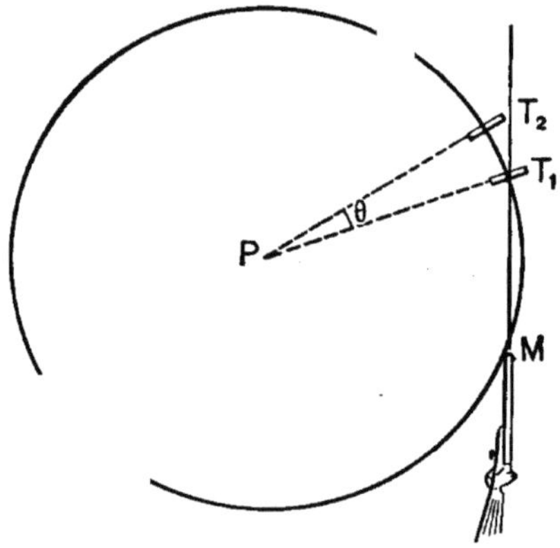

FIG. 100.—Illustrating deflection to the right.

then clear that, during the $\frac{2}{3}$ of a second that the shot is in the air, the target will move through the angle θ, amounting to $\frac{2}{3} \times 15''$, that is, $10''$ of arc; so that when the bullet strikes the target the latter is in the position marked T_2. Hence the bullet moving straight ahead according to Newton's First Law hits the target *at the right* of the bull's eye.

The student will find it interesting to prove that if the gun had been directed exactly west instead of east as shown in Fig. 100, the deflection would still have been to the right. Prove also that at the *south* pole the deflection will be always to the *left*.

(iv) Next suppose we leave the polar region and pass to a locality C whose latitude is λ. The only change is that the angular speed, *about a vertical line*, has now been reduced in the ratio of $\sin \lambda$ to unity. Thus in latitude 30° N. the drift is always to the right, but just half as large as at the north pole.

Cyclones

167. We are now prepared to consider the typical storm which takes its origin in a region of low atmospheric pressure. This low barometer may occur in consequence of various changes — let us imagine it due to excessive heating of the earth's surface by sunshine in some particular locality. As the heated air rises and flows off, the surrounding air rushes from all sides into this region of low barometer, as shown in Fig. 101. But from whichever of the four corners of the earth the air moves, observe that it will have impressed upon it the same deflection to the right that we found in the case of the rifle ball. The result is that **all storms in the northern hemisphere having a "low" center rotate in a counter-clockwise direction.** Hence also the rule of Buys-Ballot: To find the storm center stand with your back to the wind and then the center of low barometer will be on your left.

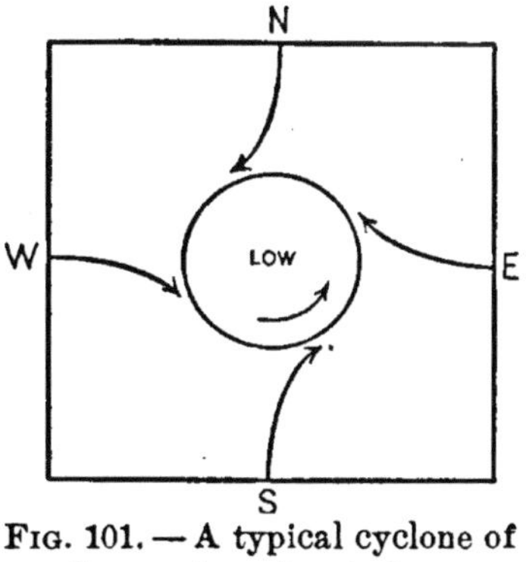

FIG. 101. — A typical cyclone of the northern hemisphere.

But some disturbances, generally accompanied by clear weather in summer and cold in winter, take their origin in regions of *high* barometer. These are called **anti-cyclones.** Here the air is moving away from the center and toward the periphery; and since the deflection is still to the right, the circulation of the air is now in the clockwise direction. The accompanying chart, Fig. 101 *bis*, shows the circulation of the atmosphere in the United States for January 2, 1904. The tracks which these great cyclones and anti-cyclones follow in sweeping over our country are interestingly shown in each copy of the *Monthly Weather Review*, published by the United States Weather Bureau. It should be noted that Buys-Ballot's rule holds equally well for cyclones and anti-cyclones.

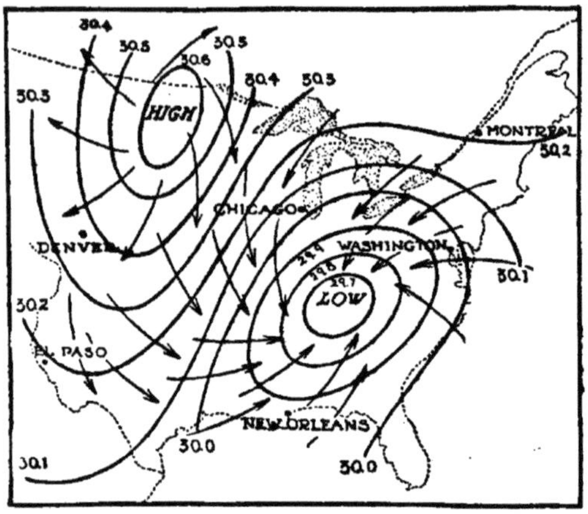

FIG. 101 *bis*. — Typical cyclone. This chart shows lines of equal barometric pressure in inches of mercury, with resulting winds, at 8 A.M. January 2, 1904. The arrows show the tendency of the wind to blow from a High towards a Low, and also the spiral motion caused by the earth's rotation, outward from a High, inward towards a Low.

168. Another point of view, and perhaps a better one, from which the deflection to the right may be regarded, is the following one, given by Professor Cleveland Abbe in terms of centrifugal force: —

"The difference between the deflection to the right in the Northern Hemisphere and that to the left in the Southern Hemisphere results from the nature of the forces that produce these deflections, and not from the way in which the observer looks at the weather map. The deflections are true natural phenomena, not mere optical delusions.

"When a body rests quietly on the earth's surface the centrifugal force, *cd* (Fig. 102), due to the diurnal rotation of the earth, gives the body a slight tendency to move toward the equator, which tendency is counterbalanced by the fact that the surface of the earth, and especially of the

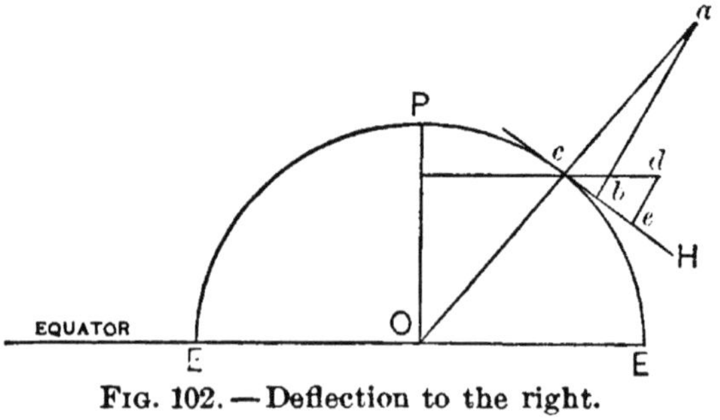

FIG. 102. — Deflection to the right.

ocean, is an oblate spheroid; the attraction of gravity, *ac*, is not perpendicular to this spheroidal surface, but is directed toward the center of the earth and its action on any body at the surface must be resolved into two components; the principal one, *ab*, is vertical or normal to the spheroidal surface and constitutes the greater part of what we call weight, the other component, *bc*, is a feeble horizontal sliding force directed toward the pole (the North Pole in the Northern Hemisphere and the South Pole in the Southern). The centrifugal force, *cd*, is directed outward in the plane of the small circle of latitude and is also to be resolved into two parts, one of which, *ed*, is normal to the surface of the spheroidal globe; it acts upward, and therefore partly counteracts the force of attraction ; the difference between it and the attraction is called *apparent gravity*, and gives rise to what is ordinarily known as the weight of a body. The other component of the centrifugal force, namely, *ce*, is parallel to the surface of the globe and is a horizontal sliding force directed toward the equator. But the earth's surface represents a state of equilibrium; therefore, the two horizontal components, respectively pushing northward and southward, just counterbalance each other, or *bc* is equal and opposite to *ce*. If the earth should rotate faster or slower, then the curvature of the spheroid would change so as to always maintain this balance between *bc* and *ce* so that bodies would have no tendency to slide either north or south.

"Now a body or a mass of water or air that is in motion east or west relative to the earth's surface is rotating around the earth's axis respectively faster or slower than the earth itself. If it has a greater velocity than the earth, it must therefore have a greater tendency to slide toward the equator; if it moves westward, as does an easterly wind, then it presses from the equator. These laws are true for both hemispheres; in both cases a west wind moving eastward presses toward the equator, which is toward the right hand for west winds in the Northern Hemisphere but toward the left hand for the Southern Hemisphere." — *Monthly Weather Review*, October, 1905.

Application of Preceding Principles to Pumps

169. Pumps of all kinds, whether for gas or liquids, whether to exhaust a vessel of the fluid it contains, or to force fluid into a vessel, are all modifications of the typical form shown in Fig. 103.

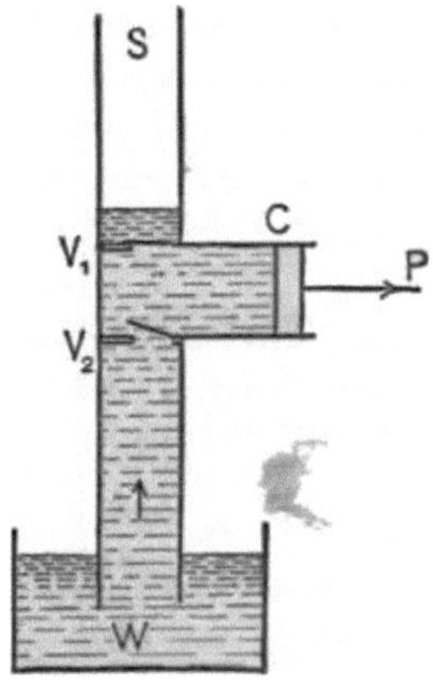

FIG. 103. — A typical pump.

S is a hollow tube dipping into a vessel of fluid *W*.

P is a piston fitting snugly into the cylinder *C*.

V_1 is a valve opening out of *C*.

V_2 is a valve opening into *C*.

1. **Force Pump.** — The figure as it stands represents the common force pump, such as that employed in the ordinary fire engine and in pumping water into standpipes. When the piston P is moved in the direction indicated by the arrow, the pressure in the region between the valves is diminished. The effect of this is to open the lower valve and close the upper one.

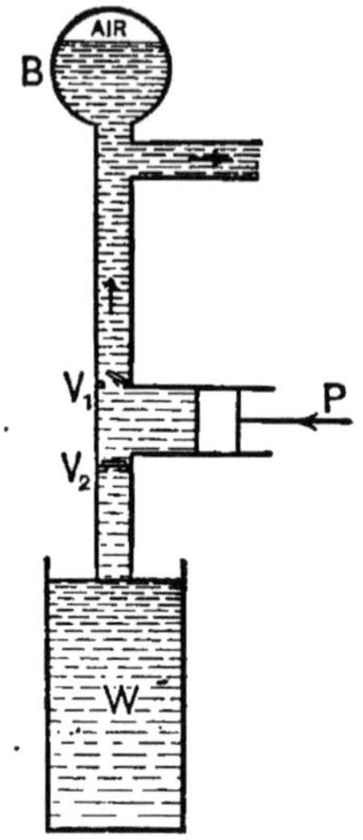

FIG. 104. — The force pump.

Not only so, but the pressure of the atmosphere on the free surface of the water in the well begins to raise the water in the tube. And if the lower valve be not too high above the surface of the well, the water will pour through the valve V_2, and partly fill the space between the valves.

If the motion of the piston be now reversed (Fig. 104), the water inclosed between the valves will be forced through the upper valve into the hose or vessel in which it is desired. Such a stream, however, would be intermittent; for in the region above the upper valve the pressure would cease, during the out stroke of the piston. To avoid this difficulty, most force pumps are furnished with an air chamber (B, Fig. 104). When the pump first begins, this chamber is filled with air. When the water is forced through the upper valve, this air is compressed; energy is stored up in it; and, when the out stroke of the piston occurs, this compressed air supplies enough energy to drive the water partially out of the bulb B, and thus keep the stream flowing until the piston begins its inward stroke. This large copper bulb on the fire-engine is familiar to every one.

2. **The Human Heart.** — The human heart is exactly such a force pump as that diagrammed in Fig. 104, only here the space between the valves is no longer furnished with a piston, but has elastic walls, which, by alternate contraction and expansion, effect the same result as the piston. When the pressure in the right ventricle of the heart (V_r, Fig. 105) is relieved, the blood from the veins rushes in by way of the right auricle Au_r; when the walls of the heart contract, this blood

is driven into the pulmonary arteries P_a on its way to the lungs. In like manner, the left side of the heart acts as a force pump, receiving fresh blood from the lungs and driving it to the extremities of the body. The elasticity of the arteries and veins takes the place of the air chamber in the force pump.

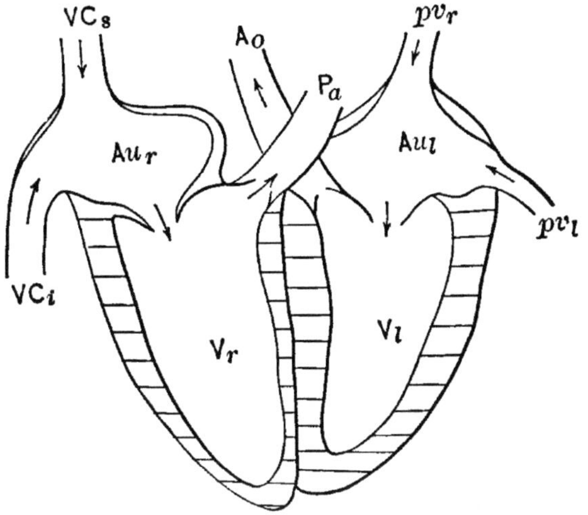

FIG. 105.—The human heart considered as a force pump.

3. **The Ordinary Lift Pump.** — Here the upper valve is placed in the piston itself, as indicated in Fig. 106. Since the pump has no air chamber, the stream is of course intermittent. On the down stroke the upper valve opens; on the up stroke it closes; the pressure between the valves is diminished; the lower valve opens and the water rushes through. On the next down stroke of the piston, it dips into the water, which is held up by the lower valve; and, on the succeeding up stroke, lifts its load of water to the spout.

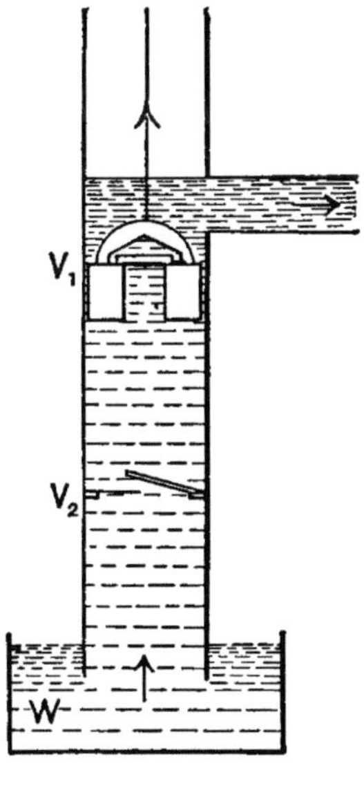

FIG. 106.—Lift pump; on up stroke.

4. **The Bicycle Pump.** — This is a force pump, in which air is the fluid transferred. The piston here also contains the valve; is the valve, in fact. The down stroke of an ordinary foot pump is represented in Fig. 107. On the up stroke the air rushes into the cylinder all around the edge of the piston; while the valve

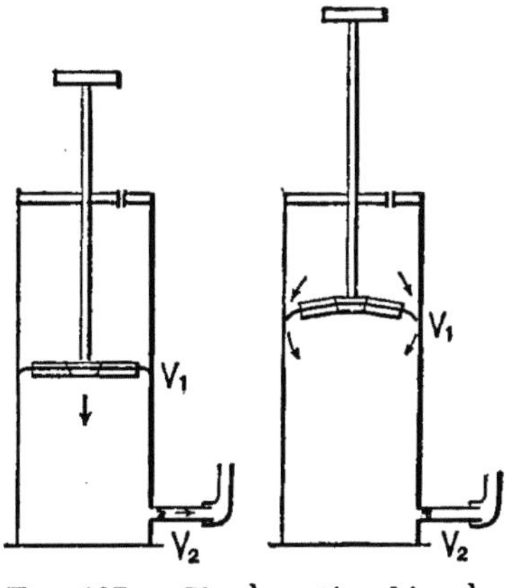

FIG. 107.—Single-acting bicycle pump.

V_2, next the tire, closes by the back pressure of the air in the tube.

Why is such a pump so hard to operate when the tire is fairly well inflated?

5. **The Vacuum Pump.**—If we reverse the valves indicated in Fig. 107, we shall have a pump which, instead of pumping air into a tire, will empty the tire. This is only the general principle of the ordinary air pump, which is one of the most important instruments of the physical laboratory. During the last half century it has received many radical modifications and improvements. Of these perhaps the most valuable is the substitution of mercury in glass for the solid piston in a brass cylinder. Mercury does not wet or soil glass, does not appreciably evaporate at ordinary temperatures, moves through a glass tube easily; and is, hence, excellently adapted to this use.

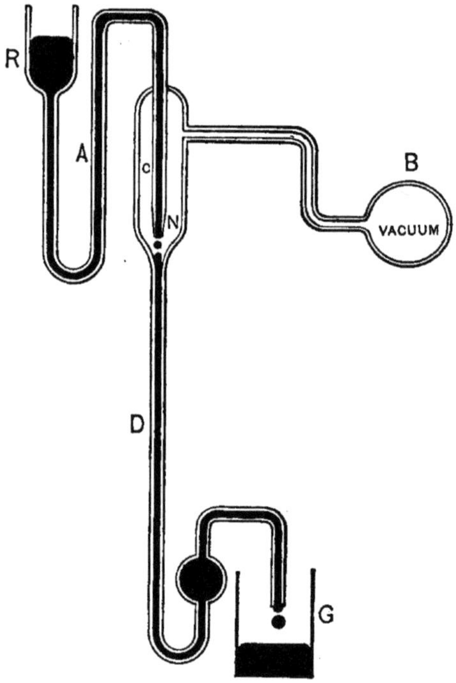

FIG. 108.—Sprengel air pump.

The action of the Sprengel air pump will be clear from the accompanying figure. The falling drops of mercury from the nozzle *N* entangle the air in the chamber *C* and carry it down the narrow glass tube. The drops of mercury act as pistons. More air rushes in from the bulb *B*, and the whole vessel is thus gradually exhausted.

The reservoir *R* is kept filled with mercury; the pressure of the atmosphere forces this over to the nozzle *N*. It is collected again in a glass vessel *G*, and poured back into the reservoir *R*. It can thus be used over and over many times.

The vertical heights of the columns *A* and *D* of mercury must each exceed the barometric height. Why? Such pumps are in daily use in the manufacture of incandescent electric lamps and "X-ray" bulbs.

6. **The Geissler-Toepler Pump,** of which there are many modifications, is shown in its essential features in Fig. 109. The large bulb *B* opens above, by means of a valve V_1 to the atmosphere. At the bottom this bulb opens, by means of the tube on the right, to the vessel which is to be exhausted. In this tube is a glass-stemmed valve V_2 which floats on mercury and prevents any air from passing back into the exhausted chamber.

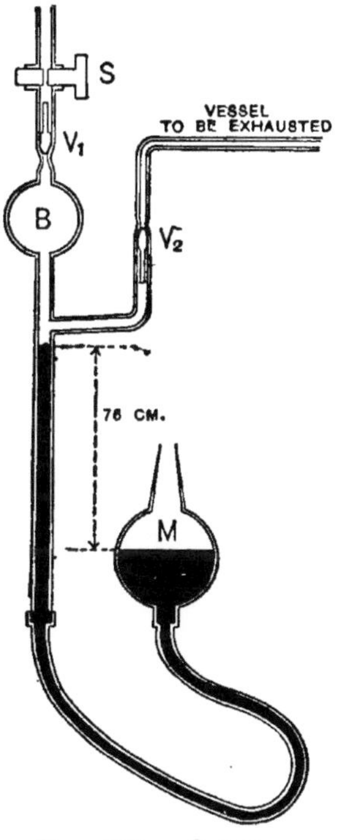

FIG. 109.—Geissler-Toepler pump.

The tube which extends vertically downward from *B* is necessarily longer than the barometric height, and is connected to a large vessel of mercury *M* by means of a strong but flexible rubber tube.

If now the vessel of mercury *M* be raised, the large bulb *B* will fill, the valve V_2 will close, and the air in *B* will be driven out through the valve V_1. If now the large vessel of mercury be lowered, the upper valve V_1 will at once close, and will remain sealed by mercury trapped between the open stopcock *S* and the seat of the valve. But as soon as the mercury falls below the junction of the two lower tubes, the air will rush from the vessel to be exhausted into the chamber *B*. On repeating this operation the bulb *B* is again emptied into the atmosphere and again filled at the expense of what air remains in the vessel to be exhausted.

7. **The Geryk Pump.**—A mechanical pump of high perfection has recently been devised in which oil is used as a sort of packing for the piston, thus leaving practically no "clearance" and producing a very high vacuum in a very few strokes. A diagram of this instrument is shown in Fig. 110. When the piston comes down upon the oil in the bottom of the cylinder, the air is all forced up through the valve in the piston, into the region above the piston.

On the upward stroke the valve not only closes, but is hermetically sealed by the layer of oil on the upper side of the piston. A pump of this type will in a few minutes produce a vacuum of 0.001 millimeter of mercury. That is to say, out of every million particles originally in the vessel only one will remain at the end of, say, ten minutes. These pumps are frequently used in tandem, the intake *I* of the second pump being directly connected with the exhaust *E* of the first pump.

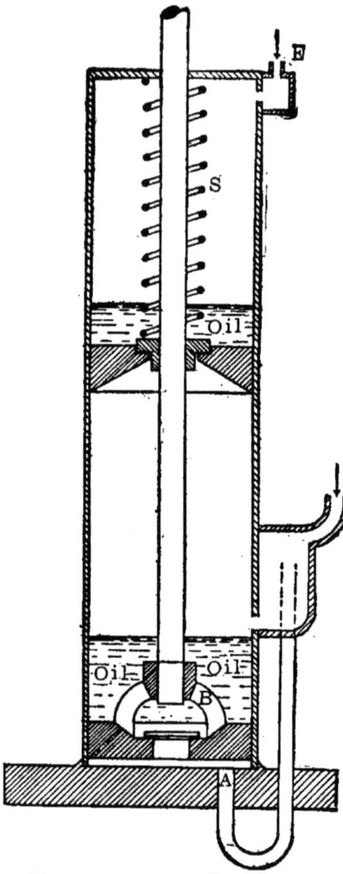

Fig. 110. — The Geryk Air Pump.

8. **The Siphon.** — When it is desired to transfer liquid from a vessel *A* at one level to another vessel at a lower level, a bent tube something like that shown in Fig. 111 is very convenient and is known as a siphon. To start the siphon it is first filled with liquid, then inverted and placed with the upper arm under the surface *A* as shown. The operation then becomes simple if we understand (i) that the pressure at *B* in the tube is the same as at *A* because they each lie in the same horizontal plane, and (ii) that the pressure increases as we go down in the liquid from *B* to *C*. But the pressure on the free surface at *C* is just one atmosphere. Hence the liquid will be continually forced out of the lower end of the tube until the free surface

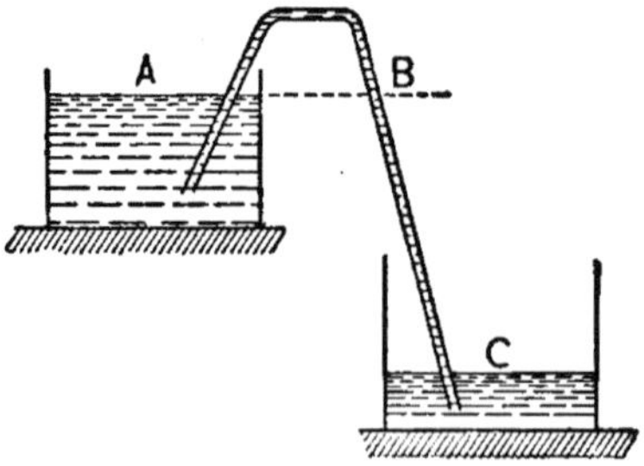

Fig. 111. — Siphon.

in each vessel stands at the same level. There are many other useful and clever forms of this instrument, all dependent upon this same principle.

Boyle's Law

170. When Torricelli discovered the proper method of measuring the pressure of the earth's atmosphere, he furnished us also a method by which gaseous pressures, in general, may be measured. The method is as follows: —

Connect the vessel containing the gas to one end of a U-tube containing mercury. The pressure due to the excess of height h (Fig. 112) of the mercury in the open arm over that in the closed arm will be measured by Dgh. If to this we add the pressure of the atmosphere B, we obtain the pressure P in the vessel.

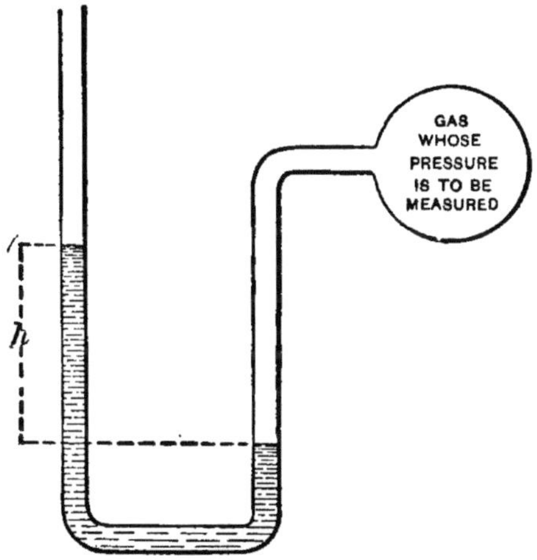

Fig. 112.

$$P = B + Dgh, \qquad \text{Eq. 67}$$

where

B = barometric pressure ;

D = density of mercury ;

g = acceleration of gravity ;

h = difference of level between the two mercury surfaces.

If the mercury in the open arm of the U-tube stands at a lower level than in the closed arm, h becomes negative, and we have

$$P = B - Dgh.$$

Shortly after the time of Torricelli's experiment, Robert Boyle, a distinguished English chemist, invented an air pump, and became greatly interested in the following question : —

Gases are easily compressed. A bladder filled with air and closed at ordinary pressure can be reduced in volume without much difficulty, *i.e.* without greatly increasing the pressure. But, "Just *how* does the volume of a given body of gas vary when the pressure is varied ?"

To discover how one quantity varies with **another,** the **general method** is **to let one quantity** vary, measure its successive values, **and also** measure **the corresponding** values **of the other** quantity.

This Boyle did. He inclosed a definite mass of air in the closed arm of a U-tube, as indicated in Fig. 113. He varied the volume V of this inclosed gas by pouring in more and more mercury, and measured the pressures corresponding.

The method employed for measuring the pressure is that which we have just described.

$$P = B + Dgh = Dg \cdot (H + h),$$

where H = height of barometer.

Now Boyle found that (i) *so long as the mass of air remained constant*, and (ii) *so long as its temperature remained constant*, **the product of the pressure by the volume of the gas was very approximately a constant.** This law was published in 1660. It may be stated algebraically as follows: —

$$PV = C = \text{constant}. \qquad \text{Eq. 72}$$

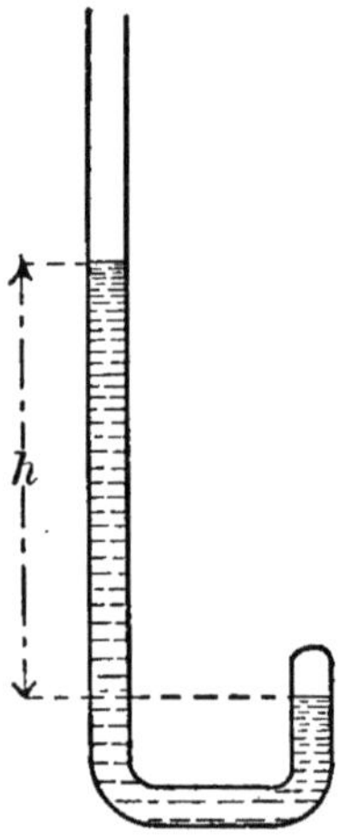

FIG. 113.—Boyle's experiment.

When the mass of gas inclosed in the short arm of the tube was doubled, it was found that, for any given pressure, the volume was also doubled; or to obtain any given volume it was now necessary to double the pressure. And in general it has been found that the product PV is proportional to the mass M of the inclosed gas, a result which may be stated as follows: For constant temperatures,

$$PV = MK \qquad \text{Eq. 73}$$

where K is a constant.

Since, however,

$$M = DV, \qquad \text{Eq. 28}$$

we may eliminate M, and obtain a result which is entirely independent of the mass of gas inclosed, namely

$$\frac{P}{D} = K = \text{constant}. \qquad \text{Eq. 74}$$

The interpretation of this equation is evidently the following: **At any given temperature, the pressure in a gas varies directly as its density, and is independent of the mass.**

171. Boyle's experiments were mostly performed with pressures greater than one atmosphere (76 centimeters of mercury); but Mariotte (1620–1684), a French physicist, shortly afterward showed that Boyle's Law holds when the air in the containing vessel is rarefied; that is, when the pressure is less than one atmosphere. This he demonstrated as follows: —

Two pieces of straight glass tubing are connected by a rubber tube, as indicated in Fig. 114. A portion of air is sealed off by mercury in the closed arm *B*. On lowering the arm *A*, the mercury in *A* falls below the level of that in *B*. Let this depression be indicated by *h*, then

$$P = Dg \cdot (H - h)$$

where, as before, *H* is the height of the barometer and *D* the density of mercury. The tube *B* is graduated so that the volume of the gas *V* can be read off directly.

In this case also the product *PV* was found to be invariable. Quantities such as *PV* are known as *invariants*. The factors which enter the product are each variable, but the product itself is a constant.

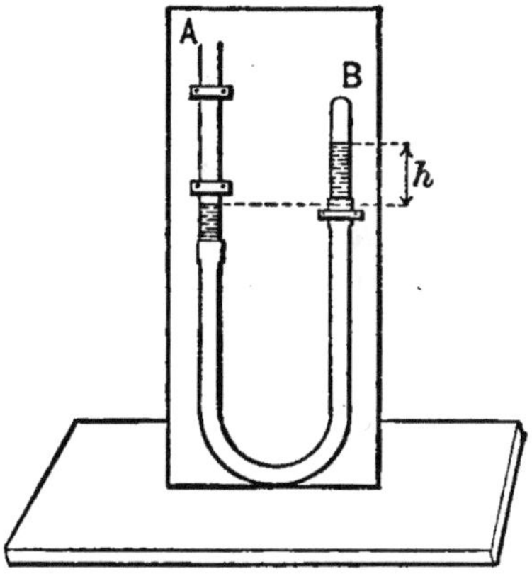

FIG. 114. — Mariotte's experiment.

172. The usefulness of this law of Boyle's can scarcely be overestimated. For, if we know the pressure corresponding to any one known volume, we know at once the value of the constant *K*, and can hence *compute* the pressure corresponding to any other volume; or, if it be desired, one can compute the volume corresponding to any other pressure. For example, suppose that a bottle whose capacity is 100 cubic centimeters contains compressed air at a pressure equal to that of 180 centimeters of mercury. Let us compute the volume which this air, at the same temperature, would occupy if allowed to escape into the atmosphere at normal pressure. Let P_1, V_1 represent the pressure and volume, respectively, of the air while in the bottle, and P_2, V_2, its pressure and volume after being released. It is required to find V_2. According to Boyle's Law, the product of pressure and volume remains constant. Hence Eq. 72 becomes

$$P_1V_1 = C = P_2V_2.$$

But $P_1 = 180$; $V_1 = 100$; and $P_2 = 76$. Hence, substituting these values, and using one centimeter of mercury as the unit of pressure, we obtain the following value for C in Eq. 72:—

$$P_1V_1 = 180 \times 100 = 18000 = C.$$

And, therefore,

$$V_2 = \frac{C}{P_2} = \frac{18000}{76} = 236.8 \text{ cm.}^3$$

Hence the 100 cubic centimeters of compressed air, if allowed to escape, would occupy 236.8 cubic centimeters.

173. A **perfect gas** is defined as one which obeys Boyle's Law exactly; but it has been found by many observers that all gases deviate more or less from the law; in other words, there are no perfect gases in nature. These deviations have been especially studied by the French physicist, Amagat, whose description of the facts, together with Boyle's original memoir, will be found in *The Laws of Gases*, edited by Professor Carl Barus (New York, 1899).

Boyle's Law has been modified by the Dutch physicist, Van der Waals, so as to describe the behavior of actual gases with much greater accuracy. Using the same notation as above, and denoting by a and b two quantities which are constant for any one gas, he writes

$$\left(P + \frac{a}{V^2}\right)(V - b) = \text{constant.} \qquad \text{Eq. 75}$$

This is the most perfect description which has yet been given for the changes in volumes which occur in gases under a wide range of pressures. The interpretation of the constants a and b will be of interest to the advanced student.

174. The graphical interpretation of Boyle's Law leads at once to the equilateral hyperbola shown in Fig. 115; or, if one employs Eq. 74 and plots pressures as a function of densities, he is then, of course, led to the straight line.

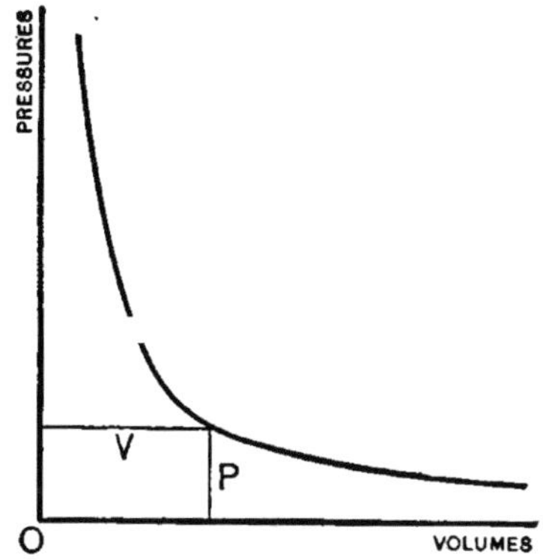

FIG. 115. — Graph of $PV =$ constant.

175. Before leaving the subject of this chapter, it is necessary to warn the beginner that there are many fundamental phenomena, such as osmosis of liquids, diffusion of gases, absorption of gases by liquids and solids, which we have not been able here to discuss.

Problems

1. What is the height of a water barometer when the mercurial barometer reads 760 mm.? Density of mercury = 13.6.

Ans. 10.336 m.

2. An atmospheric pressure of 15 lb. to the square inch is equivalent to a pressure of how many kilograms to the square centimeter?

Ans. 1.056

3. Twenty-four cubic centimeters of gas at a pressure of 71 cm. of mercury would have what volume under a pressure of 76 cm.?

Ans. 22.4

4. A cylinder 16 in. long is closed at one end by a piston. If the cylinder is filled with air at atmospheric pressure (76 cm.), and the piston is thrust down within 1 in. of the other end, what will be the pressure of the inclosed air? (It is here assumed that the temperature is the same in each case.)

Ans. 1216 cm. of mercury.

5. Two hollow spheres of radii 8 and 10 contain equal masses of hydrogen. Compare the pressures.

Ans. 1000 : 512

6. Find the lifting power of a balloon, filled with coal gas, from the following data: volume of balloon = 1000 m.^{3}; weight of balloon and car = 60 kg. A liter of coal gas weighs 1.193 g.; a liter of air weighs 1.293 g.

Ans. 40 kg.

Jones, *Examples in Physics.*

7. One wall of a room is 9 × 24 ft. If the room is closed tightly at a time when the height of the barometer is 30 in., what pressure must the wall sustain when the barometer suddenly drops to 29 in.?

Shearer, 527.

8. If 1000 cm.^{3} of hydrogen under a pressure of 20 cm. of mercury is subjected to a pressure of 84 cm. of mercury, what will be the resulting volume? *Ans.* 238.1 cm.^{3}

9. Compute the weight of air contained in a moderate-sized living room, say one which is 5 × 5 × 3 m., at the temperature of 0° C.

10. When a barometer stands at 75 cm. the volume of the empty space above the mercury is 8 cm.^{3} and its length is 13 cm. How far would the column fall if you introduced 5 cm.^{3} of air into the tube? *Ans.* 19.+ cm.

Jones, *Examples*, p. 98.

CHAPTER V

WAVES

176. The word at the head of this chapter will recall to most minds one or more of the following phenomena: the surface of a lake swept by a wind, the constantly enlarging rings produced by a stone falling into quiet water, the "swell" that follows a storm at sea, the waves which always accompany a moving boat. Most people have seen waves run along a clothes-line, or some stretched rope, when struck by a walking stick. Experience shows us that these and all similar cases have two features in common. First, there is **a disturbance at some point in a certain medium** Second, this **disturbance is propagated to** other **points of the medium.**

At a given instant the stone is dropped into quiet water; at a later instant the disturbance may have reached points several meters away. At a given instant the stretched clothesline is at rest in its position of equilibrium. When one end is suddenly depressed, this depression runs along the line at a definite rate; or if the disturbance be an elevation, it is an elevation that runs along the line. Concerning any wave motion the two important questions are (i) concerning the origin of the disturbance, and (ii) concerning the nature of the propagation.

But what is meant by "a disturbance"? Simply this, **a body is said to be disturbed when it is thrown out of equilibrium.**

But when is a body or system of bodies in equilibrium? Not necessarily when it is at rest, but when its potential energy is a minimum. A pendulum is at rest at the end of its swing, but is by no means in equilibrium. A pendulum moves most rapidly at the lowest point in its path; but there is the one point where it is in equilibrium.

A disturbance, then, implies the addition of energy to a system. And as this disturbance is propagated through the medium, some energy is in general distributed through the medium.

Definition of Wave Motion

177. We may, therefore, **define a wave motion as a state of disturbance being propagated from one part of a medium to another.**

Note that in this definition nothing is said about any up-and-down motion, such as is seen in the disturbed clothesline. One has only to replace the clothesline by a stretched spiral spring to see that waves may travel without any such up-and-down motion. Such a spring (Fig. 116) should be from two to three meters long, made of brass wire, No. 18 or 20, wound in a spiral of approximately eight millimeters diameter. Fix the ends of the spring to two rigid points *A* and *B*, so that the spring will be stretched to nearly twice its unstretched length. If this spring be grasped at any point *C*, moved in the direction of the arrow, then quickly and gently released, the disturbance thus introduced will run to and fro along the line; but there will be no appreciable up-and-down motion. This very important experiment should be carefully observed.

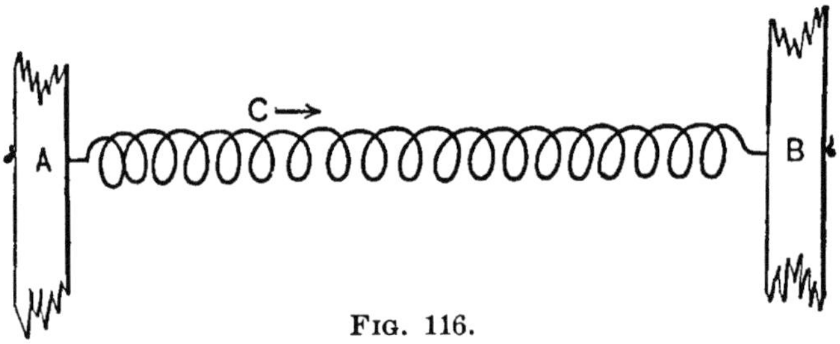

Fig. 116.

Note further that, in the definition of wave motion, nothing is said about any transfer of matter. For although in each case experiment shows that the particles of the medium oscillate through a small range, it is evidently not the medium, but the energy, which is transferred. For clearly exhibiting the distinction between the motion of the wave and the motion of the particles of the medium, there is perhaps nothing superior to the wave model used by Lord Kelvin in his *Baltimore Lectures* in 1884. This device consists of, say, 20 or 30 equal wooden bars suspended by a piano wire, as shown in Fig. 117. Since these bars are fastened so that when one of them is twisted the wire is also twisted, it is clear that a single rotation

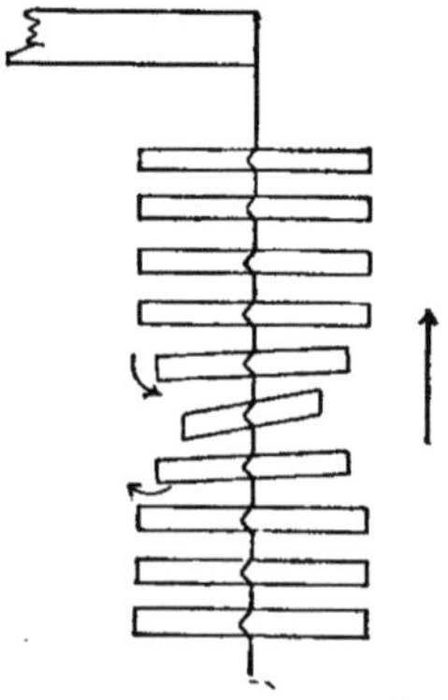
Fig. 117. — Kelvin's wave model.

given to the lowest bar will be propagated throughout the entire series, the wave moving vertically upwards, while the particles (the bars) move in a horizontal plane only.

Some Special Kinds of Waves

I. *Water Waves*

178. Of these there are several kinds, and most of them are studied to best advantage in the laboratory or in a canal.

The genesis of water waves by wind is very imperfectly understood; but we may consider that the source of these waves is anything which elevates or depresses the water at any point. Imagine OX to represent the level surface of the undisturbed water. If by any means the surface of the water be given the form indicated in Fig. 118, the operation will be equivalent to taking the water from the "trough" of the wave BC, and lifting it up into the "crest" of the wave AB. Let m be the mass of water which will just fill the trough BC; the amount of water in the crest will then also be m. Let h denote the difference in height between the center of mass of the trough and the center of mass of the crest.

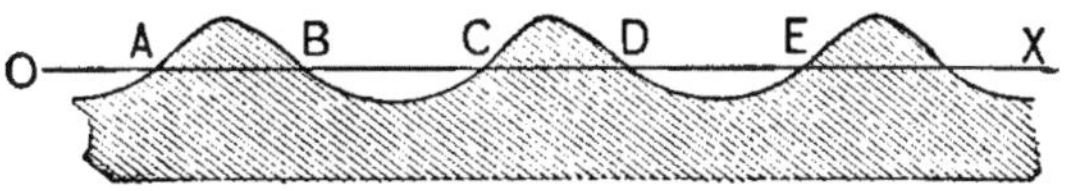

FIG. 118. — Typical sea wave.

The operation of producing a single wave in water previously quiet will require that one lift a mass m through a height h against gravity. The potential energy stored in the wave will, therefore be mgh. This potential energy tends to a minimum, *i.e.* in this case the center of mass tends to fall, and, since water has inertia, the crest in falling goes on beyond its position of equilibrium, and thus we have a **continual oscillation due to gravitation.**

In addition to this potential energy, it must not be forgotten that waves have considerable kinetic energy. Indeed, it may be shown that in wave motion practically one half the energy is kinetic and one half potential. The disturbance at any point in the water over which a wave is passing is then very similar, but not identical, with that in one arm of a U-tube partially filled with mercury and the mercury in oscillation.

Take a U-tube of a diameter not less than one centimeter. Pour it half full of clean mercury — water will answer nearly as well. Incline it and close one end with the thumb, as indicated in Fig. 119. Bring the tube into a vertical position and suddenly raise the thumb. A dozen or more oscillations follow. They are, so far as vertical motion is concerned, such as would occur at any point in an ocean of mercury, or of water traversed by waves.

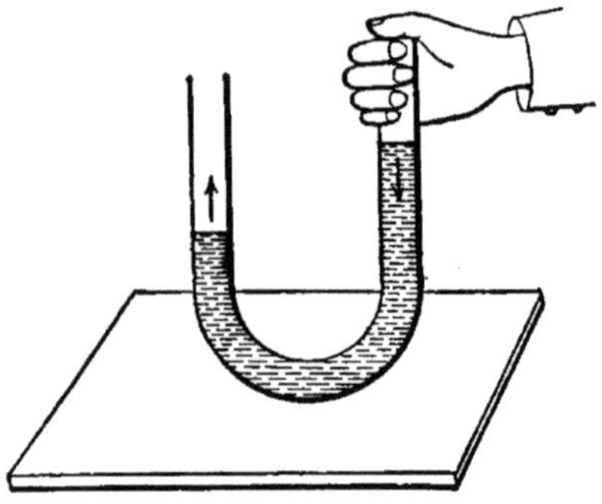

Fig. 119.

Why do these oscillations so quickly stop? Would this same cause be at work in a large lake or at sea to stop the waves?

Waves or oscillations which are made to diminish gradually in amplitude are said to be **damped**.

179. The actual motion of the water particles in sea waves seems to have been first accurately guessed by Franz Gerstner (1802), who assumed that these particles move in circular orbits, as shown in Fig. 120, where the line of centers O represents the undisturbed surface of water. When a train of waves passes over this surface, the particles P, which, in equilibrium, are situated at O, are now set into uniform circular motion.

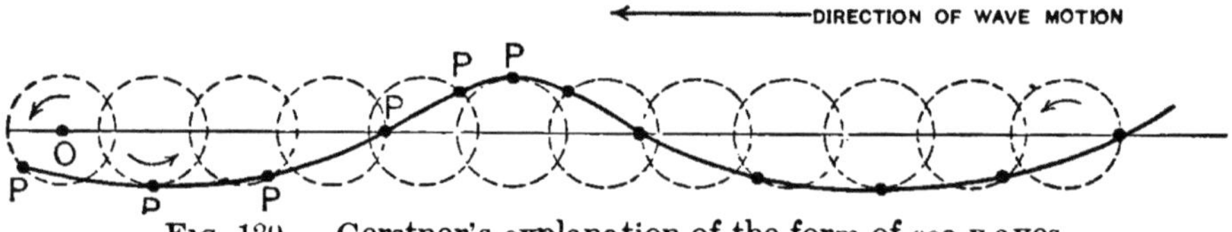

Fig. 120. — Gerstner's explanation of the form of sea waves.

The curve drawn through their positions at any particular instant is called the "locus" of these particles. This locus is the wave form at any instant. To find the wave form at any *later* instant, say when each particle has moved 90° ahead in its orbit, one has only to mark in each circle a point 90° ahead of the present position of P and then draw the locus of these new points.

In *deep* water, the motion of the particles under the surface is identical with that at the surface, except that the circular orbits become smaller and smaller as one leaves the surface. In *shallow* water, the particles move in ellipses which become flatter and flatter as one approaches the bottom. The justification for this view of Gerstner's lies in the fact that these elliptical motions can be observed on small particles scattered through the water, and also in the fact that the actual wave form is seen to be very nearly what his theory — called the "trochoidal theory" — predicts. Many of the properties of trochoidal waves are beautifully illustrated by an apparatus devised for this purpose by Professor C. S. Lyman, an excellent modification of which is made by Dr. Forbes of Columbia University.

Definition of Wave Length

180. A series of waves, such as one can start in water by rocking a boat, or by raising and lowering a floating block, is called a **train of waves.** In any train of water waves the distance between two successive crests is called a "wave length." The distance between two successive troughs is also a wave length. **But the general definition of a wave length is the shortest distance between any two particles whose motion is in the same phase.**

What would correspond to "two successive crests" in the waves produced by compression in the spiral spring?

Speed of Water Waves

181. A train of waves is traveling *over a point;* how far does this train travel during the period of one complete oscillation of the water at this point? Evidently one wave length. Let us call this wave length l, and the period of one oscillation T.

If, by any means, direct observation or computation, we can obtain l and T, we have at once the speed of the wave; for, denoting speed by V and frequency by n, we have

$$V = \frac{l}{T} = nl. \qquad \text{Eq. 76}$$

This may be called the fundamental equation of wave motion; for, since it depends simply upon the defining equation for speed, it is necessarily true for a wave motion of any type. There enters into the equation no dynamical quantity, nor, indeed, anything characteristic of the medium; it is merely a kinematic definition of wave velocity in general.

Regarding the speed of water waves as a function of the body of water upon which they travel, time permits us only to say that: —

(i) In shallow water, *i.e.* where the depth h is rather small compared with the wave length, the speed V is

$$V = \sqrt{gh}, \qquad \text{Eq. 77.}$$

where g, as usual, indicates the acceleration of gravity. Observe that in this case V is independent of the wave length.

(ii) In deep water, *i.e.* where the depth is rather large compared with the wave length l, the speed is

$$V = \sqrt{\frac{gl}{2\pi}}. \qquad \text{Eq. 78}$$

This equation describes those great waves which overtake the swiftest Atlantic liners; a simple computation shows, for instance, that a deep sea wave which is 400 feet in length travels with a speed of 26 knots an hour.

Refraction of Water Waves

182. The variation of speed with depth, and the independence of speed and wave length indicated in Eq. 77, may be easily tested in the laboratory by means of a wave trough.

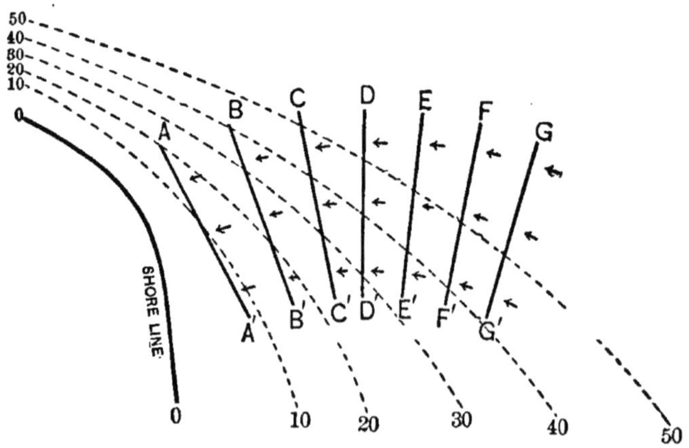

FIG. 121. — Refraction of water waves.

The equation for shallow water waves also describes very perfectly the curious fact, easily observed on the shore of nearly any lake or pond, or at the sea coast, that, whenever waves approach the shore, the crest of the wave is always more or less parallel to the shore line. In Fig. 121 let the heavy line represent the shore near the entrance to a small bay. A train of waves, *A*, *B*, *C*, *D*, etc., approaches the shore. When the waves leave the deep water, the crests have the direction indicated by $\overline{GG'}$. The nearer they approach the shore, the more they are turned about. Such a process is called **refraction.** The end *A* has travelled faster than the end *A'*. The dotted lines are called **contour lines**, and indicate the depth of the water. Every point on the contour line marked "10" has a depth of 10 centimeters; every point on the line marked "20" has a depth of 20 centimeters, etc.

Explain why the crests of the waves are rotated as they approach the shore. Since they start in parallel to each other, why do they not remain so?

Reflection of Water Waves

183. If a breakwater or solid pier (Fig. 122) is built out into a body of water, one has there an opportunity to see what happens when water waves strike a rigid body.

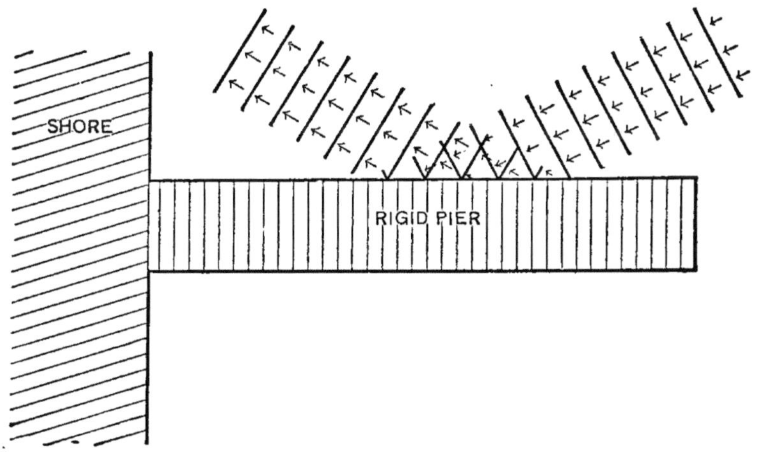

FIG. 122. — Reflection of water waves.

In such circumstances it is not difficult to see that the direction of the reflected waves is very different from that of the incident waves.

From our experience in the laboratory, we know that when a train of waves is started at one end of a wave trough, they no sooner reach the other end than they are reflected back.

These two trains of waves, the incident and reflected, may combine with each other to produce a series of waves which appear to have no motion of translation, the so-called **stationary waves**. In order that these two trains of waves should produce stationary waves, it is necessary that all the waves should be of one length. A moment's practice with a wooden paddle in a trough of water enables one to produce such waves.

A more detailed study of stationary waves will be made under the head of Waves in Strings.

Summary concerning Water Waves

1. **Medium of propagation** — bounding surface between air and water.
2. **Motion of particles** — small, in paths, nearly circular; through a part of the path the motion of the particle is in the same direction as the motion of the wave; in part of the path it is at right angles to the motion of the wave.
3. **Speed** — in deep water, depends upon length of wave, but in shallow water, it depends upon the depth.
4. Water waves are **reflected**, and thus produce **stationary waves**.

II. *Ripples*

184. In our study of liquids we found that work was required to enlarge the surface of any given body of liquid. We found that equilibrium was reached only when the liquid surface became as small as possible.

Consider any quiet body of liquid, such as a cup of tea. Its surface is in equilibrium when level, because then it has a minimum of potential energy. Any disturbance of this surface, such as is produced by drawing a spoon through it, will increase the area, and will, therefore, impart to the surface some energy.

Suppose a surface disturbed as shown in Fig. 123, and then left to itself. Not only will *gravity* tend to lower the liquid crests and fill the troughs, but the *surface tension* of the liquid will also do exactly the same thing. **Surface tension is, therefore, competent to produce waves in a liquid just as gravity is.** And when the lengths of

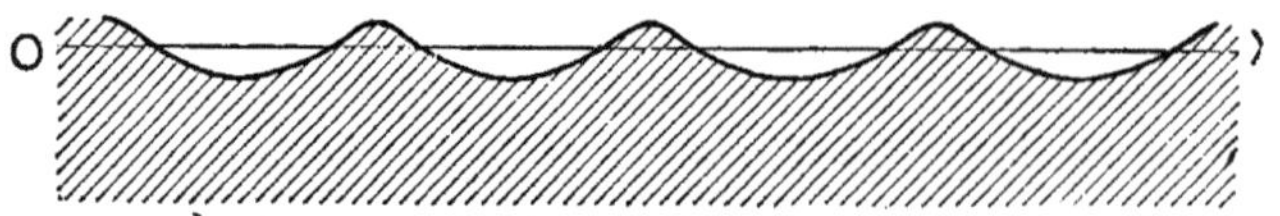

FIG. 123. — Showing that a surface disturbed by ripples is larger than a level surface.

the waves become very small, the surface tension is even more effective than gravity. So that in the case of very small waves, we may consider them as propagated by surface tension alone.

These ripples are known to every one who has watched the bow of a rowboat as it moves through quiet water. They are beautifully seen in the case of a duck swimming on a smooth pond. And what is a very curious thing, these ripples always manage to keep just a trifle ahead of the duck, however fast or slowly he swims.

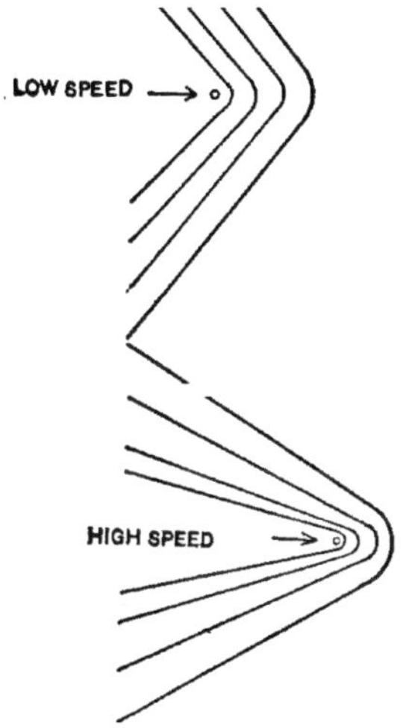

FIG. 124. — Showing that short ripples travel faster than long ones.

We may see the same phenomenon in the laboratory. When a piece of wire is drawn through a dish of water (Fig. 124), at a moderate rate, the ripples which precede it have a certain length; but when the wire is drawn more rapidly through the water the ripples become finer, *i.e.* ripples of a shorter wave length are produced. But in each case the ripples precede the moving body. It follows, therefore, that ripples of short wave length travel faster than those of long wave length. This is just the reverse of what happens in ordinary water waves; for there the longer the waves the faster they travel.

185. Neglecting the effect of gravity upon the speed of these short waves and considering only the surface tension T, it may be shown, by ordinary dynamics, that their speed is

$$V^2 = \frac{2\pi T}{lD}, \qquad \text{Eq. 79}$$

where D is the density of the liquid and l the length of the wave.

If now one takes account of gravity also, remembering Eq. 78, and that from this cause the square of speed will be increased by the amount, $\frac{gl}{2\pi}$, he has the general expression for deep-water waves, namely

$$V^2 = \frac{2\pi T}{lD} + \frac{gl}{2\pi}. \qquad \text{Eq. 80}$$

It follows, therefore, that the slowest waves are not the very short ones, nor the very long ones. The slowest of all water waves have been found to have a length of about 16 millimeters. Waves longer than this are spoken of as **water waves,** while waves shorter than this are called ripples.

In order to discover whether or not ripples are reflected, as water waves are, one has merely to produce ripples in a shallow dish of water and let the ripples proceed until they strike the edge of the dish.

Summary concerning Ripples

1. **Medium of propagation** — free surface of liquid.
2. **Speed** — increases as the wave length diminishes.
3. **Reflected** — in same manner as water waves.
4. **Motion of particles** — same as that of water waves.

III. *Tidal Waves*

186. To one who goes to the seashore for the first time, few things are more striking than the two great changes in the height of the water which occur every twenty-four hours. The water recedes until at a certain hour it reaches what is called **low tide**; not many minutes are required to see that the water is again rising. The tide is coming in; at the end of six hours after low tide, the maximum height, called **high tide**, is reached; the tide immediately begins to recede; and so on in endless procession. So gigantic is this phenomenon that one would never suspect its being a wave motion until he has formed a mental picture of the whole disturbance as reported by observers from all parts of the world.

Approximate Explanation of Tides

187. Those who have rightly understood Newton's capital discovery of universal gravitation, know that the moon attracts each part of the earth with a force which is greater as the distance between that part of the earth and the moon is less. The larger part of the earth's surface is covered by its oceans. Imagine the whole of it to be under water. In Fig. 125, consider three particles, A, B, and C, lying on a diameter of the earth, drawn in the direction joining the moon and the earth.

A is on the side of the earth next the moon; *B* at the center of the earth; *C* on the side of the earth away from the moon. *A* will be more attracted than *B*, that is, it will be attracted away from *B*; while *B*, in turn, will be more strongly attracted than *C*, that is, *B* will be attracted away from *C*. The difference between the actual attraction of the moon at any point on the earth's surface and the average attraction is called the "tide-generating force" at that point. The result of this tide-producing force is that the water at *A*, attempting to get as near the moon as possible, is heaped up, is drawn away from the solid body of the earth.

FIG. 125. — Lunar tides. Arrow at *A* indicates direction of tidal wave in passing about the earth.

To draw *B* away from *C* is equivalent to repelling *C* from *B*. So that in case the earth always presented the same side to the moon, we should have at the points *A* and *C* high tides. On the great circle which lies 90° away from these we should have low tide.

188. The above is a brief sketch of what is known as the equilibrium theory of the tides, and is a fundamental picture which each student should clearly grasp. But as a matter of fact it comes very far from accurately representing the time at which high water would occur on a rotating earth, even if we assume, as in the preceding section, that the ocean covers it to a uniform depth. For as the solid nucleus *B* is rotating on its axis, carrying the towns and cities of the earth with it, the result is evidently that every twenty-four hours each seacoast town must pass through these two antipodal tide crests, that is, through two high tides and two low tides.

This great heap of water, which is dragged over the earth's surface every day, consists, *at any one point of the earth's surface*, of a rise and fall of sea level. It is periodic as waves are; that is, the tides recur at equal intervals, — the disturbance is propagated from one part to another of the medium (the oceans); and these features show that what we have here to deal with is a wave, on a magnificent scale, traveling at a stupendous speed. At the equator the length of this wave is 12,500 miles. Its period is 12 hours and 25 minutes, half the apparent diurnal period of the moon.

189. But in order to understand the actual behavior of this great wave it is necessary at this point to make a brief digression upon the subject of

Forced Vibrations

For since this tidal wave is compelled by the moon to travel about the earth at the same rate as the moon itself, the disturbance is called a **forced vibration**. The water waves, which have been considered in previous pages, are called **free vibrations**.

In general, **a free vibration is one in which the period is determined by the properties of the vibrating body itself; while a forced vibration is one whose period is determined by some outside body acting upon it, and supplying energy to it.** Let us consider a simple gravitational pendulum, which has a point of support at A, a bob at B, and a length l, as shown in Fig. 126. If this pendulum be once set in motion and then left to itself, it will vibrate with a period T, where

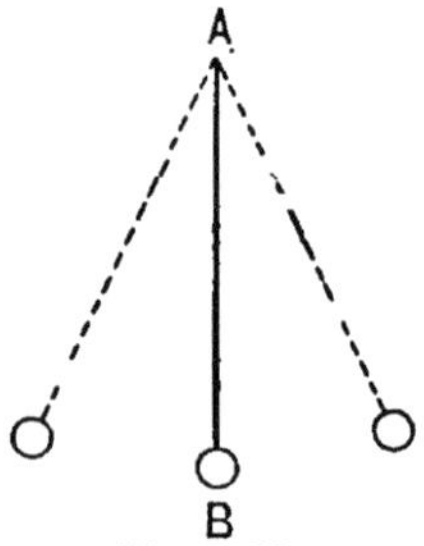

Fig. 126.

$$T = 2\pi\sqrt{\frac{l}{g}}.$$

This is known as the free **period** of the pendulum.

(i) Suppose, however, that the point of support A is moved to and fro in a horizontal direction *with a motion which has a period exactly equal to T.* The result is familiar to every one: the amplitude of the pendulum will increase until as much energy is given by the pendulum to the air and string as is given to the pendulum by the agent which moves the point A. This phenomenon is usually spoken of as **sympathetic resonance** or **sympathetic communication of vibrations**.

(ii) Let us next suppose that the point of support A is urged by a periodic force *whose period is longer than that of the free pendulum.* Here it will be observed that, as soon as the motion becomes steady, the pendulum has assumed a new period, namely, the period of the force; not only so but the phase of the force and the phase of the pendulum are the same. Thus when the force acts toward the right, as indicated in Fig. 127, so also does the pendulum bob move toward the right. The effect may be stated by saying that the pendulum acts as if it were vibrating freely but with an increased length *OB*.

(iii) If, however, we make the *period of the force at A shorter than the free period of the pendulum,* we shall again find that the actual period of the pendulum soon becomes identical with that of the force; but, unlike the preceding case, the phase of the pendulum is now just opposite that of the force, as indicated in Fig. 128. The effect is indeed that of a free pendulum with a shortened length *OB*.

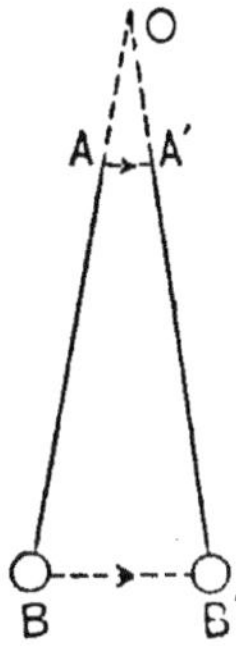

FIG. 127. Period of force greater than period of pendulum.

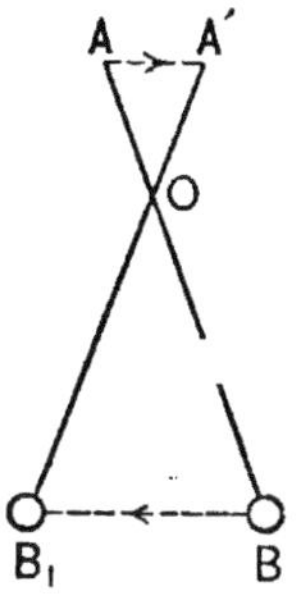

FIG. 128. Period of force shorter than period of pendulum.

190. Returning now to the subject of the tides, it will be seen that we have a tide-generating force whose period is 12 hours 25 minutes acting upon a body of water whose depth, say 3 miles, is very small compared with the length of the tidal wave. The problem is then one of waves in shallow water where the speed of the wave varies with the depth. Referring to Eq. 77, it will be seen that the oceans would have to be 13 miles deep in order that a *free* wave would sweep around the earth at the equator once in 24 hours 50 minutes. And since the actual oceans are not nearly so deep as this, it is clear that a *free* wave, even in the Pacific Ocean, would have a period much longer than that of the apparent motion of the moon.

The tides then fall under the third case which we met in the preceding section, the case where the period of the force is less than the period of the free vibration. This being so, it follows that the tides will be in opposite phase with the tide-generating force; that is, in opposite phase with the tides predicted on the equilibrium theory.

This is known as the dynamical theory of the tides, and shows that under the circumstances described, high water would actually occur where low water was expected on the equilibrium theory.

The presence of large continents of land, the varying motion of the moon, the superposition of solar tides, and many other circumstances render the actual facts tremendously complex when compared with the simple outline of lunar tides here given. Every one interested should read Sir George Darwin's popular lectures on *The Tides* (Boston, 1898).

Summary concerning Tides

1. They satisfy the conditions of wave motion.
2. They are forced vibrations.
3. Wave length — at equator — 12,500 miles.
4. Period of lunar tides 12 hours 25 minutes.
5. Period of solar tides 12 hours 0 minutes.
6. Velocity at equator 1042 miles per hour.

IV. *Waves in Strings*

191. An easy way to study these roughly is to fasten together two lengths of ordinary rubber tubing, and thus make one piece about 20 feet long. By the aid of a heavy mass, or of some more convenient means, clamp one end of the tube to the edge of a table, or even to the floor, as shown at *B* in Fig. 129.

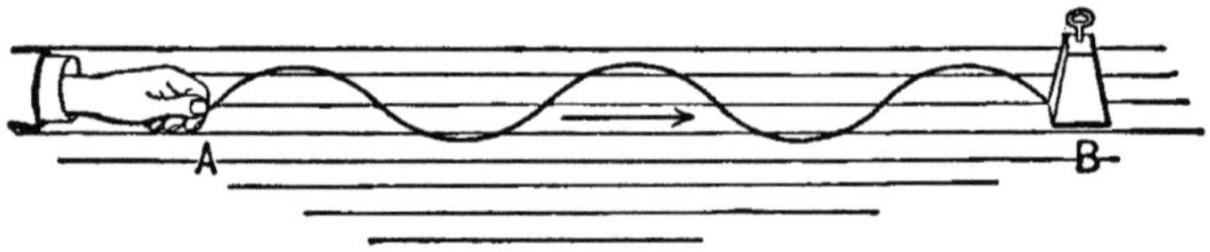

FIG. 129. — A simple method of studying waves in strings.

Taking the free end *A* in the hand, one can produce in the tube any tension he desires, and can send trains of waves down the tube in the direction given by the arrow.

Transverse and Longitudinal Waves

Consider the motion of any one point in the rubber tube. As a train of waves passes through this point, it oscillates, to and fro, in a direction perpendicular to that of the arrow. The motion of the particles in the medium is then across the direction in which the waves travel. Such waves are said to be **transverse**.

In the case of the spiral spring, alternately compressed and elongated (Fig. 130), any one small part of the spring is displaced in the *same* direction as that in which the wave is traveling. The displacement of the particles is lengthwise. Such waves are accordingly called **longitudinal** waves.

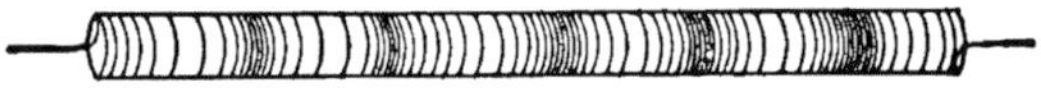

FIG. 130. — Longitudinal waves in a spiral spring.

In the short time at our disposal the only type of waves in strings which we can consider are the transverse ones.

Returning now to the rubber tube, stretched upon the floor, observe the following: —

1. That you can vary the *length* of the waves you send along the string by varying the *rapidity* with which you move your hand to and fro. Let us denote by T the time occupied by your hand in one complete vibration. During this interval the disturbance travels just one wave length l along the tube.

If we had any means of measuring T and l, we could determine the speed of the waves V; for (§ 181), —

$$V = \frac{l}{T}. \qquad \text{Eq. 76}$$

2. Observe also that you can vary the *length* of the waves by varying the force with which you stretch the rubber tube, while your hand moves to and fro at the same rate as before. From this fact it follows that the speed of the wave depends upon the tension in the string.

3. Notice that a single disturbance sent down the tube does not in general cease when the wave strikes the fixed end (B, Fig. 129). If it does, the wave has been damped out by friction on the floor; stretch the tube a little more, holding it up so as just to clear the floor. The reflected wave can now be easily seen running back in a direction just opposite that of the incident wave. Such a disturbance is called a *solitary* wave.

4. With a moment's practice one can obtain a reflected *train* of waves which interfere with the incident train in such a way as to produce stationary waves, exactly similar to the stationary water waves obtained in the water trough.

Speed of Transverse Waves in Strings

192. Imagine a perfectly flexible and uniform cord drawn through a brass tube and stretched as shown in section in Fig. 131. It is an astonishing fact that there is a perfectly definite speed with which any such cord may be drawn through the tube without exerting any pressure upon the tube.

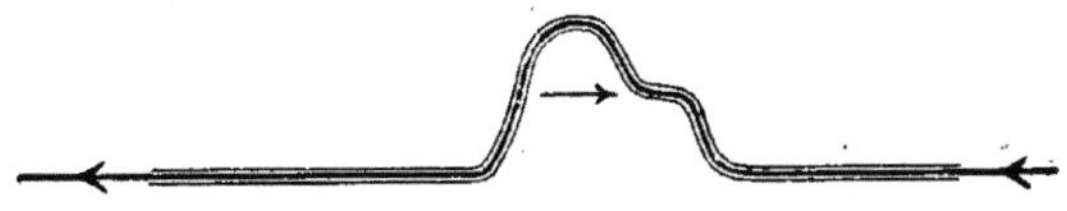

FIG. 131. — Illustrating how centrifugal force balances stretching force when a transverse wave traverses a string.

This is true whatever the curvature of the tube may be, provided only that the curvature is continuous, *i.e.* provided there are no sharp angles in the tube. And what is still more curious is the fact that this particular speed is exactly that with which a free wave will travel in this same cord. Our problem now is to determine from simple dynamics this critical velocity.

(i) First one must compute the pressure (force per unit length) which the cord would exert upon the tube, if the entire system were at rest. Let A, B, and C (Fig. 132) represent three consecutive points on the tube, and R its radius of curvature at B. Suppose that the arc ABC subtends the angle θ at the center of curvature O; then, if the tension in the string is T, the normal component of this tension at B will be $2\,T \sin\frac{\theta}{2}$; and hence

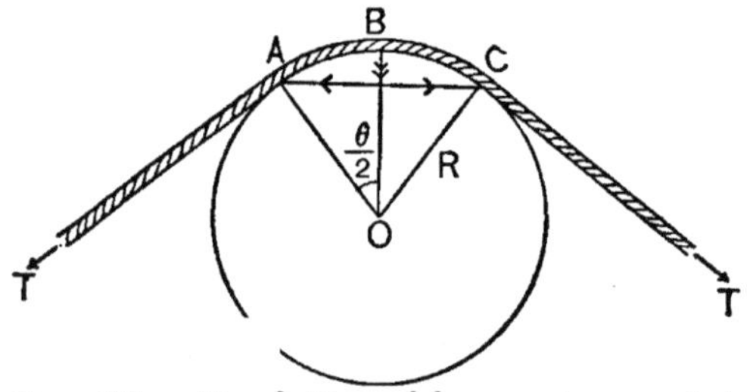

FIG. 132.—Resolution of forces at any point B on the tube.

$$\text{Pressure at } B = \frac{2\,T \sin\frac{\theta}{2}}{R\theta} = \frac{T}{R} \text{ approximately.}$$

This is the force with which each unit length of string is pressed against the tube when the cord and the tube are at rest.

(ii) Next one must compute the centrifugal force with which each unit length of the string is thrust away from the tube when it is moved through the tube.

Let μ denote the mass of unit length (linear density) of the string and V the speed with which it is drawn through the tube. Then when the string passes the point B (Fig. 132) where the radius of curvature is R, it is evident that the outward pressure on the cord will be $\mu\frac{V^2}{R}$.

It follows, therefore, that the total pressure will be zero when these two opposite pressures are equal, that is, when

$$\frac{T}{R} - \frac{MV^2}{R} = 0. \qquad \text{Eq. 81}$$

The value of the speed given by this expression,

$$V=\sqrt{\frac{T}{\mu}}, \qquad \text{Eq. 82}$$

is then the critical speed with which the string must be pulled through the tube.

But since, with this speed, there is no pressure on the walls of the tube, one may entirely remove the tube, and there remains a solitary wave traveling along the string with a definite speed, which is independent of the curvature and is a function of only the tension and the linear density.

Interference of Waves in Strings

193. Returning for a moment to the stretched rubber tube, it is easily shown that a solitary wave sent down from the hand to the fixed end is reflected and travels back to the hand. If a second wave be sent from the hand so as to meet the reflected wave, the two are said to "interfere," and the resultant disturbance at any point on the tube where they pass is the sum of the disturbances due to the outgoing and incoming wave.

The following experiment, due to Wylie, illustrates very simply the manner in which two *trains* of waves interfere.

To one of the prongs of an electrically driven tuning fork attach a light string — silk floss is excellent — and make the other end fast to some rigid object, as shown in Fig. 133.

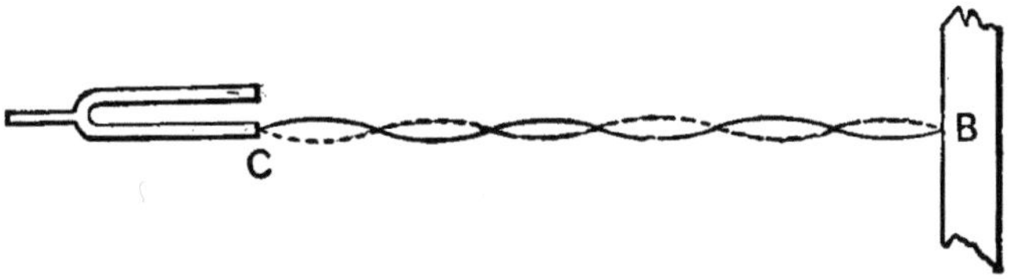

FIG. 133. — A single train of waves excited by one prong of a fork.

When the fork starts to vibrate, a train of waves will run down the whole length of the string from C to B; and if the tension of the string is properly adjusted, stationary waves will be produced *throughout the entire length of the string.*

At some point A (Fig. 134) now tie a second string of length equal to CA; attach the other end to prong D and make the tension the same in each string. Each prong of the fork will now send a wave down the cord. But the prongs of the fork, at any instant, are moving in exactly opposite directions; hence at A the trains of waves in DA and CA are in exactly opposite phases. And since they each tend to dis-

FIG. 134. — Wylie's experiment, showing two trains of waves interfering at A.

place the point A by equal amounts in opposite directions, the result is that the point A is not displaced at all. The waves DA and CA are said to **interfere completely** at A. No disturbance, therefore, can now enter that portion of the string lying between A and B.

Interference is often observed also in trains of *water* waves. Thus when the solar and lunar tides reach any point on the earth's surface in the *same* phase, they interfere and thus produce the **spring tides**. When, however, they reach any point in *opposite* phases, they interfere and thus produce the **neap tides.**

It has been suggested that the so-called tidal waves which occasionally sweep the decks of the Atlantic liners are due to the interference of two trains of running waves.

Formation of Stationary Waves in a String

194. Referring to Fig. 133, where the string is fixed at one end B, we have first to discover how the wave reflected from B is related to the wave incident at B. The condition under which reflection here occurs is that the end is fixed; hence *the reflected wave must be of such a character that when compounded with the incident wave the result will be zero displacement of the string at B.*

The simplest method for obtaining the form of the reflected wave from this principle is by use of the terms *inversion* and *perversion*. Hold a mirror perpendicular to the direction of the stretched string; the image of a wave (in the mirror) is called its **inversion**. But if the mirror is held parallel to the cord, the image of the wave is called its **perversion**.

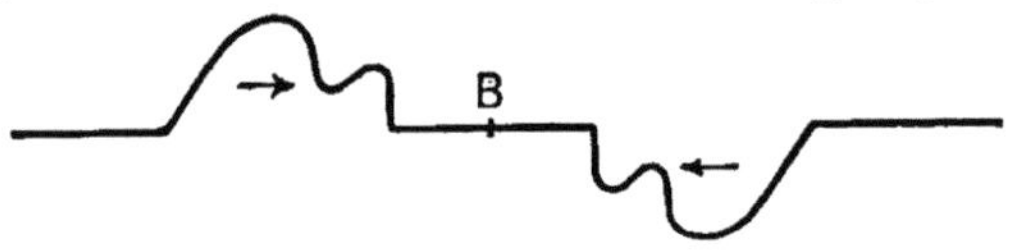

FIG. 135. — Condition of reflection from fixed end.

Let the upper wave in Fig. 135 represent the incident wave; then the wave reflected from B will be exactly the same as one which is both the inversion and perversion of the incident wave. For if we imagine the cord to be infinitely long, instead of stopping at B, and if an inversion *and* perversion approaches B from the rear, — as far behind B as the incident wave is in front, — then the two waves will evidently interfere at B in such a way as to leave the string unmoved; in other words the boundary condition will be satisfied. **The reflected wave may therefore be simply described as the perversion *and* inversion of the incident wave.**

But suppose now that a *train* of waves is reflected at B; then each particular reflected wave will, *at certain points in its return path*, meet an incident wave which is in exactly opposite phase; call this point N_1; furthermore, if at any particular instant N_1 is a point of exactly opposite phases, it will always be such a point for waves of this length. When the reflected wave has traveled half a length farther it will again meet an incident wave in precisely opposite phase; this will also be a point of no motion; call it N_2. **All such points where the interference is such as to produce no motion are called nodes.** Midway between these are other points where the two trains of waves — incident and reflected — meet in exactly the same phase: these are called **antinodes**.

With a little practice one can make the rubber tube vibrate in such a way as to show either one, two, or three, perhaps more, of these nodes. The fixed point B is of course always a node. Any portion of the vibrating string which lies between two consecutive nodes is called a **loop** or **ventral segment**.

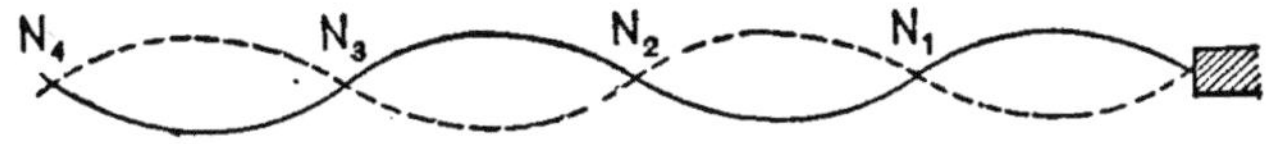

FIG. 136. — Production of nodes in a string.

The chief characteristic of stationary waves is that, *at each point* in the medium, the two interfering trains differ in phase by a constant amount, ranging from 0° to 180°. A string in which stationary vibrations have been set up may therefore be clamped at any two nodes — even two consecutive nodes — without changing the mode of vibration.

One convenient feature of stationary vibrations is that the nodes are easily observed by the eye; and since they are just half a wave apart, one can easily measure the length of the wave. Then, after eliminating V between the following equations,—

$$V = nl, \qquad \text{Eq. 76}$$

$$V = \sqrt{\frac{T}{\mu}}, \qquad \text{Eq. 82}$$

the frequency (§ 49) of the wave motion is obtainable from the directly measurable quantities l, μ, and T.

Summary concerning Transverse Waves in Strings

1. Medium is linear.

2. Motion of particles is at right angles to direction in which wave travels.

3. Speed of propagation is independent of the length of the string, the material of the string, and the shape of the wave.

Speed depends only upon (i) the stretching force T, and (ii) the linear density μ.

4. Wave reflected from fixed point is perversion and inversion of incident wave.

5. In stationary waves the distance between two successive nodes is half a wave length.

V. *Resonance*

195. When one attempts to ring a large church bell, he soon learns that he must pull the rope at certain intervals if he wishes to ring the bell with ease. Any heavy object, such as a stone suspended from a crane, can be set swinging if one's pushes are properly timed. A heavy rowboat or small sailboat, moored in quiet water, can without difficulty be set rocking, if one throws his weight first on one side, then on the other, at correct intervals. The order, "rout step," given soldiers about to cross a bridge is to prevent dangerous vibrations in the bridge.

What is more, every one knows how to discover the correct interval in each of these cases. By watching the amplitude of vibration, he at once times his efforts so that this amplitude will increase. These are all instances of the mechanical principle known as **sympathetic resonance**.

196. The following experiment should be carefully studied: Mount two heavy pendulums of any kind upon one common wall bracket or frame, as shown in Fig. 137. Heavy lead disks on $\frac{1}{4}$-inch iron rod, suspended by knife-edges at the top, do very well. The upper end of the rod should be furnished with an eye through which a piece of a three-cornered file may be thrust for a knife-edge.

The lower end of the rod should be threaded for three inches, so that the bob of the pendulum may be held up by a nut, and so that the period of either pendulum may be slightly varied at will. Paint one pendulum red, the other blue.

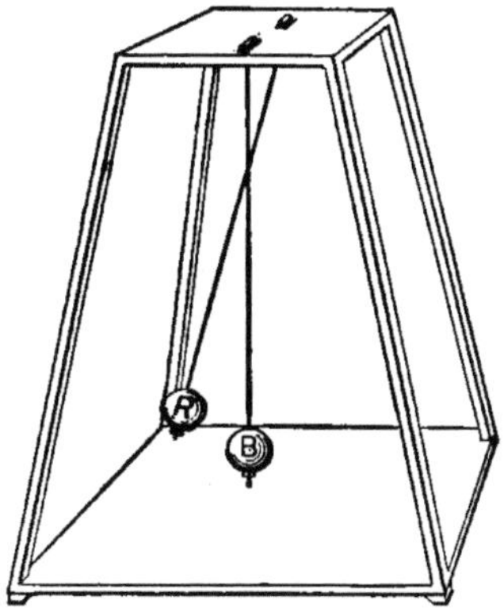

FIG. 137.—Illustrating transfer of energy by resonance.

1. Adjust the pendulums so that they have very nearly the same period. Set the blue one in vibration, leaving the red one at rest. In less than one minute, the red one will pick up sufficient kinetic energy to set itself in vibration with an amplitude not greatly inferior to that of the blue one.

Where did this energy come from? How did it get from the blue pendulum into the red pendulum?

2. Now reverse the order. Set the red pendulum in motion, allowing the blue one to be initially at rest.

3. Take a wooden mallet and tap one of these heavy pendulums in a most irregular way. The result of a dozen such blows is to disturb the pendulum very little.

Now give the same pendulum one-half dozen gentle taps, delivering each tap *just as the pendulum passes its position of equilibrium*, going away from you. How do you explain the large amplitude which these six taps produce in the pendulum?

4. Adjust the pendulums so that one is from 8 to 10 per cent longer than the other. Set the *blue* one in vibration. What is the history of the *red* one for the next five minutes? How do you explain the small amplitude and the complete stoppages of the red one?

197. A supremely beautiful example of resonance is obtained in the following simple manner described by Professor Wilberforce of Liverpool (*Phil. Mag.* October, 1894): —

At the lower end of a spiral spring suspend a mass whose moment of inertia is variable, and which can be so adjusted that its period of rotational vibration about a vertical axis is very nearly equal to its period of translational vibration along its vertical axis.

Now it may be easily shown that a spiral of wire — such as a watch spring, or even a paper ribbon — cannot be elongated without at the same time being twisted, nor can it be shortened without being at the same time untwisted.

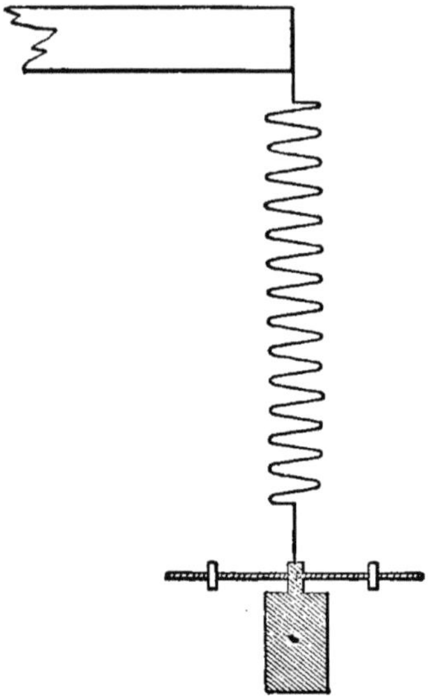

FIG. 138. — Wilberforce's spiral spring.

Accordingly, if such a spring as that shown in Fig. 138 be set to vibrating up and down, the suspended cylinder will at each vibration be given a slight torque. If these torques are timed, as we have supposed, so as to coincide exactly with the free period of rotation, the cylinder will be gradually set into rotation, until finally practically all the energy of translation has been transformed into energy of rotation. But it is equally true that a spiral spring cannot be twisted without at the same time introducing more or less of a compressional stress along the axis; nor can it be untwisted without producing a stress which tends to elongate the spring

Accordingly, when the kinetic energy of the spring has become entirely rotational, and the difference of phase between the two vibrations has changed sign (for the periods are not *exactly* equal and even if they were at some one amplitude, they would not be so at others), each oscillation of the cylinder will produce a slight up-and-down motion, and the energy will now be transferred in the opposite direction, until it is again practically all translational.

What we have here, then, is resonance between two degrees of freedom in the same body, each motion being simple harmonic.

The following dimensions for Wilberforce's spring have been found convenient: —

Height of brass cylinder (Fig. 138) .	$1\frac{39}{40}$ inches
Diameter of brass cylinder (Fig. 138)	$1\frac{1}{2}$ inches
Length of brass screw	$6\frac{1}{2}$ inches

Diameter of brass screw	$\frac{1}{7}$ inch
Weight of brass traveling nuts, each	$\frac{1}{5}$ ounce
Diameter of steel wire in spiral	$\frac{1}{16}$ inch
Diameter of spiral	$1\frac{1}{4}$ inches
Length of spiral (unstretched)	18 inches

Change of Wave Length due to Motion. Doppler's Principle

198. So far in our study of wave motion we have considered both the source of the waves and the observer of the waves to be relatively at rest. But if either the point at which the waves originate or the point at which they are observed is in motion, one with respect to the other, then the frequency of the train of waves will either be changed or appear to be changed in a definite ratio.

This idea may be most easily grasped perhaps from the following analogy, which is quantitatively correct for any ordinary train of waves.

Imagine two boys seated some distance apart upon the bank of a creek. Let the boy who is upstream have a basket of corks from which he throws into the creek n corks each second. So long as both boys remain seated the corks will float by the lower boy at the rate of n per second, *i.e.* with a frequency of n. And if the creek flows with a uniform speed V, the distance between the successive corks l will be given by the equation

$$l = \frac{V}{n}. \qquad \text{Eq. 76}$$

But let us now suppose that the upper boy starts downstream with his basket, the corks will now pass the lower boy more frequently than n per second: while if the basket of corks be carried upstream, the individual corks will float by the lower boy *less* frequently than n per second.

When both boys were at rest, there were n corks distributed over the distance traversed by the stream in one second; *i.e.* over a distance numerically equal to V. When the upper boy moves downstream with the speed v, there are n corks distributed over a shorter distance, namely, a distance numerically equal to $V - v$. Hence the wave length (distance apart of the corks) is shortened in the ratio $\frac{V - v}{V}$

If the boy with the basket had moved upstream, instead of down, the wave lengths would have been increased, of course, in the ratio $\frac{V+v}{V}$; a ratio which is not quite, but nearly, the reciprocal of the preceding whenever V is large compared with v.

The student will find it an interesting exercise to prove that if the boy with the basket remains seated, while the other boy moves upstream with a speed u, the frequency of the train will appear to be increased in the ratio $\frac{V+u}{V}$; and hence the apparent wave length is altered in the inverse ratio $\frac{V}{V+u}$.

When the boys both move up or down stream with the same speed, the problem becomes more complicated, and its discussion must be postponed. But when v is very small compared with V, it will be observed that $\frac{V-v}{V}$ is very approximately the same as $\frac{V}{V+v}$; so that in this case the effect is practically the same whether the source or the observer be in motion.

This general notion appears to have been first distinctly enunciated by Christian Doppler in 1843, and is therefore known as "Doppler's Principle." It may be stated as follows:

When a source of waves, such as a tuning fork, is approaching or receding from an observer, the length of the waves emitted is decreased or increased respectively. When, on the other hand, an observer is approaching or receding from a source of waves, the frequency of the waves appears to be increased or diminished respectively.

On the Transfer of Mechanical Energy. Waves and Currents

199. The effects of waves on the seacoast are so well known as to need no description. The energy of waves is here exhibited on a gigantic scale, — in some places undermining cliffs, in others transporting large quantities of sand or tearing away sea walls. **Wave motion is, indeed, one of nature's general methods for transferring energy.**

Every child who has played with a garden hose knows that the jet of water can be used to dig a considerable hole in the ground. In the far West these jets are employed on a large scale for mining. The process is known as **hydraulic** mining. Currents of air are employed for running windmills, thus transferring the energy of the wind to the water, which is pumped into a tank in the attic of a house. One may transfer energy to a body by throwing a stone at it or by firing a shot at it. All of these are simply special kinds of currents.

Currents of matter may be called the second general method which nature employs for transferring energy. The moving matter, in these currents, may be air, water, stones, shot, etc.

Our study of the subjects which follow, Sound, Heat, Electricity, Light, we shall find to be largely a study of the transfer of energy by one or the other of these two general methods, *Currents* and *Waves.*

Problems

1. A clothesline which weighs 1 kg. is stretched between two posts which are twenty m. apart. The tension on the line is 3 million dynes. How long will it take for a transverse disturbance started at one end to travel the entire length of the rope?

2. A stretched cord is set into stationary transverse vibration so that its frequency is 120. The nodes are 40 cm. apart. With what velocity does a wave travel in this string?

3. Some long waves coming from the deep water of a lake strike upon the edge of a shoal which is covered with a uniform depth of 8 in. How fast will the waves travel through this shallow water?

4. A train of water waves each of which is 20 ft. long passes a buoy with a speed of 24 ft. per second. If at the instant under consideration a crest is passing the buoy, what is the condition of the water at a point 85 ft. back of the buoy?

5. Compute the speed of a wave 10 ft. long when traveling over the surface of deep water.

6. How long must a wave in the Atlantic Ocean be to have the same speed as the steamer *Lusitania,* 24 knots?

7. The water in a wave tank diminishes in depth from 9 in. to 1 in. How will the speed of large waves in this tank be affected?

8. Will a train of very short ripples travel more rapidly in mercury or in water? The surface tension of water in contact with air is 75; that of mercury in air is 513.

9. A whistle emitting 1200 vibrations per second is approached by an observer at a speed of 60 ft. per second. Assume the speed of sound 1100 ft. per second. Find what number of vibrations the observer will *appear* to hear in one second.

10. How much faster will a transverse wave travel in an aluminum wire than in an iron wire of the same cross section stretched by the same force?

CHAPTER VI

SOUND

200. We have not been in this world very long before we learn to attract the attention of our friends by the use of our voice. At a very little later period we have learned to distinguish between the voices of friends. There is no longer any doubt as to whether it is mother or father that is speaking, though neither be in sight. We can soon produce sounds which are more or less loud according to our wish. The ability to "carry a tune" marks a period when we are able not only to distinguish one note from another as having a different pitch, but also to *produce* notes of different pitch. Not many years later, and we have learned to distinguish the flute from the piano, the piano from the guitar, the guitar from the human voice, and so on. In childhood we discover the echo; first perhaps that of a footfall approaching a good-sized building at night. In early youth we observe that a distant locomotive sends up a column of white steam some little while before the sound of the whistle reaches our ears; and we infer that it takes time for sound to travel the intervening distance. There is no longer any difficulty in distinguishing between a noise and a musical sound; and when several notes are sounded together, we decide at once which combination is harmonious and which is unpleasant to the ear.

Something like the foregoing is a part, at least, of our common "stock in trade" of information on the subject of sound; but the *arrangement* of these facts in our minds is perhaps quite as chaotic as the above presentation of them. The object of the present chapter is to aid the student in rearranging this large but disordered heap of valuable information into a well-ordered whole. This will involve the discovery of many new relations among many familiar facts.

201. In taking up the study of any group of phenomena which are perceived by one of our senses, we are always confronted by three problems: —

(i) The first of these is the origin and transmission of the disturbance, or stimulus, which is detected by the sense in question; a purely physical problem.

(ii) Next there is the structure and function of the sense organ which detects the disturbance; a purely physiological problem.

(iii) Then there is the purely psychological problem involved in the translation of physiological stimulation into perception.

The first of these problems, which is also the simplest, is of course the only one considered in this chapter. Any good text-book on Physiology, such as Huxley's, will discuss the second problem; and any good text-book on Psychology, such as James's, will do the same for the third.

202. It is a little difficult for the student who has only just covered the ground of General Dynamics, to realize that he has, in a certain very true sense, been over the whole ground of Physics.

The study of Acoustics — largely a study of waves in air — is, from the student's point of view, a valuable one, forming as it does a kind of connecting link between those visible motions which we have just been studying, and those invisible motions which we are shortly to consider under the heads of Light, Heat, and Electricity.

Most of the motions involved in sound can be made evident to the eye by proper experimental devices; so that one realizes here, perhaps more easily than elsewhere, that Physics is largely Applied Dynamics.

Analysis

203. The subject will be considered in the following order: —

(i) Sound a wave motion.
(ii) Stationary waves in air.
(iii) Graphical representation of sound waves.
(iv) Reflection of sound waves.
(v) Speed of sound in { Solids. / Gases.

(vi) Distinction between a noise and a musical sound.

(vii) Three features by which musical tones are distinguished from each other. { Loudness. Pitch. Quality.

(viii) Relation of tones to each other. The musical scale.

(ix) Application of preceding principles to musical instruments. { Wind instruments. Stringed instruments. Vibrating membranes.

I. Sound a Wave Motion

204. Of the multitudinous sounds which reach our ears during any single day, there are very few which we are not able to "locate" at once. And when we find the source of any sound, we find ourselves in the presence of some vibrating body.

When one string of a guitar is emitting a note, there is no difficulty in telling which one; we have only to look at the strings. A piece of thin steel — say part of the blade of a hack saw or piece of clock spring — clamped in a vice gives a note only when set in vibration. The sound disappears as the vibrations "die down." The edges of a tuning fork become hazy in appearance as it begins to emit a note.

Even when the sounds are not so distinct and definite as those mentioned, we have no difficulty in finding the vibrating body. A carriage rolling over a stony street emits many sounds; but a carriage maker will quickly detect the vibrating parts, and remove them or tighten them up, or interpose rubber in such a way that these vibrating parts no longer receive blows. The carriage then runs much more quietly.

In addition to the vibrating body, there is one other essential feature in the production of every sound; namely, an elastic medium — practically always air — between the vibrating body and the hearer.

The sound of a watch, a bell, or a clock, placed under the receiver of an air pump, is much diminished when the air is even partially removed. It is very difficult to make the sounding body quite inaudible in a vacuum, for the reason that the body must rest on some kind of support; and this support, being necessarily somewhat elastic, will transmit some vibration to the receiver, and hence to the outside air. But we are perfectly justified in inferring that, if no elastic medium whatever lay between the source of vibration and the listening ear, no sound would be heard.

For the hearing of a sound, then, two things are always required:

(i) A vibrating **body, rapidly disturbing its surrounding medium.**

(ii) An elastic **medium connecting the vibrating body and the listening ear.**

II. Stationary Waves in Air

205. One of the shortest cuts to a clear and accurate conception of sound waves is through the following series of experiments. Indeed, the whole philosophy of air waves and the propagation of sound is wrapped up in them. But the first step in understanding stationary waves in air is to understand the nature of the vibrating body which produces them. Hence we first consider the

Longitudinal Vibrations of a Rod

A round brass rod AB from 8 to 10 millimeters in diameter and from 80 to 100 centimeters in length is clamped to a table in an ordinary vise or by means of an iron clamp, as shown in the figure. A small ivory ball or glass marble or hollow sphere is suspended by two strings, as indicated, so as just to touch the end of the rod B. When rubbed with a dry cloth upon which a little powdered rosin has been sprinkled, the rod will emit a very high musical note. It is not the easiest matter in the world to get this note at first, but every one succeeds after a few trials; a fairly firm grasp and a quick but steady pull will generally accomplish the end.

Fig. 139. — Showing longitudinal vibrations of brass rod.

When this note sounds, it is observed that the ivory ball rebounds, often several inches, *showing that it has received a blow from the direction* $\overline{AB}$.

The rosin on the cloth has evidently taken hold of the rod and elongated it, then let it go, taken hold of it again, suddenly let it go again, and so on. The vibrating body here is evidently the brass rod. The motion of the particles in the rod is evidently in the direction $\overline{AB}$. This is shown by the fact that they are originally displaced in that direction by the rosin and by the fact that the momentum which they communicate to the ball is in the direction $\overline{AB}$. Such vibrations we have (§ 191) called longitudinal.

When the brass rod is elongated by a pull and suddenly let go, would you expect, from analogy with other bodies, that it would go back to its position of equilibrium and stop there, or would you expect its inertia to carry it beyond? Would you consider it necessary, for the production of free vibrations, that the vibrating body have *inertia* as well as *elasticity?* When the length of a rod is altered, as in this case, by stretching, what four factors determine the force which restores it to its original length? The present is an excellent occasion for reviewing the whole subject of elasticity.

Are the ends of the vibrating rods nodes or antinodes? The middle point must be a node because it is held fast by the iron clamp.

Character of Disturbance in Air

206. Let us now suppose that the nature of the vibrations in the rod is clearly understood. What sort of disturbance does the rod produce in the air? The displacement of the particles in the rod was so small and rapid as to be invisible; we were obliged to get our evidence by means of the ivory ball. In air, not only is the disturbance small and rapid, but the whole vibrating mass is invisible; so that, even when the disturbances are large and slow, we do not ordinarily see them.

However, Töpler of Dresden has shown that by a certain disposition of apparatus which he describes as the *schlieren Methode* it is quite possible to see these waves; while Mach of Vienna has succeeded in photographing the stationary waves which accompany a Mauser rifle ball, as shown in Fig. 140, and Wood of Baltimore has photographed sound waves under many interesting circumstances.

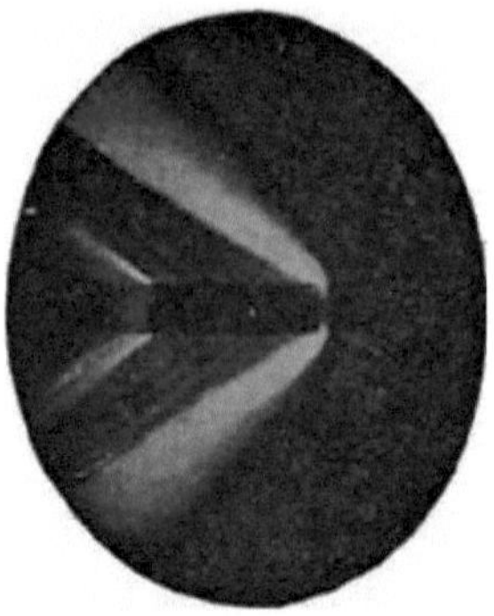

FIG. 140.— Showing the compressions and rarefactions which accompany a rifle ball in its flight.

Kundt's Resonance Tube

207. Nevertheless, the simplest manner of putting air waves into objective evidence is probably the following, which was devised in 1865 by the late Professor Kundt of Berlin: —

To one end of the brass rod described in the preceding section attach a disk of cork *B* which fits loosely into a glass tube about 30 millimeters in diameter and a meter in length.

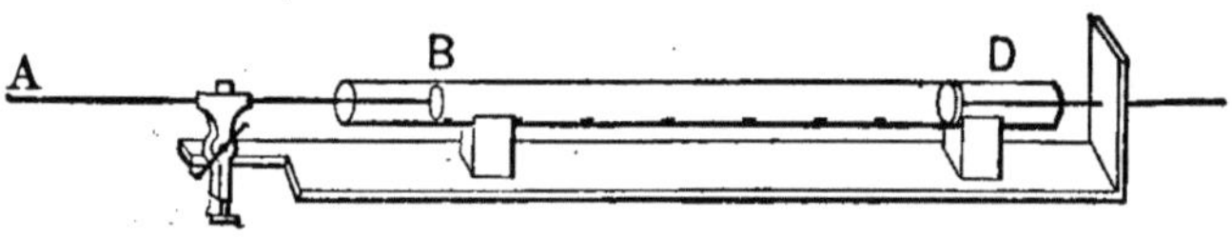

FIG. 141. — Kundt's device for studying waves in air.

By the aid of a cork and a file you can, in a moment, make enough cork dust* to lightly cover the bottom of the tube.

At the end *D* of the glass tube, fit another easily moving piston, so that the distance between the two pistons *D* and *B* can be varied. A large glass tube is more easily excited into longitudinal vibrations than the solid brass rod.

Such an arrangement is called a **Kundt's resonance tube**.

Set the brass rod into longitudinal vibration as before. If the cork piston at *B* is free and does not rub on the walls of the glass tube, the elastic air will be disturbed very much as the ivory ball was. As the rod elongates, the air between the two pistons must be suddenly compressed, for it has not time to escape around the edge of the disk — the compression is too quick for that; when the rod contracts, the inclosed air is suddenly rarefied.

This rod thus starts a train of waves down the tube exactly analogous to those which we produced in the long spiral spring (§ 191). When this train of waves strikes the fixed

* Avoid the lycopodium powder which is sometimes recommended for this purpose. The slightest bit of breath or moist air introduced into the tube will make the dust adhere firmly to the walls of the tube.

piston (Fig. 142), we have all the conditions necessary for reflection. And if reflection occurs, we shall have a train of waves starting back in the direction from *D* and *B*. We should then expect this train of reflected waves to interfere with the incident train and produce stationary waves with their characteristic nodes and loops. The nodes would be those points in the tube at which the cork dust is not disturbed; the loops would be clearly marked throughout their length by the motion of the dust.

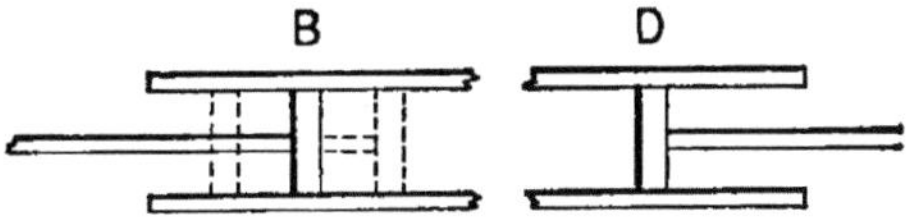

FIG. 142. — Showing action of piston in Kundt's tube.

208. The evidence for thinking that the phenomena which we have just predicted are the phenomena which actually occur is then the following: —

1. Air is a fluid. It has, therefore, no elasticity of shape. It cannot, therefore, transmit transverse vibrations.

2. There are certain equidistant points in the tube where the cork dust is not at all disturbed. (If these are not sharply marked on first rubbing the brass rod, the cork at *D* should be shifted a little.) These correspond to the nodes predicted.

3. Between these points of rest there are certain regions of very great disturbance. These correspond to the predicted loops.

4. By slightly adjusting the fixed piston, the nodes and antinodes become much more sharply marked. And this result is exactly what one might expect if the air waves had a definite wave length determined by the vibrating rod and *not* by the glass tube.

5. To these reasons for thinking sound waves longitudinal, must be added the very cogent evidence furnished by the manometric flame, which will be considered later, § 219.

From testimony of this kind one is led to think that sound consists of a wave motion in the air, and that the motion of the particles of air is to and fro in the same direction as that in which the sound is traveling, that this longitudinal wave motion can be reflected, and that two trains of sound waves may interfere to produce stationary sound waves.

What really travels through the air, then, is a series of compressions, alternating with a series of rarefactions. This is fairly represented in Fig. 143.

Just how far apart these successive rarefactions and compressions are depends entirely upon how fast the piston works. Every time the piston moves forward, a compression occurs; this compression starts out on its journey traveling at a definite rate. When the piston goes back, a rarefaction starts out. These disturbances travel in the same direction and with the same speed. The distance through which a compression (or a rarefaction) travels while the piston makes one complete oscillation to *and* fro is called a wave length. And this corresponds exactly with the definition of wave length given above (§ 180); viz. the distance between any particle and its nearest neighbor that moves in the same phase.

FIG. 143. — Showing the compression and rarefactions in Kundt's tube.

III. GRAPHICAL REPRESENTATION OF SOUND WAVES

209. It is not possible to go much farther with the discussion of air waves without an exact graphical or analytical mode of describing them. The former of these two methods is the only one here available.

As to the proper mode of representing *water* waves there can be no question. We have only to draw a vertical section through the body of water at right angles to the crest of the wave. Such a section is shown in Fig. 118. The distance of any particle above or below the level surface OX measures the displacement of the particle.

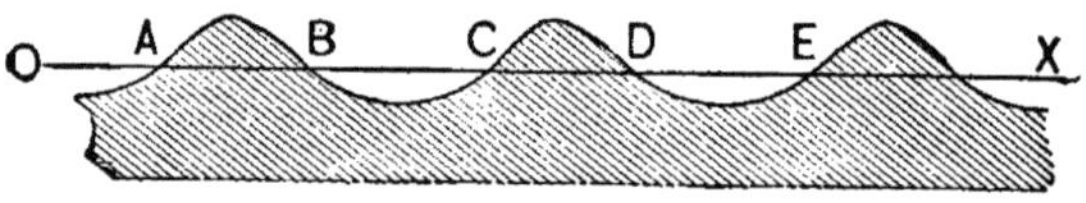

FIG. 118.

Nor can there be any hesitation as to the proper way of representing transverse waves in strings. Imagine the vibrating string at any instant to lose its flexibility; let it instantaneously become a rigid body, and we have at once a picture of the wave. Such a diagram is shown in Fig. 129.

The vertical distance of any particle from the position of equilibrium $\overline{OX}$ in Fig. 118 gives the displacement of that particle. This is the distance which is called the ordinate of the curve. From such a curve we can instantly point out those particles which are in the same phase, or in opposite phase. Be able to go to the blackboard and point out on such a curve a series of points which differ in phase by 180°; and another series which differ in phase by 90°

But suppose that we are dealing with a compression running along a spiral spring. Here the displacements of the particles all lie in the direction of the stretched spring, *i.e.* along the line $\overline{OX}$, which represents the position of equilibrium. How then shall we draw these displacements? And the same question arises when we attempt to picture a sound wave.

The whole question is this: How shall we represent the state of disturbance at any and every point in the air when a train of waves is passing through it?

The method is as follows: We agree to represent all waves in one way, namely, after the manner of the transverse wave in Figs. 129 or 136. To every particle of air in the line along which the wave travels corresponds a point on the curve. But when the wave is *longitudinal*, the ordinate of any point

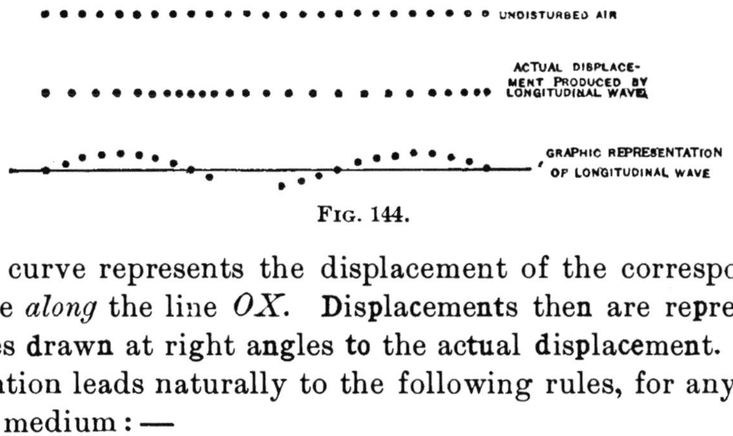

FIG. 144.

on the curve represents the displacement of the corresponding particle *along* the line OX. **Displacements then are represented by lines drawn at right angles to the actual displacement.** This convention leads naturally to the following rules, for any point in the medium:—

(i) $\left.\begin{matrix}\text{Positive}\\ \text{Negative}\end{matrix}\right\}$ ordinates indicate displacement of air particles to $\left\{\begin{matrix}\text{right.}\\ \text{left.}\end{matrix}\right.$

(ii) Where the tangent to the wave is $\left\{\begin{matrix}\text{positive}\\ \text{negative}\end{matrix}\right\}$, there is $\left\{\begin{matrix}\text{rarefaction.}\\ \text{compression.}\end{matrix}\right.$

Thus Fig. 145 represents a wave in which the displacements are all to the right, and in which the progression of the wave is also to the right. Any point *c* on the front of the wave indicates a compression. Any point *r* on the rear side of the wave indicates a rarefaction. A point exactly at the top of the wave represents that point in the medium where there is neither compression nor rarefaction, but ordinary atmospheric pressure. Such a wave might have been produced by giving the piston of a Kundt's tube a single short quick displacement to the right.

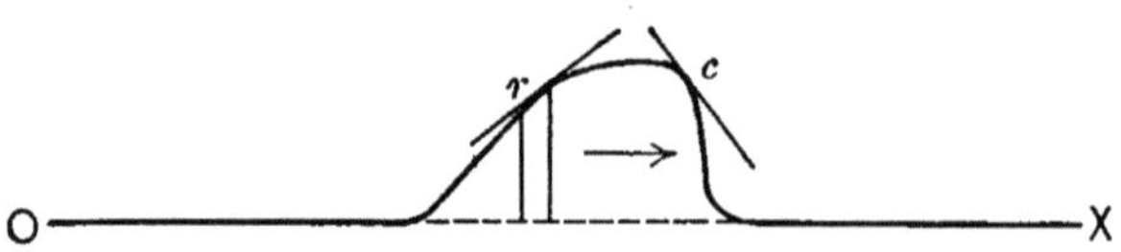

FIG. 145. — A solitary wave in air, compression leading.

Figure 146 represents a wave in air which might have been produced by displacing the piston suddenly to the left. Here the rarefaction leads and the displacement of the particles is in a direction just opposite that of the wave motion. A solitary wave of the type of either Fig. 145 or Fig. 146 is generally called a **pulse**; so also is the blow which produces it.

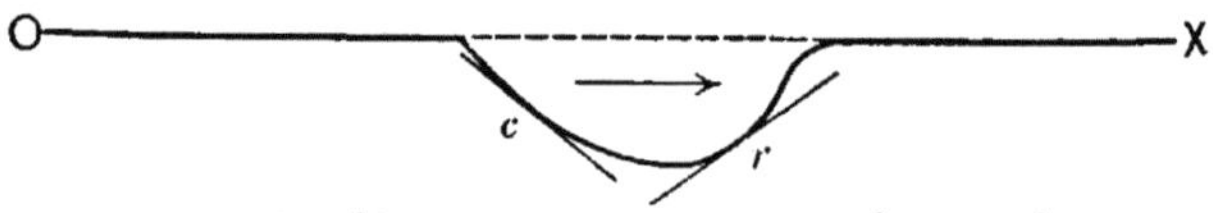

FIG. 146. — A solitary wave in air, rarefaction leading.

IV. REFLECTION OF SOUND WAVES

210. The echo from a hillside or from a large building is familiar to every one. But reflection of sound is more conveniently shown by placing a watch in the focus of a parabolic mirror, as indicated in Fig. 147. The mirror thus reflects a beam of sound, so to speak, which may be directed to any portion of the room. This also illustrates the fact that the direction of the reflected beam is determined by the same rule as in the case of a billiard ball reflected from the cushion of the table or as in the case of water waves (Fig. 122) striking a pier; namely, **the direction of the incident sound and the direction of reflected sound make equal angles with the normal to the reflecting surface.**

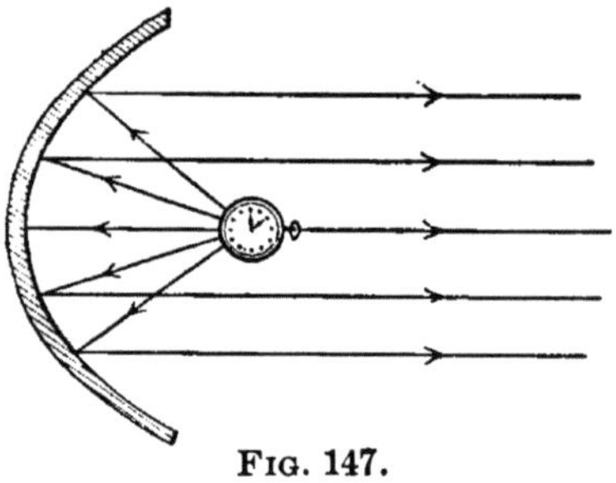
FIG. 147.

211. But in many cases

The Phase of the Reflected Wave

is also a matter of great importance — especially in Kundt's tube. Here reflection occurs at the face of a solid piston; hence *the condition of reflection is that there shall be no motion of the air particles* which lie immediately against the wall. The graphical method studied above tells us at once that the reflected wave must be both the perversion and inversion of the incident wave as shown in Fig. 135. For this is the only wave form which will compound with the incident wave and travel in the opposite direction, yet always give a zero displacement at *B*.

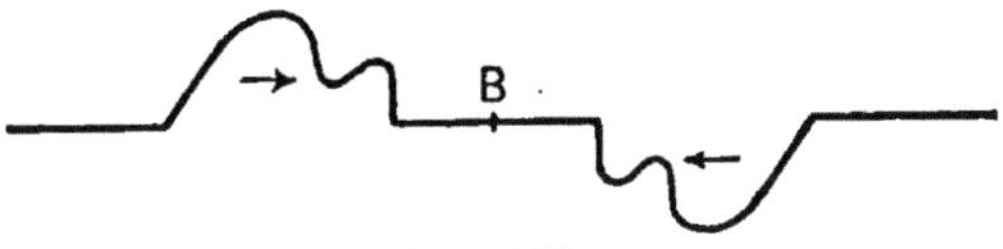

Fig. 135.

Observe, however, that the slope of the resultant at *B* is far from zero. The value of the tangent is sometimes positive, sometimes negative, sometimes zero, showing that while the *motion* at *B* is zero, the *pressure* there is continually varying.

Curiously enough the waves in the Kundt's tube are reflected also when the fixed piston is removed, and the end of the tube is left open. Here the boundary condition is changed; for now the air at the end of the tube is perfectly free to move to and fro, but the pressure is invariable, namely, that of the atmosphere.

The graphical method tells us now that the reflected wave must be such as to keep the tangent of the resultant wave always zero. From which it follows that the reflected wave is the inversion of the incident, as shown in the accompanying figure.

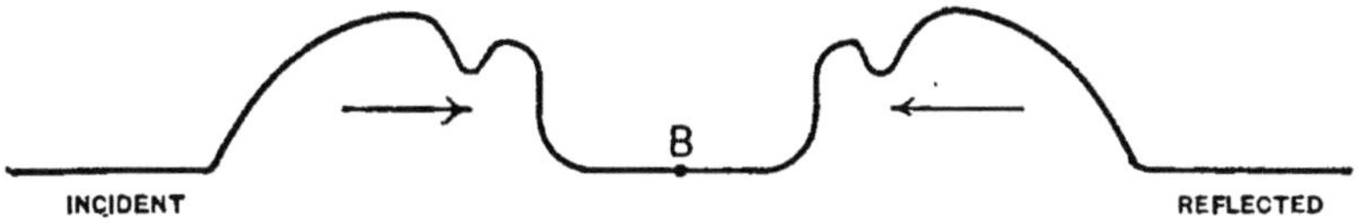

Fig. 148. Condition of reflection from open end.

212. One of the most highly instructive cases of reflection is that which occurs when a single sharp disturbance of the air is produced in front of a long flight of steps or near a fence constructed of vertical bars. From each step or bar is reflected a little wavelet; but since the distances of the steps from the ear increase in arithmetical progression, it follows that these wavelets will reach the ear as a succession of pulses—a train of waves in fact. The result is then that a single incident disturbance is reflected as a definite musical tone, whose frequency is determined by the distance apart of the bars.

About the campus of Northwestern University there is an iron fence, built of square bars placed $6\frac{1}{2}$ inches apart. On any evening when the air is acoustically clear, every footfall on the concrete walk which runs alongside the fence is reflected as a musical note whose pitch is about two octaves above middle C on the piano.

This phenomenon, which seems to have been first recorded by Thomas Young in 1813, has a very important bearing upon the explanation of what we call "white light," a difficult inquiry which must be reserved for much later study.

V. Speed of Sound

213. The interval which elapses between the fall of a distant tree or stone and the noise which reaches the ear must have been observed in primitive times. But the first accurate measure of the speed of sound was made by a commission appointed by the French Academy of Science in 1738. Some five or six stations near Paris were selected ; a cannon was placed at each station ; observers at each of the other stations recorded the interval between seeing the fire and hearing the report of one of the guns. By combining observations made in diametrically opposite directions from the gun, they eliminated the effect of the wind. The results obtained were, when reduced to a temperature of 0° C., 332.9 meters per second. These experiments were suggested largely by the fact that Newton (*Principia*, Book II, Proposition 50, Scholium) had some years previously computed the speed of sound from purely dynamical principles.

By methods whose details would here occupy too much time he concluded that the speed with which a compression or rarefaction travels through the air is given by the expression

$$V=\sqrt{\frac{E}{D}}, \qquad \text{Eq. 85}$$

where V = speed of the disturbance,
E = voluminal elasticity of the air,
D = density of the air.

In Newton's time the pressure of the air (approximately 1.014 million dynes per square centimeter) was thought to represent its elasticity. For the density of air, Newton employed the approximately correct value $\frac{1}{870}$ grams per cubic centimeter. These data gave him a value of V which was nearly 20 per cent smaller than the observed values. In the meantime, 1816, Laplace had explained this discrepancy by pointing out that the rarefactions and condensations which occur in sound waves are so rapid that there is not sufficient time for equalization of temperature between them, and that the value of E to be employed in Eq. 85 is that which represents the elasticity of air when no heat is admitted to it or leaves it. This elasticity was measured and was found to be 1.41 times greater than when the temperature of the air is maintained constant. Introducing this correction factor, which is generally indicated by γ, and denoting pressure by p, Laplace's expression for the speed of sound becomes

$$V=\sqrt{\frac{\gamma p}{D}}=\sqrt{\frac{1.41\,p}{D}}. \qquad \text{Eq. 86}$$

At this point Laplace requested the French government to make some new measures of this quantity, which they did in 1825, by a more refined application of the same method as employed by the French Academy. Their result (331.36 meters per second at 0° C.) compared very favorably with that predicted by Laplace. More accurate determinations have been made since, of which a fair average value would probably be 332 meters per second at 0° C.

Laplace's expression carries with it two interesting consequences. (i) From Boyle's Law (Eq. 74) we know that $\frac{p}{D}$ does not vary at any one temperature. *Hence as the barometer rises and falls one does not expect the speed of sound to change;* for the density is altered in the same ratio as the pressure.

(ii) But when the barometer remains constant, and the temperature changes, we should expect the density of the heated air to be less than that of the cooler, and consequently *the speed of sound to be greater in warm air than in cool.* And, indeed, the fact has been well established that, for atmospheric temperatures, the increase of velocity is about 60 centimeters per second per degree centigrade.

Speed of Sound in Solids

214. When the speed of sound in air is known, Kundt's tube offers us a beautiful method of finding the speed in the rubbed brass rod. When the brass rod is set into longitudinal vibrations we know that, whatever the length of the wave which travels through the brass rod, the two ends of the rod are antinodes; for they are free, and when compared with other parts of the rod they are places of great motion. The middle point likewise must be a node, because it is clamped, and is hence a place of no motion. Not only so, but it is the *only* node in the rod. The ends of the rod are then successive antinodes; accordingly the distance between them is one half the wave length of the disturbance in the brass.

Let us denote by N the frequency of the brass rod, and bv L its length; then, employing the fundamental equation of wave motion, § 181, —

$$\text{Speed in brass} = 2\,NL.$$

Whatever the frequency of the vibrations in the column of air, we know *it is the same as the frequency of the brass rod.* Let us call the distance between successive dust heaps d; then

$$\text{Speed in air} = 2\,Nd.$$

Hence
$$\frac{\text{Speed in brass}}{\text{Speed in air}} = \frac{2\,NL}{2\,Nd} = \frac{L}{d}.$$

$\therefore$ Speed in brass $= (L/d)$ 332 meters per second.

Accordingly, we have to measure only the distance between the dust heaps and the distance between the ends of the brass rod.

Evidently the same method may be employed with steel, glass, wood, and other substances.

Speed of Sound in Gases

215. Most useful of all is Kundt's tube in determining the speed of waves in various gases. For this purpose a glass rod (G, Fig. 149), held rigidly at its middle point by being corked and waxed into the outer tube, is better than the brass rod.

The movable piston rod at the other end must be reasonably gastight. Into each end is sealed a tube through which various gases can be introduced and removed. Measure the distance between the dust heaps when the tube is filled with air. Denote this distance by l. Now fill the tube with ordinary coal gas. The dust heaps are much farther apart than before. Call that distance L. Let N be the frequency of the glass rod. Then, since the fundamental equation of wave motion always holds, we have

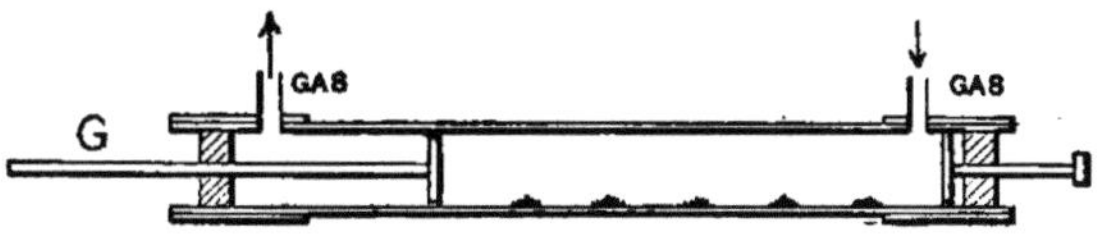

FIG. 149. — A form of Kundt's tube for use with various gases.

$$\text{Speed in coal gas} = 2\,NL,$$
$$\text{Speed in air} = 2\,Nl,$$

and hence, eliminating N, the numerical value of which we do not know,

$$\frac{\text{Speed in coal gas}}{\text{Speed in air}} = \frac{L}{l}, \qquad \text{Eq. 87}$$

where both quantities in the right-hand member are easily measurable. The process for other gases is essentially the same.

When once the speed of sound has been measured for any particular gas, Laplace's equation (86) furnishes a convenient means of determining the value of γ, a constant of great importance in the theory of heat.

Sound Waves in Three-dimensional Space

216. Hitherto we have been considering waves in air as they travel through tubes, say through a speaking tube; but in ordinary conversation we stand in an open space, surrounded

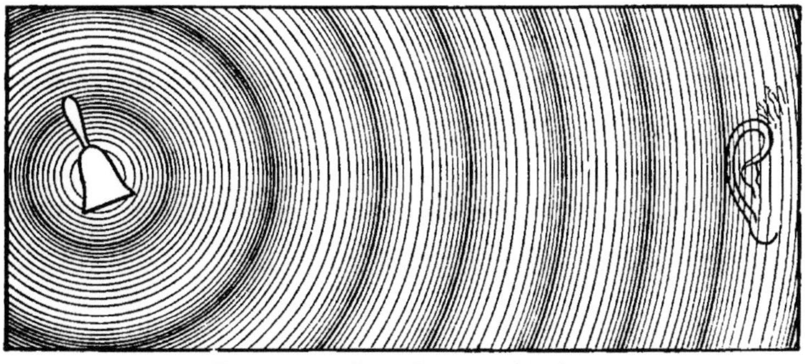
FIG. 150. — Sound waves in air.

by air on all sides; the voice is heard by those at our back, also by those in front, and by those at our side. The waves produced by our voice evidently travel away from us in all

Speed of Sound

Substance	Temp. C	Speed, meters per second	Speed, feet per second	Authority
Air	0°	332.5	1091	Kayser
Air	—45.6	305.6	1002	Greely
Alcohol	8.4	1264	4148	Martini
Aluminium		5104	16740	Masson
Brass		3500	11480	Various
Carbon dioxide	0	261.6	858	Dulong
Coppe	20	3560	11670	Wertheim
Glass from		5000	16410	Various
Glass to		6000	19690	Various
Hydrogen	0	1286.4	4221	Zoch
Illuminating gas	0	490.4	1609	Zoch
Iron and soft steel		5000	16410	Various
Oxygen	0	317.2	1041	Dulong
Water	13.7	1437	4714	Martini
Wood:				
Oak along fibre		3850	12620	Wertheim
Pine along fibre		3320	10900	Wertheim

Smithsonian Tables, pp 100, 101.

directions. As the pebble dropped into smooth water sends out rings of disturbance, the voice sends out spheres of disturbance. These expanding spheres are propagated in the same manner as waves in tubes, they travel at the same rate; the displacement of air particles is longitudinal. These spherical waves are therefore fairly represented in Fig. 150.

As an illustration of a wave motion propagated through a volume (*i.e.* a three-dimensional wave) one may take the remarkable explosion which occurred at Krakatoa on the 26th and 27th of August, 1883.

In this volcanic eruption a great column of gas was shot up to a height of perhaps 20,000 feet. When this mountain of air and other gases settled down, it sent out an air wave through the rest of the earth's atmosphere. This air wave expanded in circles. It girdled the whole earth, converging at the antipode of Krakatoa. Passing over any point of the earth's surface, it compressed the underlying air; the height of the barometer was accordingly increased; the wave of depression following had just the opposite effect. In this way all the recording barometers on the earth were affected. And by such means this great disturbance was followed through three trips from Krakatoa, near the island of Sumatra, to the antipodes. The entire circuit of the globe was performed in the almost incredibly short interval of 36 hours.

Compute the speed of this wave and compare it with the speed of sound as given in the preceding pages.

217. Preceding is a table containing the speed of sound in a few of the more important substances.

VI. Distinction between a Noise and Musical Sound

218. It is not a violent assumption to suppose that every student recognizes at once the difference between a musical tone and a noise — even though individuals may classify differently the performance of a bagpipe or of a hurdy-gurdy. But the problem which confronts us here is to distinguish between the character of two mechanical disturbances, one of which produces music, the other noise. How do the vibrations of a wagon rumbling over cobblestones, of the splash of water, or of footsteps on a pavement, differ from those of a wire in a piano? To answer this question intelligently one must first discover some means for accurately describing the vibrations emitted by various bodies.

The Manometric Flame

219. So far as concerns a picture of sound waves in the air, nothing can surpass in beauty and simplicity the "manometric flame," which was invented and perfected by Rudolph König, of Paris. The construction of the instrument will be clear on consideration of Fig. 151.

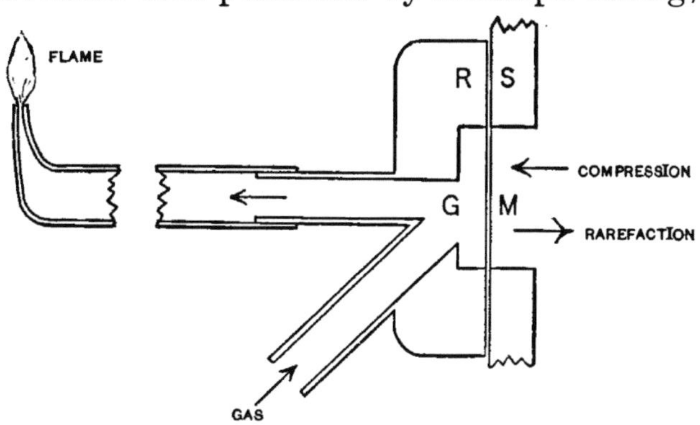

FIG. 151. — König's device to show variation of pressure produced by sound waves.

Two blocks of wood, *R* and *S*, are bored out. Between them is clamped a diaphragm of very thin goldbeater's skin. This diaphragm divides the whole apparatus into two parts. On the left-hand side of the diaphragm illuminating gas is admitted, as indicated by the arrows, and so long as the diaphragm is undisturbed, the flame is apparently undisturbed. But if the disturbance of the flame is very rapid, as is the case with acoustical disturbances generally, we might not be able to detect by the eye any changes in the flame. Accordingly a rotating mirror is placed near the flame. A crank enables the observer to give this mirror a moderate angular speed, say two or three turns a second. For every new position which the mirror assumes, the image of the flame takes a new position. The effect of rotation on the mirror is, then, to separate the images of the flame at successive instants.

The space on the right of the diaphragm *M* is one end of a speaking tube, a sort of *cul de sac*, into which all kinds of sound waves can be sent for examination and study.

The effect of a sound wave (if our previously formed conception of sound waves be correct) upon the diaphragm will be to drive it alternately forward and backward. When a compression arrives at the diaphragm, the membrane will be thrust slightly to the left. The gas space *G* will suddenly be made smaller in volume; the pressure of the gas will as suddenly rise; the flame will immediately become higher — just as if more pressure had been put on at the gas works.

But this compression in the speaking tube will be followed by a rarefaction. What is the effect of this rarefaction on the membrane? The pressure on the right diminishes, the membrane moves to the right, the pressure in the gas chamber *G* suddenly diminishes; the flame as suddenly diminishes in height.

The essential features of the instrument are, then: —

1. A membrane which responds at once to any displacement of the air particles adjoining it; *i.e.* to any wave of compression or rarefaction.

2. A gas flame so arranged as to respond to every motion of the membrane.

3. A mirror that can be rotated so as to separate the successive images of the flame, and thus allow us to detect the variations of pressure at the membrane due to any disturbance of the air.

Since now the height of the flame measures, in a general way, the displacement of the particles adjoining the membrane, we need only draw an outline of the image of the flame in the rotating mirror, and we have at once a graphical representation of a sound wave. Such are the following diagrams (see Fig. 154). They merely record the variations of pressure produced by three different trains of waves at the membrane of the manometric flame. As the name indicates, such a flame is simply a quick-acting harometer, responding to changes of pressure which occur at rates even as high as several hundred per second.

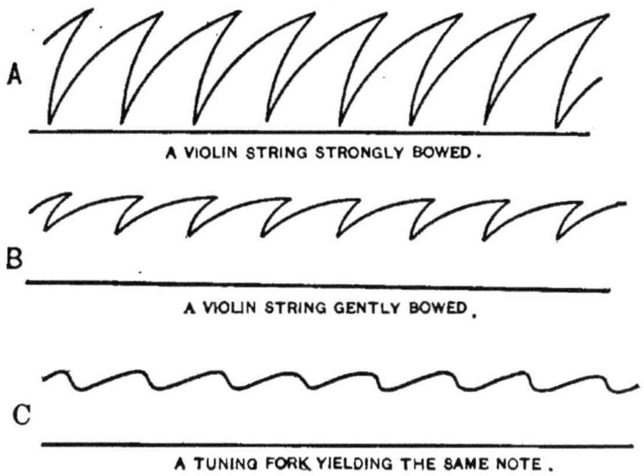

FIG. 154.

220. If now one examines with this instrument a number of noises and musical tones, he finds that corresponding to these two different sensations, the manometric flame shows two different kinds of disturbances in the air. These are exhibited fairly well in Fig. 155. The noise is made up of many small parts of trains of waves, all jumbled together. But the flame shows that a musical note is produced by *one* train of uniform waves.

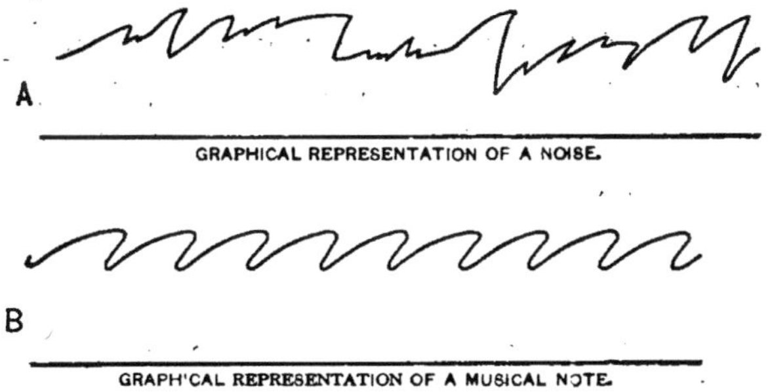

FIG. 155.

It shows also that vowels, as employed by singers, are musical tones, while consonants are essentially brief noises.

"**The sensation of a musical tone is due to rapid periodic motion of a sonorous body; the sensation of a noise to non-periodic motion.**" — HELMHOLTZ, *Sensations of Tone*, p. 8.

VII. Three Features by which Musical Tones are distinguished from Each Other

A. Loudness

221. When the key of a piano is struck gently, and again with a hard blow, we recognize that the two notes thus produced are alike except that the second is *louder* than the first. The vibrations of a guitar string when first struck are sufficiently large for us to see them; but as the vibrations diminish in amplitude, we observe that the note diminishes in *loudness*.

When a note is sung into the tube of a manometric flame, it is at once evident that the indentations in the ribbon of flame are deeper in proportion as the note is louder (see Fig. 156). That is, the louder the sound, the greater the displacement of the air particles against the membrane.

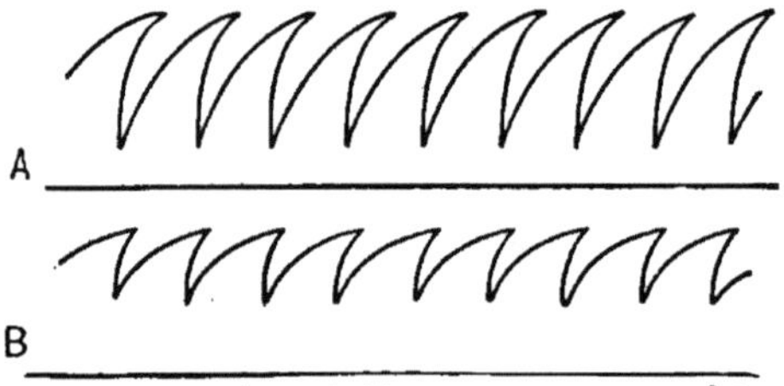

Fig. 156. — *A* and *B* are two notes alike in all respects, except that *A* is louder than *B*.

From these, and similar experiments, it is concluded therefore that **the loudness, or intensity, of a sound increases or diminishes as the amplitude of the vibrating air particles at the ear increases or diminishes.**

Since the energy of a vibrating body is, other things being equal, proportional to the square of its amplitude, loudness evidently increases with energy of vibration in the source. The variation of loudness with distance is an immediate consequence of this fact. For when a sound wave spreads out as indicated in Fig. 150, § 216, the energy emitted by the source during any minute interval of time is, at any later instant, spread out over a spherical surface which has the source at its center. Now the area of this spherical surface must increase as the square of its radius and hence the quantity of energy which is incident upon the ear at any given instant must vary inversely as the square of this distance, a fact with which every one is familiar.

Digression on the Effect of Wind upon the Intensity of Sound

222. When a spherical wave is propagated in the manner just described, it is customary to describe the continuous locus of all air particles having the same phase as the wave front. When an observer is at some distance from the origin of the sound, the wave front which strikes his ear is practically plane. Suppose that a wind is blowing in the same direction in which the sound is proceeding. One might expect the apparent speed of sound to be slightly increased; or if the wind be traveling against the sound, he would find the speed slightly diminished. But it is a matter of common observation that the audibility or loudness of sound coming from any considerable distance is also dependent upon the direction of the wind. A sound is said to "carry" better with the wind than against it.

The explanation of this phenomenon was first given, in 1857, by Stokes of Cambridge. He argued that since the upper layers of the air are always carried along more rapidly by the wind than those next the surface of the earth, a vertical plane wave front traveling with the wind will, as it advances, become more and more inclined toward the surface of the earth, as shown in Fig. 157. In this manner the sound is, so to speak, continually brought down to the ear of the observer. But if the sound be traveling against the wind, the upper portions of the advancing plane wave front will be bent backwards so that the sound is directed somewhat upwards and above the listening ear as shown in Fig. 158.

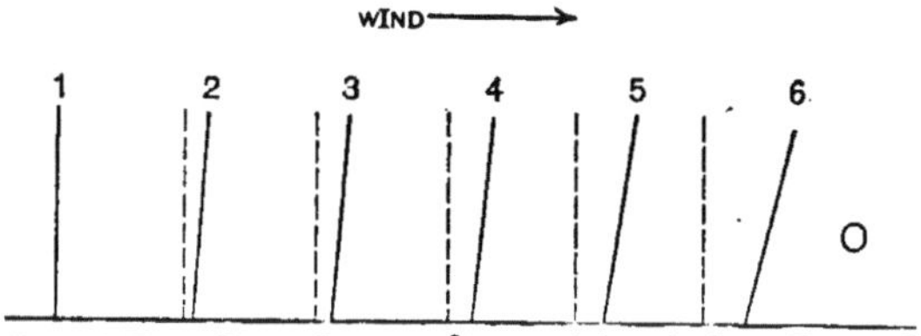

FIG. 157. — Sound traveling with wind. Dotted lines show successive positions of wave front in still air.

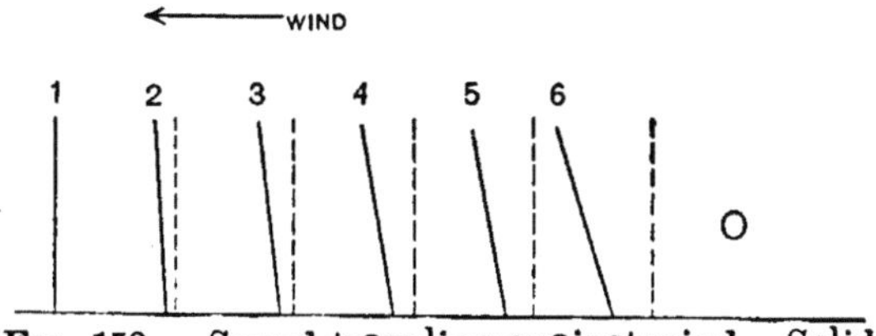

FIG. 158. — Sound traveling against wind. Solid lines indicate actual wave fronts.

If we remember that sound always travels in a direction perpendicular to the wave front, it will be clear that the observer is here left in a sort of sound shadow, all that reaches him being what little is diffracted in his direction.

223. Another factor in the loudness of sound is the presence of some large elastic body with which the vibrating body is rigidly connected. Thus a violin string when removed from any sounding board gives a thin, weak sound; but when stretched over the body of the violin, the sound is full and resonant. The explanation of this phenomenon, which is as follows, was also given by Stokes (*Papers*, Vol. IV, p. 299): —

When a solid body moves through the air slowly, the air flows from in front toward the rear, almost exactly in the same manner as if the solid were moving through an incompressible fluid; but when the motion becomes very rapid, the air in front is to some extent compressed, and equalization of pressure between the front and rear does not occur in exactly the same manner as before. If, however, the vibrating body be thin and small, as a piano wire, the equalization of pressure between front and rear is very much facilitated. But in the case of a broad flat body, such as the sounding box of a guitar, there is less chance for the lateral escape of the air and hence, corresponding to every slightest vibration of this board, a wave of condensation and rarefaction is emitted. For the air can no longer behave as an incompressible fluid unless the frequency of vibration be very much less than that employed in music. Thus, although the amplitude of the sounding board is much less than that of the string, the actual amplitude at the ear is enormously increased by its use. In the reënforcement of sound in this manner the two fundamental factors are, then, the size and frequency of the vibrating body.

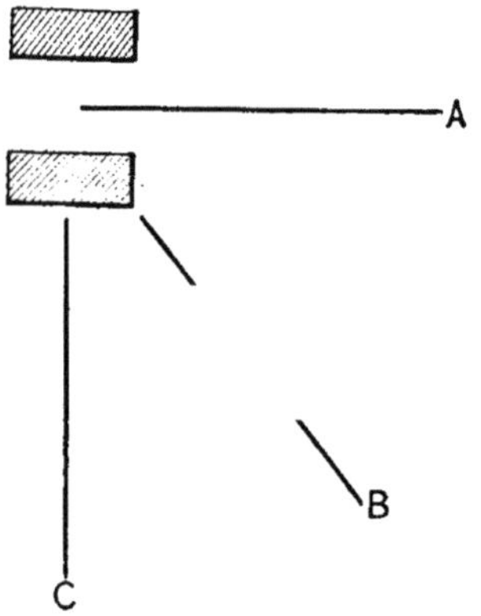

FIG. 159. — Stokes's explanation of effect of sounding board.

Stokes illustrates this principle most beautifully by holding a tuning fork near a piece of cardboard in the position indicated by the plane B in Fig. 159, where the two small rectangles represent a horizontal cross section of the fork. If the cardboard be placed in the positions A or C, the air in front of the vibrating prong has no difficulty in getting around to the rear, but when the fork is held near the edge of the cardboard, as in B, these side currents are hindered, and waves of compression and rarefaction, instead of annulling each other, go to increase the intensity of the sound.

B. Pitch

224. On clamping a piece of steel or brass spring, say a piece of clock spring, in a vise, one observes not only that the spring vibrates more rapidly as it is made shorter, but also that it *emits a higher note*. (See Fig. 160.)

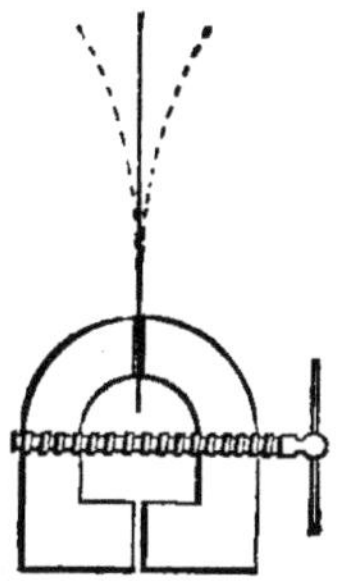

FIG. 160.—Illustrating the fact that pitch varies with frequency.

In the case of a wire, whether stretched from one fence post to another or between two fixed pins in a piano, there is no doubt but that it vibrates more rapidly the more it is stretched; we observe also that the note it emits is higher and higher the more it is stretched.

From such experience we might suspect the pitch of a note to depend upon the frequency (§§ 49, 181) of the vibrating body. But we need more direct and conclusive evidence to thoroughly establish such a conclusion. A card held against a rotating toothed wheel will furnish us just such evidence. Here we compel the card to make one vibration every time it passes a tooth. The frequency of the vibrating card is the number of teeth passed per second. When the wheel slows up, the pitch of the note produced by the card falls; when the wheel increases in speed, the pitch of the note rises.

If still further evidence be desired, try the following experiment: —

Into the speaking-tube of a manometric flame sound two notes of different pitch in rapid succession. Or, what is better still, use two manometric flames placed near each other and viewed in the same rotating mirror. It is then easily seen that when the flames are excited by notes of the same pitch, the notches in the image of each flame are at the same distance apart.

But if either flame be excited by a note of higher pitch, the notches on this flame are closer together. And if one note be an octave above another, we see that the notches are exactly twice as close together in the high note as in the low. (See Fig. 161.)

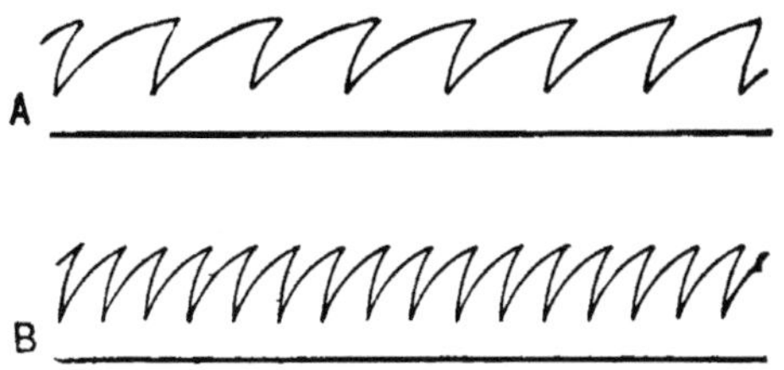

FIG. 161. — *A* and *B* are two notes alike in all respects except pitch.

Such evidence leads us to think that the pitch of any tone depends simply upon the

frequency of the vibrating body which produces it; or, what is the same thing, upon the period of the vibrating body.

For, if we denote the frequency by n and the period by T,

$$n = \frac{1}{T}$$ Defining equation for frequency.

The numerical value of any pitch is the frequency n. A tuning fork which makes three hundred vibrations per second is said to have "a pitch of three hundred."

225. There are many laboratory methods for the actual determination of the pitch of sound, most of them based upon the characteristic equation of wave motion; namely,

$$V = nl. \qquad \text{Eq. 76}$$

The following must suffice as a type of them all. Take a small whistle, such as that devised by Galton, for the source of sound whose pitch is to be measured. This source is placed at one end of a Y-shaped tube which opens into two parallel tubes. These parallel tubes, at their farther end, again unite into a Y-shaped tube, as shown in Fig. 162.

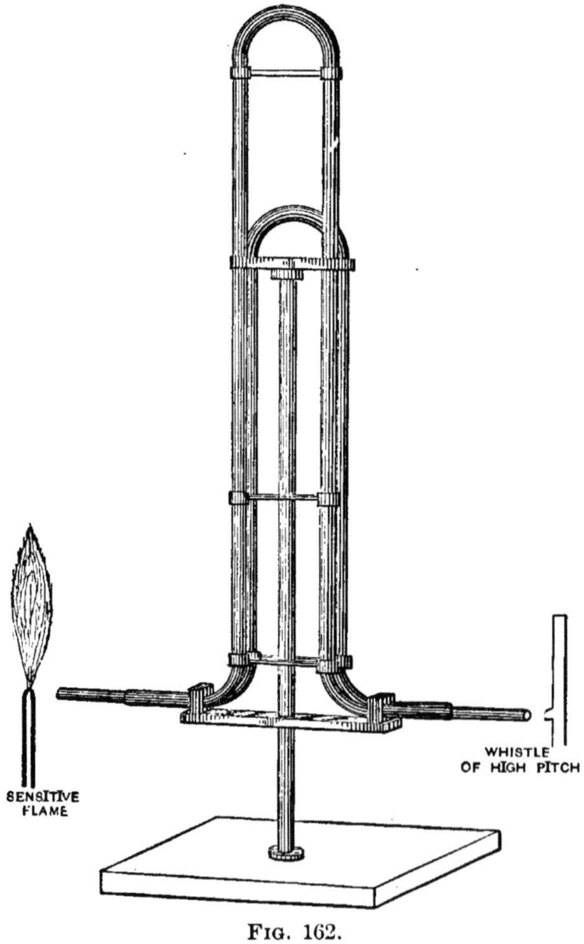

FIG. 162.

If, now, a gas flame which is just on the point of flaring be placed at the end of this second Y, it will serve as a very sensitive means for detecting any disturbance in the air at that point. The U-shaped portion of one of these parallel tubes is so arranged as to telescope upon itself, somewhat after the fashion of a trombone.

In this manner the length of path from the whistle to the flame is constant through one tube and variable through the other. At the first Y, the wave which leaves the whistle is divided into two parts. These two parts reunite at the second Y, but the phase in which they meet depends upon the difference of length between the two routes which they travel. If this difference is a whole number of wave lengths, the two waves will reënforce each other ; but if their path differs by an odd number of half wave lengths, they will interfere and produce silence. As the telescope is moved along, these silent points will be indicated by the sensitive flame. If now we measure the difference in length of the variable tube between two successive silent points, we shall have the wave length of the tone which is emitted by the whistle. Knowing the velocity of sound in air V, we may now easily compute n from Eq. 76.

One advantage of the sensitive flame is that it may be used even when the pitch of the whistle is so high as to be inaudible to the human ear.

C. *Quality*

226. But two notes may have the same pitch, and may have the same loudness, and yet be quite different. Two tenor voices singing the same note with equal loudness may be easily distinguished ; for, as we are accustomed to say, they have different qualities. Strike the same note upon a piano and upon an organ; — let them be equally loud ; no one finds any difficulty in distinguishing the piano from the organ. Two of our friends may converse in the same key and with equally loud voices, but we do not hesitate in telling their voices apart, even though they stand at our back, or perhaps in the dark. Their voices have a different quality. These differences are most strikingly illustrated in the various instruments of the orchestra, where we recognize with great ease just which instruments are being used. A guitar string when plucked at its middle emits a note of different character from that of the same string when plucked near one end.

How shall we explain these differences, so familiar to every one ? Difference of amplitude produces difference of loudness.

Difference of frequency produces difference of pitch. But what produces difference of quality?

The most direct answer to this question is obtained by means of the manometric flame. Into the speaking tube of such a flame pronounce as nearly as possible, with the same pitch and loudness, each of the vowels of the alphabet. It is observed at once that the ribbon of flame seen in the mirror is different in each case; to each particular vowel corresponds a wave of different shape. The notches in the ribbon have not the same form for any two vowels. **It is then the form of the wave which determines the quality of any note.**

It is well to remember that, while to every different quality there corresponds a different form of wave, Helmholtz has proved that the converse of this is not true. Quality depends simply on what components combine to produce the wave, not at all upon the *phases* in which they combine.

In Fig. 163 is shown the appearance of two flames each excited by a note of the same frequency and same loudness, but of different quality. In this case we see not only that the form of the wave in A is different from that in B, but we see just *how* it differs.

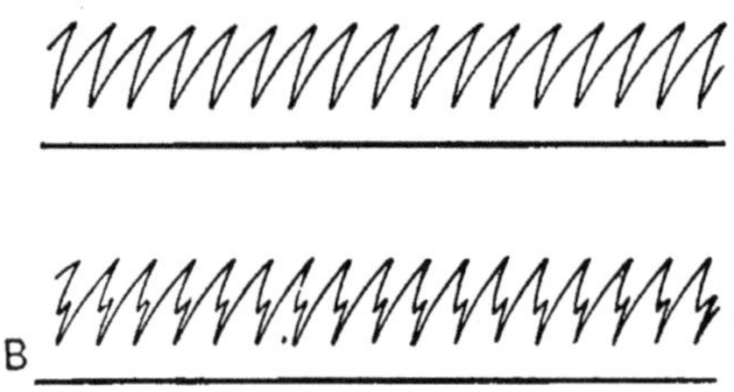

FIG. 163. — Two notes of the same pitch and loudness, but of different quality.

In B there is present a small wave having twice the frequency of that of A. So that A represents the fundamental, while B represents a note in which both the fundamental *and* the octave are present.

In the case of the vowels which we have just examined, this difference of wave form is brought about by changes in the cavity of the mouth. Pronounce the vowels *a* and *o* in succession, and observe the manner in which you change the air space inclosed within your mouth during the interval between the pronunciation of the two vowels.

In general, experiment shows that the wave form which determines the quality of any tone depends upon the presence of vibrations more rapid than the fundamental, the fundamental being defined as the lowest note which a body can give out. These other and higher notes are called **overtones** or **harmonics**.

227. The presence of these harmonics in a piano wire may be easily shown by holding down a key of, say, middle C, and then striking several times a note one or two octaves below. It will then be found that when the vibrations of the lower note have been damped out, the upper note is still sounding, owing to resonance between it and one of the harmonics of the lower note.

The resonators devised by Helmholtz (which are simply hollow spheres tuned to respond to certain definite tones), when connected with a manometric flame, form a very sensitive means for detecting those harmonics which have the same frequency as the resonators.

The existence of these overtones may easily be shown in any stretched string. Damp the string at the middle point. If now it be bowed at a point one quarter of the length of the string from either end, the string will vibrate as shown in Fig. 164, *B*, and the octave of the fundamental will be heard. Evidently, therefore, the first overtone in a stretched string is the octave of the fundamental. For (§ 215), $V = 2nL = 2n_1L_1$, where n is the frequency of the fundamental and n_1 is the frequency of the first overtone; and since $\frac{L}{L_1} = 2$, it follows that $\frac{n_1}{n} = 2$. It may be shown in like manner that the pitch of the next overtone is to that of the fundamental as 3 : 1. The third overtone is two octaves above the fundamental, and so on.

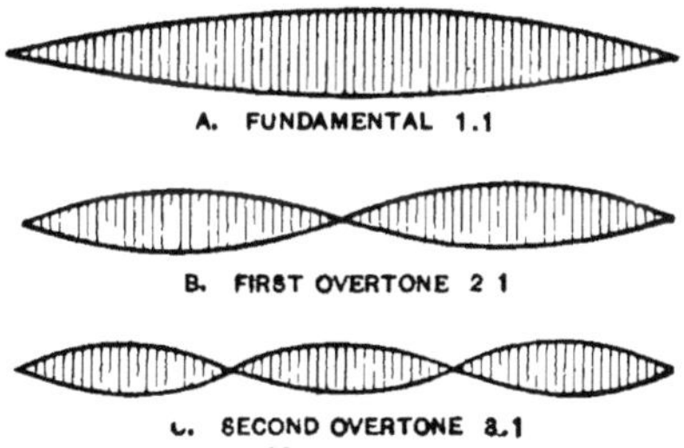

Fig. 164.

A better method still of showing these overtones is by aid of two open organ pipes (Fig. 165), one of which is an octave above the other. When both are sounded at the same time, the first overtone of the lower pipe can readily be made to beat with the fundamental of the higher pipe. It is these higher and feebler tones which give the distinguishing quality to any particular note.

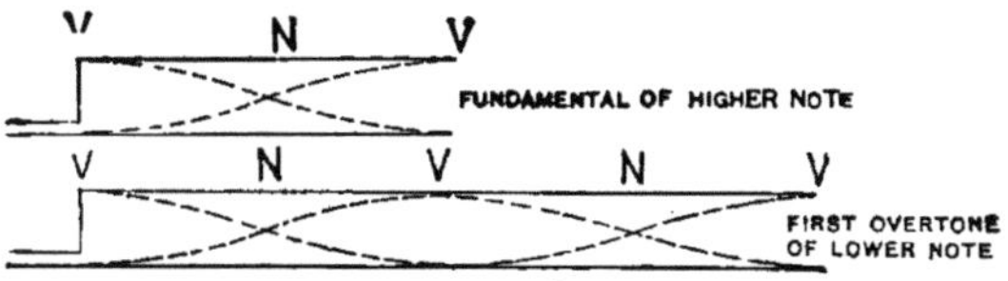

Fig. 165. — Showing wave length of overtone in one pipe equal to wave length of fundamental in other.

228. Here is perhaps the place to mention a useful distinction between the terms "note" and "tone." When, for instance, a violin string is bowed, the whole musical sound emitted is called a **note**. But, as we now see, each of these notes is made up of a series of irresolvable notes which are called **tones**. The lowest of these tones is the one which determines the pitch of the note, and is called the **fundamental**; the rest are called **overtones** or **upper partials**. If the frequencies of these overtones are exact multiples of the frequency of the fundamental, they are called **harmonics**.

Digression on Beats

229. When two notes are slightly out of unison, they have slightly different wave lengths in air. The consequence is that they reach our ears with a difference of phase which is constantly changing. At one instant, they interfere in such a way as *almost* to produce silence; at the next instant, they combine in such a way as to produce a sound louder than that of either alone.

Most people are familiar with the peculiar rise and fall that attends the sound of a large bell after the clapper has struck it for the last time. This is due to the interference of two different segments of the bell, such as A and B in Fig. 166, which travel around the rim owing to lack of perfect symmetry or homogeneity in the metal. The same phenomenon is easily shown with a glass bell or an iron ring.

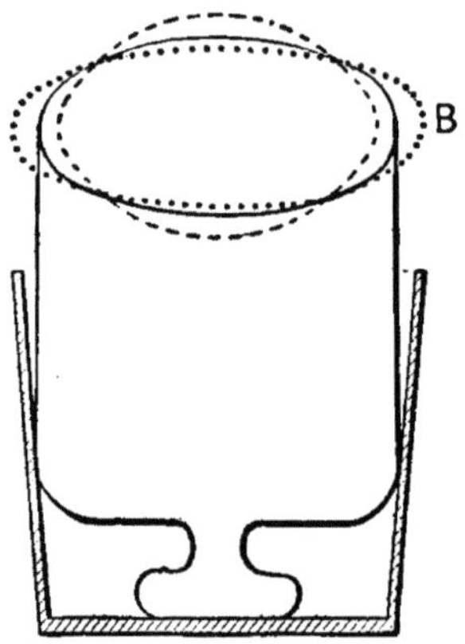

FIG. 166.—A vibrating bell jar: really equivalent to four different vibrating bodies.

Two tones which interfere in this way to produce this alternate rise and fall of intensity are said to **beat**. These alternations of loudness are called **beats**.

This phenomenon is most readily shown by means of two tuning forks which are in unison. If the forks are in exact unison to begin with, a little wax placed on the prong of either one will slow it up; for the wax adds to the inertia of the fork, and does not add to its elastic force. Consequently, whatever difference of phase may exist at any instant between the two forks, this difference will continuously change, and hence will follow a series of beats. The more wax placed on the prong, the more rapid the beats. How do you explain this? Two guitar strings are easily adjusted to unison. Tighten either one ever so little and they beat.

It is easily seen from the theory of waves in general that the number of beats per second is numerically equal to the difference in frequency between the notes beating. For the interval between two successive beats is simply the time taken for one train of waves at any point, say at the listening ear, to gain 360° (2π radians) in phase over the other train. Each train of waves travels at the same speed; but as these trains pass any one point their difference of phase is continually changing. If one note makes 500 vibrations per second and another 502 per second, it is evident that at any point the latter will gain in phase, over the first, 4π radians each second. These two notes will be in the same phase twice each second, and in opposite phase twice each second. They will therefore beat twice each second. **In general, if the frequency of one note is n and of another n', the number of beats per second will be $n - n'$.**

VIII. Relation of Tones to Each Other. The Musical Scale

230. Up to the present moment we have been studying the nature of a single train of waves in the air, the propagation of a single musical tone. We now proceed to consider some of the relations of musical tones to each other. Here we find ourselves already in the domain of **Music**.

But there are some experimental facts which must be viewed as a sort of connecting link between Music and Physics. These constitute the

Scientific Basis of Music

There are few ears which have the ability to determine, on hearing a single musical tone, just what the pitch of the note is. Frequency is a difficult thing for the human ear to determine. But there are *very many* observers who, on hearing *two* notes, can tell what the *ratio* of their frequency is. If two keys on a piano are struck, there is no difficulty in saying whether their pitches are related as 3 : 2 or 1: 2, or in some other ratio; but not more than one person in a hundred can make an intelligent guess as to the frequency (absolute pitch) of either note.

Definition of Musical Interval

231. The interval between two notes is defined as the ratio of their frequencies. Two notes having the same pitch, whatever their quality or intensity, are said to be in unison. The interval in this case is not zero, but unity. If the ratio be 2 : 1, the interval is said to be an

Octave

A note whose pitch (frequency) is 400 is, then, an octave *above* a note whose pitch is 200. A note whose pitch is 100 is an octave *below* one whose pitch is 200. The impressions produced upon the human ear by a note and its octave are strikingly similar. So alike, indeed, are the two sensations that many fairly well trained musicians sometimes mistake the one for the other. So far as the author is aware, this similarity has never been satisfactorily explained.

The Major Chord

232. Still more remarkable is the following fact concerning the human ear — a fact which has been well known for more than two thousand years, a fact which is fundamental in all music, but which, up to the present century, had baffled all attempts at explanation.

The fact is that any three tones whose frequencies are as 4 : 5 : 6 always make a pleasing impression upon the ear. This combination of notes is known as the major **triad**; or, taken together with the octave of the fundamental, is called the **major chord**. The universal judgment is that three such notes, whatever their frequencies, produce harmony when sounded together.

The explanation of this, which we owe to Helmholtz, is based upon the disagreeable impression resulting from beats.

Helmholtz likened the effect of beats on the human ear to the unpleasant sensation of a flickering light upon the eye.

When, however, the beats become *very* rapid, they cease to be disagreeable, —just as an electric light alternating in intensity ceases to be painful when the alternations are very rapid. Helmholtz showed that beats are most disagreeable when they occur at the rate of 30 per second; but they can be detected even when they run as high as 132 per second. He next showed that the physical basis for the harmony of the major chord lies in the fact that neither between notes whose frequencies bear the simple ratios 4 : 5 : 6, nor between their overtones, nor between their overtones and fundamentals, are there any disagreeable beats.

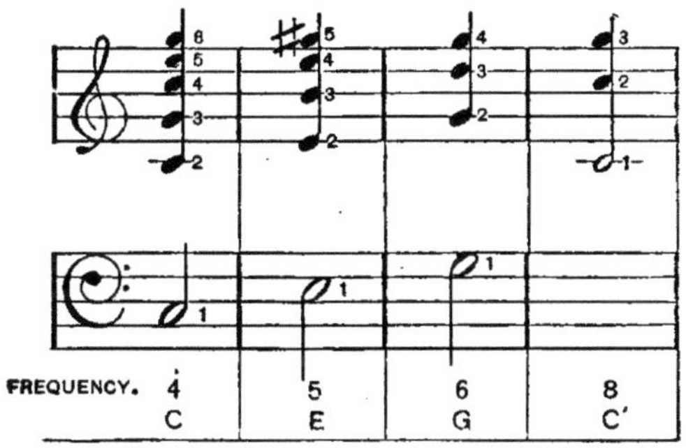

FIG. 167.—Major chord represented by half notes; harmonics represented by quarter notes.

In Fig. 167 are represented the totality of tones in the major chord up to the sixth harmonic.

Beginning with the fundamental C, it will be observed that none of the overtones are so near the fundamental or each other as to beat and produce dissonance.

Consider next the note E, whose frequency is five ; its fourth harmonic is in unison with the fifth harmonic of the fundamental. But the third and fifth harmonic of E are each more or less dissonant with the fourth and sixth of the fundamental.

When we come to the note G, whose frequency is six, we find only one harmonic, the third, which is not already present in the fundamental. This third beats with the fourth and fifth of the fundamental and distracts somewhat from the harmony. But the octave C′, it will be observed, introduces no harmonics whatever that are not already in the fundamental. This almost complete freedom from beats thus explains the remarkable harmony of the major chord.

The Major Scale

233. Musicians divide the interval between any note and its octave into a series of seven smaller intervals called **tones** and **semitones**. The notes which make up the octave are denoted by the letters of the alphabet, and are built up out of three major triads, as follows: —

$$C : E : G = 4 : 5 : 6.$$
$$F : A : 2C = 4 : 5 : 6.$$
$$G : B : 2D = 4 : 5 : 6.$$

Let us call the lowest note of the octave C, and denote its frequency by "m"; these proportions then amount to a series of equations which we may solve to find the frequency of any other note in the octave. Thus,

$$C = m.$$
$$C : E = 4 : 5. \qquad \therefore E = \tfrac{5}{4}\ m.$$
$$C : G = 4 : 6. \qquad \cdot\ G = \tfrac{3}{2}\ m.$$
$$F : 2\,C = 4 : 6. \qquad \therefore F = \tfrac{4}{3}\ m.$$
$$F : A = 4 : 5. \qquad \cdot\ A = \tfrac{5}{3}\ m.$$
$$G : B = 4 : 5. \qquad \cdot\ B = \tfrac{15}{8}\ m.$$
$$G : 2\,D = 4 : 6. \qquad \cdot\ D = \tfrac{9}{8}\ m.$$

Table of Musical Nomenclatures

Name of Note.	C	D	E	F	G	A	B	2 C
Frequency in terms of fundamental	m	$\frac{9}{8}\ m$	$\frac{5}{4}\ m$	$\frac{4}{3}\ m$	$\frac{3}{2}\ m$	$\frac{5}{3}\ m$	$\frac{15}{8}\ m$	$2\ m$
Intervals.		$\frac{9}{8}$	$\frac{10}{9}$	$\frac{16}{15}$	$\frac{9}{8}$	$\frac{10}{9}$	$\frac{9}{8}$	$\frac{16}{15}$
Name of note in Vocal music.	do	re	mi	fa	sol	la	si	do
Treble Clef								
Bass Clef								
Absolute Pitch of treble clef as used in international Scale	259	290	326	345	388	435	488	517
Scientific Scale	256	288	320	341	384	427	480	512

The preceding table indicates also the ordinary musical notation, a beautiful form of shorthand. The character employed tells the reader the time for which any note is to be sounded, while the vertical position of the note among the lines and spaces gives its pitch. The modern system of musical notation is, then, essentially that invented by Descartes in 1637 and universally employed in analytical geometry. **The bass and treble are connected by the fact that the C represented by the first line above the bass is identical with the C on the first line below the treble.** This note is called **Middle C** on the piano. Ordinarily its frequency is about 264.

234. The scale thus formed and represented in the preceding table is called the natural or diatonic scale. Between the successive notes it will be observed that there are three kinds of intervals; namely,

C–D, F–G, A–B, $\frac{9}{8}$ — called a major tone,

D–E, G–A, $\frac{10}{9}$ — called a minor tone,

and E–F, B–C, $\frac{16}{15}$ — called a major semitone.

If all music were written in the key of C, this diatonic scale would suffice. But on instruments with fixed notes, such as the piano and guitar, it would be impossible to pass to other keys without enormously complicating the instrument. To avoid this difficulty Bach (1685–1750) invented the so-called tempered scale, in which the octave is divided into twelve equal intervals, or half notes, each tone having a frequency $2^{\frac{1}{12}}$ that of the preceding. This interval is very nearly the semimajor tone. The student will now see why the black key on the piano is omitted between B and C, and also between E and F.

The names of the notes employed in vocal music are derived from the initial syllables of the following stanza, from an old Latin hymn: —

> "*Ut* queant laxis *re*sonare fibris
> *Mi*ra gestorum *fa*muli tuorum
> *Sol*ve polluti *la*bri reatum
> *Sa*nct *Jo*annes."
>
> Poggendorff, *Geschichte der Physik*, p. 802.

IX. Application of Preceding Principles to Various Instruments

Wind Instruments

235. Of these the organ pipe is perhaps the simplest and most typical. There are two kinds, *open* and *closed*. In each of them two trains of waves interfere to producé stationary waves. By an **open pipe** is meant one which is not closed at either end; by a **closed pipe** is meant one which is closed at one end only. These two types are shown in Fig. 168.

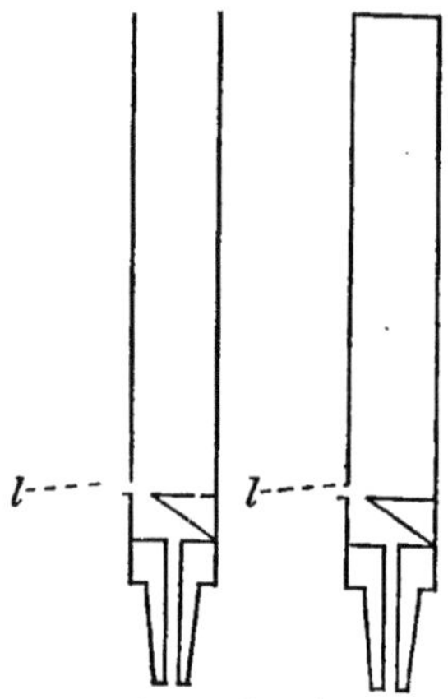

Fig. 168. — Section of open and closed organ pipes.

One train of waves is produced by blowing against the lip *l*, of the pipe; another train is produced by reflection from the upper end of the pipe.

In the case of the open pipe, reflection occurs when the incident train strikes the quiet air at the end of the pipe; in the case of the closed pipe, reflection occurs when the wave strikes against the wooden end of the pipe. In either case, the whole behavior of the air in the organ pipe is exactly that of a small portion of the air in a Kundt's tube (§ 215).

In the case of the open organ pipe, it is evident that each end is an antinode, *i.e.* a place where no compression occurs, but where great motion is always possible. Consequently, an open organ pipe is simply that portion of a Kundt's tube which lies between two consecutive antinodes, as shown in Fig. 169, where *N* denotes a node and *A* an antinode. It is evident, therefore, that the wave length of the fundamental is just twice the length of the open pipe which emits it.

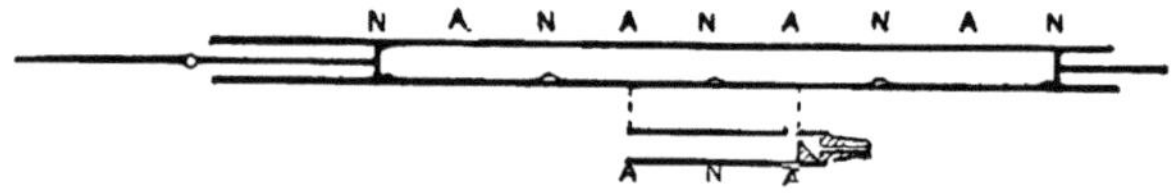

Fig. 169. — Correspondence between the open organ pipe and Kundt's tube.

As a matter of fact this rule is not quite exact; for it is found that the fundamental wave length is a little more than twice the length of the pipe. This is due to the fact that the inertia of the air surrounding the open end of the pipe has the effect of slightly increasing the effective length of the pipe, very much in the same way that the velocity of a rifle ball is found to increase even after the ball has left the muzzle of the gun. Just as the dead air about the muzzle of the gun forms a tube, so to speak, in which the gaseous products of the burning powder continue to expand, so the quiet air about the end of the organ pipe has the effect of practically extending the length of the actual circular pipe by about 0.6 of its radius, a quantity which is called the "end correction."

In the case of the closed pipe, no longitudinal motion of the air is possible at the stopped end; consequently, this end must be a node, while the lower (*i.e.* the open) end is an antinode. The correspondence is indicated by Fig. 170.

It is at once evident that the length of a closed pipe is one quarter of a wave length of its fundamental.

The student will find it interesting to prove that an open pipe yields the complete series of harmonics 1, 2, 3, 4, 5, 6, etc.; while the closed pipe gives only the odd notes 1, 3, 5, 7, 9, etc. Make a sketch, after the style of Fig. 165, showing the distribution of nodes and loops in both open and closed pipes.

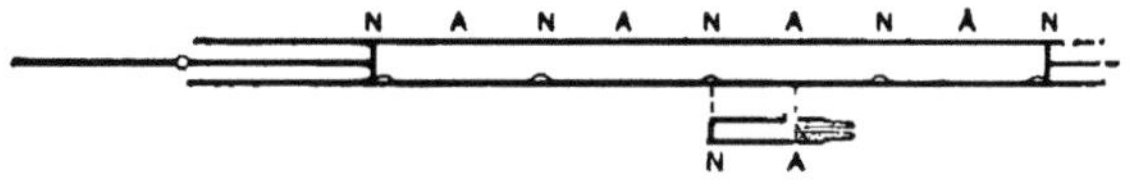

FIG. 170. — Correspondence between closed organ pipe and Kundt's tube.

A very direct means of experimentally locating the nodes and loops in an open pipe is to replace the bottom of a pill box with a thin stretched membrane, — say very thin paper, — place a little sand in the box, and lower it into the sounding tube, holding the membrane of the box horizontal. The sand distinctly announces the positions of the loops — becoming silent at the nodes.

To this class of instruments belong also the **flute**, the **clarinet**, the **cornet**, and the **trombone**. In the first three of these instruments the vibrating column of air is made to break up, by means of stops, so as to yield either the fundamental or overtones at the will of the player. The opening of a stop at any point on the tube insures an antinode at that point. While in the trombone the player controls the pitch by varying the *length* of the air column.

Stringed Instruments

236. Of these perhaps the simplest and most typical is the guitar. This instrument has six strings, three of silk covered with silver wire, and three of catgut. The range of the instrument is about three octaves. The lowest string yields the note E; the highest open string, 4 E; or when held at the middle, 8 E. Three methods of varying the pitch are illustrated in the guitar: —

1. To vary the pitch n of any one string a screw is employed to change the tension T. As the tension increases, the force restoring the string to its position of equilibrium becomes greater. The frequency therefore increases.

2. To pass from one note to another on the same string, "frets" are employed. The finger holds the string down over the fret and thus changes the length of the string.

$$V = nl = 2nL = \text{a constant.} \qquad \therefore\ n \propto \frac{1}{L}.$$

3. In passing from one string to another, the pitch of the lower strings is diminished by increasing their density. This is accomplished by wrapping them with silver wire. The strings would be too clumsy and too rigid if the same mass were given to the string by use of catgut. The wrapping of silver wire adds nothing to the restoring force, but does increase the mass to be moved, and hence diminishes the frequency.

These three facts are most briefly and definitely summarized by writing Eq. 82 (§ 192) as follows:

$$V = 2nL = \sqrt{\frac{T}{\mu}} \qquad \text{Eq. 82}$$

from which it follows that, other things remaining constant, —

(i) n varies as $\sqrt{T}$.

(ii) n varies as $\dfrac{1}{L}$.

(iii) n varies as $\dfrac{1}{\sqrt{\mu}}$.

In the **violin** the length of the string is no longer determined by the position of the fret, but is under the entire control of the player. The violin is, therefore, much more responsive to the will and to the skill of the player than is the guitar.

The **piano** is an instrument of between 80 and 90 strings, or sets of strings, each of which is employed to produce one note only. The tension and length of each string is intended to remain constant. Anything which changes the tension of a string once adjusted is said to put the piano "out of tune."

The **harp** is essentially a piano in which the strings are plucked by the fingers, instead of being struck by hammers.

Vibrating Membranes. The Human Voice.

237. Of this class, the **drum** is the well-known example. But by all odds the most interesting and most marvelous instrument of this, or indeed of any class, is that which produces the **human** voice. Here, in general, three factors enter: —

(*a*) The vibrations of the so-called vocal cords.
(*b*) The vibrations of the tongue.
(*c*) The vibrations of the lips.

The vocal cords consist of membranes situated one on each side of the throat (larynx). By changing the muscular tension of these one easily changes the pitch of his voice. By altering the form of the mouth one easily changes the overtones, and hence the quality of the tone. In this manner the various vowel sounds are produced.

Problems

1. Approaching a large building at night, I stamp my foot on the pavement and 0.6 second later I hear the echo. How far away is the building?
Ans. 99.6 m.

2. What is the pitch of sound whose wave length *in air* is 33.2 cm.
Ans. 1000.

3. Explain how the violinist varies the pitch of any one string by placing his finger upon it.

4. A well is 1800 cm. deep. How long a time will be required for a sound to travel from the bottom to the top of the well? How long will a stone be in falling from the top to the bottom of the well? How long, then, after the stone is let fall before you hear the splash?
Ans. Total interval = 1.969 sec.

5. A Kundt's tube is filled with hydrogen, and it is then found that the distance between the successive dust heaps is 3.8 times as great as when the same note is sounded in the tube filled with air. From this compute the speed of sound in hydrogen. *Ans.* 1261.6 m. per second.

6. The brass rod used to excite a certain Kundt's tube is just 1 m. long; the dust heaps which it produces in air are 99 mm. apart. What is the speed of sound in brass? *Ans.* 3353 m. per sec.

7. An open organ pipe 166 cm. long will yield a note of what pitch? *Ans.* 100.

8. A closed organ pipe 62 cm. long will yield what note?

9. Explain how it is that the speed of sound in a gas increases as the temperature rises. Show that an equation of the following form describes this phenomenon:

$$V = 332.0 + 0.6\,t^\circ \text{ meters per second}$$
$$= 332.0\,(1 + 0.00181\,t^\circ) \text{ meters per second.}$$

10. Two telegraph sounders are placed on the same electric circuit so that they click each five times a second. One of these sounders is fixed; the other I can carry off to some distance. How far away shall I have to take the movable one in order to hear its nth click at the same instant that I hear the $(n + 1)$th click of the fixed sounder? This is the "unison" method for measuring the speed of sound. *Ans.* 66.4 m.

11. In how many ways could you change the pitch of a string which now emits the note C, so that it would emit the note G in the same octave?

12. Would increase of temperature change the pitch of an organ pipe? Would it change the ratio of the frequencies of a series of pipes.
Shearer, 1265.

13. Make a diagram of the frets on a guitar and show how, starting from any one string, any other string may be tuned by the use of beats.

14. Assuming the density of hydrogen as 0.0000896, find the speed of sound in hydrogen under normal atmospheric pressure of 76 cm.

15. Assuming the density of marsh gas to be 0.000727, find the temperature at which the speed of sound in air will be equal to the speed of sound in marsh gas at 0°C.

16. A brass and an iron rod have each the same size, shape, and density Through which would you expect sound to travel more rapidly? Why? How much more rapidly?

References

TYNDALL. *Lectures on Sound* (Appleton). Price, $2.00.

BLASERNA. *Theory of Sound.* International Scientific Series. Price $1.50.

BARNES. *Practical Acoustics.* The third volume of Stewart and Gee's *Practical Physics* (Macmillan). Price, $1.10.

Ency. Brit., art. "Music," Section on Scientific Basis of Music, pp. 102–106, by Professor Bosanquet.

HELMHOLTZ. *Sensations of Tone*, trans. by Ellis (Longmans). Price, $9.00.

POYNTING AND THOMSON. *Sound* (Griffin & Co.).

TAYLOR, Sedley. *Sound and Music* (Macmillan). A popular and scholarly account of that interesting domain which lies between Acoustics and Music.

CHAPTER VII

THEORY OF HEAT

238. Our experiences with heat and cold cover a wide range. We use the words *hot*, *warm*, and *cold*, with great accuracy. And yet heat is something concerning which our two most important senses — sight and hearing — give us no direct information. Indeed there is strong evidence for thinking that this intangible, invisible, something which we call "heat" is not a substance at all, but rather a form of energy. The object of the present chapter is to discover the constant elements in the nature of heat, and especially to examine the evidence for thinking it a kind of energy and *not* a variety of matter; in short, to reduce the idea of heat to one of matter and motion, as was done in the case of sound. The subject is one which is so intimately connected with all the various parts of Physics that it is well-nigh impossible to offer any systematic analysis; but perhaps the following order of treatment will serve our purpose: —

I. Distinction between Heat and Temperature.
II. Measurement of Temperature. Thermometry.
III. Quantity of Heat. Calorimetry.
IV. Transfer of Heat.
V. Some Effects of Heat.
VI. Nature of Heat.

I. Distinction between Heat and Temperature

239. Most people will agree that a cup of water taken from a boiling kettle has, at the instant it is dipped out, the same temperature as that of the boiling water. Most people will likewise agree that for heating purposes the large kettle is much more efficient than the small cup of water. A water bag or a water bottle that holds a gallon will give out more heat and give it out longer, other things being equal, than one which holds only a pint. If the temperature of the two vessels of water are the same, the larger is said to contain more heat. If a bath tub be half filled with cold water, one can heat it more by pouring in a gallon of boiling water than by putting in a quart of boiling water; for, although the temperatures are the same, the gallon contains more heat than the quart. These simple facts are cited merely to show that popular notions concerning the distinction between heat and temperature are perfectly clear. They are also correct. But before the word "temperature" can be admitted to the rather select vocabulary of Physics, it must be defined in unmistakable English. This we now proceed to do.

Imagine three vessels of water, *A*, *B*, *C* (Fig. 171), each containing a different quantity of water. If *A* and *B* are placed side by side, in contact, and *B* thereby gains heat, — be it ever so little, — *A* has imparted heat to *B*; and *A* is said to have a higher temperature than *B*.

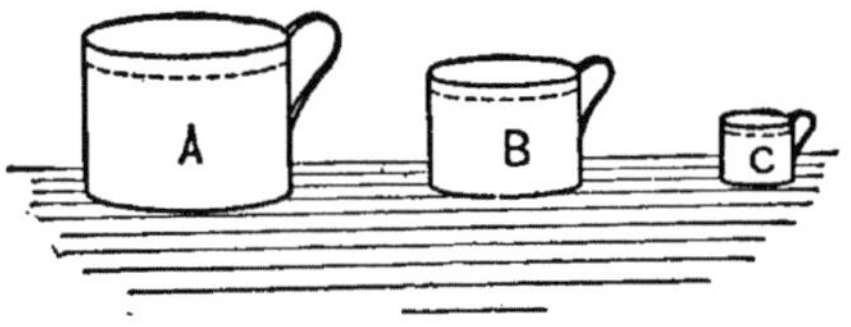

FIG. 171.

Put *C* in contact with *B*; if *B* thereby loses heat, — be it ever so little, — heat has passed into *C*; and *C* is said to be at a lower temperature than *B*.

When, in general, one body is placed in contact with another, the difference of temperature between the bodies is that which determines which way the heat flows. If the heat flows from *A* to *B*, *A* has the higher temperature; but if the heat flows from *B* to *A*, then *B* has the higher temperature.

Consider two vessels of water, *E* and *F* (Fig. 172), connected by a rubber tube. Does the water always flow from the large vessel to the small one? What *does* determine the direction of the flow? If the water flows from *E* to *F*, does the surface of the water in *F* rise by the same amount that the surface in *E* falls?

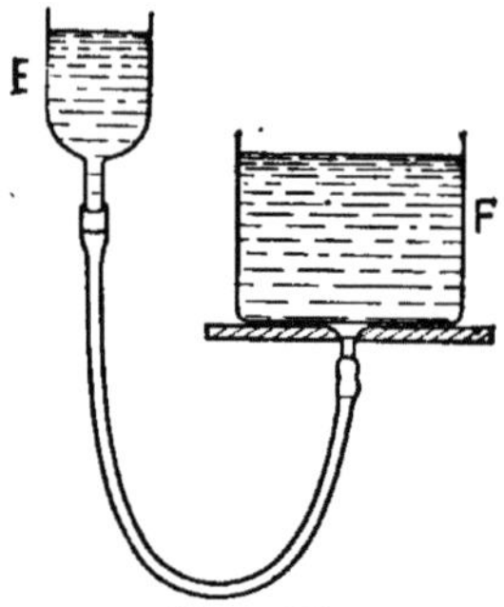

FIG. 172.

Returning now to Fig. 171, we observe that the analogy is very perfect. Whether heat flows from *A* to *C* or *vice versa* depends, not at all upon the size of the vessels, but upon their difference of heat level or, more accurately, upon their difference of temperature.

240. The following definitions will now be clear to every one who has mastered the foregoing facts: —

"DEFINITION OF TEMPERATURE. — **The temperature of a body is its thermal state considered with reference to its power of communicating heat to other bodies.**

"DEFINITION OF HIGHER AND LOWER TEMPERATURE. — *If when two bodies are placed in thermal communication, one of the bodies loses heat, and the other gains heat, that body which gives out heat is said to have a higher temperature than that which receives heat from it.*

"COROLLARY. — *If when two bodies are placed in thermal communication, neither of them loses or gains heat, the two bodies are said to have equal temperatures or the same temperature. The two bodies are then said to be in thermal equilibrium.*

"LAW OF EQUAL TEMPERATURE. — *Bodies whose temperatures are equal to that of the same body have themselves equal temperatures.*" — MAXWELL, *Theory of Heat*, Ch. II.

The point in this definition which calls for emphasis is the fact that temperature is a state of a body. "It is a condition and not a theory which confronts us."

241. In estimating temperatures, the student must be warned against drawing incorrect inferences from his sensations. For, in judging the temperature of the atmosphere, it is well known that our estimate is greatly influenced by any wind that may be blowing and by the presence of moisture in the air. If the hand be dipped into hot water and immediately afterward into tap water, the tap water will feel distinctly cold. But if a piece of ice be held in the hand for a moment, the same tap water will immediately after feel distinctly warm. The indications of thermometers such as are usually employed in the laboratory are much more independent of their previous history than are the sensations of the human hand.

II. MEASUREMENT OF TEMPERATURE. THERMOMETRY

242. The mercurial thermometer is an instrument which belongs quite as much to the household as to the physical laboratory. But the principles upon which it is constructed are not so well known as the instrument itself.

Since heat itself is invisible and intangible, and can be perceived only through its effects upon bodies, we must employ some one of these effects for the measurement of heat. **For ordinary purposes the universal choice has been change in size, which always accompanies a change in temperature.** We might take a brass rod 10 centimeters long, and then define one degree as that difference in temperature which is required to lengthen the brass rod by $\frac{1}{1000}$ millimeter. Such an instrument would be portable and would give the same reading for the same temperature; but it would be impossible to measure such a minute elongation except by means of elaborate apparatus.

If a glass bulb fitted with a hollow stem were partially filled with water, we might define one degree as that difference of temperature which would increase the volume of water by $\frac{1}{1000}$ of itself. But, unfortunately, water does not always *increase* in volume as its temperature rises. Moreover, each particular volume of the water does not correspond to one particular temperature. As we shall see presently, a body of water at 6° has the same volume as at 2°. A water thermometer would, therefore, be ambiguous in its readings. Hence water is never used for making thermometers.

243. But there is one substance, viz. mercury, which has shown itself eminently adapted to the purpose. Among the advantages of mercury may be mentioned the following: —

(*a*) It is easily prepared in a pure state.

(*b*) It does not wet glass or stick to it.

(*c*) It expands rapidly with rise of temperature, so that its changes in volume are easily read.

(*d*) To each particular volume corresponds a definite temperature.

(*e*) It does not freeze except at temperatures comparatively low, and does not boil except at temperatures comparatively high.

(*f*) Compared with water and most solutions it requires a very small amount of heat to raise its temperature through a given range.

The indications of temperature which are given by the mercurial thermometer hinge upon the fact that mercury expands with rise of temperature more rapidly than does glass. If, therefore, a glass tube having a bulb blown at one end be partially filled with mercury and immersed in a bath at higher temperature than its own, the mercury will rise in the tube. If the instrument is immersed in a bath of lower temperature, heat will flow from the mercury to the bath, and the mercury will descend in the tube. When the mercury is neither rising nor falling, it is fair to suppose that the thermometer is at the same temperature as the bath, or, at least, the part of the bath immediately about the thermometer.

Manufacture of a Mercurial Thermometer

244. The steps in the process are as follows: —

1. The selection of a piece of thick-walled capillary tubing of very uniform bore.
2. A bulb is blown on the end of this tube.
3. The bulb is filled with mercury. It is then highly heated and "sealed off."
4. The tube of the thermometer is graduated.

This last step is one which demands further consideration.

Fixed Points of a Thermometer

245. It so happens that there are two temperatures which can be easily produced and reproduced at almost any place and at any time: one of them is the melting point of ice, the other is the boiling point of water.

It has **been agreed upon by the** whole scientific world, that these **two** temperatures — the melting **point of ice and the boiling point of** water — **should be called** zero **and one hundred degrees, respectively. Together they are known as the fixed points of the thermometer.** The interval between these two fixed points is divided into 100 steps — degrees — in such a way that the apparent change in volume of the mercury between any two successive steps is exactly $\frac{1}{100}$ part of the total apparent change in volume between 0° and 100°. This is known as the "centigrade scale," and is generally indicated by writing a "C." after the numerical value of the temperature. Thus "48° C." means "48 degrees on the centigrade scale."

The first step in the graduation of a thermometer is, then, to determine the fixed points, and mark them on the tube.

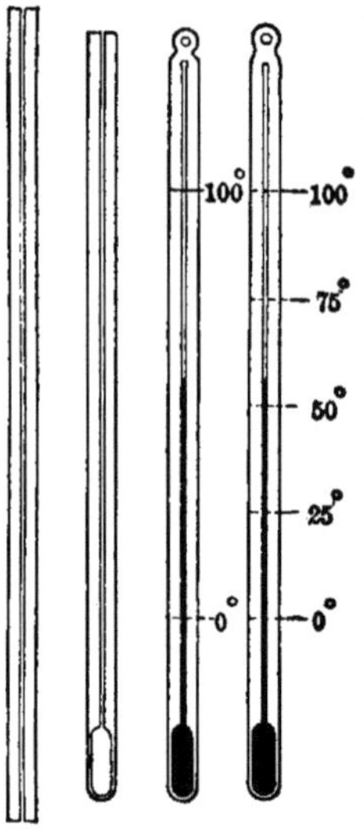

FIG. 173. — Showing four stages in the manufacture of a mercurial thermometer.

Next the 50° point is so chosen that the volume (not the length) of the tube between 50° and 100° is equal to the volume between 50° and 0°. In like manner, the 25° point must halve the volume between 0° and 50°.

The manner in which these points are determined will be considered in the laboratory.

Temperature Scales of Fahrenheit and Réaumur

246. Besides the centigrade scale which, in a modified form, was suggested by Celsius, there are two others which are widely used in everyday life. In the one proposed by Fahrenheit, the volume between the fixed points is divided into 180 parts; while the zero is placed at a point which is 32 of these degrees below the melting point of ice.

In the other, proposed by Réaumur, the volume between the freezing point and the boiling point of water is divided into 80 degrees, the former of these two points being employed as zero. It is important to notice that these three thermometers, Celsius, Fahrenheit, and Réaumur, differ in no essential respect. They are identical in principle, mode of construction, and use, the only difference being in the size of the degree employed.

247. Essential Features of a Good Thermometer.

1. It must be easily portable.
2. It must be permanent.
3. It must always give the same reading when subjected to the same temperature.
4. It must be possible for the user to test the correctness of its graduation, and to determine any errors in its graduation.
5. It must be relatively small, so that when placed in contact with a second body, the temperature of the second body will not be seriously affected.

Historical Development of the Thermometer

248. Although the thermometer is one of the oldest instruments of physical science and has a history covering three hundred years, there is even now scarcely a year goes by that does not bring either some new form of thermometer or some important improvement.

In Winkelmann's *Handbuch der Physik*, Pernet traces the development of the instrument through a series of steps which are somewhat as follows: —

(i) Some time previous to 1603, Galileo had made and used in lectures at Padua a thermoscope or indicator of temperature, which in its essential features is shown in Fig. 174.

A glass bulb with a rather long and uniform stem is inverted with its lower end placed under the surface of some colored liquid, such as wine. But before the stem is placed in the liquid the bulb must be heated to a temperature a little higher than any temperature it is afterward desired to measure. At the standard temperature, say, that of the room, the liquid rises to a certain mark on the scale and remains there, except for variations of the barometer. When taken out-of-doors the air in the bulb expands or contracts and the water in the tube thus falls with rise of temperature and rises with a fall of temperature. This simple device is surprisingly sensitive, and proved useful, even in Galileo's time, for the detection of fever.

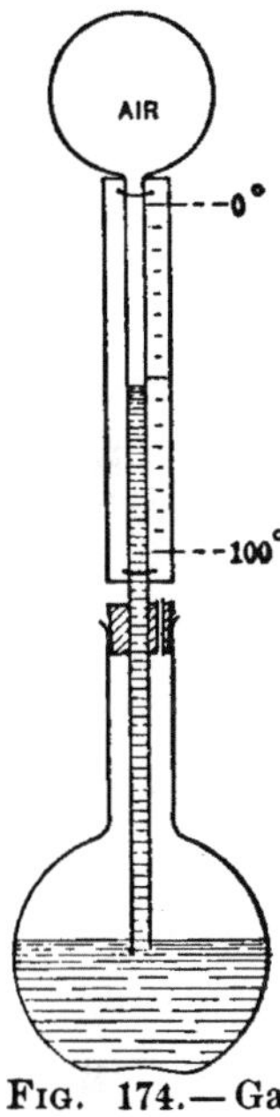

FIG. 174.— Galileo's thermometer.

(ii) As noted by Pascal in 1643 the readings of this thermometer are dependent upon the simultaneous readings of the barometer. To avoid this difficulty, Father Schoot, in 1657, sealed the liquid off from the atmosphere by bending the tube and sealing a bulb on each end as shown in Fig. 175. What this device indicates is evidently the *difference* in temperature between the two bulbs *A* and *B*. It is therefore called a *differential thermometer*.

(iii) In each of the preceding instruments a gas is employed as a thermometric substance; but in 1631 a French physician, Jean Rey, first made use of the expansion of a liquid for the measurement of temperature. A glass bulb, provided with a stem and filled with water, was used by him to distinguish between normal blood temperature and those of fever patients.

(iv) The first liquid thermometer to secure independence of atmospheric pressure was made about 1641 by sealing off the end of the glass stem. About this time the Academy of Sciences at Florence introduced the use of mercury and a scale graduated so that one end indicated the lowest temperature of winter, the other, the highest temperature of summer.

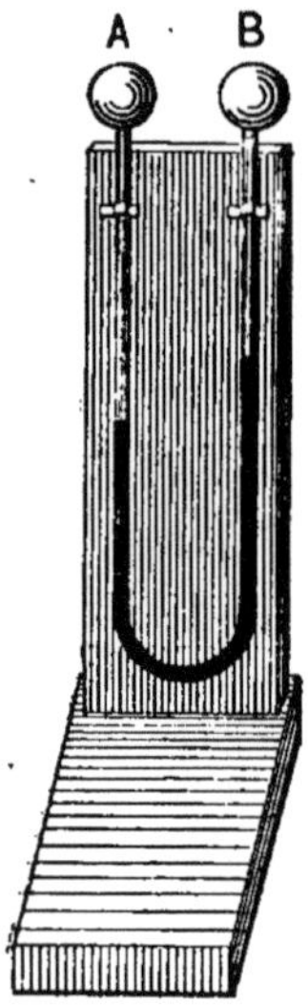

Fig. 175. — Schoot's differential thermometer

(v) These rough and inaccurate standard temperatures were partly replaced by the one excellent standard in 1664; for it was in this year that Robert Hooke, Secretary of the Royal Society of London, suggested the melting point of ice as *the* standard temperature. In the year following Huygens proposed the boiling point of water as the standard, neither of these men having yet recognized the fact that at least *two* fixed points are absolutely necessary in order to make the thermometric readings from one instrument comparable with those from another.

(vi) This important step, the use of two definite fixed points, was first proposed in 1688 by Dalance, who adopted the melting points of ice and butter as the two temperatures in question. Newton constructed a thermometer on this principle, using the melting point of ice as zero and the temperature of his own body as 12°. He also tested the uniformity of the bore in the stem of the thermometer, thus introducing a process which is now called calibration. As a thermometric substance he employed linseed oil.

(vii) The next advance is due to Fahrenheit (1686–1736), a Danish mechanician living in Amsterdam, who introduced the modern mercury thermometer with cylindrical bulb and stem, about the year 1714. For fixed points he used various temperatures, among them that of a mixture of salt and ice, 0°; melting ice, 32°; "blood heat," 96°; boiling water, 212°. Which pair were taken as standard in any particular case appears to have been determined largely by the range of temperatures for which that particular instrument was intended.

(viii) The Réaumur scale, described above, was introduced about 1730. Little interest attaches to this thermometer other than its wide use in domestic and industrial circles.

(ix) In 1742, Anders Celsius, professor of astronomy at Upsala, proposed a scale with the zero at the boiling point of water and with 100° at the melting point of ice. The inversion of this scale, which gives us the modern centigrade instrument, is due to Christin in France (1743), possibly also to Strömer, in Upsala (1750).

(x) In the meantime, many investigators had been searching for the conditions which determine the exact temperatures at which ice melts and water boils. But it was the German physicist Lambert (1728–1777) who finally placed the science of thermometry upon its modern basis by using air as a thermometric substance and proposing as an "absolute zero" that temperature at which the gas would contract so much as to bring the particles of air into actual contact with one another. Besides this he introduced many important corrections which render the air thermometer an instrument of precision.

As will be seen later (§§ 270 and 291) the absolute zero of Lambert when defined in a more practicable manner is essentially that in use to-day. Yet it was the exact measurements of the French physicist Regnault (1810–1878) upon the properties of gases and vapors that made possible a reasonably exact absolute scale of temperatures. So much for practical thermometry. The dynamical foundation upon which the absolute scale is based can be considered only after we have looked into the subject of thermodynamics.

Measurement of High Temperatures

249. The determination of temperatures which lie above, say, the boiling point of sulphur, 445° C., demands methods which constitute almost a separate science — pyrometry — and one which cannot here be considered. It may, however, be mentioned that these methods are practically all based upon one or more of the three following principles, which the student will meet later.

(i) The variation of electrical resistance with temperature, giving rise to what is known as "electrical resistance thermometry" and "platinum temperatures," since the resistance used is generally that of a platinum wire. Valuable both for moderately high and for very low temperatures; especially useful between 400° and 1200° C.

(ii) The variation with change of temperature of either (*a*) the total amount of energy radiated by a hot body, or (*b*) the energy (brightness) of a particular wave length (color) of light radiated by a hot body.

The total energy varies according to a law first determined by Stefan and Boltzmann; the partial energy according to one investigated by Wien and Planck.

These are the basis of what is known as "blackbody temperatures," since the methods are strictly accurate only for bodies which are perfectly "black." Available for the highest temperatures known.

(iii) The variation of electromotive force in a circuit made of two wires of different material when the temperature of one of the junctions of these two wires is changed, a method which leads to "thermoelectric thermometry."

Table of Reliable Melting and Boiling Points

	Centigrade	Fahrenheit
Liquid Hydrogen	– 253	– 423
Liquid Oxygen	– 182	– 295
Freezing Mercury	– 39	– 38
Melting Ice	0	32
Boiling point of Water at 760 mm. pressure	100	212
Boiling point of Aniline at 760 mm. pressure	184	363
Boiling point of Naphthaline at 760 mm. pressure	220	428
Boiling point of Diphenylamine at 760 mm. pressure	302	575
Boiling point of Sulphur at 760 mm. pressure	445	833
Melting point of Tin	232	449
Melting point of Lead	327	620
Melting point of Zinc	419	786
Melting point of Antimony	632	1169
Melting point of Aluminium	657	1214
Melting point of NaCl	800	1472
Melting point of Silver (in air)	955	1751
Melting point of Silver (in reducing atmosphere)	962	1763
Melting point of Gold	1064	1947
Melting point of Copper (in air)	1062	1943
Melting point of K_2SO_4	1070	1958
Melting point of Copper (in reducing atmosphere)	1084	1983
Melting point of Nickel	1427	2600
Melting point of Pure Iron	1503	2737

The preceding table, prepared by Dr. J. A. Harker for the Cambridge Scientific Instrument Co., in 1906, gives a fair idea of the range of temperatures employed in modern physics.

Even higher melting points than those listed in the preceding table have been measured with remarkable accuracy. Thus Waidner and Burgess, at Washington, using three different methods, obtain for palladium 1546° C., and for platinum 1753° C.; while Mendenhall and Ingersoll, at Madison, have determined the melting point of rhodium to be 1907° or 1968° C., and that of iridium to be 2292° or 2388° C., according to the value accepted for platinum, concerning which there appears to be still an uncertainty, lying somewhere between 10° and 50° *C*. The temperature of the positive crater of the electric arc lies probably between 3600° and 3900° C.; while that of the sun is probably in the neighborhood of 7800° C.

III. Quantity of Heat

250. We have already learned that temperature is merely a condition determining the direction of the flow of heat very much as pressure is a condition determining the direction of the current when two vessels of compressed air are connected. And just as we need a means of measuring the amount of air which escapes from either of the two vessels into the other, so we need a method of estimating the quantity of heat which passes from one body to another when they are brought into contact with a difference of temperature.

When a hot body is brought into the neighborhood of cooler bodies, we have excellent reason for thinking *something* passes from the hotter one because the cooler ones always rise in temperature. If a cold body is brought into the midst of hotter ones, we observe again that when the temperature of the cold body rises it is always at the expense of the hotter surroundings. Let us call this something which passes "heat," and proceed to discover what factors determine its amount. The simplest method of doing this is perhaps through the following series of experiments: —

(i) Take equal masses of water and mix them together. It will be found that the temperature of the mixture is an exact mean between the two original temperatures, whatever the masses may have been.

(ii) Now take *unequal* masses of water which we may indicate by m_1 and m_2, having temperatures t_1 and t_2 respectively. On mixing these two, it is found that the final temperature θ is obtained by multiplying the mass of each by its original temperature, adding these products, and dividing by the sum of the masses. This result seems first to have been formulated by the Russian physicist Richmann (1711–1753); it is known as Richmann's rule, and is most clearly expressed by the following equation:—

$$\theta = \frac{m_1 t_1 + m_2 t_2}{m_1 + m_2}. \qquad \text{Eq. 88}$$

If we assume that as much heat is given up by one mass of water as is gained by the other, then the experimental fact expressed by Richmann's rule shows that the amount of heat given up or absorbed by any body depends upon the mass and the change of temperature. This inference will be clearer if we rewrite Eq. 88 in the following form:—

$$m_1(t_1 - \theta) = m_2(\theta - t_2)$$

This equation may, indeed, be considered as a definition of equal quantities of heat.

Unit of Heat

251. Having once discovered a method of duplicating a given quantity of heat, the measurement of heat becomes possible so soon as a unit of heat has been agreed upon. For this purpose we have only to place equal to unity both the mass and the change of temperature in the preceding equation, a process which leads to the following widely employed definition. **Unit of heat is that quantity which will raise the temperature of one gram of water from 0° to 1° C. This unit is called a gram calorie.** Engineers more frequently employ the British Thermal Unit (written "B.T.U."), which is the amount of heat required to raise the temperature of one pound of water one Fahrenheit degree.

It has been suggested by Professor E. H. Griffiths that, in order to make the mechanical equivalent of heat (§ 289) almost exactly 42 million ergs per gram degree, *we should define the unit of heat as follows, viz. the quantity of heat necessary to raise the temperature of one gram of water from* 10° *to* 11° *C.*

(iii) So far we have been dealing with a single substance, namely, water, a most convenient material because it is easily obtained in large quantities and in a state of high purity. Now mix two liquids which differ in chemical composition but have no chemical action upon each other. It was discovered by the Scottish chemist, Black (1728–1799), that under these circumstances Richmann's rule must be modified. He found that equal masses of different substances when changing temperature by equal amounts do not give out equal amounts of heat, and was led to assign to each substance a certain constant which would indicate the specific effect of heat in raising the temperature of that particular substance. If we denote this constant by c, then Richmann's rule in its generalized form becomes

$$\theta = \frac{m_1 c_1 t_1 + m_2 c_2 t_2}{m_1 c_1 + m_2 c_2}. \qquad \text{Eq. 89}$$

This equation, which may be written

$$m_1 c_1 (t_1 - \theta) = m_2 c_2 (\theta - t_2),$$

shows that, in general, the amount of heat which a body gives up, without change of state, depends, in general, upon three factors, namely, mass, difference of témperature, and a specific constant which is now called "specific heat" and which is defined as follows:—

Definition of Specific Heat

252. **The ratio between the amount of heat required to raise the temperature of any body one degree and that required to raise the temperature of an equal mass of water one degree, is known as the specific heat of the body.**

The *numerical value* of the specific heat of any substance will be the quantity of heat (number of calories) required to change the temperature of one gram of the substance by one degree.

Let us denote by Q the quantity of heat added to a body whose mass is m and whose specific heat is c; let $t_1°$ be the initial and $t_2°$ the final temperature of the body; accordingly the change in temperature produced by Q will be $t_2° - t_1°$. Then

$$c = \frac{Q}{m(t_2° - t_1°)}, \qquad \text{Defining equation for specific heat.} \qquad \text{Eq. 90}$$

becomes the defining equation for the average specific heat between the temperatures $t_1°$ and $t_2°$.

Another useful conception is that of "heat capacity of a body" which is defined as the amount of heat in calories required to raise the temperature of that particular body one centigrade degree. Evidently, therefore, the heat capacity of a body is measured by the product of its mass and specific heat.

Various applications of Eq. 90 will be taken up in the laboratory under the head of *calorimetry.*

TABLE OF SPECIFIC HEATS

SOLIDS	TEMPERATURE IN DEGREES CENTIGRADE	SPECIFIC HEAT	LIQUIDS	TEMP.	SPECIFIC HEAT
Aluminium	20	0.214	Alcohol, ethyl	0	0.548
Brass	0	0.0883	Alcohol, methyl	5–10	0.590
Copper	0	.090	Mercury	0	.033
Diamond	10.7	.1128	Petroleum	21–58	.511
Glass, crown	10–50	.161	Sea water	17.5	.938
Glass, flint	10–50	.117			
Graphite	10.8	.160			
Ice	– 21–0	.502			
Iron	15	.109			
Lead	15	.030			

From *Smithsonian Tables*, pp. 294–296.

IV. TRANSFER OF HEAT

253. Having now defined temperature and quantity of heat in such a way that we can measure each of them, we next consider the various modes by which heat is transferred or diffused. Of these only two general methods are known, viz. (1) Conduction, and (2) Radiation.

1. *Heat Conduction*

254. The handle of a spoon may become very hot when the bowl of the spoon is placed in a cup of tea. The process by which the heat is transferred from the lower to the upper part of the spoon is called "conduction." The same spoon, when used with ices, may become very cold. Here, again, the process by which the heat is transferred, this time from the handle to the bowl of the spoon, is called conduction.

Very little, indeed, is known about the mechanism by which the particles of a solid body hand on the heat from one to another. A great deal else, however, is known about heat conduction. At any rate, we know that the process does not depend upon the visible motion of matter. Again, every one knows that he can hold the stick of a burning match for 30 or 40 seconds without difficulty; but a copper wire of the same size and shape held in a match flame for that length of time becomes unbearably hot.

An iron and a copper rod, each of the same length, say half a meter, and of the same cross section, each having one end placed in the same flame, will become unequally heated at equal distances from the junction (see Fig. 176). For if at equal distances from the flame on each rod be placed either bits of wax or matches, those on the copper will melt or burn earlier than those on the iron. A series of bicycle balls attached with wax, three centimeters apart, along each bar will melt off in such a way as to show the progress of temperatures in the bars.

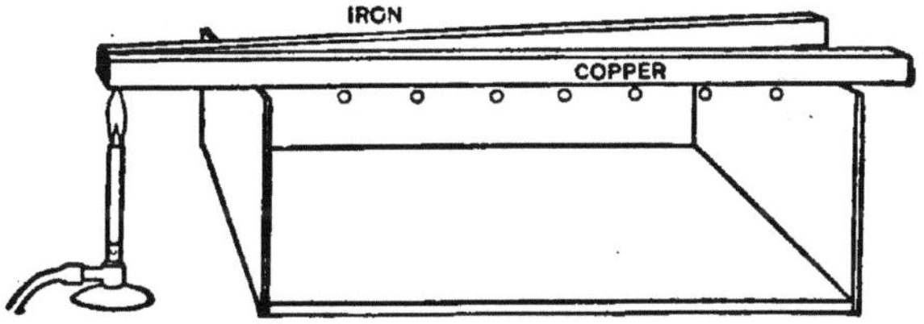

FIG. 176.—Illustrating the fact that copper is a better conductor than iron.

It is here to be borne in mind that the rapidity with which the iron and the copper change temperature at any point depends, not only upon the rate of conduction, but also upon the specific heats of iron and copper. In the present experiment, however, these two values are so nearly equal that the relative conductivities of iron and copper are fairly represented.

A good experiment to illustrate differences in conductivity is the following, the results of which are familiar to us all: Take a piece of woolen goods, a piece of glass, and a piece of iron. Put them in some cold place where their temperature will fall 50° or 60° F. below that of the human body. If these three bodies are now examined, the iron will feel very cold to the hand because it conducts off the heat of the hand most rapidly; the glass will feel moderately cold; while the woolen piece is quite comfortable, being a very poor conductor. Now place these same bodies in an oven where they will reach a temperature 50° or 60° F. above blood heat; the iron will be uncomfortably hot, because it conducts heat to the hand so rapidly, while the woolen goods feels merely warm, being unable to give up its heat quickly. Here is a case, then, where sensation is determined largely by power to conduct heat, a property which is called "conductivity."

It is a curious fact that all substances which are good conductors of electricity are also good conductors of heat. Indeed, one of the most brilliant achievements of modern electrical theory is the explanation, by the late Professor Drude of Berlin, of just how this interesting fact comes about—a matter which must be postponed until after the electron theory has been considered.

255. Among the various applications of thermal conductivity none is perhaps more important than the classical invention of the safety lamp by Sir Humphry Davy. The explosive mixture of gases which occurs in many mines and which is known as "fire damp" must, of course, be protected from naked flames. For this purpose Davy suggested that the flame be inclosed in a small cage of wire gauze. The effectiveness of this device rests upon the fact that, when the fire damp penetrates the hood and burns, the flame cannot strike back through the meshes of the gauze because the heat of combustion is carried off so rapidly, *owing to the conductivity of the iron*, that the temperature just outside is always lower than that of ignition. That this is true may be shown either by lowering a piece of flat wire gauze over an ordinary Bunsen flame (Fig. 177), when it will be seen that the flame does not pass through the meshes, or by first allowing the gas to pass through the gauze and then lighting it above. In the latter case it will be observed that the flame remains entirely upon the upper side. Since human nature is what it is, the lock on the safety lamp which prevents the miner from lighting his pipe from it is a feature quite as essential perhaps as the wire gauze.

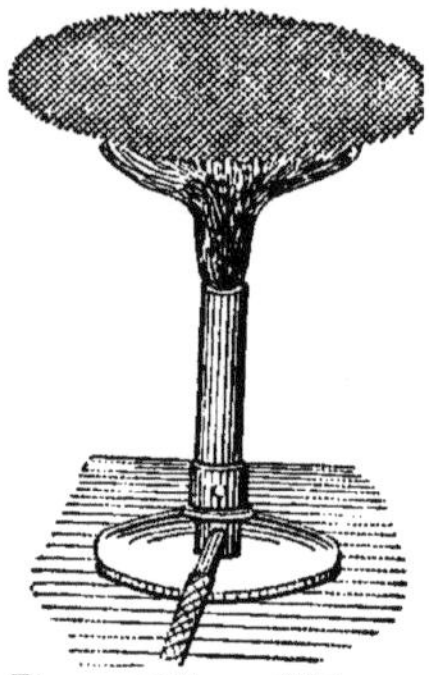

FIG. 177.—Illustrating the principle of the Davy safety lamp.

Measure of Conductivity

256. Were we to attempt to melt ice by means of a copper bar, one end of which is held in a flame at constant temperature, the other end on the ice, we should find the *rate* at which the ice is melted depends upon the dimensions as well as upon the composition of the bar. If the bar were made twice as thick, the heat would be transferred nearly twice as rapidly. But if the cross section remained constant while the bar were made twice as long, we should find the ice melting only about half as rapidly as before.

By such experiments, it has been found that the rate at which heat "flows" through a solid from one surface to another when these surfaces are at constant, but different, temperatures depends upon four things only, viz. (1) the substance from which the solid is made, (2) the distance apart of the surfaces, (3) the area of the surfaces, and (4) the difference in temperature between the surfaces. Let H denote the amount of heat which passes during the time t from one surface whose area is s to another of equal area. Then if k represents the thermal conductivity of the subtance, θ_1 the temperature of the first surface, and θ_2 the temperature of the second, we may describe the above facts by writing

$$H = k\,\frac{(\theta_1 - \theta_2)\,st}{l}, \qquad \text{Eq. 91}$$

where l is the distance apart of the surfaces.

Since all the quantities except k have already been defined, this is evidently the defining equation for thermal conductivity. The modern theory of conduction is almost entirely the creation of a single genius, the French physicist Fourier (1768-1830), whose results are contained in his *Analytical Theory of Heat.*

In actual cases which occur in the laboratory we do not generally measure the heat across a limited surface drawn in a solid, but a bar or some body of definite form is selected and the flow from one end to the other is measured. In any such case where l is considerable, as it generally is, it is clear that one has to reckon also with the amount of heat which escapes sidewise from the lateral surface of the bar. It is these "boundary conditions" which introduce many of the most difficult problems in experimental and mathematical physics.

Professor E. H. Hall has devised a clever method for experiments of this kind, in which he manages to continually send back into the sides of the bar just as much heat as escapes, thus eliminating, instead of correcting for, the surface loss of heat.

In the case of gases and liquids, it is found that the conductivities are, in general, very small compared with those of solids, as will be seen from the following table: —

TABLE OF THERMAL CONDUCTIVITIES

(Value expressed in C.G.S. units)

Solids		Liquids		Gases	
Aluminium	0.353	Alcohol	0.00042	Air	0.000568
Brass	.226	Ether	00030	Carbon dioxide	.000307
Carbon	.0004	Mercury	0201	Hydrogen	000327
Coppe	721	Petroleum	.000355	Nitrogen	000524
Cork	.0007	Water	0012	Oxygen	000563
Glass from	.0005				
Glass to	.0023				
Iron	.164				
Lead	.080				
Silver	1.096				

From *Smithsonian Tables*, pp. 197–198.

Problems

1. Give in words a definition of thermal conductivity derived from Eq. 91 by making all the quantities on the right, except k, unity.

2. Explain how the breaking of a glass tumbler by pouring in hot water may be prevented by placing in the tumbler a solid silver spoon along which the water is poured.

3. Why is it that cork handles on a bicycle are cooler in summer and warmer in winter than metal handles?

4. Why are steam pipes packed in glass wool or asbestos?

5. Three rods, one of brass, one of iron, and one of silver, have the same dimensions. Which one would you undertake to hold in a flame for the longest time without burning yourself?

6. From Eq. 91 find the dimensions of thermal conductivity. Observe that temperature will cancel out on each side of the equation.

Convection Currents

257. The toy balloons which, after being filled with hot air, may be seen rising on each Fourth of July, illustrate in an exquisite manner how the principle of heat conduction combined with the principle of Archimedes leads to a large number of our most important and vital phenomena.

The air inside the balloon is heated by conduction from the flame at the mouth of the balloon, and then rises, filling the balloon with a gas whose density is much less than that of the surrounding air. The buoyant force thus produced lifts the balloon into the upper atmosphere, where it finds a position of equilibrium determined by the temperature and pressure of the surrounding air.

If a cup of tea be too hot to drink, we may cool it by adding cream or cold water. Here the heat of the tea is transferred to the cooler liquid by the process of conduction; but the hot and cold portions of the tea do not need to be stirred; they will mix, although a little more slowly, of themselves. **This automatic mixing, which is a mere mechanical process, is called convection.**

The hot air of a chimney rises, mixes with the outside air, and gives some of its heat to the outside air. The hot air rises because its density is less than that of cold air. This process of carrying the hot air up the chimney is, again, convection. A beaker of water to which a flame is applied on one side, as in Fig. 178, becomes equally heated all through. First of all, the water just over the flame becomes hot by *conduction* through the glass. Then, by *convection*, the hot water just over the flame is displaced by the colder water, which is heavier, and therefore sinks to the bottom, as indicated by the arrows. The potential energy of the system is thus less than if the cold water remained on top.

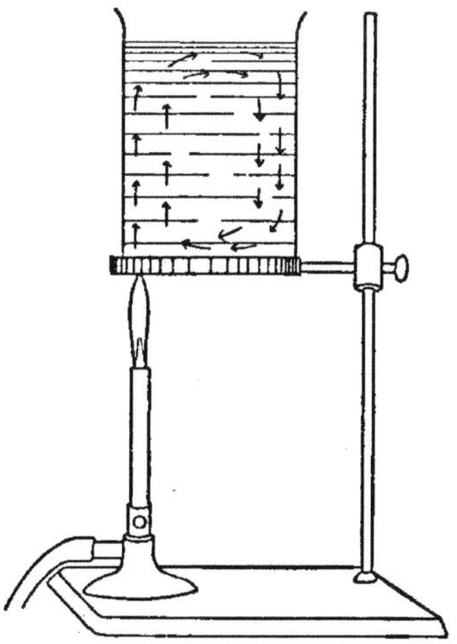

FIG. 178.—Convection currents in water.

This cold water, in turn, becomes heated by *conduction* through the glass. **These currents produced by differences of density and easily detected by small particles in the water, are called "convection currents."** They are exhibited on a gigantic scale in the equatorial regions of the earth, where the lower layers of the atmosphere, in contact with the water of the oceans or with the land, become highly heated by *conduction*, and then rise by *convection*. The air rushing in from north and south constitutes the well-known trade winds.

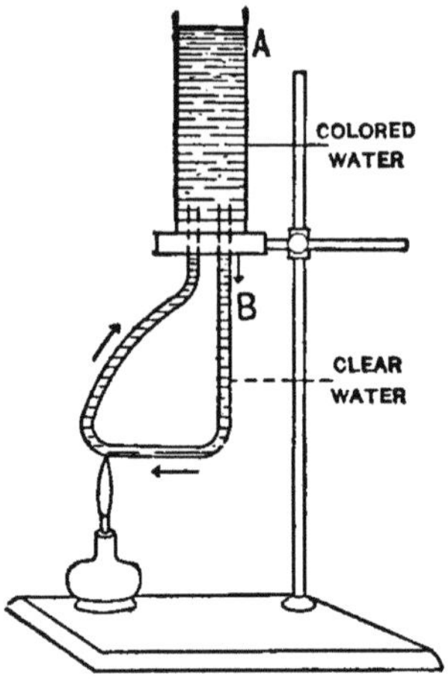

FIG. 179. — Circulation in tubes, brought about by convection currents.

Perhaps the best illustration of convection currents is obtained by the simple experiment indicated in Fig. 179. *B* is a glass tube with both ends fitting into a cork in the bottom of a lamp chimney. The chimney *A* is filled with colored water; but the tube *B* is filled with clear water. The circulation becomes very evident in a few seconds after a flame is applied to one side. Observe that here again the *heat* is communicated to the water by *conduction* through the glass, while the *water* is transferred from one part of the vessel to another by the mechanical (not thermal) process of *convection*.

From the preceding it will be evident that in fluids we have, in general, both processes, conduction and convection, always at work; while in solids, convection is necessarily impossible. When, however, heat is applied to the upper layers of a fluid, convection may be reduced to a minimum.

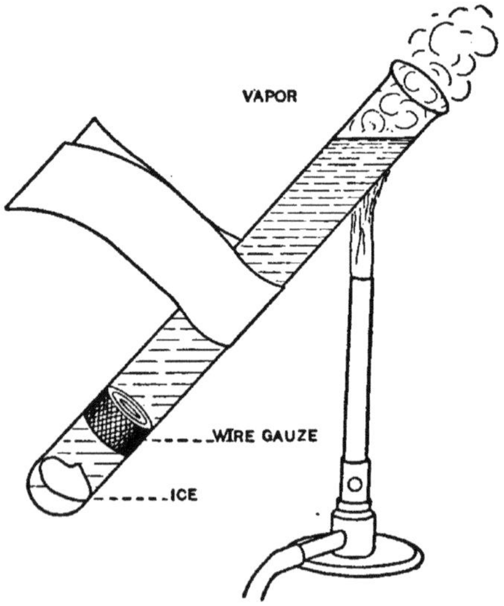

FIG. 180.— Illustrating the fact that water is a poor conductor of heat.

To illustrate, put a small piece of ice, say two grams, in the bottom of a test tube, and hold it at the bottom by means of a little coil of wire or wire gauze (see Fig. 180). The tube may be filled with water, and held over a flame until the water in the top of the tube boils. The heated portions, being lighter than the cold, remain on top, while water is such a poor conductor of heat that the ice in the bottom is not easily melted.

In all the varied processes of heat conduction, whether aided by convection or not, it will be observed that the heat is transferred from places of higher temperature to places of lower temperature. *In this respect* heat behaves like steam and all other fluids in flowing from points of higher pressure to points of lower pressure. We shall later find electricity behaving in a similar way, flowing from places of higher to places of lower electrical pressure.

2. *Radiation of Heat*

258. When the hand is held some inches at the side of, or underneath, an incandescent electric lamp, the sensation of heat is distinctly recognized. The same is true of the hand placed beneath a heated metal ball (see Fig. 181). We hold our hands before an open grate fire to warm them. How does the heat pass from the fire to the hands? Certainly not by conduction. For air is one of the very poorest conductors known (see Table, § 256). Nor is it a case of conduction aided by convection currents. For, in each of the three cases cited above, the currents due to convection are such as to cool the hand.

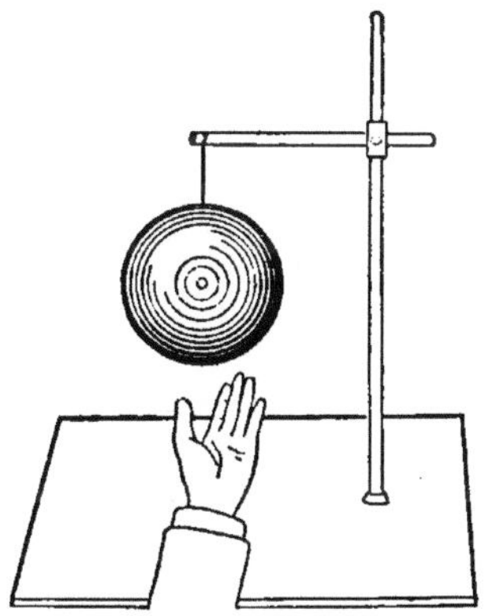

FIG. 181. — Radiation of heat downward from a hot iron ball.

Indeed, one of the earliest observations made after the discovery of the air pump was that both heat and light pass through a vacuum with the utmost ease. There is every reason for believing that the space which separates us from the sun is more nearly a perfect vacuum than any other known; yet across this vast, and so-called empty, region, the earth daily receives enormous quantities of heat. But by what process? We call it *radiation*. But what is radiation?

The question here raised can be answered (but only very imperfectly) by anticipating some of the results of optics and thermodynamics. At this point we must content ourselves with saying that there is most excellent evidence for thinking that light is a form of energy, that light consists in a wave motion, that this wave motion is propagated in a medium, called the *ether*, which permeates all bodies and fills all interstellar space, and that this wave motion travels in free space (*i.e.* space devoid of matter) at the rate of 300 million meters per second, but generally more slowly in space which is not devoid of matter.

Now the experimental evidence for thinking that radiant heat is of exactly the same nature as light is overwhelming. And all the more important laws of radiant heat and of light are included under one single and general statement of the Laws of Radiant Energy.

If, however, we once take for granted the evidence for thinking radiant heat a wave motion, we may immediately predict interference, reflection, refraction, and other properties of waves.

259. These results have been amply verified by more than two centuries of experiments, covering the work of Newton, Melloni, Stewart, Kirchhoff, Tyndall, Langley, and others. The following experiment by Newton sets forth the fact and the explanation with essential truth and clearness: —

"If in two large cylindrical vessels of glass inverted, two little thermometers be suspended so as not to touch the vessels, and the air be drawn out of one of these vessels, and these vessels, thus prepared, be carried out of a cold place into a warm one; the thermometer *in vacuo* will grow warm as much, and almost as soon, as the thermometer which is not *in vacuo;* and when the vessel is carried back into the cold place, the thermometer *in vacuo* will grow cold almost as soon as the other thermometer. Is not the heat of the warm room conveyed through the *vacuum* by the vibrations of a much subtiler medium than air, which, after the air was drawn out, remained in the *vacuum?* And is not this medium the same by which light is refracted or reflected, and by whose vibrations light communicates heat to bodies, and is put into fits of easy Reflection and easy Transmission? And do not the vibrations of this medium, in hot bodies, contribute to the intenseness and duration of their heat? And do not hot bodies communicate their heat to contiguous cold ones, by the vibrations of this medium propagated from them into the cold ones? And is not this medium exeedingly more rare and subtile than the air, and exceedingly more elastick and active? And doth it not readily pervade all bodies? And is it not (by its elastic force) expanded through the heavens?" — *Optics*, Book III, Query 18.

A strong argument for thinking that radiant heat is propagated by the same mechanism as ordinary light is found in the fact that after a solar eclipse the heat rays and light rays reappear, as nearly as can be observed, at the same instant. From this and evidence to be adduced later it appears that heat waves, light waves, and electric waves all travel with the stupendous speed of 300 million meters per second and are essentially identical except as regards wave length.

Measurement of Radiation

260. The first step toward discovering the laws of any phenomenon is the invention of some method for measuring the quantities involved. There are three principal methods for determining the amount of radiant energy which falls upon a body. **These are each dependent upon the fact that when radiant energy is incident upon a well-blackened body — such as one covered with lampblack — practically all the incident energy of wave motion is transformed into heat energy.** The measurement of radiation then reduces itself to the measurement of a rise in temperature — generally very small. Bodies which absorb all the radiant energy falling on them are said to be "perfectly black." The three methods are the following: —

(i) *Radiomicrometer.* The earliest of these instruments was the thermocouple (see § 369) and galvanometer, in which the junction of two different metals is the blackened body — a device used with great effectiveness by the Italian physicist Melloni, and greatly refined by the German physicist Rubens. But this has been surpassed in convenience by a modification due to Professor C. V. Boys, who suspends a small wire circuit of two different metals between the poles of a magnet as seen in Fig. 182. The junction marked J_1 is shielded from the radiation, while that marked J_2 is blackened and exposed to the radiation. The difference in temperature thus produced generates an electric current in the circuit and the current causes the thermocouple to rotate.

By means of this instrument, which is called a "radiomicrometer," placed at one end of a large room Boys has been able to show to an audience the radiation from a single candle placed at the other end of the room.

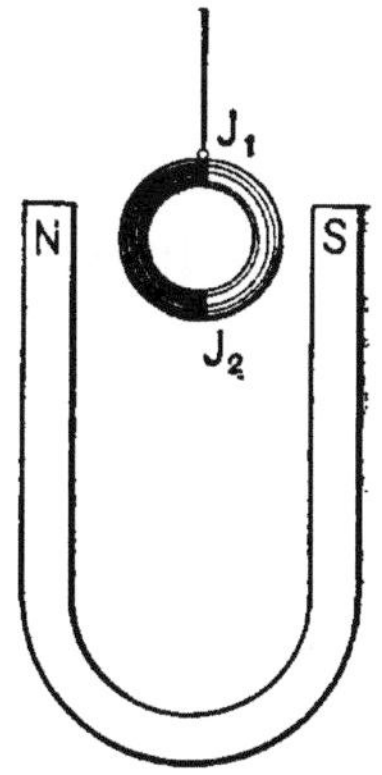

FIG. 182. — Boys' radiomicrometer.

(ii) *Radiometer.* An even more sensitive instrument is the modification of Crookes's radiometer introduced by Professor E. F. Nichols of Columbia University. This instrument consists essentially of a horizontal bar delicately suspended in a properly exhausted vessel. At each end of the bar is attached a thin glass disk blackened on one side. The radiation is admitted to the instrument through a window of fluorite which is remarkably transparent to all wave lengths. On striking the black surface, the radiant energy is degraded to heat and the black disk repelled — probably owing to the bombardment of the smallest particles of air, which, in a vacuum of this degree, are reflected freely from the disk to the wall of the containing vessel and acquire greater kinetic energy as the temperature rises. So delicate is this instrument that by means of it, Nichols and Hull, and Lebedew, have been able to accurately measure the pressure exerted by a beam of sunlight — a vastly smaller effect than that of the radiant energy which comes to us in such a beam, and which must be eliminated before the light pressure can be measured.

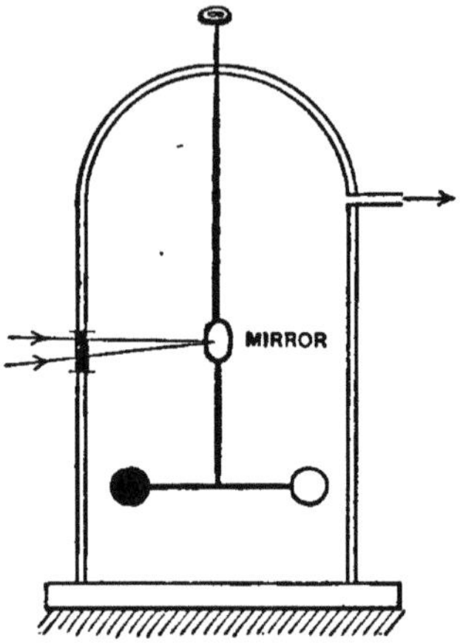

FIG. 183. — The radiometer.

(iii) *Bolometer.* Here the blackened body is a thin strip of metal which forms one arm of a Wheatstone bridge, an electrical instrument which will be met in the laboratory. Before the radiation is allowed to fall upon the wire, the galvanometer reading is zero. But the radiation heats the wire and thus increases its electrical resistance. This throws the bridge out of "balance," and the galvanometer needle is deflected through an angle which is proportional to the amount of the incident radiation. The late Secretary of the Smithsonian Institution, Professor Langley, who perfected this method, showed that differences of temperature of even less than a millionth of a degree centigrade could be detected in this way.

Laws of Radiation

261. By means of such methods as those just indicated two general results connecting radiation and temperature have been obtained.

(i) *Stefan-Boltzmann Law.* It was first suggested by Stefan, afterwards predicted from theory by Boltzmann, and has been amply verified by experiment, that **the total radiation (i.e. ra-**

diation of all wave lengths) of any perfectly black hot body is directly proportional to the fourth power of its absolute temperature.

Premising that the absolute scale of temperature begins with a zero 273° C. below that of melting ice, and denoting the total radiation by E, this law may be expressed as follows: —

$$E = C(t° + 273)^4 \quad \text{Stefan-Boltzmann Law.} \qquad \text{Eq. 92}$$

where C is a constant throughout a wide range of temperatures.

(ii) *Displacement Law.* It has also been found that if the various radiations be separated, say by a prism, so that the different wave lengths are sent off in different directions, and if the radiation of each wave length be measured, that wave length at which the radiation is most intense becomes shorter and shorter as the temperature rises, a fact which is true for all bodies. **And accurate experiment shows that this variation, in the case of black bodies, occurs in such a way that the product of the wave length λ_m of maximum radiation and the absolute temperature is constant**; or, in terms of algebra, —

$$\lambda_m(t° + 273) = \text{Constant.} \qquad \text{Eq. 93}$$

Upon laws such as these are based the optical and radiation methods for high temperature measurement described in § 249, and widely employed, not only by investigators, but also by engineers.

(iii) *Stewart-Kirchhoff Law.* Let us denote by H the amount of heat which falls upon a body at temperature $t°$ in one second: and by h denote the amount of heat which this body absorbs in one second when at the same temperature. The ratio $\frac{h}{H}$ is what Kirchhoff calls the **absorption** of the body, and is denoted by A. A body which absorbs all the heat falling upon it is said to be **absolutely black**; its absorption is unity.

Let us now denote by E the amount of heat which this same body, at this same temperature, would emit in one second in virtue of its hotness only. This quantity E is called the **emission** of the body. Now Balfour Stewart in England and Kirchhoff in Germany independently proved that

I. **The ratio of the emission to the absorption of any body depends upon the temperature only; and**

II. **This ratio is numerically equal to the emission of a perfectly black body at the same temperature.**

The quantitative form in which these laws are expressed is due to Kirchhoff. From these two statements it is evident that a piece of polished metal will not emit so strong a light as a piece of metal covered with lampblack and heated to the same temperature. For since the absorption of polished metal is less, its emission will also be less. This result is in thorough accord with what we might expect **if radiant heat** consists **in a wave motion**. For in our study of vibrating bodies (see *Resonance*, § 195), we have found that any source of vibration, *e.g.* a tuning fork or a pendulum, can absorb only those vibrations which are of the same period as its own.

A source of vibrations is a source of radiant energy. It has been proved that the red-hot carbon emits waves **of all** lengths. It might, therefore, be expected that the carbon when cool would **absorb** waves **of all** lengths, and thus appear black, *provided* the disturbances involved in the radiation of heat are ordinary mechanical disturbances such as we have been studying under the subject of dynamics and sound. Once in possession of this general law, we have the explanation of a host of important household and meteorological phenomena. For instance, land being, in general, a much better absorber than water for radiant energy, it follows that the surface of the land will cool more quickly than the surface of a lake on a clear night; conversely, it will rise in temperature more rapidly than the water during the day when the sun is shining. Indeed, the ground at the bottom of a lake or river may radiate energy through the water so rapidly as to drop the temperature of the lake bottom below the freezing point, thus producing a coating of ice which clings fast to the bottom and is known as "anchor ice."

Air being a very poor absorber of heat, the radiant energy of the sun passes directly through it with small loss, strikes the ground, is there transformed into heat, and raises the temperature of the air just over it by conduction and convection.

V. Some Effects of Heat

262. Among the multitudinous effects of heat the following two have been selected as being perhaps the most important.

1. Change of Dimensions or of Stress.
2. Change of Molecular State.

CHANGE OF DIMENSIONS

1. *Expansion of Solids*

263. The familar fact that solids as a rule increase in size as their temperature rises is illustrated in the manufacture of large guns, which are built up of coaxial cylinders, the inside diameter of each one being a trifling bit smaller than the outside diameter of the preceding. When the outside cylinder is sufficiently heated it will slip over the inside one, and, on cooling, will shrink so as to form what is practically a single piece of metal, but one which is stronger than a single casting. Wagon tires are shrunk on to wooden wheels in the same manner. On account of the changes in length caused by temperature variations, engineers generally rest one end of a girder bridge, not directly upon the pier, but upon an iron roller as shown in Fig. 184.

Coefficients of Expansion

Many accurate experiments by methods which must be studied in the laboratory have shown that the change in length is, for a moderate range of temperature, proportional to the change in temperature.

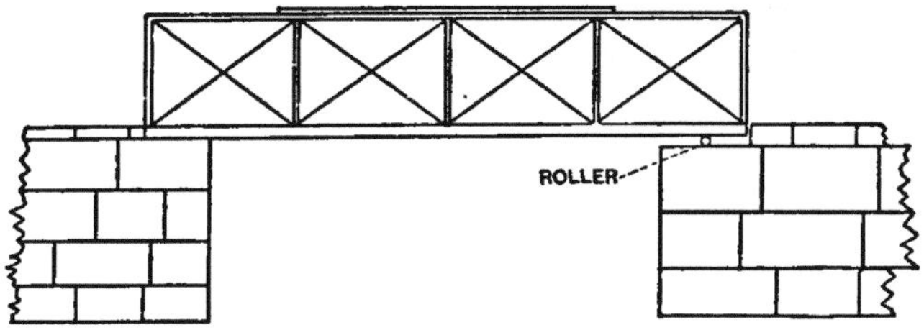

FIG. 184.—Roller providing for expansion of iron bridge.

If a bar whose length is l centimeters changes in length by e centimeters, while its temperature changes by $t°$, the change in unit length per degree is $\frac{e}{lt°}$. This increment in length experienced by each unit of length for one degree rise in temperature is called the **coefficient of linear expansion** for the material of which the bar is made.

Let us denote this coefficient by a, then

$$a = \frac{e}{lt°}. \qquad \text{Defining eq. for coef. linear expansion.} \qquad \text{Eq. 94}$$

The three quantities on the right-hand side of this equation can each be measured in the laboratory, and the coefficient a thus determined.

We are accordingly led to the following expression for the length of a bar l_t at any temperature t, when l_0 is its length at the zero of temperature: —

$$l_t = l_0(1 + at^\circ). \qquad \text{Eq. 95}$$

Observe that the value of the coefficient of linear expansion will vary from one thermometric scale to another. Measurements of the highest precision show also that this coefficient itself is a function of the temperature, thus rendering the length of a bar a function of both the first and second power of the temperature as expressed in the following equation: —

$$l_t = l_0(1 + at + bt^2). \qquad \text{Eq. 96}$$

In general, the coefficient b is very small.

264. In the case of many alloys the expansion is profoundly modified by the percentage composition. Taking advantage of this fact, the French physicist Guillaume has succeeded in discovering a nickel-steel alloy — about 36% nickel — for which the linear coefficient a has a value of less than one part in a million. Owing to the fact that the size of a body constructed of this metal is practically invariable with temperature, the substance has been called "invar." The value of its linear coefficient is as follows:

$$a = (0.965 - 0.00017\ t_0)10^{-6}.$$

The usefulness of such a metal for scales of length employed in geodetic measurements and for clock pendulums is at once apparent. —

Table of Linear Coefficients

(Centigrade Scale)

Glass	0.0000085	$= \frac{1}{120000}$
Platinum	0.0000085	$= \frac{1}{120000}$
Steel	0.000012	$= \frac{1}{85000}$
Brass	0.000018	$= \frac{1}{55000}$
Copper	0.000019	$= \frac{1}{52000}$
Aluminium	0.000023	$= \frac{1}{43000}$
Zinc	0.000029	$= \frac{1}{34000}$
Invar	0.00000096	$= \frac{1}{1000000}$

Study this table to discover why platinum wire is fused into the glass of an incandescent lamp, instead of copper or any other cheaper wire.

265. Under certain circumstances it is possible for a substance to have a coefficient of expansion which is *apparently* negative. That is, the substance may contract instead of expanding on being heated, as the following experiment shows: —

Take a piece of ordinary rubber tubing from three to five feet in length. Into each end fasten a three-way connecting tube of brass, as shown in Fig. 185. These can be wired in very easily. The upper end of the tube is attached to a support. To the lower end is fixed a heavy stretching weight. Cork up one of the openings at the top, and into the other introduce a current of steam. The rubber will contract to such an extent as to be easily seen at once by the whole class. We might be inclined to say that under these circumstances rubber has a negative coefficient of expansion. If, however, the heavy weight be replaced by a light one, we shall find that the rubber expands when the steam is introduced. From this it would seem that the *apparently* negative coefficient of expansion is due to a change of the elasticity of the rubber when heated. In other words we may say that Young's Modulus (§ 127) for rubber increases on heating so that a given stretching force produces less elongation. The above phenomenon cannot, then, be properly called "negative expansion," not at least in the usual sense of this term.

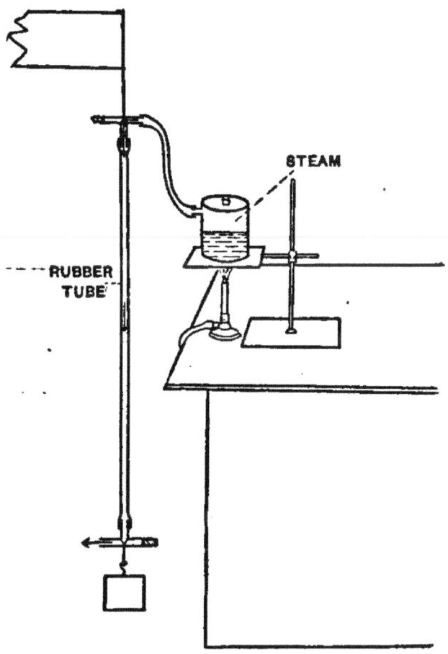

FIG. 185. — The contraction of rubber on heating.

Problems

1. Prove by squaring each side of Eq. 95 that for any solid the coefficient of superficial expansion — that is, the rate at which area increases with temperature — is twice the linear coefficient.

2. Prove in like manner that the coefficient of voluminal expansion is three times the linear coefficient.

3. Look up the subject of compensated pendulums in some encyclopedia or larger treatise on physics, and be able to explain the principle by means of a diagram on the blackboard.

4. A wagon tire 300 cm. in circumference is heated until its length has increased by 1 cm., *i.e.* by one third of one per cent. By what per cent has the area inclosed by the tire increased?

Ans. Two thirds of one per cent.

5. For every degree (centigrade) that iron is heated it increases in length by one thousandth of one per cent. When the temperature rises from 0° to 30°, how much longer will an iron bridge of 200 feet span become? *Ans.* 0.72 inch.

6. Since glass, copper, etc., expand on heating and contract again on cooling, why not employ a rod of definite length made of some one of these materials to measure temperatures? To each length of the rod would correspond a definite temperature. What practical objections?

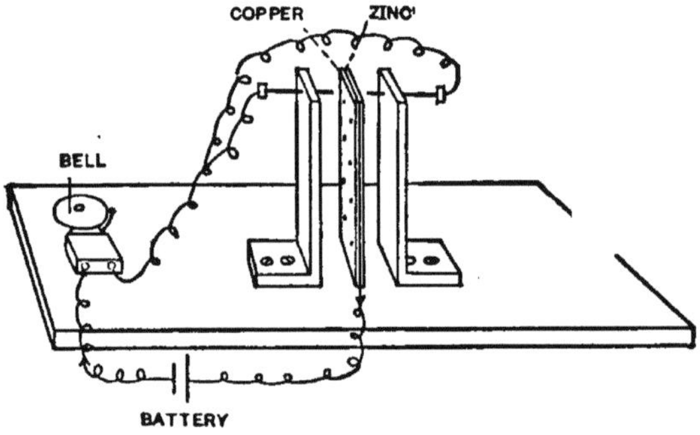

FIG. 186. — Device for announcing change of temperature.

7. Explain, by diagram on blackboard, just how a strip of zinc soldered to a strip of sheet copper, as indicated in Fig. 186, can be employed to announce (by the ringing of a bell) when a room is too hot and when too cold.

8. An iron cannon ball of radius 5 cm. is heated until its radius has been increased by $\frac{1}{50}$ of one per cent. Find by what per cent the volume of the ball has been increased.

9. Long horizontal steam-pipes are always supported in such a way that they are free to move at one end. Why?

10. A brass wire 100 ft. long is fixed at its upper end and carries a scale pan at its lower end. When a mass of one kilogram is added to the scale pan the wire is elongated by $\frac{1}{16}$ inch. To what extent must the temperature of the wire be lowered in order to restore the wire to its original length?

11. A clock with a steel pendulum is adjusted to beat seconds at a temperature of 20° C. Find how many seconds it will gain each day when its temperature falls to 0° C.

12. An iron rail 30 feet long is laid during the summer when the temperature is 30° C. Find its length in winter when the weather is 10° C. below zero.

13. An iron telegraph wire is stretched between two poles which are 400 ft. apart and has a sag of 5 ft., and a stress of 10,000 lb. per square inch. What stress will be caused by a drop in temperature sufficient to make the sag only 4 ft.?

14. Suppose the limit of accuracy in measuring the standard meter to be $\frac{1}{2000}$ mm. What variation in temperature of the bar — imagine it to be platinum — would be permissible during the measurement?

CHANGE OF DIMENSIONS

2. *Expansion of Liquids*

266. Since liquid bodies are defined (§ 122) as those which have no shape of their own, but assume the shape of the containing vessel, it is evident that a liquid body has no characteristic edges or lines, and hence no coefficient of linear expansion. But liquid bodies have characteristic volumes, and these do change when heated.

Definition of Coefficient of Expansion of a Liquid

If we denote by V_1 and V_2 the volume of a liquid at the temperatures t_1 and t_2 respectively, and if we call the coefficient of expansion b, its defining equation becomes

$$b = \frac{V_2 - V_1}{V_1 (t_2 - t_1)} \qquad \text{Defining eq. for cubical expansion.} \qquad \text{Eq. 97}$$

When, therefore, we speak of the coefficient of expansion of a liquid, we shall always refer to its **coefficient of cubical expansion, i.e. to that fraction of itself by which a volume of liquid increases when heated through one degree.**

Of all liquids, water and mercury are perhaps the two of most importance in the physical sciences. We shall consider these briefly.

Mercury

267. The very elegant method by which Regnault* determined the density of mercury at various temperatures, and hence its coefficient of expansion, has long served as a model for succeeding experimenters. The process is as follows: —

Two U-tubes, filled with mercury, as shown in Fig. 187, are connected at the bottom by a tube of small caliber; and this tube contains a little air which serves to separate the two columns of mercury. We have then really two U-tubes of mercury joined by a capillary tube of air.

* *Regnault* (b. 1810, d. 1878), a French physicist distinguished for his experimental skill.

In the portion of the tube marked C, the pressure must be the same throughout the volume of air. For if the pressure due to each of the mercury columns were *not* the same, the mercury surfaces would move one way or the other until they were equal.

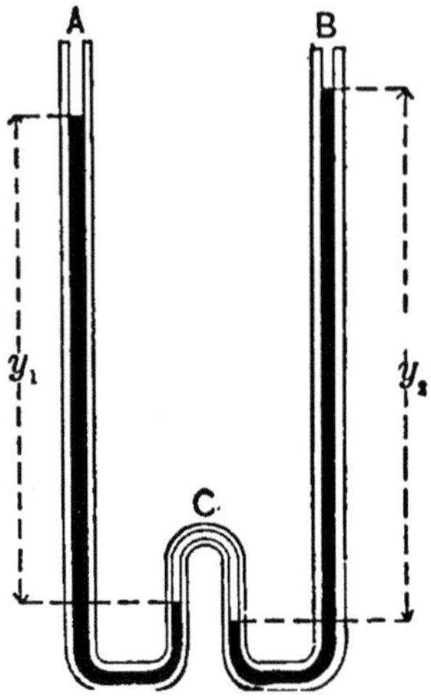

FIG. 187. — Regnault's experiment.

Suppose now that we heat the mercury in the branch B until it reaches a temperature t_2, and that in the branch A we cool to the temperature t_1. The density of the mercury in B will diminish on heating; consequently the difference of level indicated by y_2 will increase. The density of the mercury in A will increase on cooling, and hence the difference of level in A will diminish; call this difference y_1. The pressure in C due to mercury in A is, by the Third Law of Hydrostatics, y_1D_1g; while the same pressure in C due to B is y_2D_2g.

But these pressures must be the same. Hence,

$$y_1D_1g = y_2D_2g.$$

$$\therefore \frac{y_2}{y_1} = \frac{D_1}{D_2}.$$

$$\frac{y_2 - y_1}{y_1} = \frac{D_1 - D_2}{D_2}.$$

By measuring y_2 and y_1, Regnault was thus enabled to determine the fraction of itself by which the *density* of mercury was changed when its temperature varied from $t_1°$ to $t_2°$.

But how can one obtain from this fraction the value of the coefficient of expansion?

Consider any constant mass of mercury M, whose volume is changing in consequence of a variation in temperature. The defining equation for density gives

$$M = V_1D_1 = V_2D_2.$$

$$\therefore \frac{V_2}{V_1} = \frac{D_1}{D_2}.$$

$$\therefore \frac{V_2 - V_1}{V_1} = \frac{D_1 - D_2}{D_2}.$$

$$\therefore \frac{V_2 - V_1}{V_1} = \frac{y_2 - y_1}{y_1}$$

Now substituting in Eq. 97,

$$b = \frac{y_2 - y_1}{y_1(t_2 - t_1)}, \qquad \text{Eq. 98}$$

where y_1 and y_2 are differences of level corresponding to temperatures t_1 and t_2 respectively, and b is the coefficient of cubical expansion. These quantities y_1 and y_2 are, however, difficult to measure with accuracy.

Of course, we might take a single tube of mercury, Fig. 188, and measure its height at two different temperatures; but here the difficulty is that the glass vessel containing the mercury expands with the heat, so that the volume of the mercury is no longer proportional to the height of the column. In this way we should obtain only the **apparent expansion** of mercury. It **is the apparent** expansion **which** is **used in glass and** mercury **thermometers.**

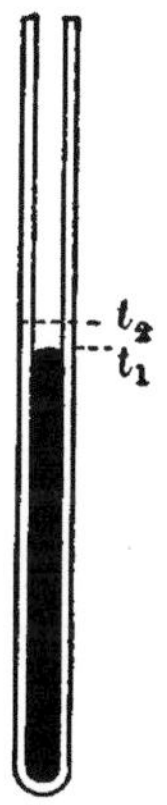

Fig. 188. Apparent expansion.

The beauty of Regnault's method is that the height of the column depends only on the temperature, and not at all upon the containing vessel. Regnault thus measured the **absolute expansion.** The value which he thus obtained for the average coefficient of expansion of mercury between 0° and 100° was 0.0001815.

The most accurate of recent values is probably that obtained at the Reichsanstalt,[1] namely, 0.00018245.

Water

268. From many points of view, water is a remarkable substance. This is strikingly exhibited in its behavior under heat. In general, warm liquids are less dense than the same liquids cold; and this is true of water except between the temperatures 0° C. and 3°.98 C. The manner in which water expands under heat is clearly set forth in the curve, Fig. 189.

The coefficient of expansion of water has recently been deter-

[1] *Reichsanstalt,* the great national physical laboratory of Germany, situated in Charlottenburg, a suburb of Berlin.

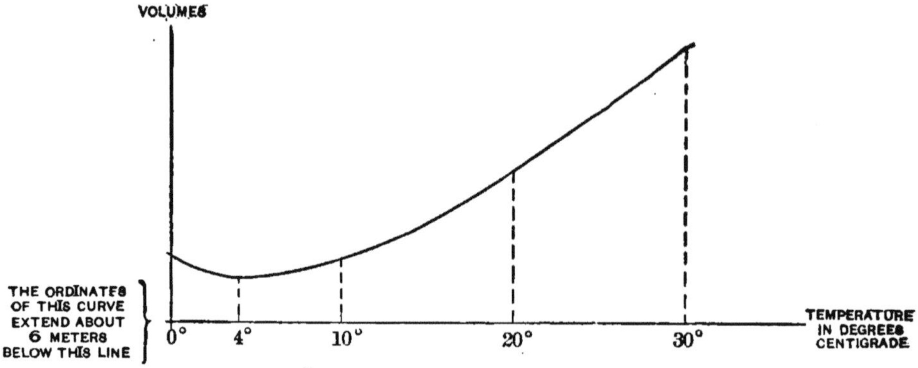

FIG. 189. — Expansion of water.

mined at the Reichsanstalt by the method which Regnault employed for mercury. The results are summarized in the following table: —

DENSITY OF WATER

TEMPERATURE CENTIGRADE	DENSITY	VOLUME
0°	0.999867	1.000132
3.98	1.000000	1.000000
10	0.999727	1.000272
15	0.999126	1.000874
20	0.998229	1.001773
25	0.997071	1.002937
30	0.995673	1.003345
35	0.994057	1.005977
40	0.992241	1.007819

The preceding figure (189) is merely a graphical description of the results contained in this table. Or, if desired, a still briefer description of the facts may be obtained in the following empirical equation,

$$V_t = V_4\left(1 + \frac{(t^\circ - 4^\circ)^2}{144000}\right), \qquad \text{Eq. } 98\tfrac{1}{2}$$

where V_t is the volume of any given mass of water at t° C. and V_4 the volume of the same mass at 4° C.

The maximum density of water at 4° C. may be demonstrated by loading a small hollow glass sphere with sand until it will just sink in water at 0°. If placed in a vessel of ice water, it will sink; if now the ice be removed the water will gradually rise in temperature, the float will also rise, and after the temperature passes 4° C. the float will again sink. The enormous economic importance of this anomaly of water is seen in the fact that it prevents our lakes and rivers from freezing except over the surface.

Problems.

1. Graham's compensated pendulum is made of a steel rod which carries at its lower end a glass cylinder containing mercury. Show that the linear expansion of the steel (which extends to the base of the cylinder) will be very approximately counterbalanced by the cubical expansion of the mercury when the height of the mercury is $\frac{1}{15}$ the height of the steel pendulum.

2. A large thermometer bulb, made of glass, and placed in melting ice, holds 200 g. of mercury. How much mercury will overflow when placed in a bath of steam at atmospheric pressure?

3. A mercury barometer in a laboratory at 15° C. reads 756.2 mm. What are the three most important corrections which must be made in order to obtain the true barometric height?

4. A glass cylinder which weighs 5 g. in air, weighs only 2.994 g. in water at 20° C. Find its weight in water at 40° C.

5. Compare the density of water at 30° as obtained
(i) from the table in § 268, with that obtained
(ii) from Eq. 98½.

Change of Dimensions or of Stress

3. *Expansion of Gases*

269. We have already seen (§ 248) that the earliest thermometers were made by inclosing a certain mass of gas in a glass bulb whose stem was sealed by immersing the end in a liquid. Such a device is at once the simplest and earliest for studying the effects of heat upon gases.

Having substituted mercury for water in the thermometer of Galileo, Amontons, in 1699, discovered that the change of pressure produced by heating through a range of 100° was the same whatever the amount of air inclosed in the bulb.

Law of Charles

By definition, the volume of a gas is always the volume of the containing vessel. Boyle's Law states the experimental fact that to each volume of gas there corresponds a definite pressure, provided the mass and the temperature remain constant. If, however, the temperature varies, the effect may be
(1) a change of volume while the pressure remains constant; or
(2) a change of pressure while the volume remains constant; or
(3) a change of *both* pressure and volume.

To determine the effect of heat on a gas, we must then measure either

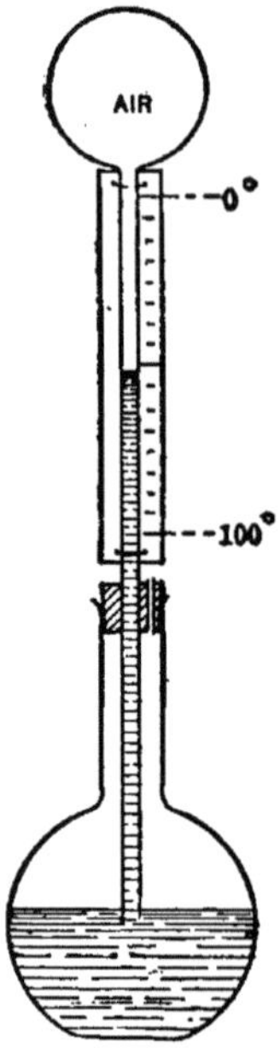

FIG. 190. — Galileo's thermometer.

(1) the ratio of increase of volume to original volume in a gas kept at constant pressure; or

(2) the ratio of increase of pressure to original pressure in a gas kept at constant volume,

in each case for a rise of one degree in temperature.

These two quantities were first roughly measured by Charles, afterwards more accurately by the French chemist, Gay-Lussac (1778–1850).

The capital discovery was thus made that:—

(*a*) These two quantities are the same for any one gas.

(*b*) They have very approximately the same numerical value for *all* gases, namely, $\frac{1}{267}$. Later and more accurate observations have shown that this value is more nearly $0.003665 = \frac{1}{273}$.

The actual measurement of these quantities is a somewhat difficult matter, requiring many precautions, and may well be reserved for the student's second or third year in physics. But the algebraic description of Gay-Lussac's *results* is a matter of the utmost simplicity. It is as follows:—

Let V_0 and V_t represent the volumes of a given mass of gas at the temperatures 0° and $t°$ respectively; then, so long as the pressure remains constant,—

$$V_t = V_0(1 + 0.003665\,t°). \quad \text{Law of Charles.} \qquad \text{Eq. 99}$$

If, in like manner, P_0 and P_t represent the pressures of a given mass of gas at the temperatures 0° and $t°$ respectively, then, so long as the volume remains the same,—

$$P_t = P_0(1 + 0.003665\,t°). \qquad \text{Eq. 100}$$

Gay-Lussac's result may be summarized by saying that the pressure of a given mass of gas at any temperature $t°$ varies directly as a factor $(1 + at)$ where a has the value $\frac{1}{273}$.

Boyle's Law

270. Boyle having already shown that the pressure varies inversely as the volume, we may combine these two results into a single expression, thus:—

$$P = \frac{C\left(1 + \frac{t^\circ}{273}\right)}{V}, \qquad \text{Eq. 101}$$

where C is a proportionality factor.

Clearing of fractions,

$$PV = \frac{C}{273}(273 + t^\circ),$$

or

$$PV = RT, \qquad \text{Eq. 102}$$

where $R = \frac{C}{273}$ and $T = t^\circ + 273$.

In other words, if we measure temperature from a zero which is 273° C. below the melting point of ice and denote these temperatures by T, we find that the pressure and the volume each vary directly as the temperature. **A gas which obeys Boyle's law perfectly is said to be an ideal or perfect gas.** Temperatures measured on this scale are called "ideal gas temperatures." Perfect gases do not occur in nature; but Lord Kelvin has proved that if they *did* occur, temperatures measured by use of any one of them would coincide with absolute temperatures, which we shall meet a little later.

The value of R in Eq. 102 is a quantity which will have different values for each particular gas, but will remain constant for any one gas. For unit mass of air the value of R is 2.872×10^6 ergs/gr. 1° C. But if pressures be measured in pounds per square foot, masses in pounds and volumes in cubic feet, then for air $R = 53.18 \frac{\text{foot-pounds}}{\text{lb. } 1^\circ \text{ F.}}$.

If we denote by m the mass of the gas being studied and consider this as a variable, then Boyle's Law may be written in its most general form as follows: —

$$PV = mRT.$$

Kinetic Theory of Gases

271. Any satisfactory theory of gases must explain at least the following observed phenomena: —

1. The Law of Boyle.
2. The Law of Charles and Gay-Lussac.
3. Dalton's Law — A mixture of gases having no chemical reaction on each other exerts at any given temperature a total

pressure which is equal to the sum of the partial pressures which would be produced by each gas separately if it alone occupied the containing vessel at the given temperature.

4. The Hypothesis of Avogadro — that "the masses of molecules are in the same ratio as the densities of the gases to which they belong," or what amounts to the same thing, *the number of molecules in equal volumes of all gases is the same.* It is here supposed, of course, that the temperature and pressure is the same for each gas.

Assumptions

272. The kinetic theory of gases, which is largely the creation of Bernouilli, Clausius, and Maxwell, begins with the following assumptions: —

(i) That a gas consists of a large number of small smooth particles which rebound with unchanged speed when they strike the wall of a containing vessel at the same temperature as that of the gas.

(ii) That the particles are so small as seldom to collide with each other.

(iii) That the particles exert no appreciable attraction upon each other.

From the first assumption it would follow that the kinetic energy of such a system ought to remain constant so long as the walls of the containing vessel remain at the same temperature.

Definitions

273. Let P = pressure of gas, measured in dynes/cm.^{2}

V = volume of containing vessel, in cm.^{3}

N = total number of particles — molecules.

$n - \frac{N}{V}$ = number of particles per cm.^{3}

m = mass of each particle.

u = square root of mean square of the speed of the particles

$= \sqrt{v_x^2 + v_y^2 + v_z^2}$ where v_x, v_y, v_z are the components of the mean velocity of all particles along the axes of X, Y, and Z, respectively.

We proceed to show that the pressure may be computed from simple dynamical principles to be

$$P = \frac{nmu^2}{3}.$$

Since the particles are supposed to move at random in all directions, the sum of the squares of the v_x's, v_y's, and v_z's are each equal to the other; and hence

$$Nu^2 = 3\,\Sigma(v_x^2) = 3\,\Sigma(v_y^2) = 3\Sigma\,(v_z^2), \qquad \text{Eq. 103}$$

or, the mean value of $v_x^2 = \frac{1}{3}u^2$. Eq. 104

The fundamental hypothesis of the gas theory is that the pressure on any wall of the containing vessel is due to the momentum which is delivered to that wall per second by the gas particles. Imagine the vessel to have two parallel walls perpendicular to the axis of X and these walls to be a centimeters apart. The Y and Z components of the velocity being each parallel to these walls will contribute nothing to the pressure on these walls.

Limiting our consideration now to a single particle, this particle will travel to and fro from one of these walls to the other $\frac{v_x}{2a}$ times per second. At each impact the velocity of the particle changes by $2v_x$ and its momentum by $2mv_x$. The momentum imparted to the wall per second by this particle is therefore $2mv_x\, \frac{v_x}{2a}$ or $\frac{m}{a}v_x^2$. But momentum per second means average force; so that the expression just obtained gives us the mean force which a particle rebounding from one of these walls to the other would exert upon each of them.

Now consider N particles; then if the total force on each of these walls be denoted by F, we shall have

$$F = \frac{m}{a}Nv_x^2.$$

But since $v_x^2 = \frac{1}{3}u^2$, it follows that—

$$F = \frac{mNu^2}{3a}. \qquad \text{Eq. 105}$$

If s is the area of the wall over which this force is exerted, the pressure will be

$$P = \frac{F}{s} = \frac{Fa}{V}.$$

Substituting the value of F from Eq. 105, we have

$$P=\frac{1}{3}\frac{Nmu^2}{V} \qquad \text{Eq. 106}$$

$$=\frac{1}{3}nmu^2,$$

which is the fundamental equation of the kinetic theory of gases, a result which we now proceed to interpret.

Laws of Boyle and Charles

274. If now we make the common and perhaps reasonable assumption that the kinetic energy resident in any mass of gas is the kinetic energy of translation of the small particles and that the ideal gas temperature (§ 270) is proportional to the mean kinetic energy of all the particles, we have $T \propto \frac{1}{2}mu^2$, and

$$T=\frac{Cmu^2}{2}, \qquad \text{Eq. 107}$$

where C is an unknown proportionality constant.

Let R be a constant such that

$$R=2/3C \times \text{total number of particles} = 2N/3C.$$

Then it follows from Eq. 106 that

$$PV=RT,$$

which is precisely the law of Charles and Boyle as derived by experiment.

Dalton's Law

Total pressure = Sum of partial pressures.

275. This follows at once, since we assume that the particles are so small as not to interfere with one another; hence the pressure would vary as the total number N of particles in the vessel, or

$$P \propto N.$$

Avogadro's Hypothesis

276. For any two gases, we may write from our fundamental equation

$$P_1=\frac{1}{3}n_1m_1u_1^2$$

and

$$P_2=\frac{1}{3}n_2m_2u_2^2.$$

Since the temperatures and pressures are the same for each gas, we have

$$T_1 = T_2 \therefore \tfrac{1}{2} m_1 u_1^2 = \tfrac{1}{2} m_2 u_2^2$$

and $$P_1 = P_2 \therefore \tfrac{1}{3} m_1 u_1^2 n_1 = \tfrac{1}{3} m_2 u_2^2 n_2.$$

But these equations can be true simultaneously only when

$$n_1 = n_2,$$

which is Avogadró's hypothesis.

V. Effects of Heat (*continued*)

2. Change of Molecular State

277. There are probably not many people whose memories reach back to years so early that they recall their surprise at first seeing a piece of ice melt in their tiny hands and apparently disappear. The change from water to steam which occurs in a boiling teakettle is equally familiar. These two instances of **melting** and **boiling** are so typical of many similar cases met with in nature that they deserve close attention.

Melting

278. (i) In the first place, it is well known that water and ice have the same chemical composition and practically the same chemical properties. A similar statement concerning solid and liquid paraffine would be true.

(ii) In the second place, a glass of water so long as it con tains ice and is stirred does not become either hotter or colder on standing. The ice may melt away; but, so long as there is any ice left, the water will remain approximately at what we call "the temperature of melting ice."

(iii) Thirdly, the temperature of melting ice can be changed by placing the ice under pressure. The most convenient method of showing this is perhaps the following: —

Choose a block of ice having a section of from 100 to 200 square centimeters. Support it, as indicated in the figure, upon two stools or boxes. Over the ice tie a loop of steel piano wire about one half millimeter in diameter. Hang a mass of about 25 kilograms on the wire. In this way you can subject the ice to a pressure of several hundred kilograms per square centimeter. Compute the area of the wire in contact with the ice. The weight of the mass attached divided by this area will roughly give the pressure. The block of ice will, in general, be not far from its melting temperature; but the melting will all occur at the surface until pressure is put on the wire; it will then be observed that the ice just underneath the wire melts also and lets the wire cut its way through the block. But the water behind the wire freezes at once, so that the ice remains a solid block although it has just been cut through. The effect of pressure is, then, to produce melting, *i.e.* to lower the freezing point of water.

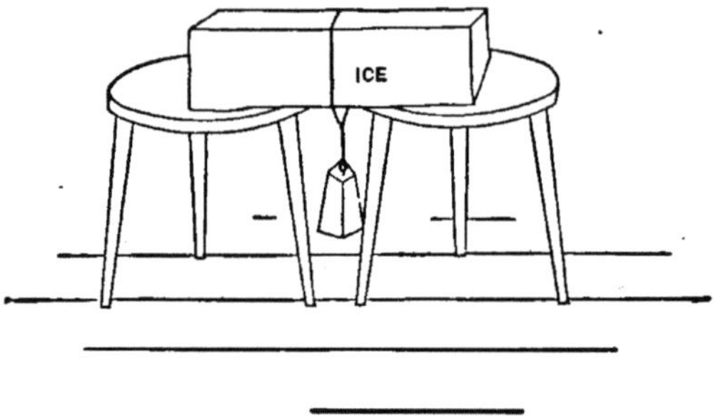

FIG. 191.—Bottomley's experiment showing effect of pressure on melting point of ice.

This process of melting under pressure and again freezing is known as "regelation." In this principle lies the explanation of the fact that cold snow will not "pack" into a good snowball; also of the fact that glaciers apparently "flow" down a mountain side as if ice were a very viscous fluid. In front of points of great pressure the ice melts; the pressure being by this means transferred to some other point, the water again freezes, but in a new configuration. In this manner the glacier gradually adapts itself to the shape of the valley as it proceeds downward.

Our experience with ice and all other solids may be summarized in the following statement: —

Experimental Law of Melting

"**The pressure** remaining **the** same, **there is a definite melting point for every solid; and (provided the** mass **be** stirred) **however much heat be** slowly **applied, the** temperature of **the** whole **remains at the melting point** till **the last particle** is **melted.**" — TAIT, *Heat*, Ch. VIII.

Boiling

279. (i) In like manner, the chemical composition and the chemical properties of steam do not differ from those of water.

(ii) However much you boil the water in the teakettle, its temperature does not change after boiling has once begun.

One can easily satisfy himself of this with an ordinary mercury thermometer.

(iii) But the temperature of boiling water is very much changed by a change of pressure.

This is most easily proved by boiling in a kettle of water a bottle half filled with water. If this bottle be corked while still boiling, it may be removed with so little air inclosed in it that the pressure on the inclosed water soon becomes greatly less than that of the atmosphere. Under these circumstances, the water in the bottle will continue to boil long after it has reached a temperature not uncomfortable to the hand.

In the laboratory this phenomenon is most commonly shown by boiling over a Bunsen flame a Florence flask half filled with water. Cork the flask while boiling. Remove it at once to a support previously prepared and cool with cold water. Violent boiling will occur during the cooling process. To show that to each different pressure corresponds a definite boiling point, a very simple method is that indicated in Fig. 193. Through the cork of a large-necked flask insert a thermometer, a U-tube that can be partially filled with mercury and used as a pressure gauge, also an escape pipe *e*.

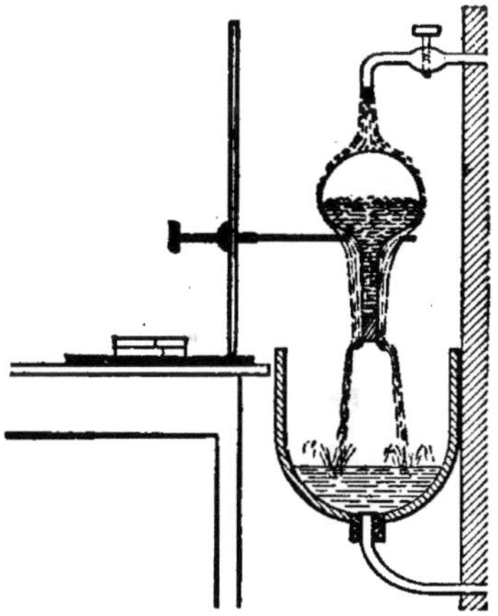

FIG. 192. — Franklin's experiment showing boiling point lowered by diminution of pressure.

The escape pipe is fitted with a tip of rubber tubing and a pinchcock. By a little manipulation of the pinchcock one can produce and maintain any desired height *h* in the pressure gauge. In like manner, to observe the effect of diminished pressure, one can connect the escape pipe *e* with an air pump. For each different *h*, the thermometer registers a different boiling point. It will thus be seen that the boiling point of water is much more sensitive to changes of pressure than is the melting point of ice. Not only so, but the effects of pressure in these two cases are exactly opposite: the melting point of ice is lowered by pressure; the boiling point of water is raised by it.

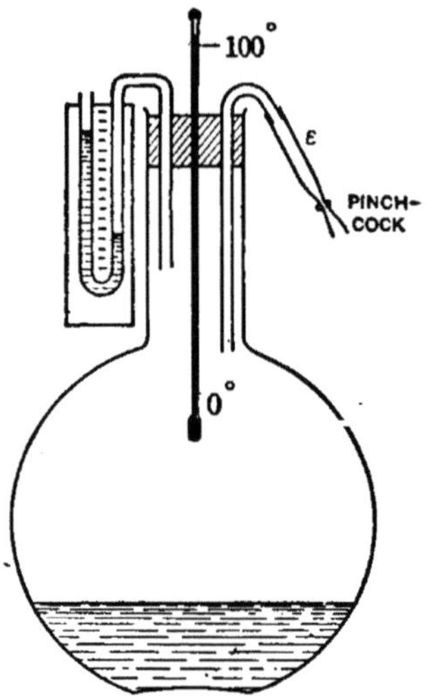

FIG. 193. — Boiling point raised by increase of pressure.

Our experience concerning boiling may be summarized as follows: —

"**The pressure remaining the same, there is a definite boiling point for the free surface of every liquid; and (provided the mass be stirred) however much heat be applied, the temperature of the whole remains at the boiling point till the last particle is evaporated.**"

— TAIT, *Heat*, Ch. IX.

280. It can be shown that liquids are undergoing evaporation at all temperatures. But the region over the free surface of a liquid may contain so much vapor from the liquid that condensation will occur as rapidly as new vapor is formed. The atmosphere over the liquid is then said to be **saturated.**

The process of **boiling** occurs when the temperature has reached a point where vapor is formed in the interior of the liquid, and rises to the top. But unless the pressure of this hot vapor were equal in pressure to that at the free surface of the liquid, the bubble of vapor would collapse in the liquid.

The temperature of boiling (what we ordinarily call the boiling point) is then simply that temperature at which the saturated vapor of the liquid exerts a pressure equal to the pressure at the free surface of the liquid.

PRESSURE OF SATURATED WATER VAPOR

TEMPERATURE CENTIGRADE	PRESSURE IN ATMOSPHERES	TEMPERATURE CENTIGRADE	PRESSURE IN ATMOSPHERES
0°	0.006	90°	0.691
10°	0.012	100°	1.000
20°	0.023	110°	1.415
30°	0.042	120°	1.962
40°	0.072	140°	3.576
50°	0.121	160°	6.120
60°	0.196	180°	9.929
70°	0.306	200°	15.380
80°	0.466	220°	22.882

The preceding table, due to Regnault, shows us very clearly just how the boiling point of water varies with the pressure.

Distinction between a Gas and a Vapor

281. With such apparatus as that indicated in Fig. 193, adapted of course to accurate laboratory measurement, it is readily shown not only that the pressure of a saturated vapor of any given substance depends upon its temperature, but also that it depends upon nothing else. Every student should clearly grasp the idea that the saturated vapor pressure does *not* depend upon the volume of the vapor. The simplest and clearest illustration of this is perhaps the following: —

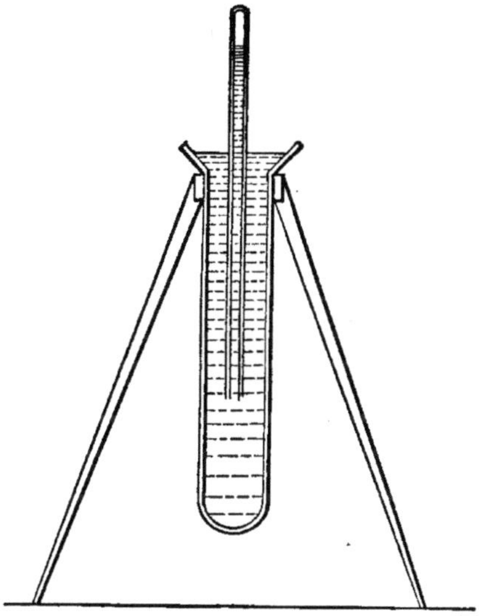
FIG. 194.

Fill a barometer tube with mercury and invert it over a deep basin of mercury as shown in Fig. 194. At the top of the tube there will be a small Torricellian vacuum. With a penfiller, having its tip bent as indicated in the accompanying figure, a few drops of sulphuric ether may be introduced into the bottom of the barometer tube. The ether at once rises to the surface of the mercury and the Torricellian vacuum becomes filled with the vapor of ether. The result is that the mercury column falls from a height of 76 to a height of perhaps 36 centimeters; for the pressure of this vapor is determined by the temperature of the room, which is probably about 20° C. The entire volume of the tube above the mercury is filled with saturated ether vapor.

FIG. 195.

If now the barometer tube be raised or lowered in the mercury bath, it will be observed that the height of the *mercury* column inside the tube does not vary, so long as the temperature remains constant. See Fig. 196, where the same barometer tube is shown in three different positions. In other words, the vapor pressure does not vary with the volume. How does the mass of *vapor* vary as the tube is raised and lowered?

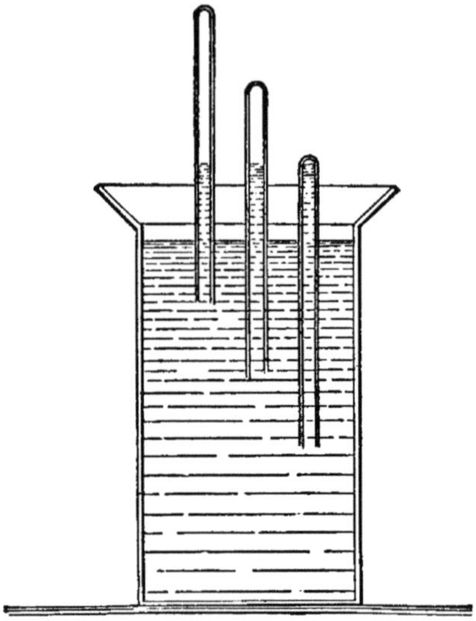

Fig. 196. — Illustrating the fact that vapor pressure is independent of volume.

Now take a similar tube, fill it with mercury, and invert it so as to form a barometer as before; but into the Torricellian vacuum introduce now a few bubbles of air, instead of ether. The behavior of the mercury column as the tube is raised and lowered is now totally different. When the volume is halved the pressure is now doubled and so on; in short, the gas obeys Boyle's Law and the apor does not.

This entire question as to the difference between a gas and a vapor was cleared up about 1869 by Andrews at Glasgow, who showed that two conditions must be fulfilled in order to condense a gas into a liquid, viz.: —

(i) The temperature must fall below a certain critical point, and

(ii) The pressure must, for each particular temperature, rise above a certain definite value.

He illustrates this by plotting the pressure as a function of the volume for a given mass of carbonic acid gas, at various temperatures. When the compression occurs at a constant temperature, the curve which connects pressure and volume has the form shown in Fig. 197 and is called an "isothermal curve." The horizontal portion such as that indicated by *ab* in the figure represents

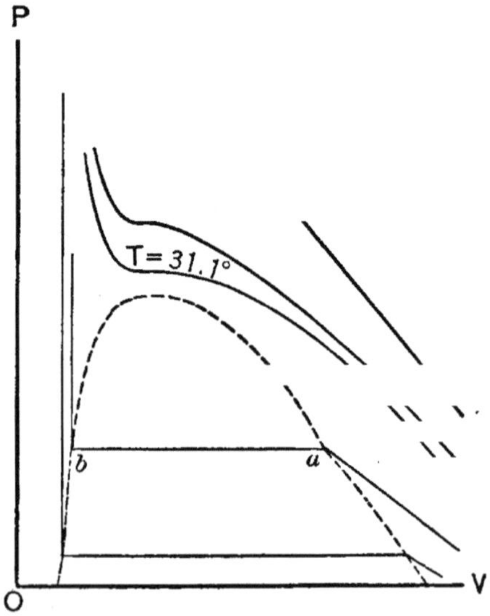

Fig. 197. — Isothermals for carbonic acid gas.

a state in which the substance is partly gaseous and partly liquid. Thus at *a* the pressure assumes a constant value and retains it until the condensation reaches the point *b*. It is this constancy of pressure which, as we have seen above, indicates saturation of the vapor.

Now Andrews discovered that after reaching a certain temperature (in the case of carbonic acid gas, about 31° C.) the isothermals ceased to have any region of saturation, *i.e.* they no longer showed any horizontal portions.

In other words, when condensation from a gas to liquid takes place above a certain temperature, no free surface occurs in the liquid; the change is a perfectly continuous one.

FIG. 197 *bis*. The critical tube.

282. A very satisfactory mode of illustrating the passage of a liquid from the region below to the region above the critical temperature is by use of what is called a "critical tube," that is, a thick-walled glass tube a little less than half filled with ether, and closed at both ends. When this tube is suspended in a test tube filled with melted paraffin and is heated still further with a Bunsen burner, it will be observed that presently the liquid expands very rapidly, the meniscus flattens, and finally disappears. The ether now fills the whole tube and has become a gas. When the flame is removed, a hazy cloud forms in the middle of the tube, a meniscus reappears, rapid contraction occurs, the critical temperature has again been passed, and the ether has now returned from a state of gas to a state of liquid and vapor.

283. The temperature of the isothermal at which the substance ceases to show a free surface during condensation is known as the critical temperature. The point on the isothermal at which the horizontal portion just disappears is called the **critical point**. The pressure and the volume corresponding to this point are called the **critical pressure** and **critical volume** respectively.

We are now in a position to define "vapor" and "gas." **"Vapor" is the name given to a substance in the gaseous condition when *below* the critical temperature; while the term "gas"**

is confined to substances in the gaseous condition when *above* the critical temperature.

TABLE OF CRITICAL DATA

SUBSTANCE	SYMBOL	CRITICAL TEMPERATURE	CRITICAL PRESSURE IN ATMOSPHERES	DENSITY AT CRITICAL POINT. WATER AT 0°=1
Air		– 141°	39.6	
Carbon dioxide .	CO_2	30°.92	77	0.65
Carbon monoxide	CO	– 141°.1	35.9	
Ether	$C_4H_{10}O$	197°	35.8	0.208
Hydrogen .	H_2	– 220°	20.0	
Nitrogen .	N_2	– 146°	35.0	0.45
Oxygen . .	O_2	– 118°.8	50.8	0.60
Water	H_2O	365°	200.5	0.429

— From Landholt and Börnstein's *Tables.*

284. Many meteorological occurrences with which we are most vitally concerned, as well as some of the most exquisite of natural phenomena, such as geysers, are merely special cases of change of state brought about by change of temperature. Thus *dew*, *fog*, *clouds*, and *rain* are four forms of water which make their appearance when the air is saturated with water vapor at any temperature and afterward more or less cooled below this temperature. The manner in which the necessary cooling is brought about is not always the same.

In the case of dew, the bodies upon which the moisture is deposited are cooled mainly by radiation; these bodies in turn cool the adjacent layers of air below the point of saturation, or "dew point" as it is sometimes called. The result is that the moisture formerly in the air is now deposited over the surface of the stone, plant, or whatever the body may be. As we have already learned (§ 280), there is, for each particular temperature, a certain pressure which water vapor can sustain; when the amount of vapor in the air is at any time greater than what is sufficient to supply the pressure, the excess of moisture is naturally precipitated.

In the case of clouds and rain, the conditions which bring about the cooling of the air are, mainly:—

(i) The intermixture of two masses of air at different temperatures;

(ii) A land breeze bringing the moist air of the ocean over the cooler land; and

(iii) Convection currents which lift masses of moist air into higher altitudes and thus allow them to cool by expansion.

Kelvin and Aitken have shown that there is also another condition for rainfall not dependent upon change of temperature, namely, the presence in the air of small particles of dust (or, as we now know, ions) upon which the moisture can condense. The student is now prepared to follow the explanation of this fact, which he will find discussed in one of the larger text-books, such as that of Watson, Ames, or Hastings.

285. In the case of geysers, which were first satisfactorily explained by the great German chemist, Bunsen, in 1847, we have a crevasse, generally in an old lava bed, the heat of which is sufficient to raise the temperature of the spring water, which feeds into this crevasse, above the boiling point. The water in the lower part of the crevasse (which has usually become closed up on the sides so as to resemble a tube) is under considerable pressure, owing to the superincumbent water. Its temperature therefore rises above 100° *C*. If now any thin, horizontal layer of water in this tube becomes displaced — tilted perhaps — so that a part of it whose temperature is, say, 102° C., gets into a region where the boiling point is only 101°.9 C., the result is that some of the water is transformed into steam. This steam pushes some of the water out of the crevasse and thus slightly diminishes the pressure on the water below. This diminution of pressure is followed by a much larger conversion of water into steam, which in expanding, expels a large quantity of water, until the entire mass is cooled to the normal boiling point or below. Much of the water is wasted and does not fall back into the tube; this waste is supplied by springs. The above process of heating and explosion is then repeated; and so on, indefinitely, so long as the sources of heat and water remain. In the geysers of the Yellowstone Park this period is found to vary from a few minutes to several days.

In Fig. 198 is shown a model suggested by Bunsen, consisting of a sheet-iron tube, the upper end of which is soldered into a basin to catch the water when falling back. Three ther-

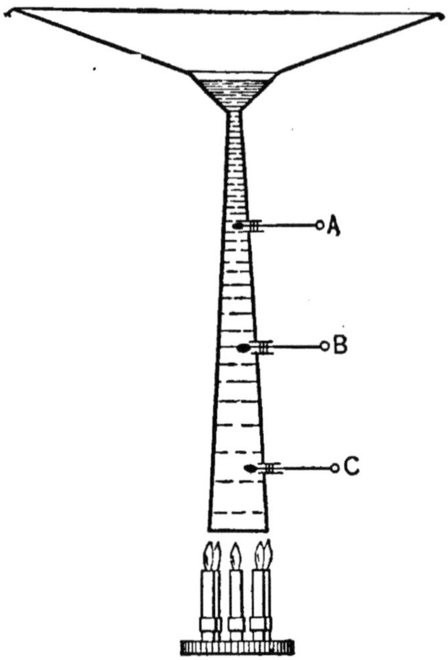

FIG. 198. — Bunsen's model of geyser.

mometers, *A*, *B*, *C*, corked into the side of the upright tube, prove that the geyser goes into action only when the temperature conditions are those described above. A triple or quintuple Bunsen flame at the bottom takes the place of hot lava in nature.

Heat of Fusion

286. It has been found by experiment that **a definite amount of heat is required** just **to melt one gram of a substance,** *i.e.* to change it from a solid to a liquid **without changing its temperature.** This quantity of heat is, in general, different for each different substance and is called the **Heat of Fusion** of the substance. For the last hundred years this quantity has been known as the "Latent Heat" of the substance. This name is, however, rather misleading; for, as we shall presently see, it is exactly during the present century that men have learned that heat is a form of energy which, when employed to melt a body, disappears as heat *of any kind*, and assumes some other form of energy concerning which we know comparatively little. There is possibly one justification for the name in the fact that when a substance freezes, instead of melting, this internal energy is returned in the shape of heat. Before water at 0° C. can turn into ice it must give off a very considerable amount of heat.

If H calories be required just to melt m grams of any substance, the heat of fusion of this substance F is measured as follows: —

$$F = H/m. \qquad \text{Defining equation for heat of fusion.} \qquad \text{Eq. 108}$$

In the case of ice, the heat of fusion is found to be very nearly 79 calories per gram.

Heat of Vaporization

287. In a strictly analogous manner, experiment shows that **it takes a definite amount of heat** just **to change one gram of**

any liquid **into vapor** without **changing the temperature.** This quantity, for any particular substance, is called its **Heat of Vaporization.**

In the case of water boiling under standard atmospheric pressure (760 millimeters of mercury), it is found that 536 calories are required to evaporate one gram without changing its temperature. The heat of vaporization of water is, therefore, said to be 536 calories per gram.

VI. The Nature of Heat

288. In our study of the effects of heat, we have nowhere considered the question as to what this something is which changes the size of a body, alters its temperature, melts and vaporizes it.

Up to the beginning of the present century it was thought to be a form of matter somewhat like air, only of much greater tenuity; for in this way it was easy to explain the fact that, in conduction, the amount of heat leaving one system of bodies is always equal to that entering the other. This substance, which was given the name of "caloric," was supposed to be so "thin" that it would penetrate solid bodies with the utmost ease. It was supposed to be without weight because hot bodies do not weigh any more than the same bodies when cold.

That there is no such substance as "caloric" and that heat is a form of energy was established through the efforts of a large number of investigators. But the picture will perhaps be essentially correct if we say that this accomplishment is mainly due to the joint labors of Rumford, Davy, and Joule.

Count Rumford (1753–1814), who was by birth an American, but by residence an Englishman and a German successively, was greatly impressed by the fact that in the boring of cannon the metal chips came out very hot. In some experiments which he tried in the arsenal at Munich he found that the heat generated by a blunt tool was proportional to the time during which the tool was driven and not proportional to the amount of abraded metal. The caloricists explained the heat which made its appearance in the chips by saying that the specific heat of brass in a finely divided condition was less than that of solid brass. But Rumford showed by experiment that this was not true; and he suggested that this "something" which a system of bodies can continue to furnish in unlimited quantities can hardly be a material substance.

Sir Humphry Davy (1778–1829), the English chemist, showed that by rubbing together two pieces of ice, in a room whose temperature was below the melting point of ice, he could not only melt the ice, but obtain a product of abrasion, so to speak, namely, water, whose specific heat, so far from being less, is more than twice as large as that of ice. Rumford's experiments were in this manner thoroughly confirmed.

First Law of Thermodynamics

289. It remained for Mr. James P. Joule (1818–1889), an English physicist, to discover the exact relation between heat and mechanical energy. He arranged a paddle wheel in a vessel of water in such a way that the paddle wheel could be kept in rapid rotation by means of a descending clock weight. By placing a thermometer in the water, he found that the more the water was stirred, the warmer it became. **But, what is more important, he measured the work done by the descending weight and the corresponding heat produced by the paddle in turning in the water, and found that in every case the ratio between these two quantities was the same.**

Let us denote by W the work done in raising the mass m (Fig. 199) through the height h; then

$$W = mgh.$$

Barring a slight correction for friction at the bearings, this is also the amount of work which the descending weight does on the paddle wheel.

Let us denote by M the mass of the water stirred, and by $t_2^\circ - t_1^\circ$ the rise in temperature produced by the stirring. Then the heat acquired by the water H is

$$H = M(t_2^\circ - t_1^\circ).$$

The capital discovery of Joule was that for every unit of mechanical energy which *disappears* as the weight descends, a definite and **constant** quantity of heat *appears* in the stirred water; *i.e.* H is proportional to W.

In terms of algebra,

$$\frac{W}{H} = \frac{mgh \text{ ergs}}{M(t_2^\circ - t_1^\circ) \text{ calories}} = J = \text{a constant}.$$

By the use of this equation, Joule determined the numerical value of this constant, *J*, and found it to be very nearly 42,000,000.

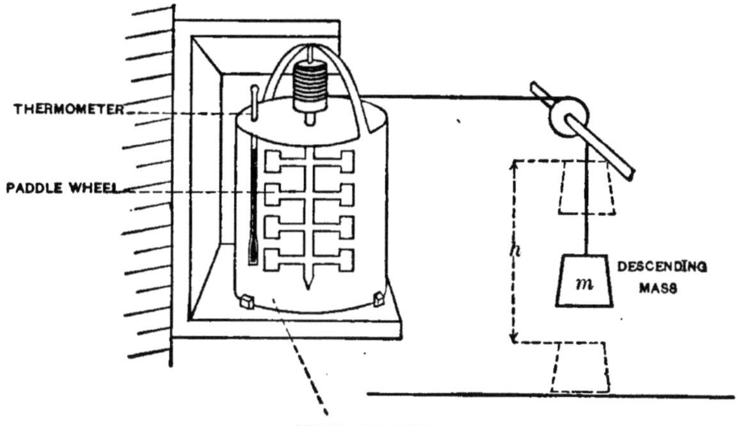

FIG. 199.—Joule's experiment to determine the mechanical equivalent of heat.

The physical meaning of this constant, which is called the **Mechanical Equivalent of Heat,** is as follows, viz. **42** million **ergs of** work must **be done in order to produce one calorie of heat; in** like manner, **when one** calorie of **heat disappears, some other form of** energy **equivalent in amount to 42** million ergs **of work always appears.**

Joule expressed his results in terms of foot-pounds, by saying that the amount of work which will raise the temperature of a pound of water one degree Fahrenheit is 772 foot-pounds. It is in this form that English engineers always use the mechanical equivalent of heat. The same result may be expressed in the metric system by writing

$$J = 424 \text{ gram-meters per calorie.}$$

The inference, then, is, since work disappears as heat appears, and *vice versa*, that **heat is a form of energy. The principle expressed in the equation**

$$W = JH, \quad \text{Eq. 109}$$

is known as the First **Law of** Thermo**dynamics.** The transformation of heat into forms of energy other than mechanical will be discussed in following chapters.

Having shown heat to be a form of energy, we see that **the first law of** thermo**dynamics** is merely a special **case of the Law of the Conservation of Energy.**

Second Law of Thermodynamics

290. About the middle of the nineteenth century Clausius and Kelvin established the general principle that it is **impossible for a self-acting machine, unaided by any external agency, to convey heat from a body at one temperature to another body at a higher temperature.** Kelvin showed, by an argument which must be reserved for advanced study, that it is impossible by use of the most perfect imaginable heat engine to convert into mechanical work more than a certain definite fraction of the heat in any body. Let us suppose an engine working under the most favorable circumstances and receiving an amount of heat H_1 at a temperature of t_1. Every heat engine disposes of a part of the heat which it receives to a body at some lower temperature t_2. Let us denote by H_2 the heat given out by the engine to the refrigerator whose temperature is t_2. Then Kelvin showed that, if temperatures are measured on the ideal gas scale, the maximum possible value for the ratio between useful work and heat taken in — a quantity generally called "efficiency" — is $\frac{t_1 - t_2}{t_1}$. Or, putting this same result in terms of algebra, we have

$$\text{Efficiency} = \frac{H_1 - H_2}{H_1} = \frac{t_1 - t_2}{t_1}. \qquad \text{Eq. 110}$$

This efficiency equation may be considered as a quantitative expression for the second law of thermodynamics.

Absolute Scale of Temperatures

291. Still another way of putting this result is to say that the efficiency of the most perfect heat engine working under the most favorable conditions is a function only of the temperatures of the source and refrigerator respectively.

One of the important applications of the second law of thermodynamics is the determination of temperature by a method which is independent of the particular thermometric substance employed. Such temperatures are said to be "absolute." An outline of the method by which this is done is as follows: Eq. 110 may be transformed by simple algebra so as to read

$$\frac{H_1}{H_2} = \frac{t_1}{t_2}. \qquad \text{Eq. 111}$$

With clear vision, Kelvin saw that the ratio $\frac{H_1}{H_2}$ is independent of the material, and being a function of the temperature alone, may be used as a measure of the ratio of two temperatures on a new and absolute scale.

The experimental determination of the numerical value of this ratio $\frac{H_1}{H_2}$ for a substance working between the boiling and freezing points of water was made by Kelvin and Joule, and found to be $\frac{373}{273}$ This is, of course, also the value of $\frac{t_1}{t_2}$ for the same two fixed points. In order to definitely determine either of these two temperatures on the absolute scale, it remained only to discover another equation between them. This we have in the universal convention that the difference between the melting point of ice and the boiling point of water shall be 100°, or, calling t_1 the temperature of steam and t_2 the melting point of ice,

$$\left.\begin{aligned} t_1 - t_2 &= 100^\circ, \\ \text{and} \quad \frac{t_1}{t_2} &= \frac{373}{273}, \end{aligned}\right\} \qquad \text{Eq. 112}$$

Eliminating t_1 and t_2 respectively between these two equations, we have

Temperature of melting ice $= t_2 = 273°$ on the absolute centigrade scale, and

Temperature of boiling water $= t_1 = 373°$ on the absolute centigrade scale.

It thus appears that **the zero on the absolute** scale is **identical with that on the ideal gas** scale.

From Eq. 110 it follows that **the absolute** zero is **simply the temperature of a refrigerator at** which **a perfect** engine would **convert into** mechanical **work *all* the heat** supplied **to** it.

Observe that the *size* of the degrees here employed is determined by Eq. 112, and that they are centigrade. If $t_1 - t_2$ had been fixed at 180°, the size of the degrees would have been Fahrenheit and the absolute zero would then have been 461° below that of melting ice.

Heat Engines

292. We have seen how Joule transformed mechanical energy into heat by means of his raised weight and paddle.

The heat engine is a machine devised to accomplish the reverse operation, viz. to transform heat into mechanical energy. The most important of these engines are included in the three following groups: —

(i) *The reciprocating steam engine.* This is the ordinary steam engine, of which the common railroad locomotive is an example. Its essential features are the following: *boiler*, *H*, in which heat, furnished by combustion of fuel, is employed to evaporate water; a *cylinder*, *C*, in which slides a close-fitting *piston head*, *P*, and a *piston rod*, *R*, as indicated in Fig. 200; a steam chest, *S*, containing an automatic valve which admits

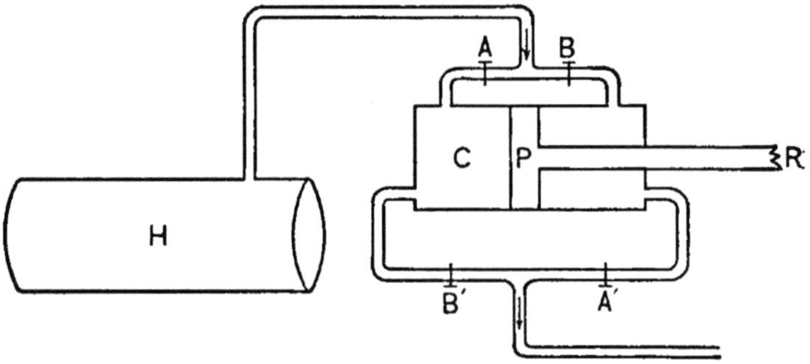

FIG. 200.

the steam alternately to one end and to the other of the cylinder. By means of a connecting rod this to-and-fro motion of the piston is transformed into a circular motion of the shaft and pulley in a manner which will be clear from an inspection of Fig. 200 *bis*. This form of engine is very largely the creation of the Scottish engineer, James Watt.

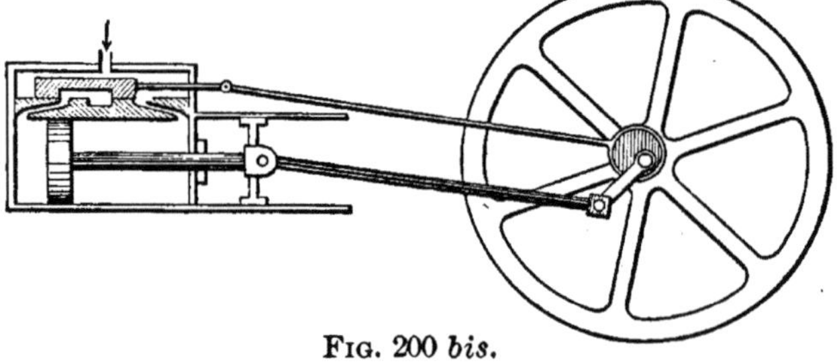

FIG. 200 *bis*.

In marine engines (and in some others also) the exhaust steam is collected, condensed into water, and again returned to the boilers. The same water is thus used over and over again. The steam is not used up, but is merely a "working substance," a carrier to transfer the energy of the combining coal and oxygen in the furnace to the machinery which is doing mechanical work.

The student who would know more of the history or the

theory of the steam engine should consult the very readable article of Professor Ewing on this subject in the *Encyclopædia Britannica;* also the biographical sketch of James Watt there given.

(ii) *The steam turbine.* When water escapes from pressure, its energy becomes kinetic — $\frac{1}{2}mv^2$. If a stream of water, say a slender jet, is directed against a flat surface, such as a door, it exerts a pressure upon this surface. If the door gives way and moves in the direction of the force, work is done upon it — Fs — and this work is performed at the expense of the kinetic energy of the moving water. For many years the miners of the west used a form of water-wheel — the hurdy gurdy — made up of flat vanes attached to a shaft. This wheel was made to run by playing a jet of water on the vanes. The Pelton water-wheel of the present day employs the same principle, but by use of *curved* vanes is able to utilize a much larger proportion of the kinetic energy of the water.

When steam escapes from pressure the velocity factor of its kinetic energy is very great. Consequently a jet of steam directed against a resisting surface exerts upon it a force similar to that of the wind, with which we are all familiar; but the force due to a jet of steam is likely to be of far greater intensity.

A rough sort of steam engine may be made by running steam through an ordinary Pelton water-wheel; while the De Laval steam turbine is simply the Pelton wheel properly designed for the use of a light and expansive medium — steam.

FIG. 201.— The De Laval steam turbine.

The disadvantage of the De Laval wheel is that it must run at an excessively high speed in order to absorb a large fraction of energy from the steam which has freely expanded and has thereby acquired an enormous speed.

We have seen that in the ordinary reciprocating steam engine the steam does work by expanding the volume of the cylinder in which it is confined.

In the Parsons turbine this principle is also made use of to some extent in order to absorb the energy from the steam and yet to keep the speed of the steam down to about four hundred feet per second.

To accomplish this the cylinder (stator), instead of being fitted with a piston, is provided with a coaxial shaft (rotor) which is covered with small blades *inclined to the axis of the shaft*, on the same principle that the blades of a windmill are inclined to the direction of its shaft. Corresponding to each movable blade on the rotor, there is a *fixed* blade on the *inside* of the stator.

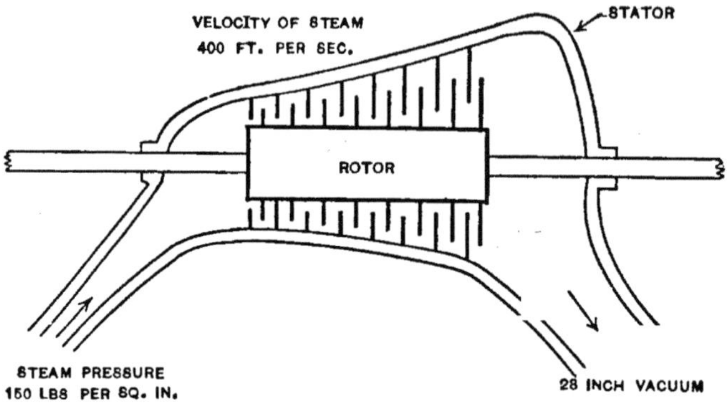

Fig. 201 *bis.* — Longitudinal section of Parsons turbine.

Accordingly we may think of the passage of the particles of steam between these blades as similar to their passage through a long rectangular pipe one side of which is constantly receding from the other and is therefore absorbing energy. The impulse principle — that of the De Laval turbine — is also employed in the Parsons to abstract still another portion of the energy of the steam.

This type of engine is especially adapted for the smooth running of steamships and dynamos. The steam turbine owes its initial development largely to the English engineer, Parsons.

(iii) *The internal combustion engine.* The two preceding engines derive their energy from fuel which is burned outside the cylinder. But there has recently been perfected a type of machine which burns its fuel (either gas or a spray of oil) right inside the cylinder. When the proper fuel is used, the products of combustion are entirely gaseous, so that the cylinder easily clears itself. These are the little engines so familiar to us on automobiles. In most of them there are four steps through which the "working substance" is taken. These are indicated in Fig. 202. These four steps constitute what is known as the "Otto cycle," after the German engineer, Otto, who did much to perfect the modern gas engine. In these engines the piston does work only when moving in one direction; they are therefore said to be "single acting," to distinguish them from the ordinary reciprocating steam engine, which does work both on the forward and on the backward stroke, and is hence said to be "double acting." The use of intake and exhaust valves *I* and *E*, shown in Fig. 202, will be sufficiently clear from the diagram. The intake valve admits the explosive mixture — say air and gasoline vapor; compression then occurs; explosion is then produced by an electric spark; the products of combustion, CO_2, etc., are then swept out through the exhaust valve *E*. These are the four steps in the Otto cycle.

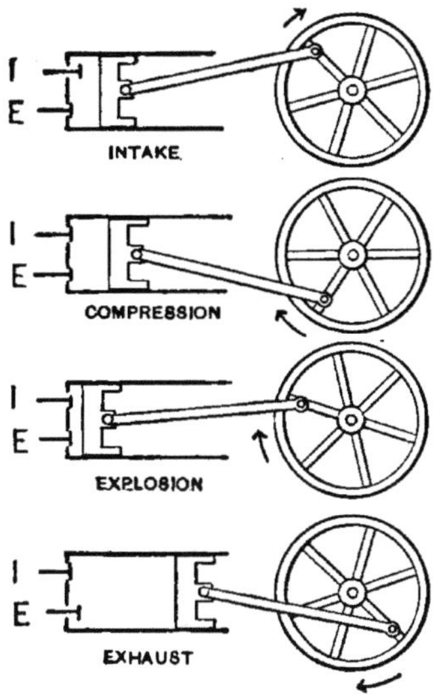

FIG. 202.

References

MAXWELL. *Theory of Heat* (Longmans, Green & Co.), pp. 343. Price $1.25. First four chapters especially valuable for the beginner.

TAIT. *Heat* (Macmillan), pp. 368. Price $2.00. Exceedingly lucid. Covers the whole subject.

TYNDALL. *Heat as a Mode of Motion* (Appleton), pp. 529. Price $2.00. A series of popular lectures, marked by their wealth of experimental illustration.

HELMHOLTZ. *Popular Lectures, First Series* (Appleton). Price $2.00. Seventh lecture recommended.

KELVIN. Art. "Heat" in *Ency. Brit.* pp. 554-565 are especially recommended to beginners.

PRESTON. *Theory of Heat* (Macmillan), pp. 719. Price $5.00. A treatise recommended to the advanced student as an excellent compendium of both experimental and mathematical results.

EWING. *The Steam Engine and Other Heat Engines* (Cambridge Press). Price, $3.75.

REEVE. *Thermodynamics of Heat Engines* (Macmillan). Price $2.60.

BUCKINGHAM. *Theory of Thermodynamics* (Macmillan).

STODOLA. *The Steam Turbine*, trans. by Loewenstein (Van Nostrand).

Problems

1. The average reading of the barometer at the top of Mount Hamilton (the Lick Observatory) is 65 cm. At what temperature, therefore, does water boil on Mount Hamilton? *Ans.* 95° C. approximately.

2. At the top of Pike's Peak water boils at a temperature of 86°. What, therefore, is the average reading of the barometer on Pike's Peak?

Ans. 45.68 cm.

3. The pressure of the atmosphere is approximately 15 lb. weight to the square inch. The boiler of an Atlantic liner carries a steam pressure of 300 lb. to the square inch. What is the temperature of the water in this boiler? See Regnault's table. *Ans.* 212°.3 C.

4. How do you explain the fact that as a vessel of sap from the maple tree is boiled away the percentage of sugar in the solution left behind continually increases?

5. Explain how one can warm his hands by rubbing them together.

6. Explain the fact that a piece of iron is heated by hammering. What kind of energy is here transformed into heat?

7. Explain the fact that when a piece of copper wire, at any point, is rapidly bent first in one direction, then in the other, it becomes hot.

8. How is it that when the tire of a bicycle is being pumped up,

(*a*) the barrel of the pump becomes warm?

(*b*) the rubber hose connecting the pump and the tire becomes warm?

9. At a place where the acceleration of gravity is 980, a bullet whose mass is 8 g. is allowed to fall from a height of 1000 cm. What will be the kinetic energy of the bullet in ergs just before it strikes the ground? To what fraction of a calorie will this mechanical energy be equivalent?

Ans. 7,840,000 ergs = 0.187 calorie.

10. Suppose, in the preceding problem, that when the bullet strikes the ground one half its kinetic energy goes to heat the bullet, the other half to heat the ground. If the specific heat of lead is $\frac{1}{30}$, by how much will the temperature of the bullet be raised? *Ans.* 0°.34 C.

11. (*a*) The height of the American Falls at Niagara is 50 m. How much kinetic energy will a gram of water acquire in falling through this distance? ($g = 981$.) *Ans.* 4,905,000 ergs.

(*b*) What will be the equivalent of this energy in heat units?

Ans. 0.117 calorie.

(*c*) By how much then will the temperature of the water at the bottom be increased over that at the top? *Ans.* 0°.117 centigrade.

12. To raise a mass of 10 lb. through a vertical distance of 8 ft. (80 foot-pounds) requires how many ergs of work?

Ans. 1079 million ergs.

13. Assuming that the specific heat of iron is 0.12, find the number of heat units required to raise the temperature of 300 g. of iron from 20° C. to 360° C.

14. Why is it that alcohol or ether poured on one's hand gives him the sensation of cold?

15. How much heat must be given to 12 g. of water at 40° C., in order to make it boil?

16. Why are damp clothes apt to make one chilly?

17. Ten lb. of water at 0° C. are mixed with 40 lb. of water at 50° C. Find the temperature of the mixture.

18. Give five illustrations of the transformation of some form of energy into heat.

19. What is meant by a "land breeze"? By a "sea breeze"? How do you explain these?

20. Write an examination paper of five questions on the subject of heat.

21. A harvesting machine, made partly of wood and partly of iron, is allowed to stand overnight in an open field. In the morning, which parts will be covered with dew, the iron or the wood? Why?

22. What are clouds and how are they produced?

23. Why is it that in winter ice which is partly covered with dirt or ashes melts sooner than clean ice?

24. Why are teapots often provided with wooden handles?

25. Fifty cubic centimeters of water are heated from a temperature 15° C. to 35° C. Find the change in volume thus produced.

26. A steel rail 10 m. long will change its length by how many millimeters when heated from 0° C. to 50° C.?

27. Why do you pour hot water on the neck of a bottle when you wish to loosen the glass stopper?

28. Explain how the temperature of a greenhouse, not heated artificially, rises above that of the surrounding air.

29. What is the theoretical efficiency of a steam engine whose boiler is at a temperature of 160° C. and whose condenser is at 40° C.?

30. Suppose the temperature of combustion in the cylinder of a gasoline engine to be 1600° C. and the temperature of the exhaust 350° C. Find the highest possible efficiency.

31. How would you proceed to graduate an electrical resistance thermometer?

32. Sketch the isothermals of carbon dioxide and show how they differ in form above and below the critical point.

CHAPTER VIII

MAGNETISM

293. Any body which attracts iron filings is said to be magnetized, and is called a magnet.

One kind of iron ore is found to be already magnetized as it comes from the earth. If a piece of this ore (which mineralogists call magnetite or loadstone) be dipped into a box of iron filings, it is seen that certain regions of it attract the filings very strongly, while the remaining parts attract scarcely at all.

Those parts which attract the iron filings most strongly are called magnetic poles.

LEADING FACTS OF MAGNETISM

294. The following ten facts, presented in essentially the order in which they were discovered, are perhaps the most important phenomena connected with this subject: —

1. **If a piece of magnetized iron or a piece of loadstone be freely suspended, it will always set itself so that a certain direction in it makes a fixed angle with the geographical meridian, i.e. with the north and south line.**

The Mariner's Compass

This fact is illustrated by the ordinary compass and constitutes its fundamental principle. A slender piece of steel is magnetized by drawing it over a magnet or loadstone, and is then mounted on a sharp pivot so that it is free to turn in any direction. This idea of the compass, which dates from a time certainly not much later than 1200 A.D., is the first important discovery in the science of magnetism.

In the case of a long, slender magnet, such as is used for a compass needle, there is, in general, one pole near each end. **That pole which turns toward the north, when the magnet is free** to rotate, is called the north-seeking pole, or simply the north pole; that pole which turns toward the south is called the south pole.

The direction of the line joining the poles is called the magnetic axis of the magnet. This definition of "axis" will later need a slight revision, since the poles have not yet been defined as mathematical points.

The direction assumed by the axis of a freely suspended magnet is called the magnetic meridian.

The angle between the magnetic and geographical meridians is called the magnetic declination. It is generally expressed in degrees, minutes, and seconds. In making an accurate determination of the magnetic declination, one must be careful not to assume that the magnetic and geometric axes of the needle coincide.

In actual measurement any uncertainty as to the position of the pole is eliminated by using a thin flat compass needle and by observing first with one face up, then with the other. Thus in Fig. 203, if GG be the first position of the compass needle

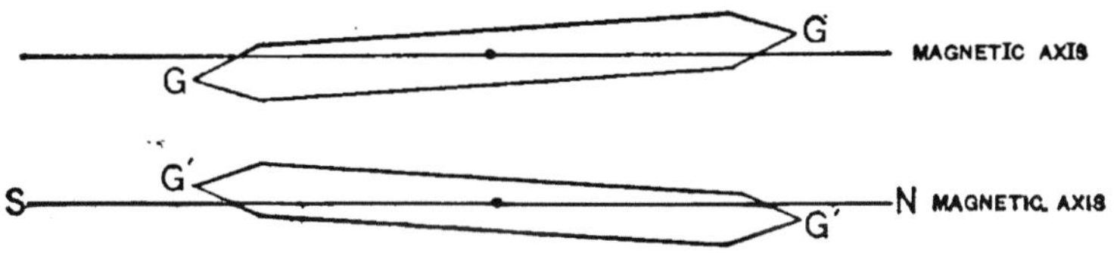

FIG. 203. — Illustrating the difference in direction between geometric and magnetic axes.

and $G'G'$ its second position, we shall find that while the magnetic axis remains fixed in space whichever face of the needle is up, the geometric axis does not. But the directions of the geometric axis GG and $G'G'$ are those which are actually read upon the graduated circle. Hence the *difference* of the two readings, before and after the needle is reversed in its bearings, will give us twice the angle by which the magnetic and geometric axes diverge. Not only so but the *mean* of these two readings will give us the correct direction of the magnetic meridian. One thus avoids the necessity of locating the magnetic axis as a mathematical line in the body of the needle.

A little later it will be clear how we may avoid the necessity of locating the magnetic *pole* as a mathematical point, situated on the magnetic axis at a certain distance from the axis of rotation.

Every student should read, in the *Encyclopædia Britannica*, art. "Compass," the description of a beautiful needle devised by Lord Kelvin in 1876 and now used on practically all the sea-going vessels of the world, a simple device which eliminates practically all divergence between magnetic and mechanical axes.

295. 2. The Discovery of Columbus. Columbus used the compass on his first voyage to America. Up to this date it had been supposed by European navigators that while the suspended magnet in the compass did not always point exactly north and south, it did always point in the same direction; *i.e.* they supposed the angle between the geographical and magnetic meridians to be constant. But Columbus, much to his own astonishment, and much to the dismay of his sailors, found that as he proceeded west, the needle pointed to a part of the sky which lay more and more to the west, and thus discovered **that the magnetic declination varies from one point to another on the earth's surface.** This we may call the second important discovery in magnetism.

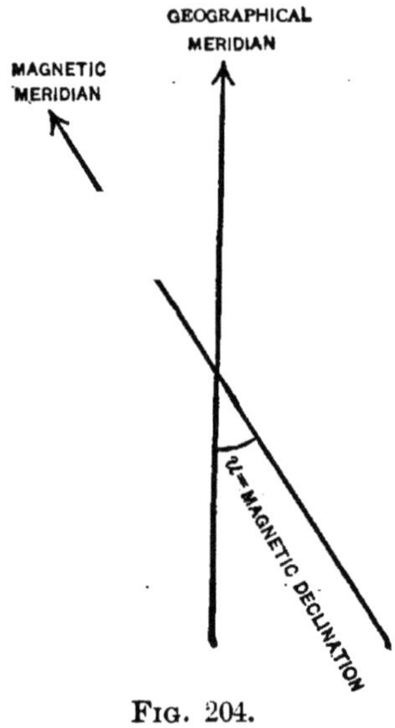

FIG. 204.

Another way of putting it would be to say that what Columbus discovered is that the general statement contained in § 294 is true only for a single geographical location; or more strictly, for the locus of all geographical points for which the magnetic declination is the same.

Curves drawn in such a way that the magnetic declination is constant for every point on them are called "isogonic curves." The accompanying map will show the trend of these isogonic lines in various parts of the world.

MAGNETIC DECLINATIONS (1900)

Augusta, Me.	16° W.	New Orleans	5° E.
Boston	12° W.	Kansas City	8° E.
New York	9° W.	Omaha	10° E.
Cleveland	2° W.	Denver	14° E.
Knoxville, Tenn.	0°	San Francisco	17° E.
Chicago	3° E.	Tacoma, Wash.	23° E.

THE DECLINATION OF THE COMPASS, FOR THE YEAR 1910.

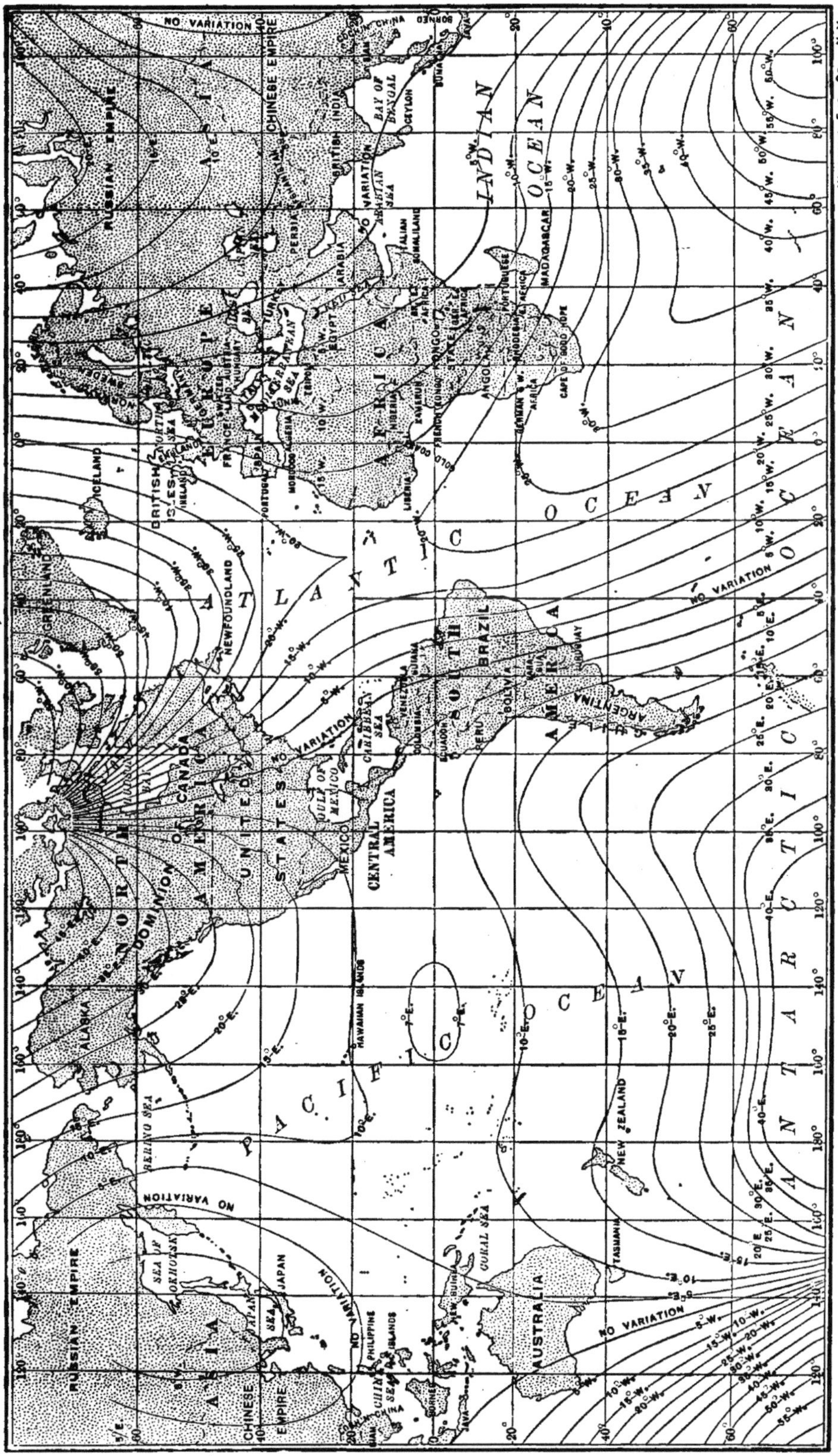

FIG. 205.—Map showing the direction in which the compass needle points at various parts of the earth's surface.

The preceding table shows the values of the declination in different parts of the United States, to the nearest degree. W. and E. indicate west and east declinations, respectively.

The Magnetoscope

296. In the laboratory one has frequently occasion to use a compass needle of some sort. A most convenient and easily prepared form is the following: A wooden block about one decimeter square is bored halfway through by an auger (extension bit) which has the same diameter as the glass chimney of an Argand lamp. The chimney is then waxed into the block, and a small steel magnet is suspended by a fiber through a cork at the top. Such an instrument is not disturbed by currents of air in the room. It is called a **magnetoscope**. Why?

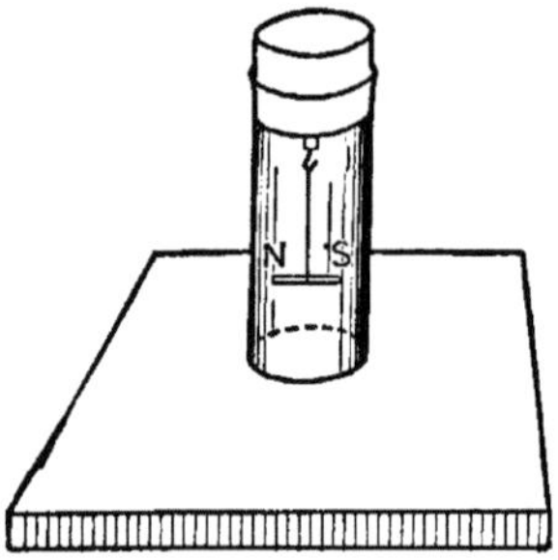

FIG. 206. — Simple form of magnetoscope.

297. 3. The **magnetic dip**. Not long after the time of Columbus it was found by Hartman (b. 1489, d. 1564) that if an iron needle be suspended so as to lie *horizontal* before magnetization, it takes up a different position after magnetization. Imagine the needle before it is magnetized to be horizontal. After magnetization it is found that the north end "dips" down as if it had become heavier than the south end. But Norman (1576), the inventor of the dipping needle represented in Fig. 207, weighed the iron needle and found that it was no heavier after magnetization than before. As we shall see later, this dip is due to the fact that the earth itself is a great magnet. **Only the direction, therefore, and not the weight of the needle is changed by magnetization.** In the southern hemisphere it is the south end of the magnet which "dips."

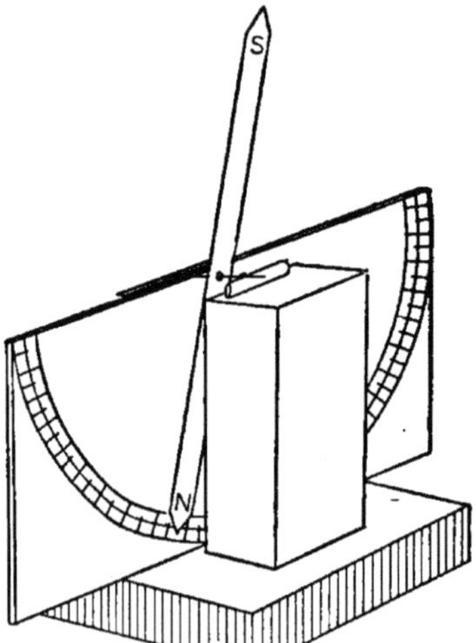

FIG. 207. — Dipping needle.

In the United States it is customary to attach a small weight to the south end of a surveyor's compass needle in order to counteract the effect of the dip.

The phenomenon of magnetic dip might be called the *third* important discovery in this subject.

298. 4. Like poles repel each other; unlike poles attract each other. The evidence for this may be shown in a variety of ways; perhaps the most convenient method is the following: —

Mount two magnets, as shown in Fig. 208. Determine the north end of each, and slip bits of paper over them as indicated. When one magnet is brought near the other, the repulsion between the two north poles or between the two south poles is very evident; likewise the attraction between a north and a south pole.

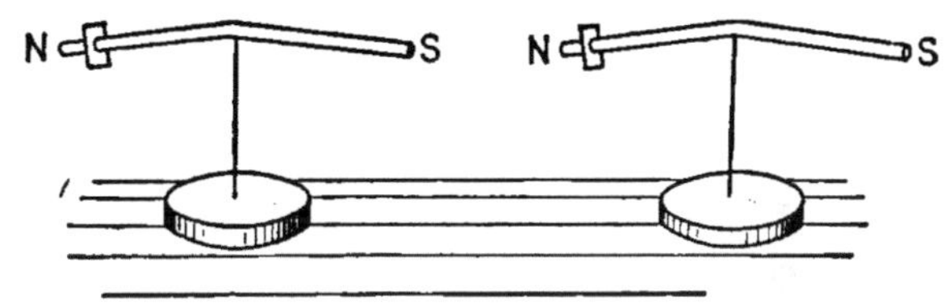

FIG. 208. — Showing attraction between north and south poles.

Norman magnetized a sewing needle, and floated it on water. By bringing into the neighborhood another magnet, he had a delicate means of showing attraction and repulsion of the poles.

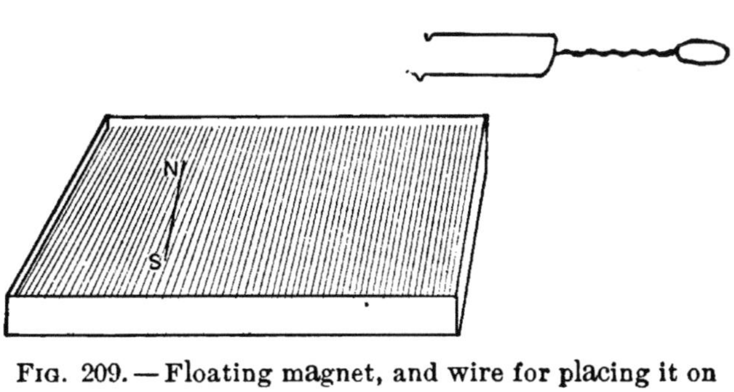

FIG. 209. — Floating magnet, and wire for placing it on the water.

Concerning the *amount* of this attraction or repulsion, it was first shown by the French electrician Coulomb (1736–1806), and afterwards accurately verified by the German astronomer Gauss (1777–1855), that the attraction of one pole for another, *in air*, varies inversely as the square of the distance separating them and directly as the products of the pole strengths — a quantity which will be defined a couple of pages hence. If we denote this distance by r, and the pole strengths by m and m', respectively, we may describe Coulomb's result by writing: —

$$F = k\frac{mm'}{r^2}, \quad \text{Coulomb's Law.} \quad \text{Eq. 113}$$

where, for air, k is a mere constant of proportionality which may be determined later.

The Magnetic Field. Lines of Force

299. If a sheet of white paper be laid over a magnet, and iron filings be sprinkled over the paper, one obtains very striking evidence that the region about the magnet is different from other parts of space. The iron filings arrange themselves in curved lines; and these curved lines remain the same, however many times we repeat the experiment.

"We next introduce a method of conceiving and describing magnetic actions which was invented and much used by Faraday. Since a magnet acts upon a magnetic needle placed anywhere in the surrounding space, we call that space the magnetic field of the magnet. Neglecting the earth's magnetism, we may map out this field as follows: Conceive any plane drawn through the axis of the magnet and place it so that this plane shall be horizontal. Then, at any point in this plane, place a very small magnetic needle, and note the direction which its axis assumes under the action of the magnet; then proceed to move the center of the needle in the direction in which its north pole points, and continue the motion so that, at each point, the center is following the direction indicated by the north pole.

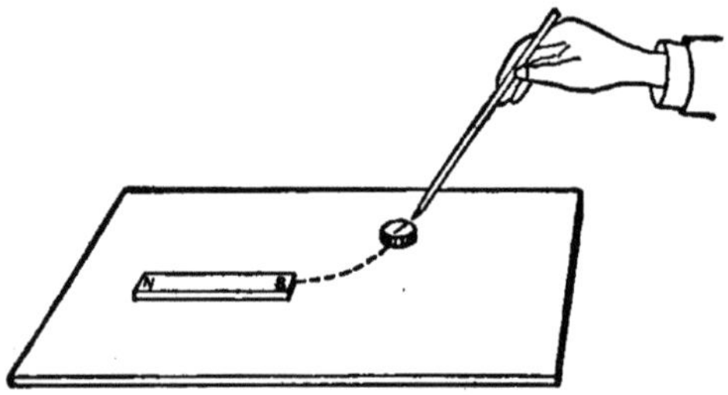

FIG. 210. — Method of mapping magnetic fields.

"The line thus traced will at last cut the surface of the magnet at some point lying toward the south pole; and if we continue the line backward by following the direction continually indicated by the south pole of the needle, it will cut the surface of the magnet at some point lying toward the north pole. Such a line is called a **line of magnetic force**; and since one such line can be drawn through every point in the plane, and any

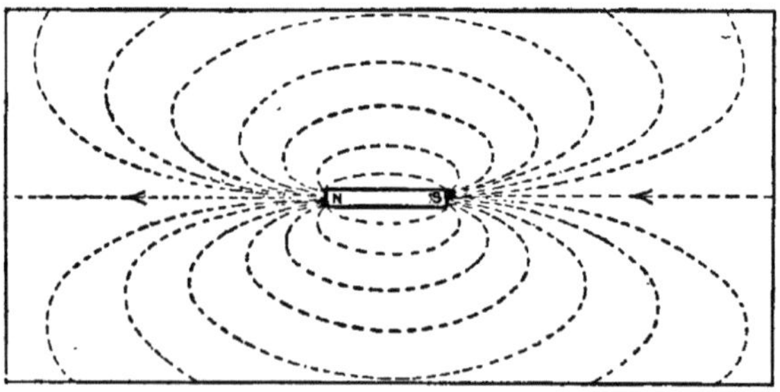

FIG. 211. — Field about a straight magnet.

number of planes can be drawn through the axis of the magnet, we can conceive the whole magnetic field filled with such lines."— CHRYSTAL, *Ency. Brit.*, art. "Magnetism."

Figure 210 indicates the process of mapping these fields.

Figure 211 shows the field about a strong bar magnet. The effect of the earth's field is here negligible.

Figure 212 represents a small portion of the field due to the earth alone.

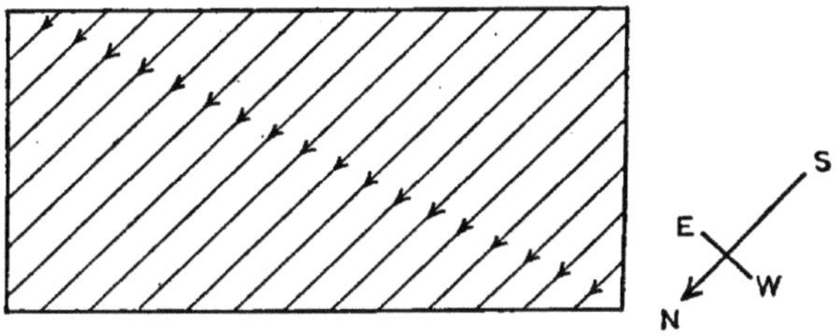

FIG. 212. — Uniform magnetic field due to earth alone.

A magnetic field as actually mapped in the laboratory will always be the resultant of the earth's field and the field of the magnet.

Intensity of Magnetic Field

300. So far we have considered only the fact that a magnetic field has, at every point, a certain definite direction. If, however, a small horseshoe magnet be brought into the field of a large one, and held between the thumb and forefinger, the observer will easily "feel" that the little magnet experiences a *force* driving it toward one pole or the other of the larger magnet, always tending to move into the strongest part of the field. Observe that in those parts of the field where the lines of force are nearly parallel to each other, the force is very minute.

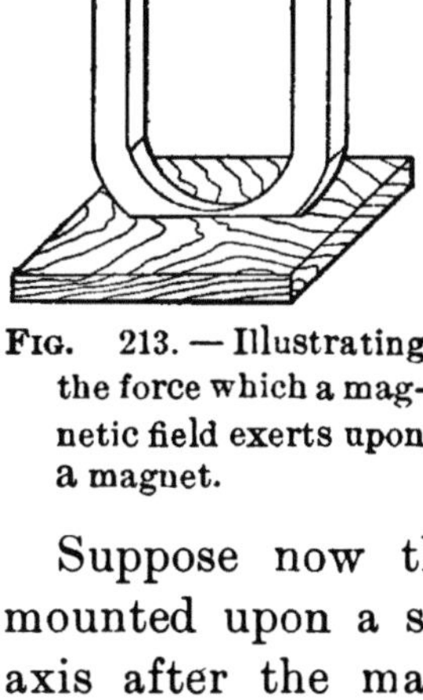

FIG. 213. — Illustrating the force which a magnetic field exerts upon a magnet.

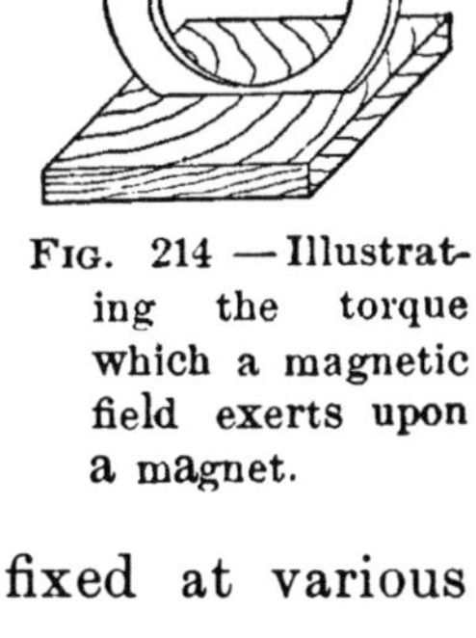

FIG. 214 — Illustrating the torque which a magnetic field exerts upon a magnet.

Suppose now that the little magnet be mounted upon a sewing needle as a vertical axis after the manner shown in Fig. 214. The needle may well be placed in a block of wood so that the rotating system may be fixed at various heights above the poles of the larger magnet. Now the *force* of translation is counteracted by the reaction of the needle, and one can feel in a very convincing manner the *torque* which a magnetic field in general exerts upon a magnet placed anywhere within it.

Having determined that a magnetic field exerts both a force and a torque upon any magnet placed within it, we next inquire what is meant by a strong or weak magnetic field — a question which can be clearly answered only after the "strength of a magnetic pole" has been defined. For this latter purpose the law of Coulomb, $F = k\frac{mm'}{r^2}$, is employed.

Definition of Unit Magnetic Pole

301. To begin with we *assume* that the value of the constant k in Eq. 113 does not depend upon the medium in which the poles are placed, *so long as the medium is the ordinary atmospheric air;* in other words, we assume air as our standard magnetic medium. Next let us *assume* that the two poles which are attracting each other are equal and are placed at a distance of one centimeter apart. **If under these conditions the two equal poles be chosen of such a size that they attract or repel each other with a force of one dyne, they are said to be unit magnet poles, or poles of unit strength.**

The chief advantage of this definition, which may at first glance appear rather arbitrary, is that it makes the constant k disappear from Coulomb's Eq. 113. For having assumed $m = m' = 1$ when $F = r = 1$, it follows that in air k must always be unity, a unit vector, indeed; whence follows

$$F = \frac{mm'}{r^2} \qquad \text{Eq. 114}$$

which now becomes not only the law of Coulomb, for air, but also the defining equation for the unit magnet pole.

Measure of Magnetic Intensity

302. We are now, for the first time, in a position to define what is meant by the intensity of a magnetic field, namely, the ratio of the force which a magnet pole experiences in that field to the strength of the pole.

Let R denote the intensity of field, and F the force with which the pole m is urged. Then

$$R = \frac{F}{m}. \quad \text{Defining equation for intensity of magnetic field.} \qquad \text{Eq. 115}$$

Field intensity, being a ratio of a vector to a scalar quantity, is itself a vector quantity, and is therefore to be resolved and compounded according to the rules for vector addition. From Eq. 114 it follows that a field of unit intensity is one which will act upon a unit magnet pole with a force of one dyne. By international agreement (Paris, 1900) this unit is called, after the eminent German astronomer, a "gauss."

We are now provided with a clear and simple definition for the intensity of field; but since we have as yet no practical method for measuring pole strengths, Eq. 115 cannot be called a laboratory equation.

Just how this difficulty is avoided will be cleared up in the following sections when we come to consider the earth's magnetic field.

303. 5. The earth itself a great magnet. We have seen that at any one point on the earth's surface the compass needle points in a definite direction, generally not far from north. We have seen that, in the northern hemisphere, the north end of the needle dips; while in the southern hemisphere, the south end dips.

But this is exactly the manner in which a small compass needle would behave in the presence of a large iron sphere which had been magnetized. Accordingly Dr. Gilbert* was led to suggest that the earth itself was a large magnet. And since his time the theory has been confirmed by navigators who have located the north magnetic pole of the earth not far from Baffin's Bay, and the south magnetic pole somewhere between Australia and the south geographical pole.

If we map the magnetic field of the earth by means of compass and pencil as indicated in the preceding paragraph, we shall find that **the lines of force due to the earth are, in any limited region, practically parallel to each other.**

* *Dr. William Gilbert* (b. 1540, d. 1603), the leading man of science in England during the reign of Queen Elizabeth.

304. 6. The earth's action a couple. In a preceding section (15) we have considered the distinction between a motion of translation and one of rotation. To a system of forces which produce rotation only the name **couple** has been given. **Norman offered the following three experiments to show that the earth does not exert any force of translation upon a magnet,** or, as we might say, three reasons for thinking that the force which the earth *does* exert upon a magnet is a *couple.*

"First, he weighed several small pieces of steel in a delicate balance, and then magnetized them, but could not detect the slightest alteration in their weight, 'though every one of them had received virtue sufficient to lift up his fellow.'

"Secondly, he pushed a steel wire through a spherical piece of cork, and carefully pared the latter so that the whole sank to a certain depth in a vessel of water and remained there, taking up any position about the center indifferently. After the wire was magnetized very carefully, without disturbing its position in the cork, it sank to the same depth as before, neither more nor less, the only difference being that now the wire set itself persistently in a definite fixed direction parallel to the magnetic meridian, the north end dipping about 71° or 72° below the horizon.

"Thirdly, he arranged a magnetized needle on a cork so as to float on the surface of water, and found that, although it set in the magnetic meridian, there was not the slightest tendency to translation in any direction. .

"He concludes that there is no force of translation on the magnet, either vertical or horizontal." — CHRYSTAL, *Ency. Brit.*, art. "Magnetism."

Recalling the fact that a couple is made up of two parallel and equal forces acting in opposite senses, but not in the same straight line, we conclude that the earth's action is a couple precisely because its lines of force, throughout the limited region of any single experiment, are parallel.

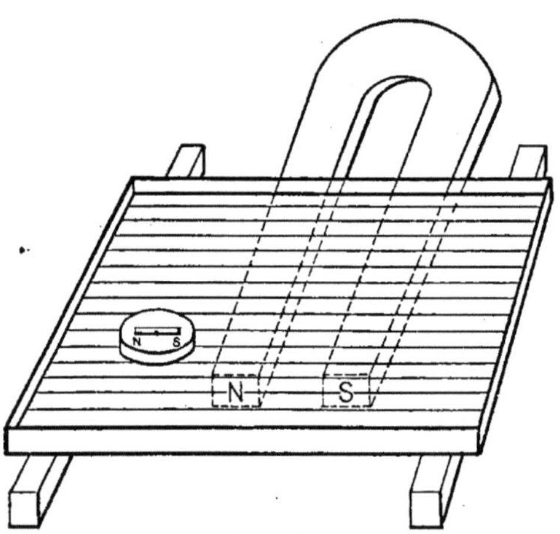

FIG. 215.

If the dish of water, shown in Fig. 215, be supported over a rather strong permanent magnet, and if the magnetic needle be floated on a small cork, it will be seen that this artificial field exerts both a couple and a force upon the floating magnet, continually drawing the magnet back toward the strongest part of the field. The explanation of this lies in the fact that here the force acting upon the *north* pole of the floating needle is *not* exactly opposite in direction to that acting upon the south pole. Hence the resultant of these two forces cannot be zero. The advanced student will find that the floating magnet here simply moves in such a way as to make its potential energy a minimum.

Digression on the Measurement of the Earth's Magnetic Field

305. On both historical and scientific grounds the following problem is one worthy of rather careful study; for not only is it the first problem in which the absolute system of units was introduced into the science of electricity and magnetism, but it is a thoroughly typical example of the manner in which physical science deals quantitatively with phenomena which are very imperfectly comprehended.

As a preliminary it is necessary to define

Magnetic Moment,

a quantity of frequent use in magnetic measurements. **The magnetic moment of any magnet is defined as the product of its pole strength m by the distance l between its poles.** This quantity is generally denoted by M, and is defined by the following equation: —

$$M = ml. \quad \text{Defining equation for magnetic moment.} \qquad \text{Eq. 116}$$

Horizontal Intensity of the Earth's Field

306. We have already learned, in a general way at least, how the declination and the dip of the compass needle may be measured. In order to complete the determination of the earth's field, it remains only to measure the horizontal component H of the total intensity T. See Fig. 216.

Confining our attention now to the parallel field which belongs to the earth, let us imagine a magnet suspended in this field by means of a delicate fiber. The magnet will oscillate to and fro about its position of equilibrium and will then come to rest, pointing along the magnetic meridian. If the magnet lies in a horizontal plane, the only directive force acting upon it is the horizontal intensity H.

307. To determine the numerical value of H the two following steps are needed: —

(i) *To obtain the torque, acting upon the needle in terms of its period.* When the suspended magnet is rotated through an angle θ about its suspending fiber as an axis, there will be a force $+mH$ acting upon the north pole of the magnet, and a force $-mH$ acting upon its south pole.

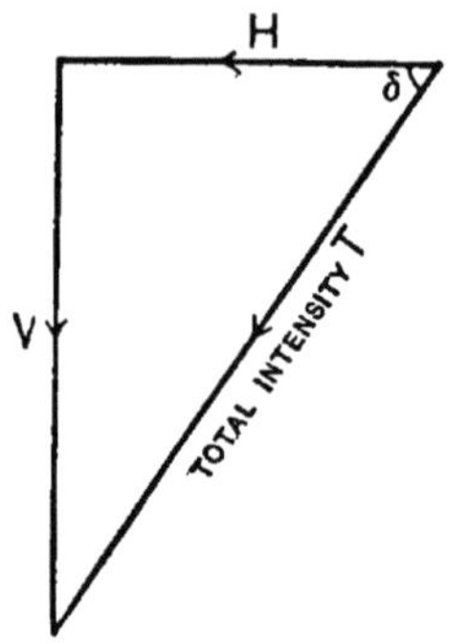

FIG. 216. — Illustrating the relation between the four magnetic elements, namely, horizontal, vertical, and total intensities, and angle of dip δ.

If we denote by l the distance between the two poles, $l \sin \theta$ will be the perpendicular distance between the lines of action of the two forces $+mH$ and $-mH$. The couple L acting upon the needle will therefore be $mHl \sin \theta$. And since the sign of the couple which tends to restore the needle to its position of equilibrium is always opposite to that of the angular displacement θ, we may write

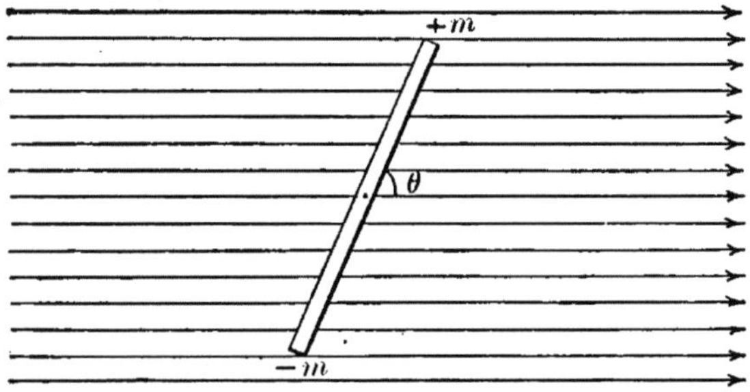

FIG. 217. — Illustrating the couple which acts upon a magnet suspended in the earth's field.

$$L = -ml \cdot H \sin \theta = -MH \sin \theta. \qquad \text{Eq. 117}$$

When θ is very small, we have with sufficient accuracy

$$L = -MH\theta. \qquad \text{Eq. 118}$$

Dividing each side of this equation by the rotational inertia I of the magnet, we obtain for the angular acceleration A,

$$A = \frac{L}{I} = -\left(\frac{MH}{I}\right)\theta. \qquad \text{Eq. 119}$$

It thus appears that the angular acceleration of the vibrating system is, at every instant, proportional to the angular displacement and opposite in sense. The behavior of the magnet therefore satisfies the criterion (§ 53) for simple harmonic motion. And since the period of any simple harmonic motion is $2\pi\sqrt{-\left(\frac{\text{Displacement}}{\text{Acceleration}}\right)}$, it follows that the period of vibration of this magnet is

$$T = 2\pi\sqrt{\frac{I}{MH}}. \qquad \text{Eq. 120}$$

Now T and I are easily measurable quantities; but these data do not enable us to determine either of the two unknown quantities M and H. Equation 120 merely gives us the product MH in terms of T and I. Hence the necessity for the second step, namely,

(ii) *To obtain the deflection of the needle in terms of M and H.* Gauss has shown how this relation may be obtained by a very simple experiment.

Having suspended another, and much smaller, magnet with a small mirror rigidly attached to it, so that the direction of the magnetic field in which it is placed may be at once indicated, one takes the magnet *NS* (Fig. 218) whose period he has just measured, and fixes it as nearly as possible at right angles to the magnetic meridian and at a distance r east or west of the suspended needle O.

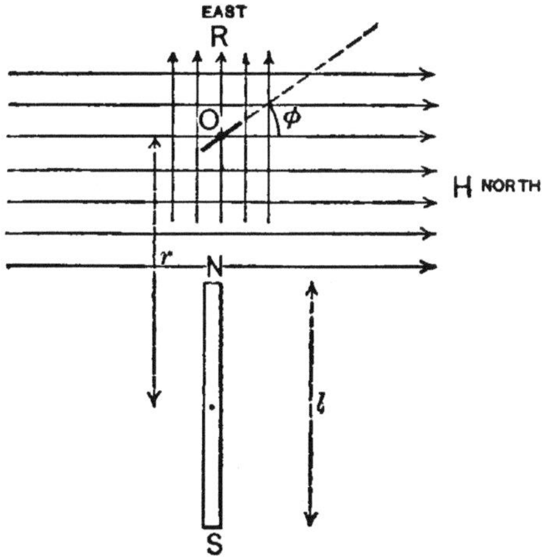

FIG. 218.—Showing the superposition of an artificial field upon the earth's field, and the consequent deflection of the suspended needle.

Before the deflecting magnet *NS* was brought into the neighborhood, the suspended needle pointed north and south along the magnetic meridian. If there were no magnetic field due to the earth, the suspended needle would now point east and west due to the field R of the deflecting magnet; but since both fields are present, the suspended needle sets itself along the *resultant* of the two fields as indicated in the figure. Now if we knew the intensity of each of these fields, we could easily compute the direction of the resultant; for, since these two fields are at right angles to each other,

$$\frac{R}{H} = \tan\phi, \qquad \text{Eq. 121}$$

where ϕ is the deflection, *i.e.* the direction of the resultant referred to the earth's field.

Conversely, if we know the direction ϕ and the intensity R we can compute H. This latter we now proceed to do. The magnetic field at O, due to the pole N alone, is by Coulomb's law (§ 301) $\frac{+m}{\left(r-\frac{l}{2}\right)^2}$; while the field at O due to S alone is $\frac{-m}{\left(r+\frac{l}{2}\right)^2}$. Hence the total field R due to the entire magnet NS is given by

$$R = m\left[\frac{1}{\left(r-\frac{l}{2}\right)^2} - \frac{1}{\left(r+\frac{l}{2}\right)^2}\right] = \frac{m \cdot 2\,lr}{\left[r^2-\left(\frac{l}{2}\right)^2\right]^2}.$$

Or, if we choose r so large that l^2 may be neglected in comparison with r^2

$$R = \frac{2\,Mr}{r^4} = 2\,\frac{M}{r^3}. \qquad \text{Eq. 122}$$

Substituting this value of R in Eq. 121 and solving for H/M, one has

$$\frac{H}{M} = \frac{2}{r^3 \tan\phi}. \qquad \text{Eq. 123}$$

Since r and ϕ are easily measured, we have here a laboratory equation for the ratio H/M. Our second step is now complete.

308. Between Eqs. 120 and 123 it is now possible to eliminate M and obtain the horizontal intensity of the earth's field, thus

$$(MH)\left(\frac{H}{M}\right) = H^2 = \frac{8\,\pi^2 I}{T^2 r^3 \tan\phi}; \qquad \text{Eq. 124}$$

or to eliminate H, and obtain the moment of the deflecting magnet, thus

$$(MH)\left(\frac{M}{H}\right) = M^2 = \frac{2\,\pi^2 I r^3 \tan\phi}{T^2}. \qquad \text{Eq. 125}$$

It should be observed that this method of Gauss avoids the necessity of measuring either the pole strength m or the distance between the poles l. The neat manner in which this is done is to combine these two quantities into a product M and then eliminate M by means of a second equation which is obtained, of course, by means of a second (deflection) experiment.

The advanced student will find that some of the approximations here introduced are not permissible in work of high accuracy, hence certain corrections must be made to Eqs. 124 and 125; but the philosophy of the subject is not thereby changed.

Returning from this digression, we now proceed to the next fundamental fact in the science of magnetism.

309. 7. Magnetic quality disappears at red heat. Any substance which is attracted or repelled by a magnet is said to possess **magnetic quality**. It has been found that the only substances which exhibit magnetic quality to any great extent are iron, nickel, and cobalt. Iron is the only really important magnetic substance known. But if a piece of iron or steel be magnetized and then heated until red-hot all over, it will be found to have lost its magnetization. As soon as the iron has cooled again, however, it will be found to have regained its magnetic quality, for it will now attract *either* end of a compass needle. There is, then, a marked difference between a piece of iron which is magnetized and one which is not. The iron which is not magnetized still possesses *magnetic quality* at all ordinary temperatures; but it does not possess any *magnetic poles* or any *magnetization.*

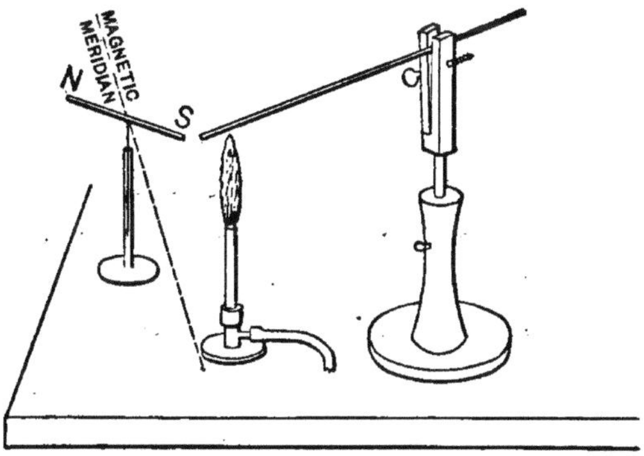

FIG. 219 — Illustrating the disappearance of magnetic quality at red heat.

The evidence for thinking that magnetic quality disappears at high temperatures is simply shown as follows: —

Mount a compass needle some five or six inches long on a single pivot as shown in Fig. 219. Bring near it, in a wood clamp, a piece of unmagnetized iron wire, say, five or six inches long. The compass needle will magnetize the wire and will be deflected from the magnetic meridian, so that one end of it will swing about and touch the wire. If now a Bunsen flame be brought underneath the iron wire until it becomes a bright red, the compass needle will "let go" and will swing back into the meridian, showing that the hot iron has lost its magnetic quality. On removing the flame, the iron will be seen to cool and one end of the compass needle to swing around into contact with the iron, showing that the wire on cooling has regained its magnetic quality.

310. 8. Magnetization a molecular property. The evidence furnished by the preceding experiment, together with the fact that a magnet — say a piece of magnetized watch spring — may be broken into an unlimited number of pieces (Fig. 220), each of which remains a complete magnet, with north and south pole, leads us to think that, whatever magnetism may be, it is something which belongs to the smallest particles of which the body is made up. The experiment of breaking up the magnetized watch spring should be tried by each student for himself. The fragments are easily examined by use of the magnetoscope.

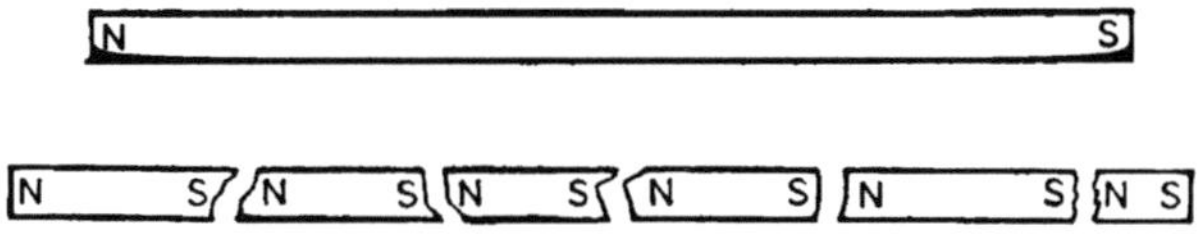

FIG. 220. — A piece of magnetized watch spring broken into fragments, each fragment remaining a complete magnet.

A tube of iron filings, represented in Fig. 221, may be magnetized by placing it across the poles of a strong permanent magnet. It may be again demagnetized by shaking. And it is not improbable that when we heat a magnet red-hot we produce such violent motions in its smaller parts as to continually "shake out" any magnetization that might be introduced.

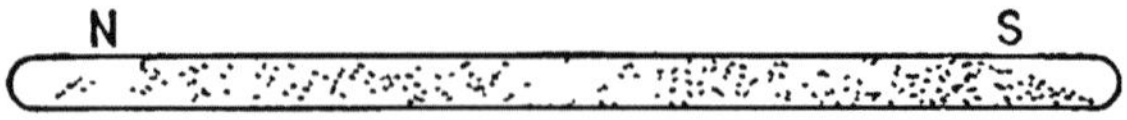

FIG. 221. — A tube of iron filings magnetized.

A long wire of soft iron — say 70 centimeters in length and 1 or 2 millimeters in diameter — may be strongly magnetized; but if it receive even a slight jar, its magnetization will almost entirely disappear. This also would lead us to think magnetization not a property of the surface of the body, nor of the body as a whole, but of its innermost structure; that is, its molecular structure. This we may call the *eighth* important discovery in magnetic science.

In the *Philosophical Magazine* for September, 1890, Professor J. A. Ewing, the English engineer, explains practically all the phenomena of magnetism and electromagnetism (see § 312) in terms of this molecular hypothesis, and illustrates them with an exquisite model in which molecules are represented by small compass needles.

311. 9. **The phenomena of magnetic induction.** — If we bring a piece of soft iron near a magnet, we shall find that the iron acquires magnetic poles; *i.e.* it exhibits magnetization as well as magnetic quality.

A wire nail is made of soft iron, and hence shows this phenomenon very easily, as indicated in Fig. 222.

If the iron nail be tethered at its lower end, it will remain suspended in mid air. That it is a magnet is at once shown by the fact that its lower end will attract small nails or iron filings.

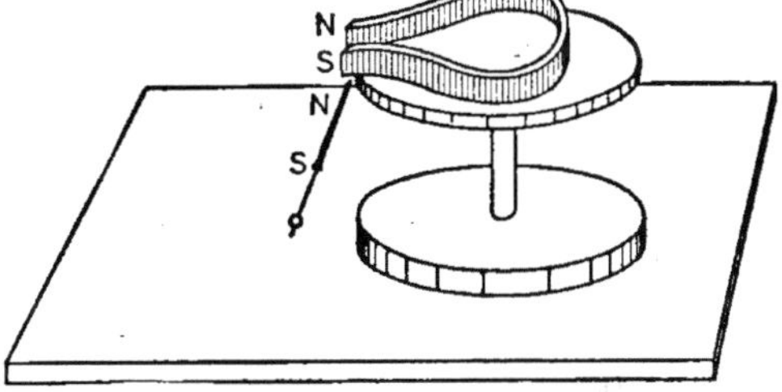

FIG. 222. — A wire nail becomes a magnet when brought into a magnetic field.

A wire nail held near either end of a magnetoscope will attract the needle. But if the magnet *NS* be brought up near the nail, the latter will attract one pole of the magnetoscope and repel the other.

One of the very best illustrations of magnetic induction is the following: Take a strip of "tin plate" so called (which is really sheet iron plated with tin), having dimensions something like 1 inch by 12.

If this strip has not been magnetized, either end of it will attract the north pole of the needle in a magnetoscope. Next hold the strip with one end down and bend it to and fro sharply between your thumb and forefinger at various points throughout its length. On examining it with the magnetoscope it is observed that the end which has been held down in the earth's field now repels the north pole of the needle. The bending apparently gives freedom to the particles of iron, and they probably arrange themselves in a new way under the influence of the earth's field. If now the other end of the strip be held down during the bending process, the poles of the strip will be reversed. Explain this by a diagram.

To magnetize a piece of iron, therefore, we need not touch it with a magnet; we need only bring it into the neighborhood of a magnet, i.e. into a magnetic field.

This process is called magnetic induction, more properly magnetization by induction.

It is here all-important to note that the original magnet loses none of its own magnetism by inducing magnetism in another body.

If we assume that a magnet is made up of small particles, or molecules, and that each particle is a magnet, then the phenomenon of magnetic induction is easily explained as follows: A piece of iron is made up of molecular magnets which are arranged in no particular order, arranged, in fact, in the utmost disorder, as indicated in Fig. 223. But when this iron is brought into a magnetic field, — when, for instance, it is brought near a magnet, — the north poles of these little magnets will each tend to point in a definite direction, and the south poles each in an opposite direction. The result will be that the little magnets will be brought more or less into line, producing a north pole at one end of the iron and a south pole at the other, as indicated in Fig. 224. The iron is then said to be magnetized by induction. All magnets, in fact, are produced by induction.

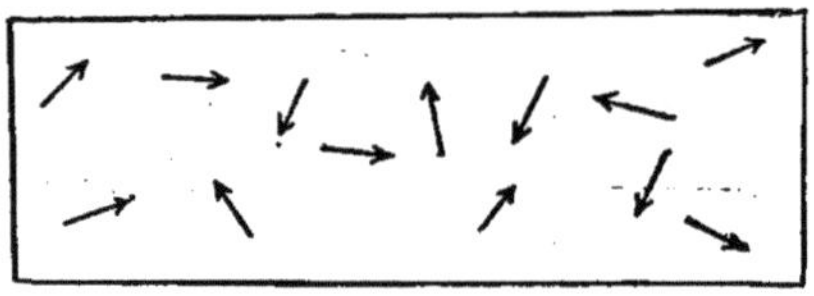

FIG. 223. — Probable arrangement of particles in a piece of ordinary iron.

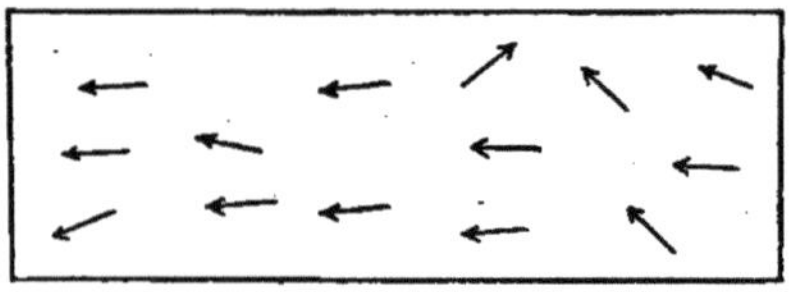

FIG. 224. — Probable arrangement of particles in a piece of magnetized iron.

The Electromagnet

312. One of the most remarkable and fundamental facts of electrical science is that every electric current is surrounded by a magnetic field; or in other words, every electric current carries with it a magnetic field. Accordingly a piece of iron brought into the neighborhood of an electric current will in general become magnetized by induction, just as it would if brought near a permanent magnet. An arrangement of this kind, which is called an "electromagnet," is the most important case of magnetic induction. Economic applications of the electromagnet, such as occur in the telegraph, telephone, dynamo, and motor, form an important chapter in the engineering achievements of the nineteenth century. Some of these will be studied in a later chapter.

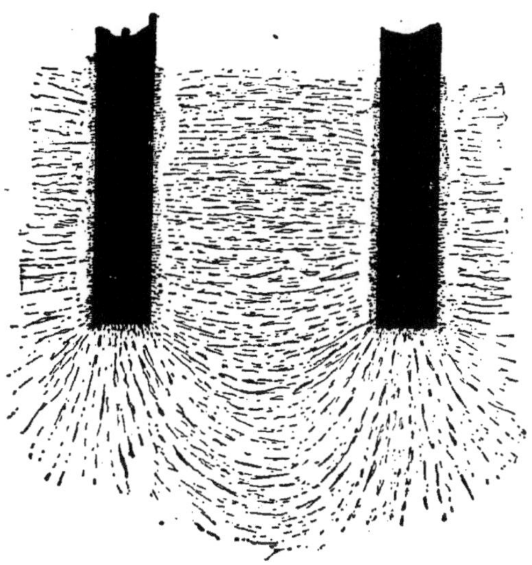

FIG. 225.

313. 10. Iron has great conductivity for lines of force. When a piece of iron is placed in a magnetic field, not only does the iron become magnetized, but the trend of the lines of force in the field becomes greatly altered. This alteration is always such as one might expect if the lines of force had a decided preference for the iron. They appear to go out of their way, so to speak, for the sake of passing through the iron. Figure 225 represents the field between the two poles of a horseshoe magnet. Figure 226 represents the same field with a small block of iron placed between the two poles.

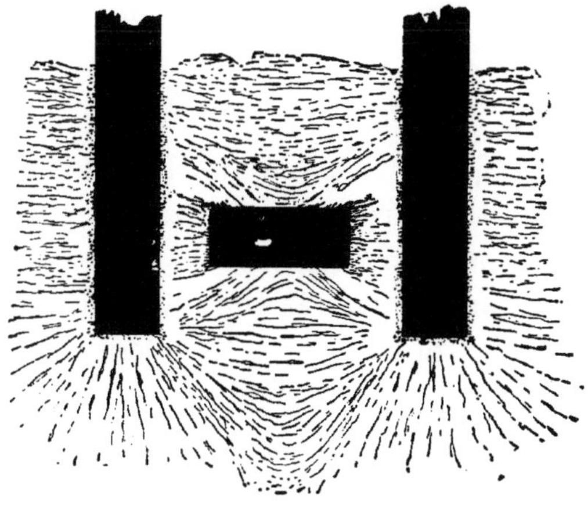

FIG. 226.

If an electric current be passed through a spiral of wire, it will be found that the interior of the coil will be a magnetic field. The lines of force will pass from one end of the helix to the other. But if an iron rod be slipped into the helix, the number of lines of force passing through will be multiplied many times. In this case many new lines of force are created by the introduction of the iron.

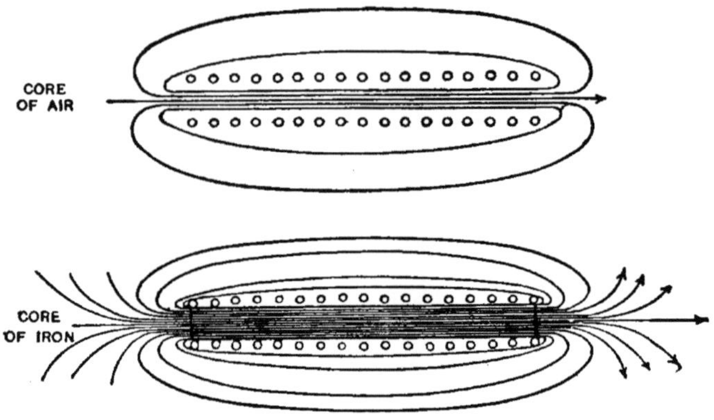

FIG. 227.—Illustrating the great conductivity of iron for lines of magnetic force.

If B is the number of lines of force passing through each unit of area of the cross section of the iron rod and H the number of lines passing through the same unit of area when the helix was filled with air, then we may describe the preceding facts by writing

$$B = \mu H, \qquad \text{Eq. 126}$$

where μ is a proportionality factor for which Lord Kelvin proposed the name *magnetic permeability*. This factor μ is, however, not a constant, but a variable depending upon H, and in the case of iron ranges in value from 1 to 2000 when measured on the C.G.S. system.

Magnetic Variations

314. Besides that mentioned in § 295 there is another exception to the general statement that the direction of the magnetic needle remains fixed. There are several distinct changes constantly taking place. These "variations," as they are called, may fairly be reckoned among the most instructive cosmical phenomena with which we are acquainted. The two most important of them are perhaps the following: —

1. *The diurnal variation.* In the northern magnetic hemisphere during the early morning hours the north-seeking end of the needle is found to be swinging to the eastward; this it continues to do until 7 to 9 A.M., some time during which interval it has in general reached its most easterly position. After hovering near this extreme for a while, it turns about, and begins to march westward. About 10 to 11 A.M. it passes

through the average position for the whole 24 hours, and by 1 to 3 P.M. has reached its extreme westerly position. Then it once more turns to the eastward, recrosses the mean position in general some time between 6 P.M. and midnight, and gradually returns to the position from which it started in the morning. The average total change in the course of a year between the morning and afternoon extremes is greatest in the United States in northern Minnesota, where it amounts to as much as 14′, while in southern Florida it is as low as 4′; these quantities vary with the seasons, being about one third greater in midsummer and a like amount less in midwinter.

2. *The secular variation.* In addition to this daily oscillation, the *mean position* about which the needle swings is itself slowly changing. The annual amount of this secular change varies not only for different localities, but also with time. In this country the north-seeking end has *at the present time* (1907) practically no secular variation for points along a line passing approximately through the western end of Lake Superior and the extreme eastern portion of Alabama. For places east of this line-of-no-change the mean position of the north-seeking end of the needle is now moving to the westward, and for places west of the said line, to the eastward. The *present* annual rate of secular change in the region of the Pacific Coast states is about 4′ and in New England about 3′. This change is of great importance to surveyors and map makers. If suitable corrections are not made for the secular variation, a boundary line whose direction has been determined by means of the compass needle will, when resurveyed some years later, have a direction *apparently* different from that previously found.

References on Magnetism

PERKINS. *Electricity and Magnetism* (Henry Holt & Co.), Chapters VIII and IX.

STEWART AND GEE. *Practical Electricity and Magnetism*, Vol. II, pp. 19–26.

CHRYSTAL. Art. "Magnetism," *Ency. Brit.*

HENDERSON. *Practical Electricity and Magnetism*, Chapter VI. A modern and valuable guide for advanced work in the laboratory.

EWING. *Magnetization of Iron and Other Metals.* Undoubtedly the best treatise on the subject in the English language.

MASCART. *Traité de Magnétisme Terrestre.* Gauthier-Villars: Paris, 1900.

CHAPTER IX

ELECTROSTATICS

315. If a sheet of dry writing paper held against the wall be given a few quick flaps with a piece of flannel, the paper will adhere to the wall quite strongly. The rubber handle of a fountain pen, when rubbed over the sleeve of one's woolen coat, acquires the property of attracting small bits of paper or wood; glass rubbed on silk behaves in the same way.

The substance earliest found to have this property was a fossil resin (ordinary amber) found on the shores of Greece.*

These facts have been known ever since the time of Thales, 600 (?) B.C.; and for two thousand years following him they constituted practically the whole of electric science, barring some knowledge of lightning and of the electric eel, which no one then imagined to have any connection with the amber phenomenon.

Bodies that have acquired this property of attracting small bits of paper are said to be electrified, or to have an electric charge.

First Law of Electrostatics

316. In more recent times, however, it has been found that *any* two different substances when rubbed together, or even when brought into contact, become electrified.

But before we can examine the evidence for this statement it will be necessary for us to devise some means by which we can detect the presence of an electric charge. One method is that indicated above, viz. the attraction of bits of thread or straw, but this is not a very sensitive test; much smaller charges can be detected by the gold leaf **electroscope**, an instrument which we shall now describe.

* The Greeks called this resin *electron* (ἤλεκτρον). And hence later Dr. Gilbert called this phenomenon *electric*, and gave us our word *electricity*.

The Electroscope and Proof Plane

This device is essentially a metal rod supported on a cork of sealing wax or, better still, of sulphur, and carrying two leaves of aluminium or gold foil suspended from its lower end, the leaves being inclosed in a glass vessel, as shown in Fig. 228.

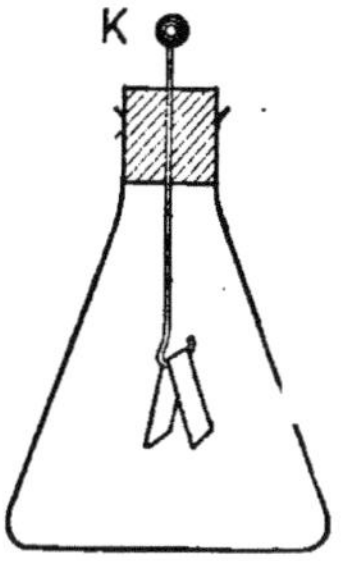

FIG. 228.—Electroscope made of Erlenmeyer flask and aluminium foil.

If, now, any electrified body be brought into contact with the knob *K* or, as we shall see presently, even into the near neighborhood of *K*, the leaves of gold foil will diverge. If the body is not electrified, the leaves will not diverge.

We have thus a simple and exceedingly delicate means for detecting electric charges. The theory of the electroscope we shall consider presently.

Try touching the tip of a camel's-hair brush on the table top and then to the electroscope. Rub the brush gently over your own hair and again bring it to the electroscope. **Since electrification is produced by such slight means, we are led to think that any two different substances brought into contact become electrified.** This is possibly the most fundamental fact of electrostatics, and may be called the **First Law of Electrostatics.**

But we sometimes wish to examine the electrification of a body which is too large to carry to the electroscope, or of a body which we do not wish to move. In such a case, it is most convenient to wax a penny to the end of a glass or rubber rod (see Fig. 229). We may use the glass rod as a handle, and then by touching the penny alternately to the body and to the electroscope, we may determine whether or not the body is charged. Such a device is called a **proof plane.**

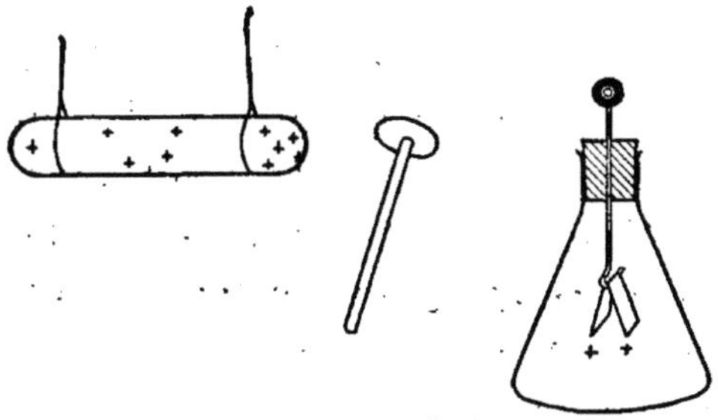

FIG. 229.—Proof plane.

317. Sensitive as is the form described above, the gold leaf electroscope was modified by C. T. R. Wilson of Cambridge in 1901 so as to be much more efficient.

He supports a rigid strip of brass *B* from a small block of sulphur *S*. On the side of this flat brass rod a narrow strip of gold leaf is attached by its upper end. The whole system of sulphur, brass strip, and gold leaf is suspended by a rod *T*, which is rigidly attached to the containing vessel by means of a block of ebonite.

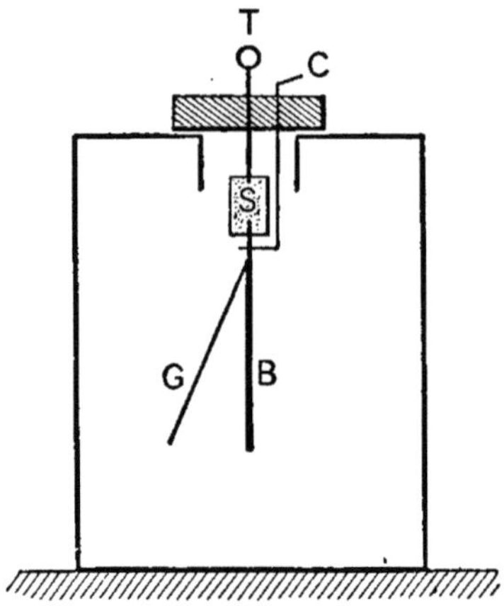

FIG. 230. — C. T. R. Wilson's gold leaf electroscope.

Sulphur is such a superb nonconductor that, in reasonably dry air, an electroscope of this kind will hold its charge for several hours. In order to communicate a charge to the suspended system, a small wire *C* is passed through the hard rubber block on top of the containing vessel and then bent twice at right angles. When the gold leaf is once charged, the contact between it and this wire may be broken by simply rotating the upper end of *C*.

CONDUCTORS AND NONCONDUCTORS

318. A *second* great discovery in electrostatics was made by Stephen Gray in London (d. 1736). He found that when a glass rod is electrified by rubbing, the electrification "leaks off" if the rod is held in the hand, or if the rod is supported by cotton thread, or metal, or a moistened string of any kind. By connecting an electrified glass rod to a distant point by means of a wet string, he was able to transfer the electrification through a distance of many meters.

He found also that if a body containing an electric charge was supported on glass or rubber, or suspended by a dry silk thread, the charge disappeared very slowly. **Those substances which lead off the charge quickly are called conductors; those which prevent the charge from escaping are called nonconductors or insulators.** A conductor supported upon a nonconductor is said to be "insulated." But there are some substances, such as carbon and most acids, which occupy an intermediate position, and can hardly be called either conductors or nonconductors. The following list of substances, mostly from Perkins's *Outlines of Electricity and Magnetism*, p. 12, is arranged in the order of their ability to conduct electricity.

Substances	Relative Conducting Power
Silver	1,000,000,000
Mercury	16,000,000
Gas carbon	400,000
Sulphuric acid, dilute . .	160
Gutta-percha	0.000000000004

Two Kinds of Electrification

319. A *third* great discovery was made in electrostatics by a French soldier and electrician, Dufay (b. 1698, d. 1739). He found that if a light electrified body, say a pith ball, be suspended by a silk thread, other electrified bodies behave always in one of two ways when they are brought near the suspended body. Some charged bodies will attract the pith ball, and some will repel it. The electrification which appears on glass when rubbed with silk behaves in one way ; the electrification which appears on gutta-percha (or a rod of sealing wax) when rubbed with flannel behaves in the opposite way Dufay called these two kinds of electrification *vitreous* and *resinous*, respectively ; but the following names suggested by Benjamin Franklin are in some respects happier, and have been adopted by the whole world : —

Definitions. 1. **That kind of electrification** which makes its **appearance on** glass when **rubbed with** silk is called **positive.**

2. **That kind of electrification** which **appears on** sealing **wax or gutta-percha** when **rubbed with flannel** is therefore **called negative.**

It will be observed that the first of these definitions is a purely arbitrary one. The fact of nature is that there are two kinds of electrification, but which one we shall call positive and which one negative is a matter of choice. The scientific world, has, however, agreed upon the definitions given above. The advanced student will find evidence for thinking that if the nomenclature of electrostatics had been postponed until the present day, these two names, positive and negative, would have been interchanged.

Figure 231 may aid the student to bear in mind not only the present convention, but also Dufay's evidence for thinking there *are* two kinds of electricity.

It is important here to observe that the pith ball is always repelled by the glass rod when it has been charged from the glass rod, while it is always repelled by the sealing wax if it

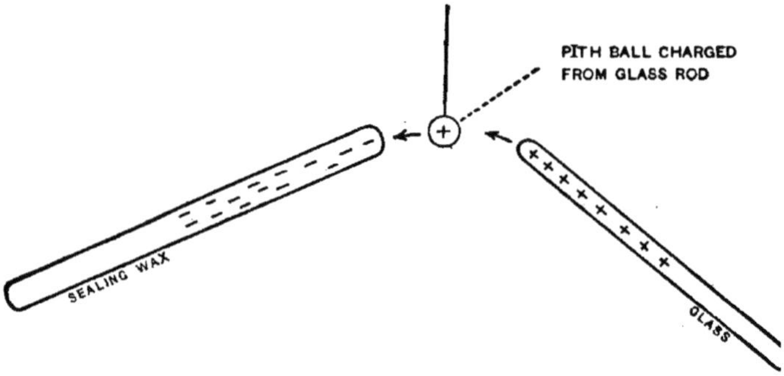

FIG. 231. — Differences in behavior of the two electrifications.

has been charged from the sealing wax; in other words, like electrifications always repel each other.

We now see why it is that the gold leaves of the electroscope always diverge when they are charged. For whichever of the two electrifications is given to the electroscope, the leaves will each be charged in the same way, and will, therefore, repel each other. It is important also to observe that the uncharged electroscope gives the same indication, *i.e.* a divergence, for a positive charge as for a negative one. How, then, can we determine when we are dealing with a positively charged body and when with a negatively charged body?

Experimental Distinction between Positive and Negative Charges

320. First charge the electroscope with a known kind of electrification, say positive. Then if the positive charge on the electroscope be increased, the divergence of the leaves will increase; if, however, we give to the electroscope a negative charge, the divergence of the leaves will diminish.

If we know the original charge on the leaves to be positive, we will then have no difficulty in saying just what the sign of any other charge is; for if the unknown charge increases the divergence, it must be positive; but if it diminishes the divergence, it must be negative.

SECOND LAW OF ELECTROSTATICS

321. If we assume now that two different substances when brought into contact become electrified, and that bodies having charges of the same sign repel each other, while bodies having charges of opposite signs attract, the next inquiry is, *How* do these bodies attract (or repel) each other? This question, it will be observed, is, for electrical charges, exactly the same one which Newton asked concerning the behavior of pieces of ordinary matter.

The French physicist Coulomb (b. 1736, d. 1806) answered the question for electricity by a series of experiments which are somewhat difficult to repeat, though very simple in idea. He showed that the experimental facts are as follows: —

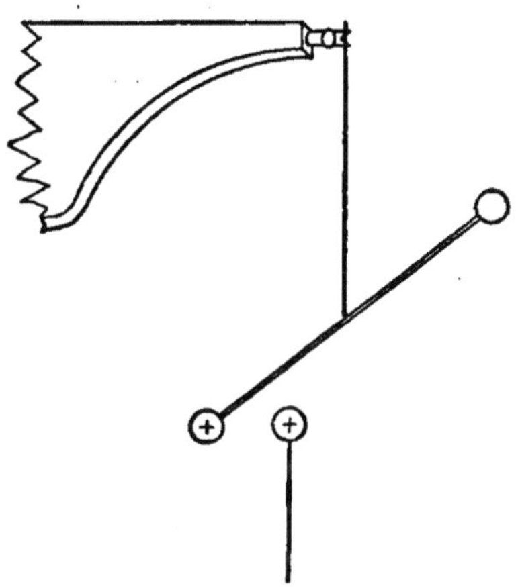

FIG. 232.

(i) The repulsive (or attractive) force between two quantities of electrification (Fig. 232) is directly proportional to the product of the two quantities. Equal quantities of electricity can always be obtained by exactly duplicating any process which produces electrification.

(ii) The force which one charged particle (or a body acting as a particle) exerts upon another varies inversely as the square of the distance separating the bodies.

(iii) In addition to the above two facts discovered by Coulomb, it was shown by Cavendish and Faraday that the force which one charged particle exerts upon another depends also upon the material of the nonconductor which separates them. This nonconducting material is generally called the medium or the dielectric.

A summary of these three facts, which we may call the *Second Law of Electrostatics*, is more elegantly made in terms of algebra, as follows: —

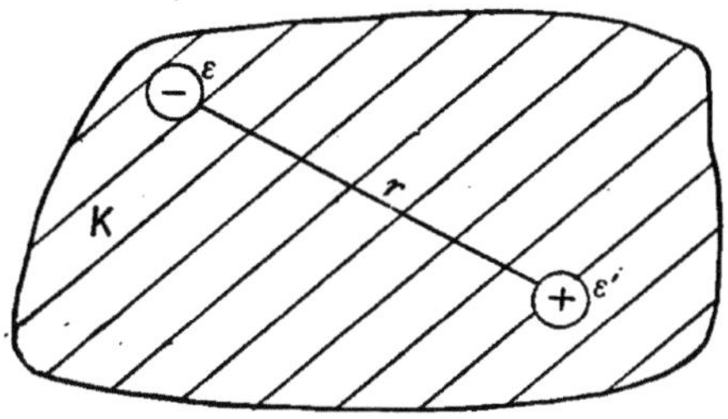

FIG. 233.

Let e and e' (Fig. 233) denote the electric charges of the two particles, r the distance which separates them, and K a quantity which expresses (measures) the effect of the nonconducting medium; then we know by experiment that

$$F \propto e; \ F \propto e'; \ F \propto \frac{1}{r^2}; \text{ and } F \propto \frac{1}{K}.$$

Hence
$$F \propto \frac{ee'}{Kr^2},$$

or
$$F = C\frac{ee'}{Kr^2}, \qquad \text{Eq. 127}$$

where C is a constant of proportionality. This quantity K is known as the *dielectric constant*, and is assumed arbitrarily to have the value of unity when the dielectric is a vacuum. But its value for air (1.0005) differs so little from unity, that for all ordinary purposes we may put $K = 1$ in air. For various kinds of glass K ranges from 3 to 10; while for ebonite and paraffin its value generally lies between 2 and 3.

The essential thing here is to understand that the force which one charged particle exerts upon another depends upon the medium in which the particles are situated, as well as upon the distance between the particles.

Definition of Unit Charge

322. Before Eq. 127, which describes the law of inverse squares discovered by Coulomb, can be of any value for purposes of computation, it will, of course, be necessary to evaluate the constant C. This is done as follows: Let the two charges indicated in Fig. 232 be taken equal to each other, and let them be placed at a distance of one centimeter apart *in air* The size of either charge we have still left undetermined. In terms of algebra, we have assumed only the following :—

$$e = e'; \ r = 1; \text{ and } K = 1.$$

If now, in addition to these assumptions, we agree to call e unity when $F = 1$, it follows, by substituting these values in Eq. 127, that $C = 1$. And this is precisely the reason for choosing a unit charge of this size, namely, that the constant of proportionality is thus made to disappear from the equation, giving us the working form

$$F = \frac{ee'}{Kr^2}. \qquad \text{Eq. 128}$$

A unit charge of electricity is, then, one of such a size that it will repel with a force of one dyne an equal charge of the same sign placed at a distance of one centimeter in air.

Compare this law of force with those which hold for gravitation and magnetism. The fundamental importance of this law will be realized when it is understood that this is the first bridge which connects the two great domains of electricity and mechanics. The connection is, however, only a quantitative one; no light is shed upon the nature of an electric charge by thus defining, in terms of a force and a distance, a certain charge with which other charges are to be compared. At the same time, this is a most essential step toward discovering various facts which do illuminate the nature of electrical phenomena.

When the consideration of electrical currents is taken up, we shall find that the unit of electrical quantity employed by engineers—the *Coulomb*—is 3000 million times as large as the electrostatic unit above defined.

An Electric Field

323. Since, in general, a force is exerted upon one positively charged body in the presence of another positively charged body, it is clear that we shall have to do work upon one of these charged bodies if we move it nearer to the other one. And, in general, if we move either body in any direction, either we shall have to do work upon it, or it will do work for us.

A region in which work must be done to move an electric charge from one point to another is called an electric field.

The intensity of the field in any direction, at any point, is measured numerically by the amount of work, in ergs, which must be done to move a unit quantity of electricity through unit distance along the given direction in the neighborhood of that point.

Electric intensity is, of course, to be measured in the same fundamental units—centimeters, grams, and seconds—as are mechanical forces and mechanical work.

In accordance with this definition, a unit electric field is one in which a unit charge is acted upon with a force of one dyne.

If, therefore, a body carries a charge of e units in an electric field whose intensity is R, the force F acting upon this body will be given by

$$F = eR,$$ Defining equation for intensity of electric field. Eq. 129

which is the defining equation for R, and which is a strict analogue of the equation (115) which defines the intensity of a magnetic field.

Since F is a vector quantity and e a scalar, it follows that R is a vector. Hence in order to specify an electric field it is quite as necessary to state its direction as its intensity

Lines of Electric Force

324. This fact led Faraday to introduce into electrical science the conception of "lines of force," a most helpful mode of description. He imagines that in any region where there are charged bodies these "lines of force" run from the positive charges to the negative ones. At any point in space the line of force takes the direction of the electric field at that point. Not only so, but the closeness (density, if you like) with which the lines of force are drawn at any point in the field is proportional to the intensity of the field at that point. In this manner, Faraday and Maxwell gave a simple, lucid, and accurate graphical description of an electric field.

As one of the simplest possible illustrations consider a positively charged metal sphere placed on a table in a room. Here the *negative* charge lies, as we shall presently learn, upon the walls of the room. If now the region about the charged sphere be tested by a bit of linen thread, say 2 inches long, supported at its middle point by a silk fibre, it will be found that the bit of linen acts as a compass needle for the electric field and that the electric field is radially distributed about the sphere. Hence the lines of force are to be drawn in a radial direction as in the accompanying figure.

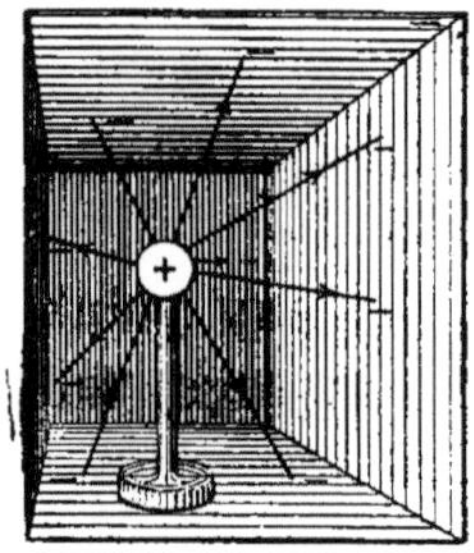

FIG. 234 — Lines of force representing the electric field about a charged sphere.

As another illustration consider two equal parallel plates, A and B, Fig. 235, one charged positively, the other holding an equal negative charge. In such an arrangement, the electric field is limited almost exclusively to the region between the plates; hence also are the lines of force. And

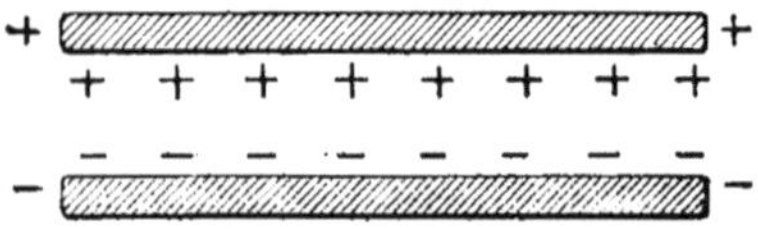

FIG. 235. — Showing distribution of charges over parallel plates.

since the field is also uniform, it follows that its representation must be very like that indicated in Fig. 236. The analogy between electric and magnetic lines of force is obvious.

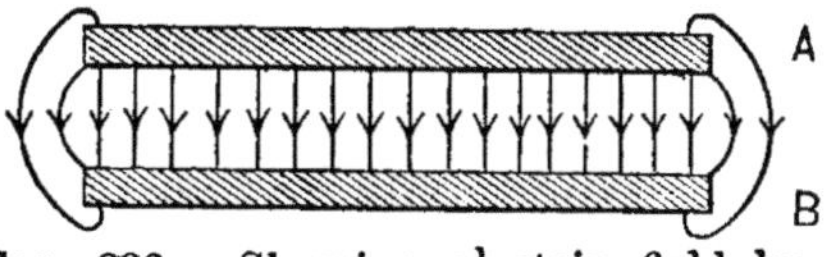

FIG. 236. — Showing electric field between parallel plates in terms of lines of force.

CAUTION

The student is here warned not to imagine that electrical forces differ in any respect from the other forces which he has studied and will study. They produce the same effects upon bodies as gravitational forces, they are measured in the same units (dynes) as gravitational forces, and they are no more mysterious than gravitational forces. An electrical force of 13 dynes acting upon a charged body will produce the same acceleration in the body as any other force of 13 dynes.

A SECOND METHOD OF PRODUCING AN ELECTRIC CHARGE. ELECTROSTATIC INDUCTION

325. For charging bodies there are other methods besides friction. One of the most interesting of these is the following, which appears to have been discovered by Stephen Gray.[1] He found that one may charge a body, provided this body be a conductor, by simply bringing it into the field of another charged body.

Let A (Fig. 237) be a charged body. A tin can on a block of paraffin, or a hollow brass sphere supported on a glass stem, will answer well. Let B indicate an insulated conductor which is not charged. A block of wood covered with lead foil and supported on a glass or rubber stem will answer well.

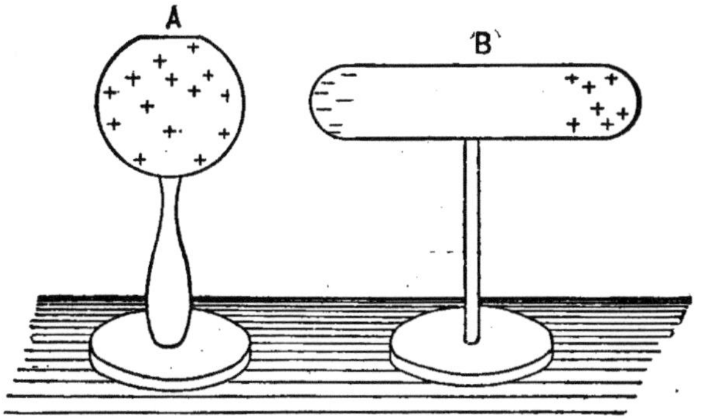

FIG. 237. — The phenomenon of electrostatic induction.

Bring B into the neighborhood of A — say within three or four

[1] Poggendorff, *Geschichte der Physik*, p. 839.

centimeters of it—and examine *B* with the proof plane and electroscope (§ 316) as follows: —

1. Touch the proof plane to the charged body *A*. Carry this charge to the electroscope. You will get a distinct separation of the gold leaves.

2. Touch the proof plane to the end of *B* which is nearest *A*, and carry the charge to the electroscope while the leaves still diverge. The divergence of the leaves diminishes, showing that the charge on this part of *B* is opposite to that on *A*.

Repeat the process until the leaves have completely collapsed and have again diverged. They are now charged with the same kind of electrification that is on the near end of *B*.

3. Touch the proof plane to the end of *B* that is most remote from *A*. The leaves begin to collapse again, showing that the two ends of *B* have charges of different signs.

This experiment should always be repeated until the evidence is perfectly clear for thinking that when an uncharged conductor is brought into the field of a charged body, the uncharged body becomes electrified exactly as indicated in Fig. 237 above. **The charges on *B* are said to be induced, and the phenomenon is called electrostatic induction.**

At this point it is all-important to note that the induced charge

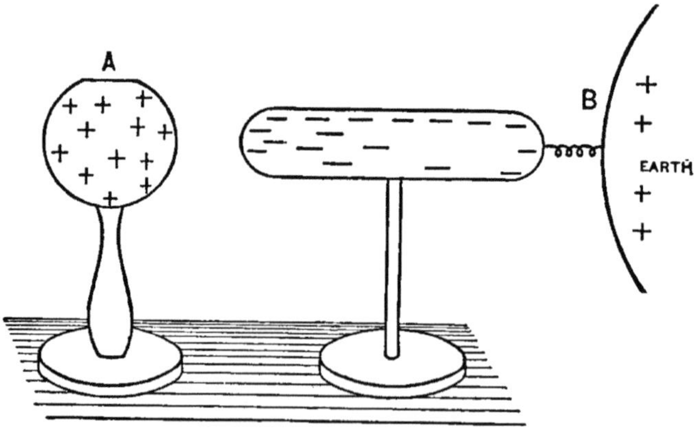

FIG. 238.—Charging by induction.

on *B* does not at all diminish the charge on *A*. It is produced by means of the charge on *A*, but not at the expense of the charge on *A*.

326. Up to this point we have considered each of the conductors, *A* and *B*, insulated. Imagine now the body *B* to be connected with the earth, which we know to be a good conductor. We have then

AN INTERESTING SPECIAL CASE. CHARGING BY INDUCTION

Our conductor *B* (Fig. 238) has now been very much enlarged; indeed, it includes the earth itself, so that *B* is at least 8000 miles long. There is, therefore, only one end of *B* which is near enough for us to examine, and at this end we find, as before, an electrification which is opposite that on *A*.

Suppose now we remove the wire. The body *B* is again reduced to its former size; but it will now be charged negatively while *A* is charged positively. **We have thus learned how to charge a body without using friction and without diminishing the original charge on *A*.**

The Electrophorus

327. A simple illustration of the foregoing principle as well as one of the most convenient instruments in the laboratory for producing electric charges quickly and easily is the electrophorus invented by the Italian physicist, Volta. It consists of a plate of hard rubber, rosin, or some substance which can be electrified by friction and a metal disk *C*, which is provided with an insulating handle *H*, and which is called a "carrier." The rubber plate *P* generally rests upon a sheet of metal *S*, which is called the "sole plate."

Suppose the plate to have been rubbed with catskin and therefore to be negatively electrified. Then when the carrier is made to approach the plate of ebonite the induced charges will be distributed as shown in Fig. 239. When the carrier is placed upon the plate, it will be supported by slight rugosities so that it does not receive a charge from the plate. If now the finger be touched to the carrier, the negative charge will run off to earth and the positive charge remains bound. When, however, the carrier is removed, by the handle, to a distance from the plate, the bound charge becomes free, and available for charging other bodies. Evidently this process may be repeated as many times as one likes.

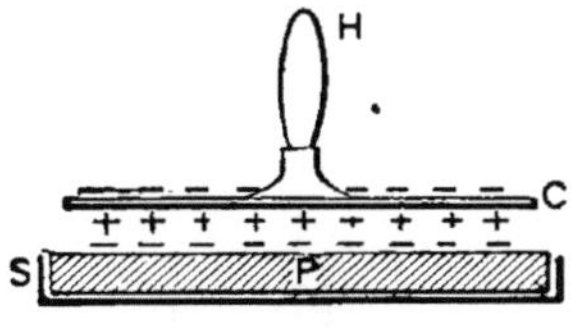

FIG. 239. — Electrophorus of Volta.

Machines have been recently perfected by the joint labors of Holtz, Toepler, and Wimshurst by means of which one can start with an indefinitely small charge on a body *A*, and produce an indefinitely large charge on *B*. These machines are based on the simple principles which we have just been studying; but they are somewhat complicated in their action, and the student beginning physics is advised to first observe the structure and behavior of one of these machines and then consult some of the larger text-books or encyclopedias for its description and explanation.

Third Law of Electrostatics

328. Let us now modify the electroscope, described § 316, either by fitting it with a hollow metal vessel, as indicated in Fig. 240, or by connecting it, by use of a light wire, to a hollow insulated vessel, as diagrammed in Fig. 241

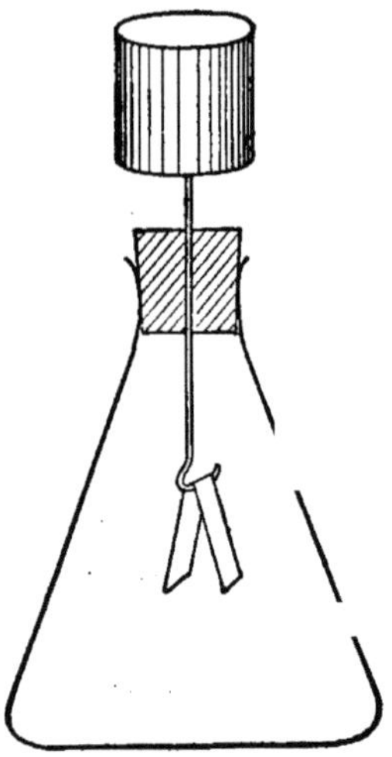

Fig. 240. — Equality of positive and negative charges.

Any charged body introduced into this can will affect the gold leaves even more than if it were brought up near to the electroscope itself; for, indeed, the can is now a part of the electroscope. And, what is most remarkable, it matters not in what part of the can the charged body is placed, the effect on the gold leaves is just the same.

We are now prepared to try the following fundamental experiment: Take two nonconducting glass rods, *A* and *B*, Fig. 242. About the end of one of them fasten a short pocket or tube of woolen goods so that it will not easily slip off. On the side of this pocket sew another similar pocket of woolen cloth into which the second glass tube will slide with gentle friction. By merely inserting the glass rod into the pocket and twisting it about, one is easily able to electrify the lower end of the free glass rod positively and the woolen pocket negatively.

Observe then that —

(i) No effect is produced upon the electroscope so long as both the rods are held in the can. When they are electrified by twisting the glass rod in the pocket, no divergence of gold leaves follows.

(ii) But if either the free rod or the one that carries the woolen cloth be removed, the gold leaves diverge, showing that both the cloth and rod are charged.

(iii) The charge on the glass is opposite in sign to that on the woolen cloth.

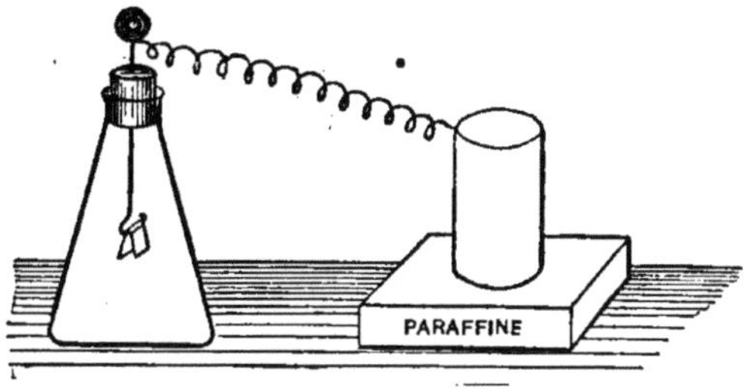

FIG 241.

The same series of experiments may be tried with a piece of silk and a glass rod.

In each case we have two oppositely charged bodies such that their combined effect on the electroscope is zero. The same phenomenon is observed, indeed, whatever two bodies are rubbed together. The inference is, therefore, that whenever we produce by friction a certain electric charge, we always produce at the same time an equal charge of opposite sign. This is also true in the case of induction. If we induce two opposite charges on a conductor and then remove the inducing charge, we find that the induced charges will just neutralize each other; all evidence of electrification disappears, showing that the two induced charges are equal and opposite. This general truth, expressed as follows, may be called the *Third Law of Electrostatics:* —

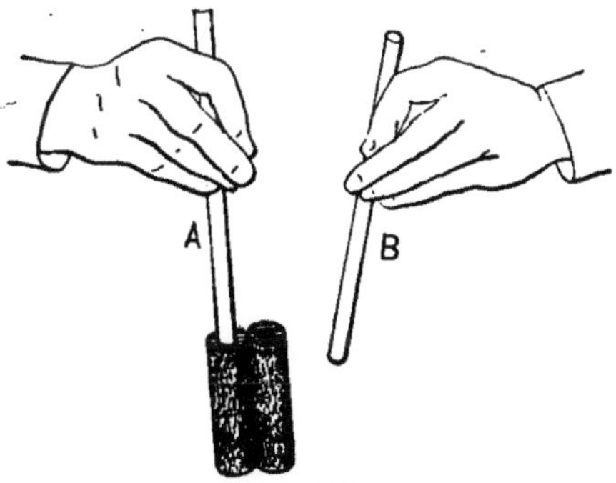

FIG. 242.

When electrification **is produced by friction, by induction, or by any** other **means, the** positive **and negative charges so produced** are **always** equal.

FOURTH LAW OF ELECTROSTATICS

329. It is also interesting to examine, by means of a proof plane and electroscope, the various parts of the surface of an irregularly shaped conductor, such as that represented in Fig. 243.

The proof plane, after being brought once into contact with a flat portion of the body, does not produce nearly so great a divergence of the gold leaves as when touched once to either end of the body. The sharper the end (*i.e.* the greater the curvature of the surface), the greater the charge which the proof plane carries away.

We see from experiments of this kind that the electrification is not uniformly distributed over a conductor. The sharper any end or point, the more the charge tends to collect there. When

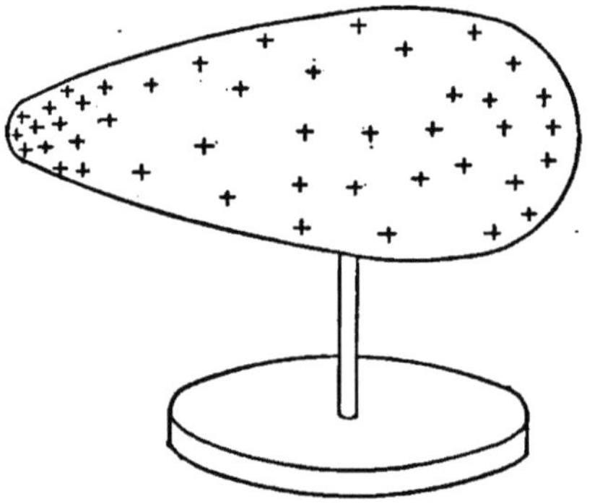

FIG. 243. — Electric charge accumulates near sharp end of body.

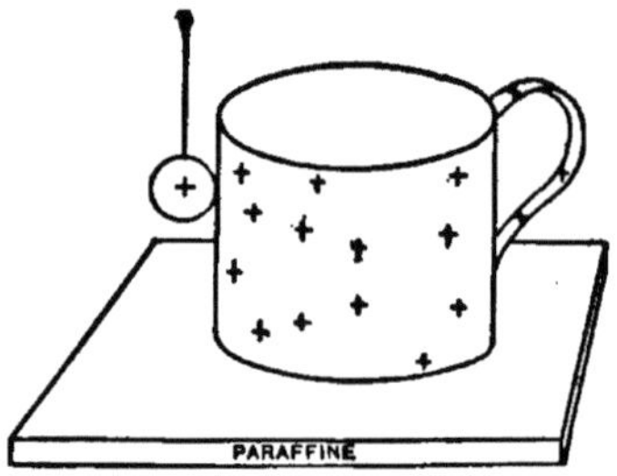

FIG. 244. — Charge resides on outside of conductor.

we examine the inside and outside of a tin cup, in the same manner, we find that practically all of the charge resides on the outside.

In general, it has been found that the manner in which electrification distributes itself upon a conductor depends only upon the shape of the body and upon the presence of surrounding charges.

To describe *each* of these various distributions would be an infinite task; but we may always be certain that **the charge on the outside of any closed conductor is distributed in such a way that it will produce no electric field — no electric force — inside the closed conductor.**

Even in the case of a conductor such as a tin pail which is not quite closed, the electrification is very nearly all on the outside, and the intensity of the electric field inside is very small.

Accordingly, if we wish to protect any instrument from electrical disturbances, we have only to place it inside an electrical conductor which is closed or nearly closed.

Take a small electroscope and place it inside a roll of wire screen, as shown in Fig. 245. If the meshes of the screen are reasonably fine, there is scarcely any electrical disturbance you can produce outside the screen which will affect the sensitive gold leaves inside.

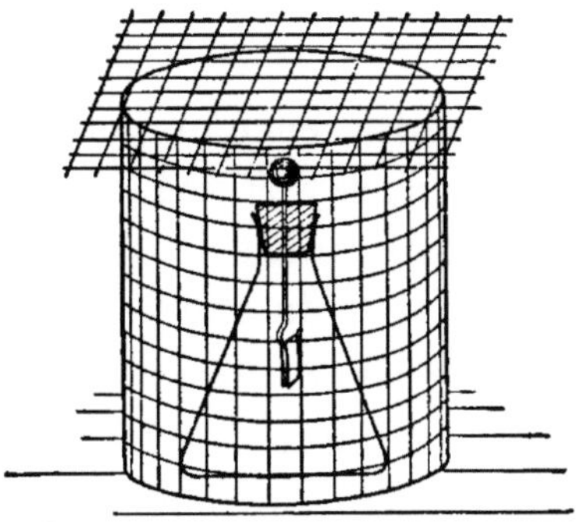

FIG. 245. — Electroscope protected by a network of wires.

This great fact that any electrical disturbance outside a closed conductor has no effect on the region inside the closed conductor was thoroughly established by Faraday. We may call it the *Fourth Law of Electrostatics.*

330. With this principle in mind, Maxwell suggested that the best method for protecting houses from lightning is to place them in a network of conductors (Fig. 246), which shall be as nearly as possible equivalent to a closed conductor. Benjamin Franklin, in 1760, first suggested the use of single metallic rods, connected with the earth, as a means of protection against lightning.

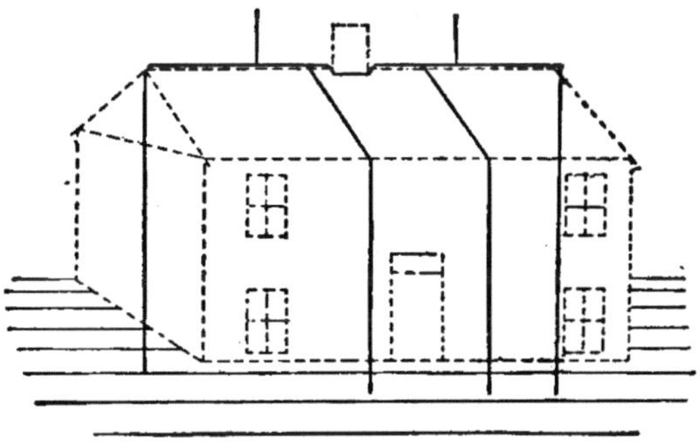

FIG. 246. — Maxwell's method of protection against lightning.

The Idea of Electric Potential

331. One of the principal ambitions of every student of physics should be to obtain a simple and unitary view of the entire subject and likewise of each subdivision of it. Few conceptions — perhaps none — have ever been introduced into the treatment of electricity that have been more helpful, in the way of simplification, than that of "potential," an idea which is due to the celebrated French astronomer, Laplace. The name is due to the English physicist, Green.

Definition of Electric Potential

The difference of potential between any two points in an electric field, say *P* and *S* in Fig. 247, is defined as the work required to carry a unit of positive electricity from *P* to *S* against the electrical forces in the field.

It is important to observe that in this definition all forces, such as friction, which are always opposed to the direction of motion are especially ruled out; the only force considered is an electrical one, which, as Coulomb has shown, depends upon the distance only. Since this is true, a little consideration will show that the work done on the unit charge in passing from *P* to *S* is independent of the path pursued between these two points. For, suppose that less work is done when the charge is carried along the route *A* than along the route *B*; and imagine the charge to be carried along the entire circuit *PASBP*. If the work done along the path *PAS* is positive, that along the path *SBP* will be negative; in other words the field will do work upon the charge as it passes from *S* to *P* along *B*. And since more work is done along the path *B* than *A*, it follows that we shall, by going through this cycle, get more work out of the field than we have spent upon it; and every time this cycle is repeated we shall gain a definite amount of energy. But since it is impossible to create energy (§ 100), we must conclude that our hypothesis is wrong and that the same amount of work will be required to carry a charge from *P* to *S* along any path whatsoever.

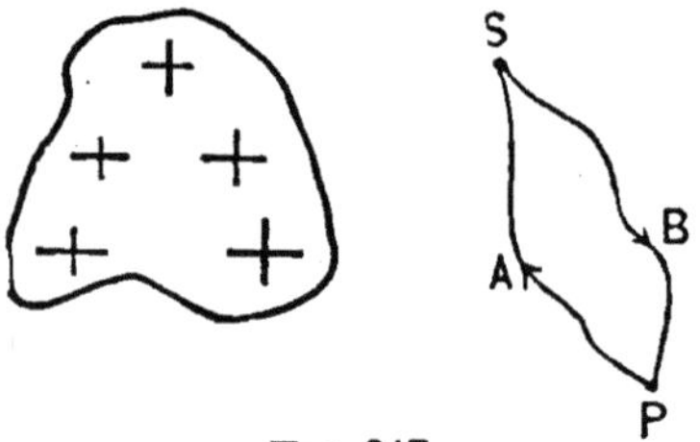

FIG. 247.

Potential Difference a Scalar Quantity

332. From the definition just given it will be seen that difference of potential is a ratio between a certain amount of work and a certain electric charge. Its defining equation, if we denote potential difference by *P.D.*, work by *W*, and charge by *Q*, is

$$P.D. = \frac{W}{Q}; \quad \text{Defining equation for difference of potential.} \qquad \text{Eq. 130}$$

and since W and Q are each scalar quantities it follows that potential difference is also a scalar.

Special Case

333. *Find the difference of potential between any two points P and S when the electric field is due to a charge Q on a small sphere at O.*

For the sake of simplicity we may assume (what will not detract from the generality of the result) that the three points O, P, and S, all lie in the same straight line, as indicated in Fig. 248. Let us now suppose the distance PS to be divided into a large number, say n, of small equal parts, $\overline{r_P r_1}$, $\overline{r_1 r_2}$, $\overline{r_2 r_3}$, . . . $r_{n-1} r_S$. Remembering that electric intensity is the force on unit positive charge, we have

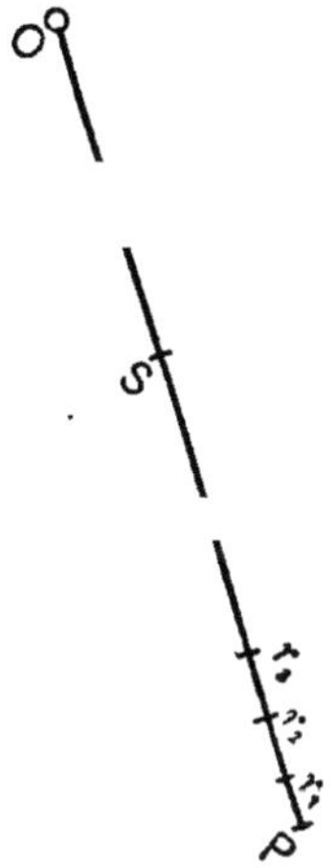

FIG. 248.

$$\text{Electric intensity at } P = \frac{Q}{r_P^2},$$

$$\text{Electric intensity at } r_1 = \frac{Q}{r_1^2},$$

$$\text{Electric intensity at } r_2 = \frac{Q}{r_2^2},$$

etc.

$$\text{Electric intensity at } S = \frac{Q}{r_S^2}$$

Hence

$$\text{Average intensity over distance } \overline{r_P r_1} = \frac{Q}{r_P r_1} \text{ very approximately.}$$

$$\text{Average intensity over distance } \overline{r_1 r_2} = \frac{Q}{r_1 r_2} \text{ very approximately.}$$

etc.

Hence the work in carrying the positive unit

$$\text{from } r_P \text{ to } r_1 = \frac{Q}{r_P r_1}(r_P - r_1) = Q\left(\frac{1}{r_1} - \frac{1}{r_P}\right),$$

$$\text{from } r_1 \text{ to } r_2 = \frac{Q}{r_1 r_2}(r_1 - r_2) = Q\left(\frac{1}{r_2} - \frac{1}{r_1}\right),$$

etc.

$$\text{from } r_{n-1} \text{ to } r_S = \frac{Q}{r_{n-1} r_S}(r_{n-1} - r_S) = Q\left(\frac{1}{r_S} - \frac{1}{r_{n-1}}\right).$$

Let us now take the sum of these small quantities of work, and it will be observed that all the intermediate terms cancel out, and we have for the total work done in carrying a unit charge from P to S, the quantity

$$Q\left(\frac{1}{r_S}-\frac{1}{r_P}\right),$$

which is the solution of the problem.

Potential at any Point

334. Let the initial point P be chosen at such a great distance (say at infinity) from the charge Q that $\frac{1}{r_P}=0$; then the intensity at P is also zero, and we have an expression for the amount of work done in bringing the positive unit from an infinite distance to the point S, namely $\frac{Q}{r_S}$.

The potential at any point in an electric field is accordingly defined as the amount of work which must be done in carrying a positive unit charge from an infinite distance to that point. This quantity may be denoted by V_S, and for a single charged particle may be defined algebraically as follows: —

$$V_S=\frac{Q}{r_S}. \qquad \text{Eq. 131}$$

It is important to observe that the theorem just proved applies not only to a field produced by a charged particle, but also to a body which acts as if its charge were all concentrated in a single point. A charged sphere acts in this way, namely, as if its entire charge were concentrated at the center of the sphere.

If there be several charges in the field, then the total potential at any point S is obtained by adding the separate potentials at the point S due to each separate charge. Potentials, it must be remembered, are scalar quantities. Hence follows the general theorem that **the potential at any point S, in an electric field, is obtained by adding the quotients of each charge divided by its distance from the point S.**

For purposes of computation, the following algebraic expression of this theorem is convenient: —

$$V_S=\Sigma\left(\frac{Q}{r_S}\right). \qquad \text{Eq. 132}$$

When this expression is positive, it means that work must be done *upon* the positive unit to bring it from infinity to the point S; but when negative, the meaning is that work will be done *by* the positive unit in coming from an infinite distance to the point S.

Zero of Potential

335. Ordinarily it is not necessary to go to infinity in order to reach a point where the electric intensity is zero. For the earth is such a tremendous body that its potential is not appreciably affected by any operations which we may carry on in our laboratories and factories. **Hence the earth is chosen as a body of zero or standard potential on exactly the same principle that the mean sea level is chosen as a surface of zero or standard height.**

Equipotential Surfaces

336. One of the best methods for describing an electric field is simply to join together all those points in it which have the same potential as any selected point P. In general the locus of these points will be a surface; and this surface is said to be "equipotential."

Accordingly an equipotential surface is defined as one on which the potential has the same value at every point.

Thus if the charge which produces the field is situated on a single particle, and if the charge remains constant, the potential at any point S will depend only upon the distance of S from the particle; and hence, for all points which lie on the surface of a sphere drawn about the charged particle as a center,

$$\frac{Q}{r} = \text{constant.}$$

Any one of the circles shown in Fig. 249 represents, therefore, a section, by the plane of the paper, of an equipotential surface (a sphere) due to a point charge at Q.

Since there is no change of potential over any one of these surfaces, it follows, from the definition of potential, that no work will be done in carrying a unit charge from any one point to any other point on the same equipotential surface. But in taking a positive unit from the surface passing through P to the surface passing through S, one must do $V_S - V_P$ units of work. The simplicity of this expression for work shows the convenience and importance of the idea of potential.

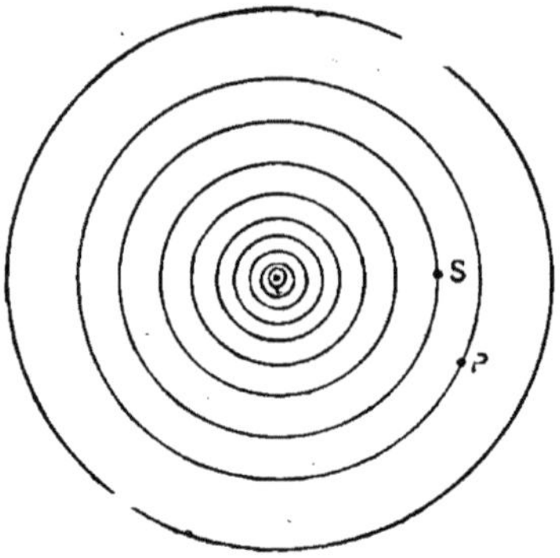

FIG. 249.—Equipotential surfaces due to a point charge.

When the charge resides upon a body other than a particle or a sphere, the equipotential surfaces become, of course, more complicated than those shown in Fig. 249.

The work done in carrying a unit charge from one point to another in an electric field may also be regarded as the product of the average intensity of the field R and the distance s between the two points.

Therefore

$$V_S - V_P = Rs, \qquad \text{Eq. } 132\tfrac{1}{2}$$

where s is the distance between P and S. See Fig. 249.

From this it follows that between any two equipotential surfaces the intensity R will be greatest in the direction in which the distance s is least.

337. And this leads us back again to the idea of

Lines of Force

For "the direction of an electric field" at any point is merely the direction in which the electric intensity is greatest, that is, the direction in which the distance between any two equipotential surfaces drawn near that point is least. **The direction of the electric intensity at any point is therefore normal to the equipotential surface passing through that point.** This result would also follow immediately from the definition of an equipotential surface as one along which no work is required to carry an electric charge.

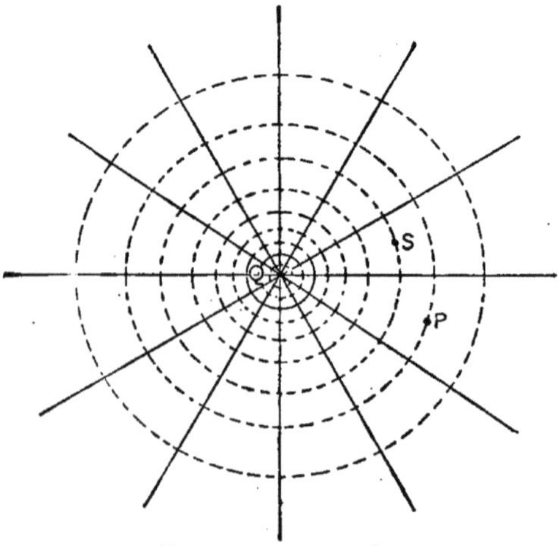

FIG. 250.—Illustrating the relation between lines of force and equipotential surfaces.

When, therefore, a field has once been described by drawing its equipotential surfaces, it becomes a matter of great ease to also describe it in terms of lines of force. For we have merely to draw the normals to these surfaces at every point. Thus in Fig. 250 the equipotential surfaces are indicated by dotted lines; the normals which are shown solid are, therefore, the lines of force.

Surfaces of Conductors are Equipotential

338. A little consideration will show that when a conductor is insulated and charged it will be an equipotential surface as soon as the electric charge comes to rest, which it does almost instantaneously. For if the component of the electric intensity along the surface of the conductor had, at any point, a value other than zero, the electricity at this point would move in the direction of the force exerted upon it. But, by hypothesis, the charge is at rest. **Any conductor holding a charge at rest is therefore an equipotential surface.** Indeed, this is the one practical method for "manufacturing" equipotential surfaces in the laboratory.

At this point the student will find it helpful to show by means of a diagram just how one can arrange a conductor (i) so that it will have a positive charge while it remains at zero potential, and (ii) so that it will have zero charge while it is at positive potential.

The advanced student will discover that the entire science of electrostatics reduces itself almost to a special case of the potential theory.

Measurement of Potential

339. The simplest means for obtaining a quantitative indication of potential difference is perhaps the gold leaf electroscope. But why, some one asks, does not the divergence of the gold leaves measure quantity instead of potential? For it is evident, at once, that the force which drives the gold leaves apart is directly proportional to the amount of charge on the leaves. The reply to this query is that if the electroscope be isolated and entirely disconnected from any other body, the divergence of the leaves *does* measure the *quantity* of charge on the instrument *as well a its potential.* If, however, the electroscope be connected, *by means of a rather long* wire, with another and much larger body, it is clear that the amount of work done in bringing up a unit charge from infinity (or from the earth) to the instrument will depend, not at all upon the charge of the second body, but *only* upon the quantity of electricity *on the electroscope;* in other words, the potential of the system made up of the electroscope, the wire, and the second body is shown directly by the divergence of the gold leaves.

Wherever the electroscope be placed, the connecting wire insures that its potential shall be the same as that of the body; for this wire renders the entire system a single metallic conductor.

If while the electroscope be still connected with, say, the tin cup in Fig. 251, another electrified body or conductor be brought into the immediate neighborhood of the instrument, its

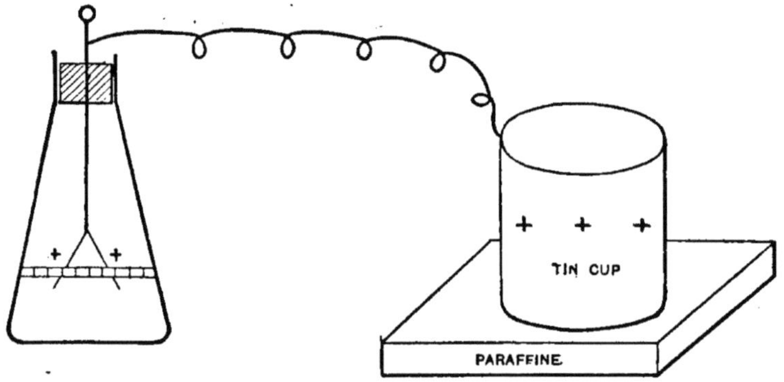

FIG. 251.—Illustrating the use of the electroscope to measure potential.

divergence may change very much, but its potential will remain constant. The reason the divergence may change is precisely *because* the potential of the gold leaves does remain constant. For if a unit charge be now carried up to the electroscope from infinity, the charge on the neighboring body *B*, in Fig. 252, will either aid or oppose the motion; and hence the charge on the leaves will need to be either larger or smaller, respectively, than before, in order to maintain their potential at the same value.

Therefore, in order to use an electroscope as an instrument for measuring potentials, it is only necessary to connect the body and the electroscope by a wire which is long in comparison with the dimensions of the body, and to see that no third body influences the gold leaves.

It need hardly be added that the size of the electroscope must be small in comparison with that of the body whose potential is to be measured, otherwise the original charge on the body would be appreciably diminished and its potential thereby reduced.

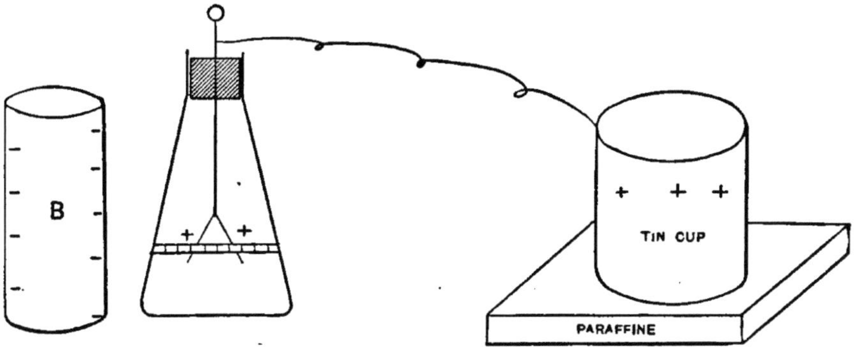

FIG. 252. — Illustrating the effect of a neighboring body on the reading of an electroscope.

The Idea of Electrical Capacity

340. Having a clear understanding of what is meant by electrical quantity Q and electrical potential V, the meaning of electrical capacity becomes a very simple matter; for the **electrical capacity** of any conductor is defined as **the ratio of its charge to its potential when all other conductors in the neighborhood are connected to earth.**

The capacity of a conductor is *not*, therefore, "the amount of electricity it will hold when it is full," but its numerical value is the same as the amount of electricity which the conductor holds when its potential is unity; in other words, it is the constant ratio which experiment shows to exist between the charge and the potential of the conductor.

If capacity be denoted by C, its definition may be given in algebra as follows: —

$$C = \frac{Q}{V}. \quad \text{Defining equation for electrical capacity.} \qquad \text{Eq. 133}$$

This equation will be found exceedingly useful in computing any one of the three quantities involved, and may indeed be regarded as the fundamental equation of electrostatics.

Capacity of a Sphere

341. As will be learned by the advanced student, the capacity of a body depends entirely upon its geometrical form and upon the dielectric constant (§ 321) of the medium. To compute

2 A

the capacity of a conductor from geometrical data is in general a rather difficult task; but in the case of some simpler forms the process is easy. Thus in the case of the sphere let us assume for electrical forces what Newton proved for gravitational forces, namely, that in the case of a spherical shell the forces at all points, except on the inside, are the same as if the entire charge were concentrated at the center of the sphere. For the potential of any point on the surface of a sphere of radius R and charge Q, we may then write

$$V=\Sigma\left(\frac{q}{R}\right)=\frac{1}{R}\Sigma\,(q)=\frac{Q}{R}, \qquad \text{Eq. 132}$$

where q denotes any small element of the charge.

And since in all cases we have $V=\frac{Q}{C}$, it follows that in the case of a sphere, in air,

$$C=R. \qquad \text{Eq. 134}$$

This result serves to illustrate what is meant by saying that capacity is a purely geometrical quantity.

Capacity of Two Concentric Spheres

342. With very little trouble it may be proved that in the case of concentric spheres, such as diagrammed in Fig. 253, the capacity C of the system is

$$C=\frac{RR'}{R-R'}. \qquad \text{Eq. 135}$$

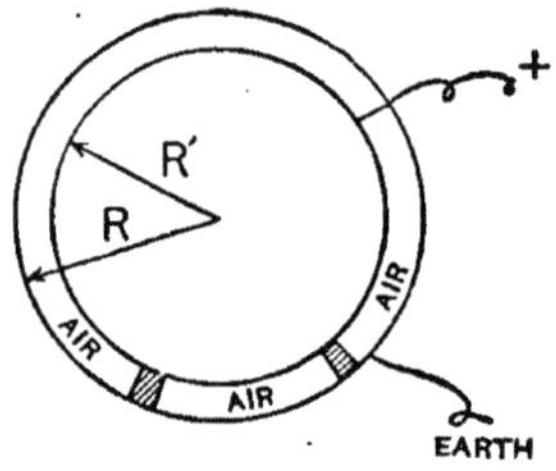

Fig. 253.—Condenser made up of two concentric spheres.

Such an arrangement is frequently called an electrical "condenser" One of the most common forms of condenser is a bottle which is lined both inside and outside with tin foil. This device, first used in Leyden, is called a "Leyden jar."

Capacity of Two Parallel Plates

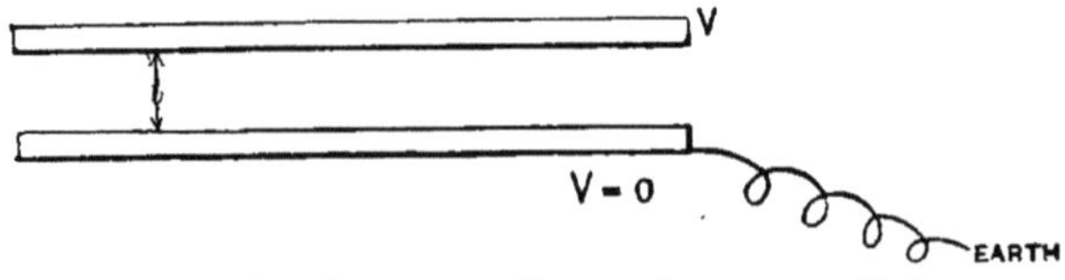

Fig. 254.—Condenser made up of two parallel plates.

343. With equal ease it may be shown that in the case of two parallel plates, in air, each of area

A and separated by a small distance t, the capacity of the system is

$$C = \frac{A}{4\pi t}. \qquad \text{Eq. 136}$$

Effect of Medium on Capacity

344. If the medium between the two spheres or between two plates has a dielectric constant K which is different from that of air, then these expressions for capacity become

$$C = \frac{KRR'}{R - R'} \text{ and } C = \frac{KA}{4\pi t}, \text{ respectively,}$$

equations which are useful in determining the value of K by experiment.

Energy of a Charged Conductor

345. If for a moment we define recent science as dating from the middle of the nineteenth century, one might, perhaps with fairness, say that the all-important problem in recent physical science is the determination of the amount and transformation of energy involved in its various phenomena.

In the earlier periods of science, stress was laid mainly upon the forces involved; but, in view of the fact that the energy of any isolated system — electrical as well as mechanical or thermal — remains constant, the energy expression is now a matter of the utmost importance.

To obtain the energy of a charged conductor is not a difficult matter; for, by the law of the conservation of energy, we have only to measure the amount of work done in bringing the charge up and putting it on the conductor.

Suppose we start with an insulated conductor whose potential is zero; let it be charged by conduction from some source. We may think of any finite charge which is given the conductor as brought up in very minute charges, one at a time; each one of these will slightly increase the potential of the conductor, so that when the charge is complete the final potential of the conductor will be, say, V.

Had the potential of the conductor remained at its initial value (zero), the work required to bring up any small charge would have been vanishingly small, since the potential of any conductor is the amount of work required to bring up a unit charge from the earth to that conductor.

If, on the other hand, the conductor had been at potential V during the whole of the charging process, then the amount of work required to bring up any small charge q would have been qV.

But, as a matter of fact, the potential of the conductor varied directly as the amount of charge placed upon it, so that **the mean value of the potential during the charging process was** $\frac{V}{2}$. **If then the total charge be called** Q, **the amount of work required to charge the conductor is** $\frac{1}{2}QV$; **and this by the principle of the conservation of energy represents the energy of the charged conductor.** In virtue of the defining equation for capacity, $Q = CV$, this energy expression may take any one of the three following forms, each of which is useful, namely:

$$\left.\begin{matrix}\text{Energy of}\\ \text{Charged Conductor}\end{matrix}\right\} = \tfrac{1}{2}QV = \tfrac{1}{2}CV^2 = \tfrac{1}{2}\frac{Q^2}{C}. \qquad \text{Eq. 137}$$

When C, Q, and V are each measured in C.G.S. units, it follows, of course, that Eq. 137 gives the energy in terms of ergs.

When a Leyden jar or other conductor is discharged, this expression gives us a measure of the heat which is produced in the spark gap and in the wires which lead to it. The first to measure this heat was Dr. Kinnersley (contemporary and friend of Franklin), who inclosed two knobs between which the discharge took place in a glass bulb which was essentially the bulb of an air thermometer. The lead wires from the two coatings of the jar passed through the glass walls of the bulb.

The Oscillatory Discharge

346. It was first observed by Joseph Henry (1842) that, when a Leyden jar is discharged through a coil of wire so as to magnetize a sewing needle placed inside the coil, sometimes the needle is magnetized in one way, sometimes in the opposite way, even though the inner coating of the jar be always charged with electricity of the same sign and the coil connected in the same way. See Fig. 255.

This fact Henry explained by supposing that the charge oscillated from one coating to the other, but became a little smaller at each oscillation. Accordingly the last discharge which was strong enough to reverse the magnetization of the needle was the one which determined which end of the needle would be the north pole and which the south.

Later (1857) Feddersen examined the spark of a Leyden jar discharge by means of a rotating mirror, such as that which is used for viewing a manometric flame (§ 219), only turning much more rapidly, and found that in this mirror the image of the spark looked something like a string of beads, thus furnishing irrestistible evidence that the spark was, under his conditions, a periodic phenomena.

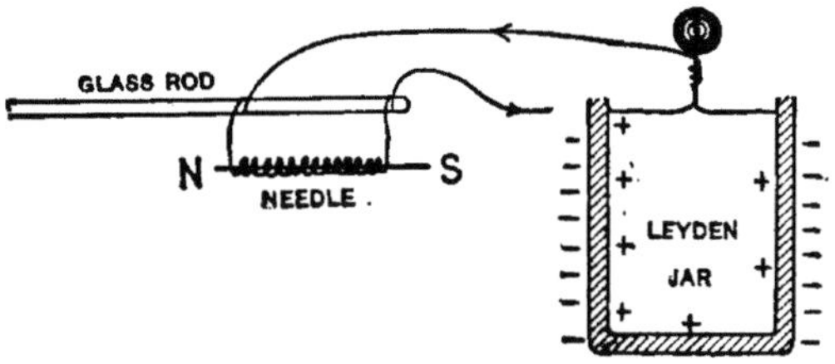

FIG. 255.—Illustrating Henry's experiment.

In more recent times, C. V. Boys and many others have photographed this oscillatory discharge.

The advanced student will be interested in following Lord Kelvin's beautiful prediction of this periodic phenomenon, made in 1855 before Feddersen had proved it by experiment. For this see *Ency. Brit.*, art "Electricity," p. 81. Here Kelvin proves, what has since been verified by experiment, that the period of the discharge is

$$T = 2\pi\sqrt{\frac{L}{1/C}} = 2\pi\sqrt{LC} \qquad \text{Eq. 138}$$

where C is the capacity of the jar and L a quantity which represents the *electrical* inertia of the circuit and which is known as *self-inductance.* The reciprocal of capacity represents therefore, by analogy with the pendulum, spiral spring, and other vibrating systems, the electrical elasticity of the circuit. But this is a matter which can be cleared up only by more advanced study.

Electrical Resonance

347. If two Leyden jars which are alike in size and shape so as to have practically the same period of oscillation be provided each with a wire frame such as that shown in Fig. 256, then one of them can be set into electrical vibration by means of a discharge at the other. The wire frame serves to make the period definite and the same for each jar.

The spark gap G in the circuit B is very small, say $\frac{1}{2}$ millimeter; while that in A is larger, say $\frac{1}{2}$ inch. When the spark of the electric machine passes across S, a periodic disturbance is produced which sets up surgings of electricity in B, on the same principle that one pendulum will set into motion another of equal period, if there be any elastic connection between the two.

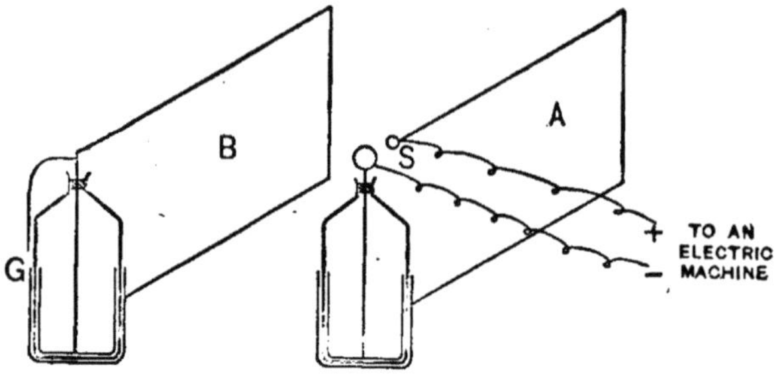

FIG. 256. — Lodge's experiment.

When these surgings become sufficiently large, the charge "slops over," so to speak, and a small spark is seen at G. The effect of the discharge in the first jar is to send through the medium which separates the jars a properly timed electrical impulse, first in one direction and then in the other; this impulse separates the two electricities and produces the spark. The disturbance in the medium is made up of electric waves such as are used in wireless telegraphy. This form of the experiment is due to Sir Oliver Lodge.

Problems

1. An insulated conducting sphere whose radius is 7 is charged until its potential is 5. Find the amount of its charge. *Ans.* 35.

2. Two insulated conducting spheres whose radii are 3 and 8 respectively are charged so that their potentials are 5 and 10 respectively. Find the common potential which these two spheres will assume when connected by a wire whose capacity may be neglected. *Ans.* $8\frac{7}{11}$.

3. In the preceding problem find the energy of each of the two charges before the spheres are connected. Then compute the energy of the final charge. Explain the difference.

4. A small sphere of rosin carries a charge of 2 electrostatic C.G.S. units and is placed in an electric field whose intensity is 25 C.G.S. units. Find the force in dynes with which the sphere will be urged.

5. Find the attraction between two small balls, 4 cm. apart in air, one of them having a positive charge of 12 units, the other a negative charge of 5 units.

6. At each of two corners of an equilateral triangle whose sides are 60 cm. is placed a charge of 40 C.G.S. units of electricity. Find the intensity of the electric field at the third corner.

7. A conducting sphere of 20 cm. radius carries a charge of 80 units. If it be made to share its charge with another insulated sphere whose radius is 5 cm., what will be the ratio of the charges on the two spheres?

8. Find the electric potential at a point 40 cm. distant from the center of a hollow sphere of 4 cm. radius charged with 60 C.G.S. units of electricity.

Suppose the radius of the hollow sphere to be 100 cm., how will the result be affected?

9. Find the "dimensions" of electrical quantity, electrical potential, and electrical capacity.

10. Find the capacity of a condenser made up of two concentric spheres having radii of 100 and 98 respectively, and separated by air.

11. Two condensers of the same dimensions as in the preceding problem are exactly alike except that in one case the space between the spheres is filled with an oil whose dielectric constant is 2. How will a charge of 100 units divide itself between these two condensers?

12. Which will require the greater work; to charge a condenser of capacity 10 to a potential 5, or a condenser of capacity 30 to a potential 3?

13. A charged insulated disk is connected with a gold leaf electroscope. Describe and explain the change of divergence in the leaves as a second disk held in the hand is made to approach the first.

14. Explain the difference between joining up Leyden jars "in series" and "in parallel."

15. A parallel plate condenser in air has its plates separated by a distance of $\frac{1}{2}$ mm. What change in its capacity will be produced if the plates are separated to a distance of 2 mm. and the intervening space filled with an oil whose dielectric constant is 3?

16. Draw the lines of force which represent the distribution of electricity on a Leyden jar, the outer coating of which is connected to earth.

17. How much power would be required to charge a condenser having 3000 C.G.S. units of capacity to a potential 200 C.G.S. units in $\frac{1}{5}$ second?

18. A condenser whose capacity is C is charged with a quantity Q. What diminution of energy will occur when one half this charge is led off to earth?

19. What must be the radius of a sphere in order that a charge of 80 units may raise its potential 16 units?

20. How would you proceed to find the difference of potential between two points, knowing their distance apart and the average intensity of the field between them?

21. Prove that an equipotential surface can intersect neither itself nor any other equipotential surface.

22. What is the curve which exhibits the relation between the charge and the potential of an insulated conductor?

23. What is the intensity of field between two plane condenser plates which are $\frac{1}{10}$ cm. apart and differ in potential by 200 C.G.S. electrostatic units?

CHAPTER X

ELECTRIC CURRENTS

348. We now proceed to study some of the remarkable properties which are acquired by a wire and the region about it when that wire is made to connect two charged conductors. Suppose we select for our two charged bodies the inside and outside coatings of the Leyden jar which we have been using in our study of electrostatics.

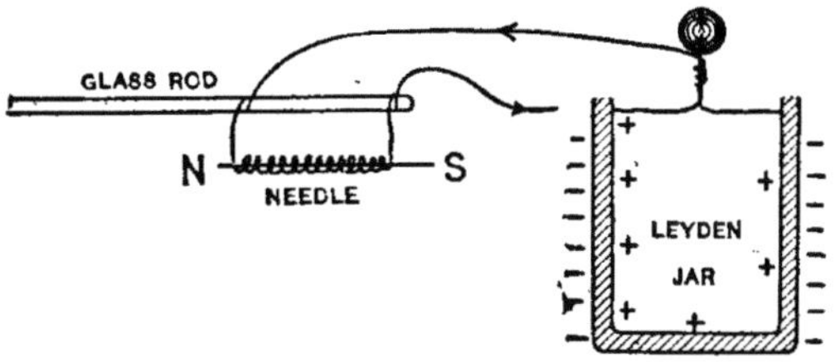

FIG. 257. — An electric charge passing along a coiled wire magnetizes a sewing needle.

On winding a copper wire about a sewing needle, as shown in the figure, no effect is produced on the needle; but if we take the copper wire and (handling it by means of a glass rod) connect one end to the positive coating and the other to the negative, we shall find that when the discharge occurs the needle becomes magnetized; that is to say, this little coil of copper wire has suddenly acquired the ability to magnetize a piece of steel. After the spark has occurred and the two conductors are discharged it will be observed that the wire no longer has the ability to magnetize steel. **A wire which possesses this property of carrying with it a magnetic field is said to have an electric current flowing through it.**

An even simpler manner of illustrating this so-called electric current is to connect the carrier and the sole plate of a Volta electrophorus with the two lead wires of a galvanometer, say a Rowland-D'Arsonval, as indicated in Fig. 258.

The reader is here asked to take it for granted, for a moment, that a galvanometer is merely an instrument which will indicate the presence of an electric current. First put one terminal of the galvanometer in connection with the sole plate. Then, when the electrophorus has been electrified by friction, place the carrier, *by means of its handle*, upon the plate. The carrier now contains two charges, a negative one on top which is free to run off as soon as connection is made with the ground or sole plate, and a positive one which is held bound by the negative charge on the ebonite. When the second terminal of the galvanometer is now connected to the carrier, this negative charge passes to earth *via* the galvanometer and sole plate, giving a deflection of the galvanometer in one direction. When, however, the carrier is lifted off the plate, the *positive* charge will become free and will flow off through the galvanometer, giving a deflection in the opposite direction. Either of the two preceding experiments will suffice to illustrate the connection between the subject we have just left and the one we are just approaching.

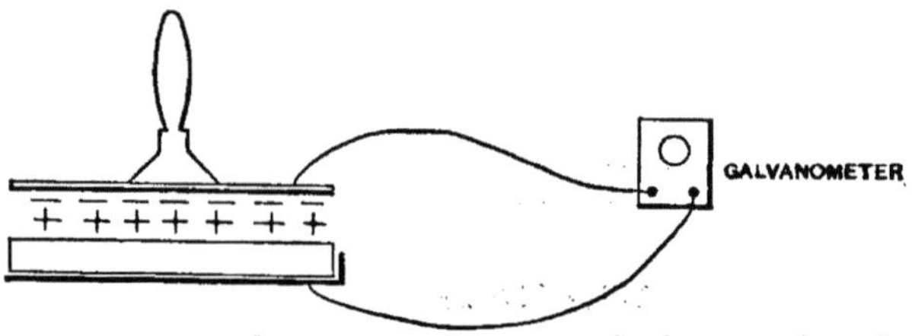

FIG. 258. — Showing the current which is produced when an electrophorus is discharged.

As we shall presently see, there are many other and more convenient ways in which electric currents may be produced in wires.

METHOD OF STUDYING ELECTRIC CURRENTS

349. It will be remembered that, in taking up the study of magnetism, we examined a number of magnetic phenomena, and employed a compass needle, a magnetoscope, as it is sometimes called, to detect the presence of a magnet. It is thus not difficult to say whether or not any particular piece of steel is a magnet; but it is very difficult to say just what "magnetism" is.

It is precisely so with electrification; we were able to detect its presence very readily by means of the gold leaf electroscope, though no one has yet been able to find out what "electricity" is. What copper or iron is no one knows, although very much else is known about copper and about iron. "Electricity" is, then, just as mysterious as copper, but no more so.

We shall now consider some of the phenomena connected with electric currents; but, as before, we shall first need some ready method of detecting the presence of an electric current in a wire. This once obtained, we shall be in a position to study some of those remarkable properties of electric currents which have proved so useful during the nineteenth century.

These will be grouped under three heads, viz.: —

I. The Production of Electric Currents.
II. The Measurement of Electric Currents.
III. Some Effects of Electric Currents.

Preliminary Concerning the Galvanometer

350. As we have already seen and shall see many times again, every wire conveying an electric current is surrounded by a magnetic field, which is just like that surrounding a magnet, only the lines of force about the wire have a direction different from those about the magnet.

This magnetic field is such that, if the wire be placed parallel to a compass needle, the needle will be deflected so that it will not remain parallel to the wire. This is the fundamental discovery made by the Danish physicist Oersted, in 1820. Such an instrument is the one most commonly used to detect the presence of a current, and is called a **galvanometer.*** Almost immediately after a current ceases to flow in the wire, the magnetic field about the wire ceases to exist, and the wire no longer affects the compass needle. **But practically as soon as the current begins to flow** the needle tends to set itself at right angles to the direction of the current. **This type of current indicator is therefore called a "moving magnet" galvanometer.**

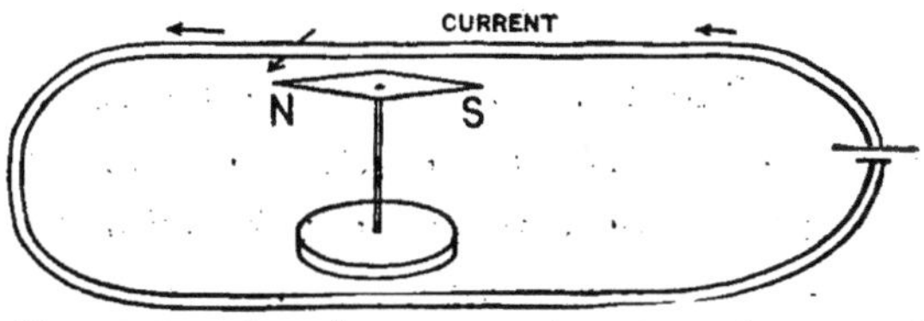

Fig. 259. — A galvanoscope in which the magnet is movable, while the wire conveying the current is fixed.

But if the compass needle is fixed and the wire movable, then the *wire* tends to set itself at right angles to the compass needle. The student will observe that this is merely another

* The first galvanometer was a freshly prepared frog's leg, which twitched whenever a current passed along the sciatic nerve and muscles of the hind leg.

instance of the great general fact that action and reaction are equal and opposite.

In practice a compass needle is not so convenient as the U-shaped magnet shown in Fig. 260. The movable wire is a coil which can rotate about a vertical axis. This device, first used by Kelvin and D'Arsonval, is called a "moving coil" galvanometer.

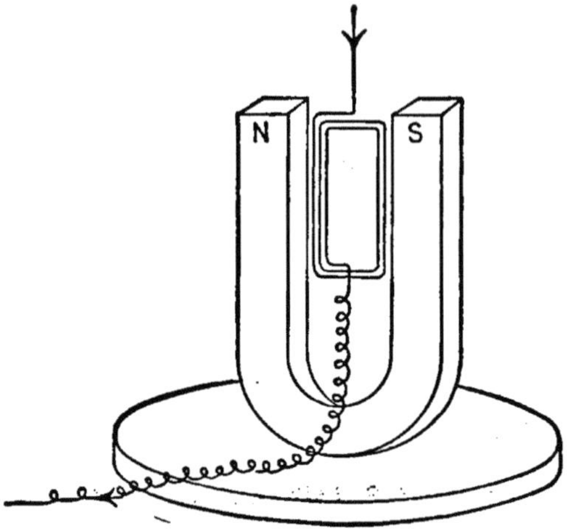

FIG. 260. — A galvanoscope in which a part of the wire conveying the current is movable, while the magnet is fixed.

We have seen (§ 299) that a magnet is surrounded by lines of force, leaving its north pole and running to its south pole. A simpler and much more useful statement of the fundamental principle of the galvanometer is this: Consider any closed electric circuit, *i.e.* a series of conductors, joined end to end so as to form a closed path.

(i) **The principle of the moving magnet galvanometer, then, is that whenever an electric current flows through any circuit, a compass needle in the neighborhood tends to set itself so that the greatest possible number of its lines of force will pass through the circuit;**

(ii) **While the principle of the moving coil galvanometer is that the portion of the circuit which is movable sets itself always so as to include as many as possible lines of force from the magnet.** In either case, the electric current announces its presence by producing a rotation, *i.e.* an angular displacement.

From what precedes, it will be clear that each of these two forms of galvanometer indicates a current by the fact that, in general, the current exerts a turning moment on the magnet, or, what amounts to the same thing, the magnet exerts a turning moment upon a part of the circuit conveying the current.

The advanced student will discover that this behavior of the compass needle and the moving coil is merely a special case of the general principle that a body at rest tends to move into a position where its potential energy is a minimum.

I. Methods of Producing Electric Currents

351. We now proceed to consider the three most important methods of producing electric currents, viz.: —

(*a*) The method of Galvani and Volta — the so-called Voltaic cell.

(*b*) The method of Faraday — cutting lines of magnetic force.

(*c*) The method of Seebeck — the thermoelectric couple.

(*a*) The Voltaic Cell

352. The simple and well-established facts of the Voltaic cell are as follows: —

At the very close of the eighteenth century (1799) it was discovered by the Italian physicist Volta that all conductors of electricity can be divided into two classes. This division is based upon the following experiments: —

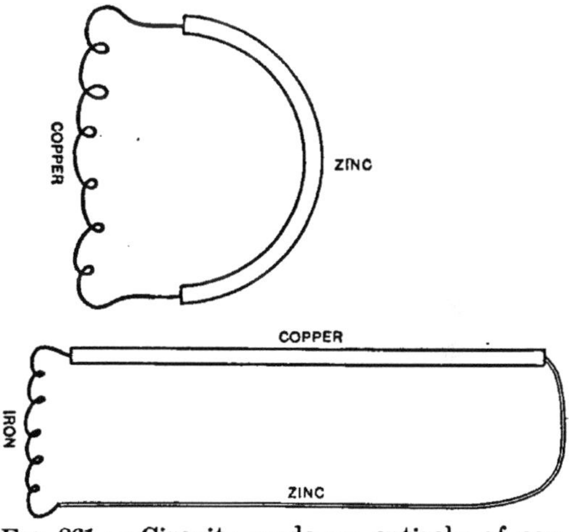

Fig. 261. — Circuits made up entirely of conductors of the first class.

If we make a closed circuit out of several different metals, *i.e.* if we make an endless chain in which each link is composed of a different simple substance, such as zinc, copper, or iron, we see that no electric current is produced. **All substances which when joined together at the same temperature in any order, as in Fig. 261, produce no current, are called conductors of the first class.**

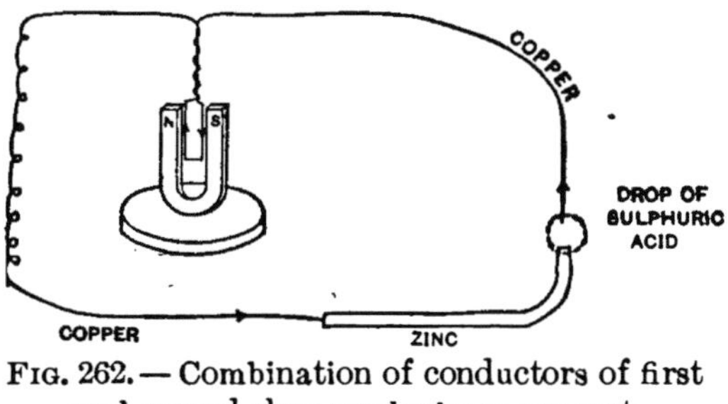

Fig. 262. — Combination of conductors of first and second class producing a current.

Volta found, however, that if into a circuit such as the above he introduced one link composed of a *compound* substance, such as brine (a solution of table salt) or dilute sulphuric acid or copper sulphate, he then always obtained an electric current. **Conductors of this kind, which Volta called conductors of the second class, always undergo chemical decomposition when introduced into a circuit containing two different metals and always yield a current.** The modern name for a conductor of the second class is electrolyte, *i.e.* any substance which is decomposed when a current passes through it. Conductors of the first class practically include only carbon and the various metals.

Definition of a Voltaic Cell

353. **It has been found by experiment that no two conductors when joined together will produce a current so long as they are at the same temperature.** We may use any of the three combinations indicated in Fig. 263, but no current will thus be obtained. But the following combination, suggested by Volta and named after him the "Voltaic Cell," will always give a current: —

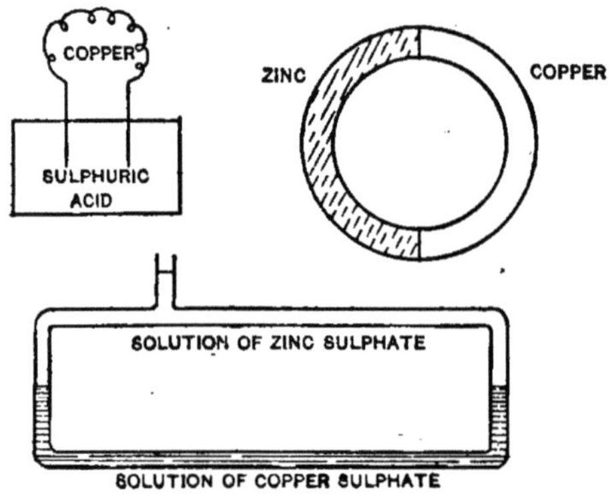

FIG. 263. — No two conductors can act as a Voltaic cell.

The Voltaic cell is defined as three or more conductors in series, each conductor being made of a different substance, and not all belonging to the same class.

A Typical Voltaic Cell

354. One of the simplest forms of these cells is the following: Take a strip of sheet copper, say one centimeter by ten; solder it, end to end, to a similar strip of zinc. Take a small beaker of dilute sulphuric acid, say one part strong sulphuric acid to ten parts of water. Complete the circuit by bending the copper-zinc strip into a U-shape and dipping the ends into the acid, as shown in Fig. 264. Bubbles of gas are given off from each metal, but much more freely from the copper.

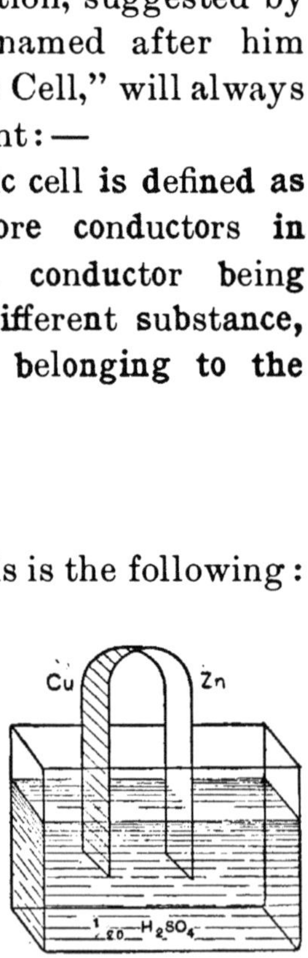

FIG. 264. — Typical Voltaic cell.

Here the zinc and copper are the conductors of the first class: the dilute sulphuric acid is a conductor of the second class. On dipping the metals into the acid, what evidence presents itself that the acid undergoes decomposition and hence belongs to conductors of the second class?

Now take two strips, one of zinc and one of copper; immerse them in dilute sulphuric acid and connect them by wires to a galvanometer, and determine whether or not an electric current is flowing while the gas is being given off. If the strip of zinc be well cleaned and have a little mercury rubbed over it until it is thoroughly amalgamated, the gas will no longer be given off from the zinc. But under these circumstances it is found that the zinc is still "eaten up" by the acid. Zinc sulphate is formed and dissolved in the solution, which becomes less and less acid. The copper remains intact, while hydrogen gas is set free at the copper strip. The amalgamation has merely the effect of preventing a local waste of energy, and thus secures the transformation of a larger per cent of chemical energy into electrical energy.

These phenomena are summarized by the chemist as follows:—

$$Zn + H_2SO_4 = ZnSO_4 + H_2.$$

The student who is familiar with the elements of chemistry will translate this equation into words as follows: The changes in this cell take place in the following proportion, viz.: 65 grams of zinc unite with 98 grams of sulphuric acid to form 161 grams of zinc sulphate and liberate 2 grams of hydrogen.

Some Definitions

355. Closed circuit. A series of conductors joined end to end so as to form an endless chain is said to be a closed circuit. It may or may not have an electric current passing through it.

Open circuit. When a closed circuit is cut or interrupted at any point, it is then called an open circuit.

Poles. If there be a Voltaic cell in the circuit, the two ends thus left free by opening the circuit are called the **poles** of the cell. The pole which is connected with the copper plate, or with the plate whose electrical behavior is like that of copper toward zinc, is called the **positive pole.** That pole which is connected with the zinc plate, or with a plate whose electrical behavior is like that of zinc toward copper, is called the **negative pole.**

Direction of Current. When the circuit is closed by a wire joining the positive and negative poles, the direction of the current in this wire is said to be from the positive to the negative pole. We may hereafter, therefore, speak of the current as leaving the positive pole and entering the negative pole. It may be shown by the following experiment, due to Volta, that the direction of the current as above defined is the direction in which the positive electrification travels. Take an adjustable parallel plate condenser such as that shown in Fig. 265 and join the positive pole of a Voltaic cell to one plate, the negative pole to the other plate. Connect one pole, say the negative, to earth also. Place an electric key between the copper pole and the movable plate of the condenser. Then, if an electroscope be joined to the movable plate, it will be observed that this plate always acquires a positive charge when the key is depressed. In order to make the charge as large as possible the condenser plates should be brought very close together before the key is depressed (in order to increase the capacity of the system) and then separated after the key has been released. In like manner, it is found that when the

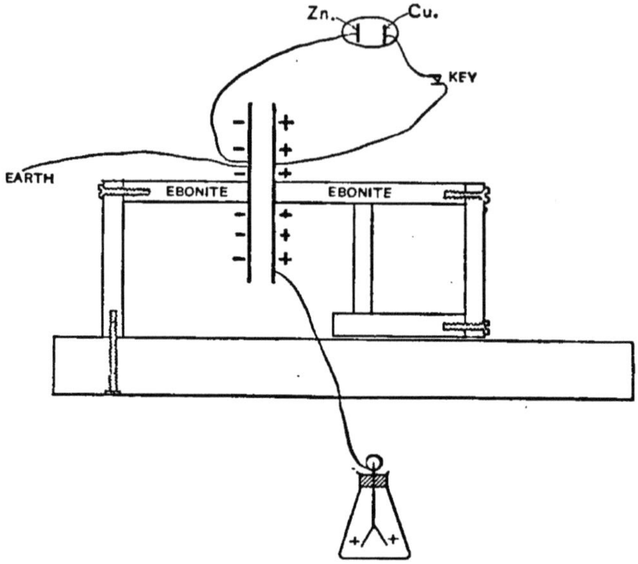

Fig. 265. — Illustrating the reason for calling the copper pole of a cell positive.

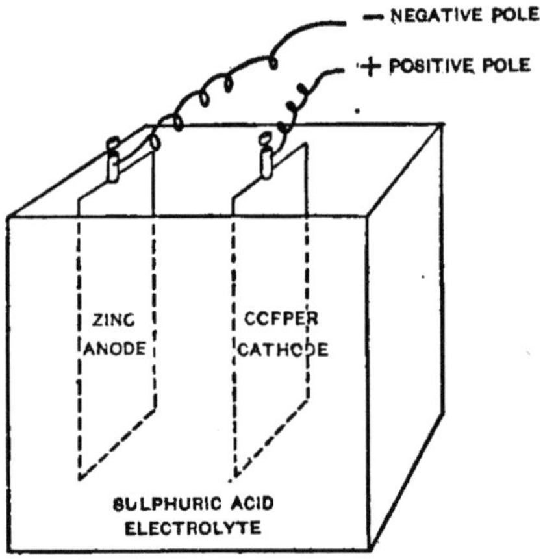

Fig. 266. — Various parts of a Voltaic cell.

poles of the cell are reversed, the charge on the electroscope is also reversed.

Electrodes. We have already seen that at least one member in the chain of conductors which makes up the circuit of a Voltaic cell must be an electrolyte, practically always a solution. The two conductors which connect with this liquid are called **electrodes.**

The electrode by which the current enters the electrolyte is called the **anode**; and that electrode by which the current leaves the electrolyte is called the **cathode.** In the typical cell which we have just studied the zinc plate is the anode and the copper plate the cathode. Etymology of these two words? Very often a current from some outside source is passed through a cell containing an electrolyte; in such a case both electrodes may be made of the same substance.

Battery. A number of Voltaic cells connected together is called a **battery.**

The eight definitions just given are so frequently and so continually used in all departments of electrical science, that the student should master them here and now, once for all.

Three Practical Forms of the Voltaic Cell

356. 1. The Gravity Cell. This cell is composed of a copper plate immersed in a saturated solution of copper sulphate, and a zinc plate immersed in a solution of zinc sulphate, generally arranged essentially as indicated in Fig. 267.

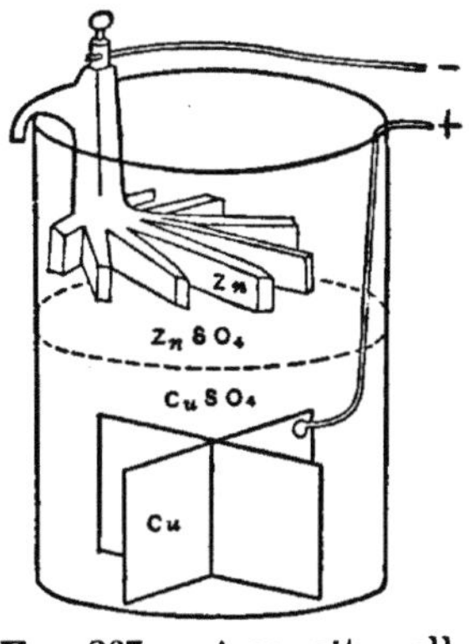

FIG. 267.—A gravity cell.

The copper sulphate is somewhat denser than zinc sulphate, and hence remains in the lower half of the jar, while the zinc sulphate, when carefully poured in, floats on top. Hence the name, "gravity" cell.

One of the chief merits of this cell is, that so long as the circuit is open there is not much waste of material taking place in the cell. When, however, the circuit is closed and the electric current begins to flow, we observe that the zinc plate grows smaller, that the copper sulphate grows less and less, while the zinc sulphate increases and the copper electrode becomes covered with a coating of fresh metallic copper.

In the shorthand of chemistry these phenomena are described quantitatively as follows:

$$Zn + ZnSO_4 + CuSO_4 + Cu = 2\,ZnSO_4 + 2\,Cu$$

The metallic zinc has disappeared; the metallic copper has increased, at the expense of the copper sulphate. Accordingly, if such a battery is to furnish a current continuously, it must now and then be supplied with a new zinc plate and with a fresh solution of copper sulphate.

The source of energy is here evidently the zinc, which is oxidized — "burned" if you like — to form zinc sulphate.

This cell is well adapted to work in which small currents are needed very frequently or for a long while at a time.

357. 2. The Leclanché Cell. In this cell the two conductors of the first class are zinc and carbon; the electrolyte is a solution of ammonium chloride, sometimes called "sal ammoniac." With the carbon is mixed some dioxide of manganese, and both are held in a porous jar or small canvas bag. The current easily passes through the pores of the jar or the meshes of the bag. The carbon plate here evidently corresponds to the copper plate of the gravity cell. Here again the zinc is used up whenever the poles are joined and the current flows. The sal ammoniac also diminishes. Accordingly, from time to time, the zinc and sal ammoniac must be renewed.

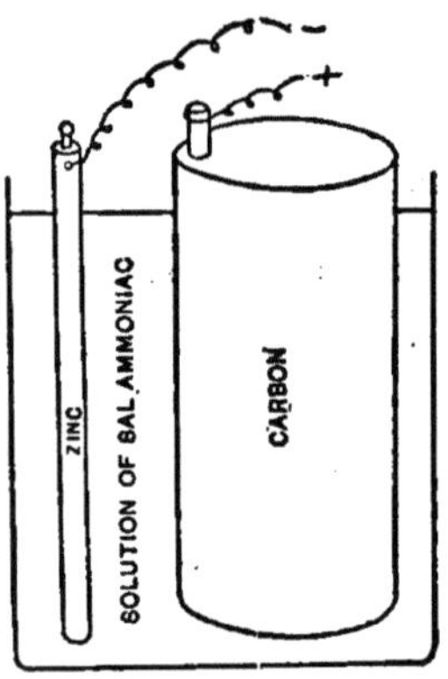

FIG. 268. — A Leclanché cell.

The student has probably already asked himself why the oxide of manganese is mixed with the carbon. The answer is found in the fact that, if carbon alone is used, the carbon becomes covered with a layer of hydrogen gas soon after the circuit is closed; because the ammonium chloride (NH_4Cl) breaks up, giving off ammonia (NH_3) and depositing free hydrogen (H) at the carbon electrode. The chlorine unites with the zinc.

This coating of hydrogen makes the carbon plate cease to act as a carbon plate and to behave like a metallic plate; it also offers more resistance than carbon to the passage of the current. **This is called polarization, which may be defined as a tendency to send a current in a direction opposite to that furnished by the cell.**

The effectiveness of a cell is always diminished when polarization occurs. The manganese dioxide, however, uses up the hydrogen and keeps the carbon plate clean. The purpose of the manganese dioxide is, then, to prevent polarization. It does this by furnishing oxygen which unites with the hydrogen to form water; and once in union with oxygen the hydrogen is harmless.

In the language of chemistry, the action of the Leclanché cell is described as follows: —

$$Zn + 2\,NH_4Cl + 2\,MnO_2 = ZnCl_2 + 2\,NH_3 + Mn_2O_3 + H_2O.$$

This same equation also describes what goes on in the so-called "dry cell" now in such common use.

The Leclanché cell is excellently adapted to furnishing a current for a small time, or for intermittent use, as in the case of a door bell or call bell. It will remain on open circuit with still less waste than occurs in the gravity cell.

358. 3. The Storage Cell. In this cell the electrodes are of lead and lead peroxide, respectively; the electrolyte is dilute sulphuric acid. The lead plate is the negative one; the lead oxide is the positive one. When the circuit is closed, the lead peroxide plate gives up a part of its oxygen, while the lead plate, the negative electrode, becomes oxidized, until finally the two electrodes become very much alike, and the current therefore becomes less and less, other things being the same. The battery is now said to be discharged. For, as we have already seen (§ 353), two electrodes which are alike never give a current when immersed in any one electrolyte.

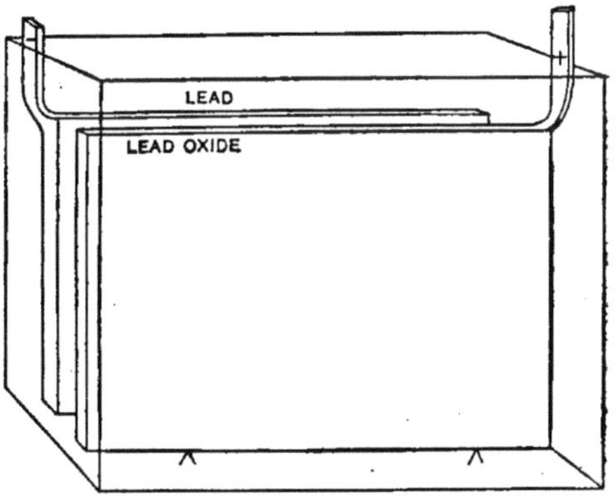

FIG. 269. — A storage cell.

But in order to put the cell again in good working shape, it is necessary only to pass an electric current through the cell in a direction opposite to that obtained when the cell was in use. **This process of regeneration is known as charging.** The charging simply restores the oxygen to the positive plate, and removes the oxygen from the negative plate, leaving it spongy metallic lead. **When the cell is used as an electric generator, the process is known as discharging.**

The following two features distinguish the storage battery from most other Voltaic cells: —

(*a*) The fact that the plates are prepared, *i.e.* made chemically different, by electrical means.

(*b*) The fact that they are capable of yielding currents which are enormous compared with those from other forms of batteries.

The chemical behavior of the lead storage cell is essentially as follows: —

(i) During the charge, at the anode or "positive grid,"

$$PhSO_4 + SO_4 + 2\,H_2O = PbO_2 + 2\,H_2SO_4.$$

(ii) During the charge, at the cathode or "negative grid,"

$$PbSO_4 + \overset{+}{H}_2 = Pb + H_2SO_4.$$

The ion SO_4 of the first equation and the ion $\overset{+}{H}_2$ of the second result from the decomposition of the sulphuric acid by the charging current.

(iii) During the discharge, we have at the positive plate,

$$PbO_2 + H_2SO_4 + \overset{+}{H}_2 = PbSO_4 + 2\,H_2O.$$

(iv) During the discharge, at the negative plate,

$$Pb + SO_4 = PbSO_4.$$

Observe that while the effect of the charging current is to deliver hydrogen at the negative plate, the effect of the discharging current is to produce hydrogen at the *positive* plate.

Edison has recently devised a storage cell in which the anode is spongy iron, the cathode nickel peroxide, and the electrolyte caustic potash.

In charging, the iron salt is reduced; in discharging, oxidized.

A storage cell is often spoken of as an accumulator, not because it stores up electricity but because it stores up chemical energy, just as the mainspring of your watch stores up mechanical energy in order to give it out again as needed.

It will be observed that all three forms of cells just described have this one common feature, viz., they consist of two different conductors of the first class, joined in series with one or more different conductors of the second class.

THE COMMUTATOR

359. In most of our work with batteries, whether in the classroom or in the laboratory, we shall need to change the direction of the current in some part of the circuit. This can be done in many ways ; but the device represented in Fig. 270 is typical of them all. The block *ABCD* is made of wood and has four holes bored about halfway through it. These holes, placed one at each corner of a square, act as cups to hold mercury. Into the sides of the block are inserted four wires *E*, *F*, *G*, *H*. Each of these wires connects with the mercury in the cup nearest it. The poles of the battery are joined to two diametrically opposite cups, say *B* and *D* ; the wires from the rest of the circuit, say from a galvanometer, are joined to the remaining two cups *A* and *C*.

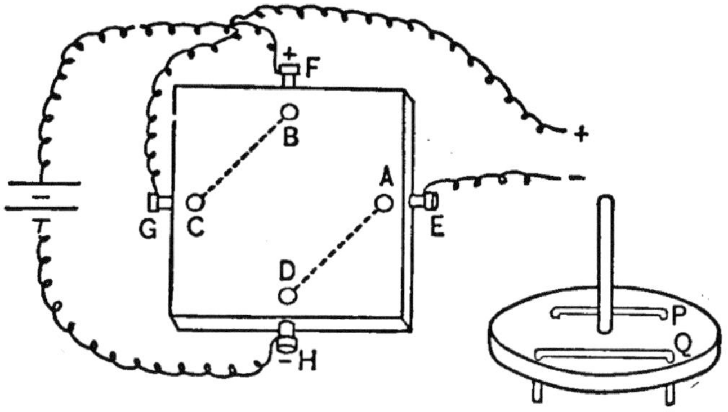

FIG. 270.— A typical commutator.

For closing the circuit, two short thick copper conductors, *P* and *Q*, are mounted on a movable block, as shown in the figure. These conductors *P* and *Q* may be placed in the mercury cups in such a way as to connect *B* with *A*, and hence *C* with *D* ; or they may be placed so as to connect *B* with *C*, and hence *A* with *D*.

In the one case, the current flows from *E* to *G* through the galvanometer ; in the other case, from *G* to *E* through the galvanometer. To change the direction of the current in the galvanometer circuit, we have, therefore, only to lift the top block and rotate it through 90° in either direction. **An instrument which may be thus used to reverse the direction of a current is called a commutator.** To "break the current" we have merely to lift off the top block. **Any device which is thus employed to interrupt a current is spoken of as a key or switch.**

(*b*) INDUCTION OF ELECTRIC CURRENTS

360. We return from this digression on the commutator to consider the second of the three important methods mentioned above (§ 351) for producing electric currents.

The reader has already seen (§ 299) that the region about

a magnet is endowed with properties totally different from other portions of space. This magnetic field, as it is called, does not appear to the naked eye to be different from any other portion of space; but we know that it is different, for a magnet brought into such a field behaves in a manner peculiar to this magnetic field. As has been pointed out, Faraday described such a magnetic field by conceiving it to be filled with lines of force — purely imaginary lines — drawn, at each point in the field, in the same direction in which the axis of a magnet would set itself if placed at that point. The earth itself being a great magnet, the whole of the space inhabited by the human race is a magnetic field. In these fields the lines are distributed just far enough apart to make the number which passes through unit area, drawn perpendicular to the direction of the field, numerically equal to the intensity of the field.

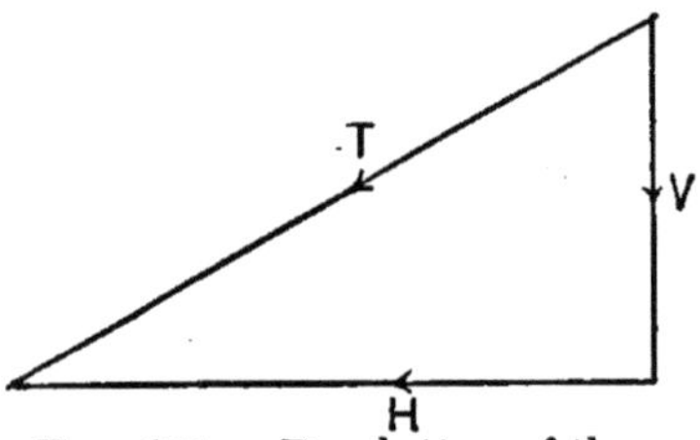

FIG. 271. — Resolution of the Earth's Magnetic Field into two components.

The lines of force which describe a magnetic field being directed, *i.e.* vector, quantities, they can be resolved into components. We accordingly speak of the vertical component and the horizontal component of the earth's field, as represented in Fig. 271, where T represents in direction and in amount the total magnetic force of the earth on any magnet pole; V represents in direction and amount the vertical component of T on the same pole; and H the horizontal component of T on the same pole.

361. If the preceding is clear, we are now in a position to consider the beautifully simple method by which Faraday, in the autumn of 1831, first succeeded in producing currents without the use of the Voltaic cell and without the use of heat.

If we consider any closed circuit (say a piece of copper wire bent into the form of a circle or square) placed in a magnetic field, we may think of it as containing or inclosing a certain number of lines of force, just as a fence might inclose a group of trees.

In Fig. 272 six lines of force are represented as being inclosed by the wire circle. If by any means the magnetic field were made twice as strong, we should represent this change by drawing the lines of force twice as close together, so that now the circle would inclose twelve instead of six lines. Or, if the copper circle were carried parallel to itself to another region where the magnetic field is not so strong, then we should describe this change by drawing *fewer* liues of force through the circle.

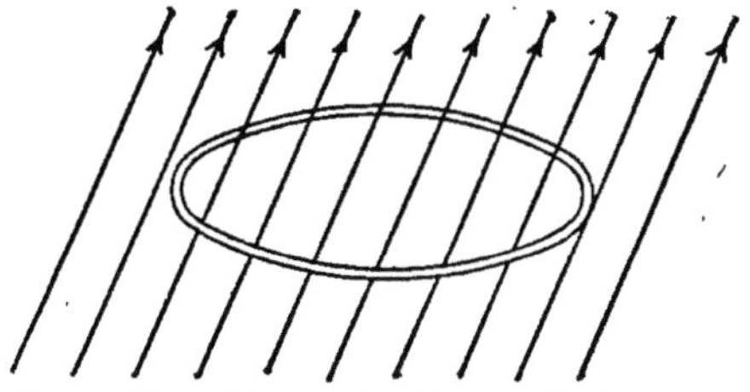

FIG. 272. — Magnetic lines of force inclosed by an electric circuit.

Faraday's discovery may then be described in a quantitative manner as follows: **When, by any means whatever, the total number of lines of force passing through any circuit is changed, an electric current is produced in that circuit. Such a current is called an induced current.**

Various Illustrations of Induced Currents

362. *Experiment* 1. Connect the terminals of a D'Arsonval galvanometer* with a copper wire, so that with the suspended coil a closed circuit is formed. Take a single loop in this circuit, and thrust through it one pole of a permanent magnet. Such a magnet carries its magnetic field (*i.e.* its lines of force) about with it. And when these invisible lines are thrust through the loop, we observe that for the instant a current passes in the galvanometer. See Fig. 273. Now remove these lines by removing the magnet; a current passes, but now in the opposite direction. **Observe that in either case the current lasts only so long as the motion of the magnet continues.**

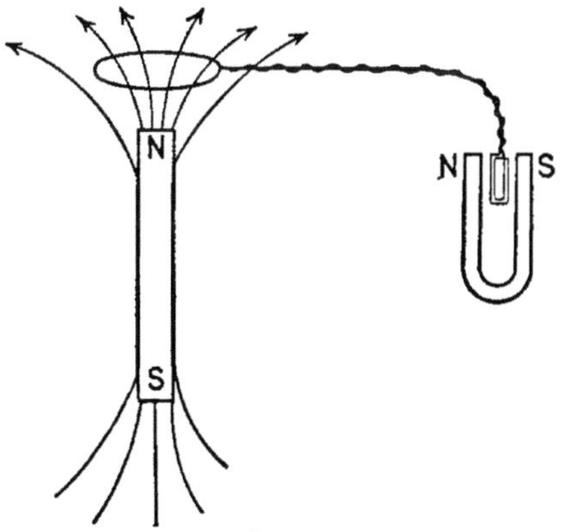

FIG. 273. — An electric current produced by thrusting lines of force through a circuit.

* In the presentation of the phenomena of induced currents a fairly sensitive D'Arsonval galvanometer, costing from $20 to $40, is great economy of time and energy. The Rowland form answers nearly every purpose for which a galvanometer is required in the laboratory or in demonstration work.

363. *Experiment* 2. Now support the magnet on some convenient stand, and slip the loop of wire over one end of the magnet. See Fig. 274. A current is again observed. If we slip the coil off, the *direction* of the current is reversed.

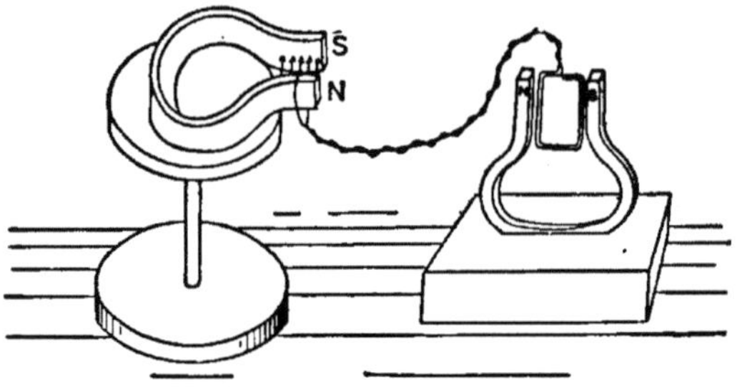

FIG. 274.—An electric current produced by moving a coil in a magnet field.

If the loop of wire be slipped on and off quickly, a current will be produced which flows first in one direction, then in the other.

(i) A current which changes its direction in a regular and continuous manner is called an "alternating" current.

(ii) But if the current, during certain intervals of time, ceases entirely, it is said to be "intermittent."

(iii) Currents which flow continuously in one direction are said to be "direct."

From these two experiments, it is evident that it makes no difference whether it is the loop of wire or the lines of force that are moved. **The relative motion of the wire and the magnetic field changes the number of lines of force inclosed, and hence produces the current.**

364. *Experiment* 3. Since we live on the surface of a large magnet, it becomes interesting to see how we must move a closed circuit in the earth's magnetic field in order to produce an electric current.

If a large loop of wire be made in one part of a closed circuit containing a galvanometer, we have only to hold this loop alternately in an east-and-west and in a north-and-south vertical plane. When the plane of the loop is vertical and north-and-south, no magnetic lines from the earth pass through it; they pass along each side of it. But if the coil be suddenly rotated about a vertical axis so as to lie in an east-and-west

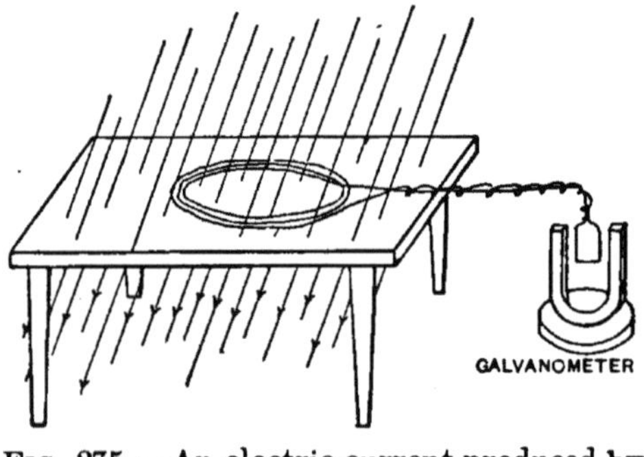

FIG. 275.—An electric current produced by capsizing a coil of wire in the earth's magnetic field.

plane, the earth's lines of force will thread through it, and we observe in the circuit an induced current. If we rotate the coil of wire back into a north-and-south plane, the current is in the opposite direction. **If in either of the preceding experiments we leave the total length of wire unchanged and take two turns of wire instead of one, the deflection of the galvanometer is observed to be twice as great as before.**

If a coil of wire which lies flat on the table (Fig. 275) be quickly turned over, the lines of force passing through it will be changed by twice the original number. For if in the first case there were N lines passing through in, say, a *positive* direction, when the coil is turned over there will be N lines passing in a *negative* direction, *i.e.* the change will be

$$N-(-N)=2N.$$

365. *Experiment* 4. But there are still other ways in which we may alter the number of lines of force passing through a circuit. We have seen that a wire conveying a current is surrounded by a magnetic field. If we bring a loop of our galvanometer circuit near such a battery current, we shall find that the magnetic lines from the battery current will in general thread through the galvanometer circuit, and will thus produce an induced current. In Fig. 276 lines of force marked 3, 4, 5, and 6 each pass through the galvanometer circuit as well as the battery circuit. When the motion ceases the current ceases.

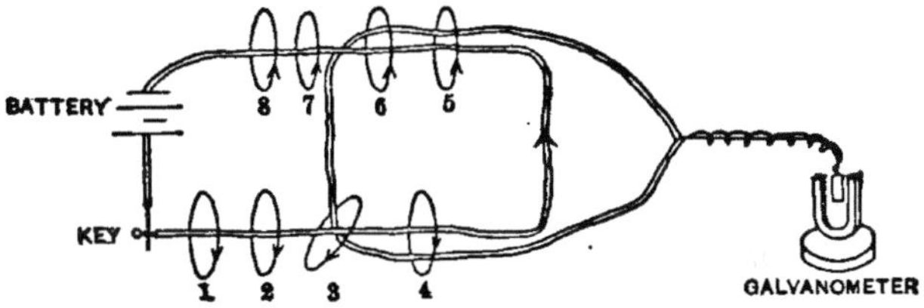

FIG. 276. — An electric current produced in a loop of wire by making or breaking a battery circuit in the neighborhood of the loop.

If instead of bringing the loop and the battery current near together, we simply place the circuits side by side and suddenly open or close the battery circuit, we observe in each case an induced current in the galvanometer circuit. To interrupt or break the battery current is equivalent to carrying it off to a very great distance. To close the battery circuit is electrically equivalent to bringing it suddenly into the neighborhood of the galvanometer circuit. Indeed, anything which increases the battery current — as, for instance, shortening the wire in the circuit — is equivalent to moving it nearer to the galvanometer circuit; while anything which diminishes the battery current has the effect of motion in the opposite direction.

Primary and Secondary Current

366. The two circuits of which we have just been speaking are so frequently used in physics that special names have been given to them. **The circuit whose current furnishes the magnetic lines of force is called the primary circuit; the one in which the current is induced is called the secondary circuit.**

The facts of the case may be stated more generally still by saying that any variation of the primary current will, in general, induce a current in the secondary. For as the amount of current in any circuit varies, the number of lines of force about that circuit also varies.

It has been found that when the primary circuit is wrapped about a core of iron, it furnishes an enormously greater number of magnetic lines of force than would be furnished if the wire were wound on, say, wood or glass tubing. This is the electromagnet which was invented and so named by Sturgeon in 1822, and which has proved to be an all-important factor in modern electrical machinery.

The Transformer

367. *Experiment* 5. Take an iron ring such as that shown in Fig. 277; wrap it with two coils of wire; pass an alternating current through one of them. It will then be observed — perhaps by means of an incandescent lamp — that a current is passing through the other coil. **A pair of electric circuits, each insulated from the other, but having the region between them filled with iron, is called a transformer.** Whenever the current in the primary circuit P is made to vary, an induced current appears in the secondary circuit S. Conversely, if a current be passed through the circuit S and made to vary, a current will

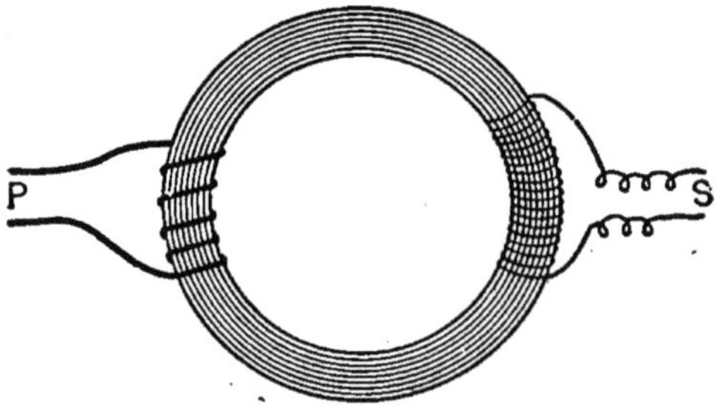

FIG. 277. — A transformer.

be induced in the circuit *P*. An instrument of this kind is in wide use for lighting houses by means of alternating currents. The coil *P* consists, in this case, of many turns of fine wire, while the coil *S* consists of a few turns of heavy wire. The result is that a small alternating current of high pressure passed through *P* induces a large current of low pressure in S.

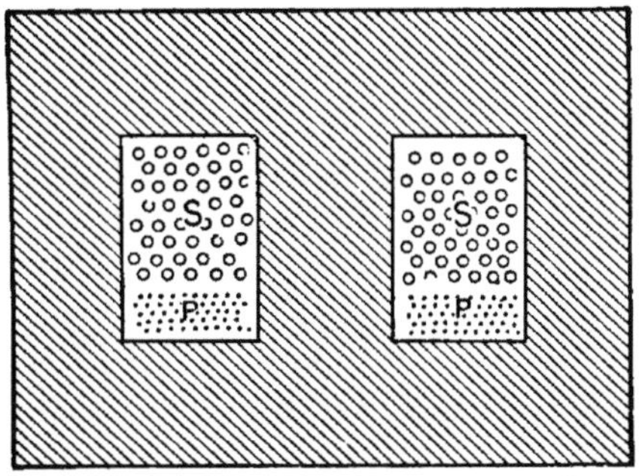

FIG. 278. — Transformer cores are made of a pil of iron plates having this shape. e

A particular kind of transformer called an "induction coil" — **is one in which both coils are wound, one on top of the other, on a cylindrical bundle of iron wires.** This instrument has a wide variety of uses in the laboratory and in medical practice. For description see § 406.

The Dynamo

368. *Experiment* 6. Take an electromagnet such as that represented in Fig. 279. Between the poles of this magnet mount a single turn of wire so that it can be rotated about an axis which is perpendicular to the magnetic lines of force. When the terminals of this single turn of wire are joined to the terminals of a galvanometer, it is observed that during each rotation of the wire loop a current passes through the galvanometer, first in one direction, then in the other. An instrument of this kind is called an "electric generator." When a portion of the current in the loop is shunted off as shown in Fig. 279 so as to supply the electromagnet, the machine is called a "dynamo."

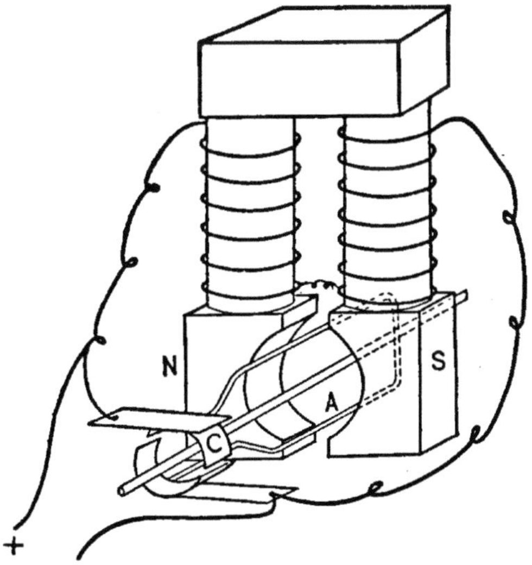

FIG. 279.

The dynamo is then a machine consisting of a strong electromagnet which we may call the primary, and a movable coil which we may call the secondary. This movable coil is rotated by some mechanical means, such as a steam engine, so that the number of lines of force passing through the coil is continually altered. The distinguishing feature of this machine — the details of which must be left for the engineering student — is that it furnishes its own primary current.

The coil *A* (Fig. 279) which is rotated in the magnetic field is called the "armature"; the electromagnet *NS*, which constitutes almost the entire weight of the modern dynamo and which furnishes the magnetic field, is called the "field magnet." The current produced by the armature is, in general, an **alternating current**, *i.e.* one which flows first in one direction, then in the other, changing, perhaps a hundred times a second. But sometimes this armature is fitted with a device *C*, which changes the connections of the armature coils with the using circuit whenever the direction of the current in these coils changes. The result is that the current in the using circuit is direct, although the current in any armature coil is alternating. This device is essentially a **commutator**, such as we have already studied (§ 359), only here the commutator is automatic. A machine in which the armature is not provided with a commutator, but has two collector rings, as shown in Fig. 280, is called an alternating current generator.

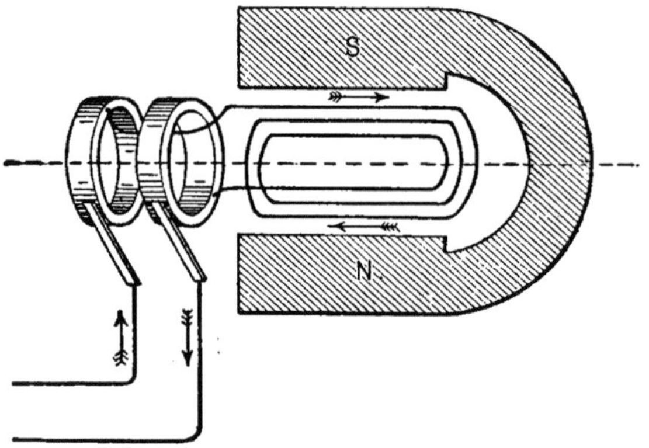

FIG. 280. — Collector rings of alternating current generator.

Engineers find it frequently more convenient to rotate the field magnet and allow the armature to remain stationary. But this is a mere mechanical variation and does not in any way change the electric problem. For a brief general theory of Dynamo and Motor see § 410.

(c) Thermoelectric Currents

369. The two methods just described for producing electric currents are practically the only ones employed in the industries. But there is a third method which deserves notice, partly on account of its wide application in science and partly because of its simplicity and directness, namely, the method in which the energy of the electric current is obtained directly from heat. This mode of generating electric currents was discovered by Seebeck and reported to the Berlin Academy of Sciences in 1822. We have already learned that the essential feature of a Voltaic cell is that it will maintain at a different potential the two conductors of the first class which have their respective ends immersed in a conductor of the second class. It has also been noted (§ 352) that, when two conductors are joined together so as to form a circuit at uniform temperature, whatever tendency there is at one junction to produce a current is exactly counterbalanced by an opposite tendency at the other junction. See Fig. 263.

But Seebeck discovered that when one of these junctions is at a different temperature from the other, the equilibrium is destroyed and a direct continuous current flows in the circuit so long as this difference is maintained. From this it appears that when a metal is heated, its electrical properties are changed just as definitely as if its chemical composition were altered. Not only so, but different metals are differently affected by the same change of temperature. These thermoelectric currents, as they are called, are in general very minute; but the modern galvanometer is so extraordinarily sensitive that even these small effects are easily observed and measured. As noted above (§ 249) the thermocouple is already a standard instrument for measuring temperatures. The reason for this is that the strength of the thermoelectric current depends, other things remaining the same, upon the difference of temperature between the heated and the cooled junction.

The important fact appears to be that at any single junction a change of temperature produces a change of potential difference. And the important constant connected with any particular couple is its **thermoelectric power, which is defined as the rate at which this potential difference changes with temperature.**

The radiomicrometer devised by Professor C. V. Boys and the thermogalvanometer of Duddell (§ 400) are instruments of such great interest that they should be looked up in some of the larger text-books.

Leaving this brief, superficial, and merely qualitative description of electric currents, we pass to their quantitative consideration. The electrical properties which we have just been tudying were mainly discovered in the *first* half of the nineteenth century; but the electrical measurements to which we now proceed were chiefly perfected during the *second* half of the last century.

II. The Measurement of Electrical Quantities: Electromagnetism

370. Electrical measurement is almost a complete and independent science by itself, so well established are its standards and methods. Yet the entire structure is built essentially upon two fundamental principles, one of which connects the difference of potential in any circuit with the rate at which the magnetic field in this circuit is varying, while the other connects the current strength with the strength of the magnetic field produced by the current. The advanced student will find that these two principles can be described in a general and quantitative way only by the two fundamental equations of Maxwell and Hertz. Here, however, we are not seeking after the utmost generality, but rather after a clear grasp of a few special and concrete but exceedingly important cases. For these particular cases lead naturally to a clear understanding of the two fundamental principles.

The Idea of Electromotive Force

371. The essential feature of the Voltaic cell is that it will maintain a difference of potential either between its terminals if the circuit be open, or between any two points on a circuit which is closed.

But Faraday showed that, without touching the closed battery circuit, it was possible to exactly counteract this potential difference and make the current disappear by simply thrusting lines of magnetic force through the circuit in the proper sense and at the proper rate. The introduction of the lines of force is therefore equivalent to the introduction of a potential difference; the effect is in each case to make the electric current flow; and in each case the effect is said to be produced by an "electromotive force," just as the motion of a stream of water might be said to be due to a "hydromotive force." And just as in the case of a basin of water one can make the water flow around in the pan by dragging a towel in a circle over the bottom, without any difference of level, so also in the case of an electric circuit one can make the current flow, without any difference of potential, by merely introducing lines of force. In other words, while a difference of potential always acts as an electromotive force, it does *not* follow conversely that all electromotive forces are differences of potential. Electromotive force, which we shall hereafter write "E.M.F.," is the more general term.

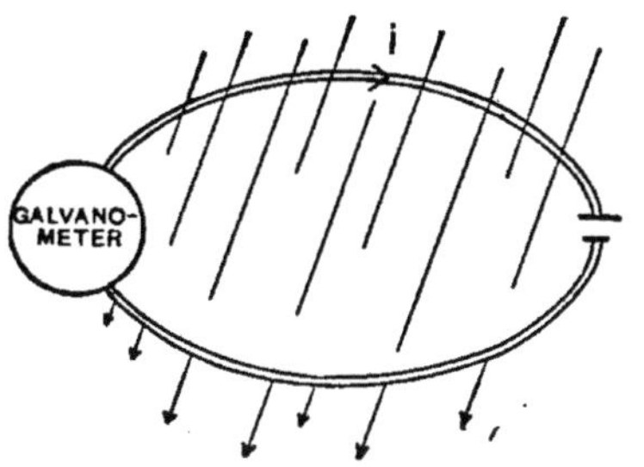

FIG. 281. — Showing potential difference of a battery may be counterbalanced by introducing lines of force.

Concerning the E.M.F. which is produced by magnetic induction, the three following experimental facts should be verified by each student for himself in the laboratory: —

(i) Other things being the same, the amount of E.M.F. varies directly as the number of lines, say N, introduced into the circuit.

(ii) Other things remaining the same, the E.M.F. varies inversely as the time occupied by their introduction.

(iii) To produce a current in any given sense, lines of force must be thrust through the circuit in a sense which is just opposite to that of the lines produced by this current. Thus in Fig. 281 the lines of force due to the current i would be directed downward; and hence any lines of force which are thrust downward through the circuit will produce an induced current which is opposite in sense to i and may be made to just annul i.

372. In 1845 F. E. Neumann formulated these facts in the following simple manner: Let N_1 and N_2 denote the total number of lines of force threading any given circuit at the instants t_1 and t_2 respectively, and k a constant of proportionality. Then for this circuit

$$\text{E.M.F.} = -k\left[\frac{N_2 - N_1}{t_2 - t_1}\right]_{t_2 \doteq t_1} \qquad \text{Neumann's equation.} \qquad \text{Eq. 139}$$

This brief expression, which says that the E.M.F. of any induced current is to be measured by the rate at which lines of force are cut, is the first of the two general principles mentioned at the outset. In a modified form it is also the engineer's formula for computing the E.M.F. of a dynamo which he is designing. The negative sign means that the sense of the induced E.M.F. and the sense of the lines thrust through the circuit are related as the rotation of a left-handed nut to its direction of travel on the screw.

Unit of Electromotive Force

373. About one half a century ago it was agreed by the British Association for the Advancement of Science and assented to by the remainder of the scientific world that we should take as unit E.M.F. that which is produced in any closed circuit when it cuts one line of force per second. This definition has the effect of making $k = 1$ in Eq. 139, so that the defining equation for E.M.F. is

$$\text{E.M.F.} = -\left(\frac{N}{t}\right), \quad \text{Defining equation for E.M.F.} \qquad \text{Eq. 140}$$

where N is the number of lines of force added during the interval t. We have already learned how to measure the number of lines passing through unit area, *i.e.* the intensity H of a magnetic field; and N is, of course, simply the product of H by an area.

This unit of E.M.F. is called the electromagnetic C.G.S. unit; but it is entirely too small for practical purposes. Hence the International Electrical Congress has agreed to adopt the following "practical unit," which is called the "volt" in honor of the Italian physicist Volta.

The "international volt" is defined as that electromotive force which is induced in any closed circuit cutting lines of magnetic force at the rate of one hundred million per second.

Or in terms of arithmetic, —

1 volt $= 10^8$ electromagnetic C.G.S. units of E.M.F.

Standard of Electromotive Force

374. In order to measure electromotive forces in practice, it remains only to establish a standard E.M.F. with which others may be compared. This was much facilitated by the discovery of Latimer Clark that a Voltaic cell made up of zinc rod in zinc sulphate for the cathode and mercury adjoining mercurous sulphate for the anode gave a remarkably constant E.M.F. when properly used. Accordingly the International Electrical Congress at Chicago in 1893 adopted $\frac{1000}{1434}$ of the E.M.F. between the poles of Clark's cell as representing the volt "sufficiently well for practical use." In other words, the E.M.F. of the Clark cell is 1.434 legal volts. For rough work one may consider the gravity cell as having an E.M.F. of one volt; and the Leclanché as giving a volt and a half.

The Definition of Unit Current

375. The second great principle in the accurate study of electric currents rests upon the fundamental discovery of Oersted that about every electric current there is a magnetic field of force.

In the same year in which Oersted found that a current will deflect a magnet, Biot and Savart showed that, in the case of a very long straight wire,

(i) The magnetic field at any point about the wire varies inversely as the distance r of that point from the wire. See Fig. 282.

(ii) The intensity R of this magnetic field varies directly as the current strength i.

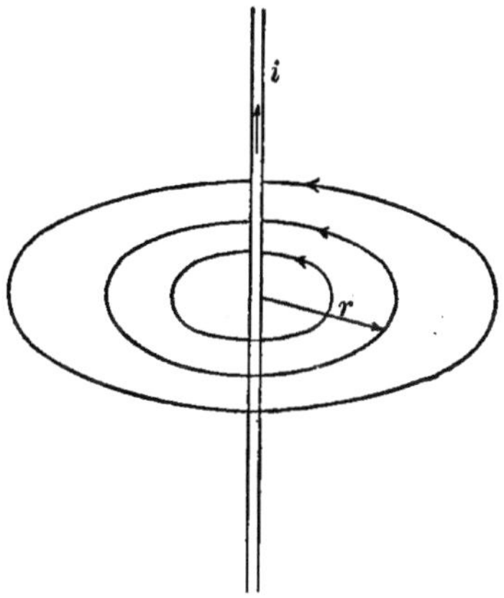

Fig. 282. — Illustrating Biot and Savart's Law.

(iii) The direction of the current and the direction of the lines of force are related as the direction of translation and rotation in a right-handed screw. See Fig. 282.

The facts may be summarized by writing

$$R = K\frac{i}{r}. \quad \text{Biot and Savart's Law.} \qquad \text{Eq. 141}$$

376. But a much more convenient form of circuit is one where the wire is bent into the form of a circle. In this case, it can be shown, by the apparatus indicated in Fig. 283, that

(i) the magnetic field R *at the center of the circle* varies directly as the current strength in the wire.

(ii) The magnetic field *at the center of the circle* varies inversely as the radius of the circle.

(iii) The direction of the field *at the center of the circle* and the direction of the current in the wire are related in the manner of the right-handed screw.

These facts are illustrated by Poynting as follows: Wind on a frame three circles of insulated wire such as those marked 1, 2, and 3 in Fig. 283. Make the radii of "2" and "3" just twice that of "1." If the same current be passed through each of the three circles, but so that the sense of the current in the small circle is opposed to that in the two larger circles, then there will be no effect upon the magnetic needle suspended at the center. For the larger radius of the outer circle is just compensated by the fact that there is here just twice as large a current acting upon the needle. Let us denote the current by i, the magnetic field at the center of the coil by R, and the radius of the circle by r, then a summary of the above facts becomes

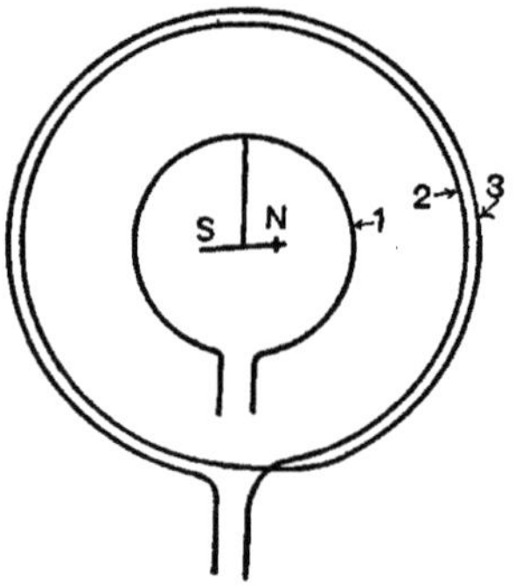

FIG. 283. — Poynting's apparatus.

$$R = k\frac{i}{r} \qquad \text{Eq. 142}$$

Now it has been agreed by the scientific world to give the constant k an arbitrary value of 2π, and then call this the defining equation for the electric current, namely, —

$$R = \frac{2\pi i}{r}. \quad \text{Defining equation for current.} \qquad \text{Eq. 143}$$

This is equivalent to saying that unit current is one which when passing about a circle of unit radius will produce at the center of the circle a magnetic field of intensity 2π. This definition of unit current makes the constant K in Biot and Savart's Law (Eq. 141) equal to 2; so that for a long straight wire $R = \frac{2i}{r}$. When the centimeter, gram, and second are used as fundamental units, this current is called the "electromagnetic C.G.S. unit of current."

The "international ampere," which is the practical unit of current, is one tenth of the C.G.S. unit.

377. Suppose now that one takes a positive unit magnet pole and carries it once around (*i.e.* along) one of the circular lines of force which surround a long straight wire and which are depicted in Fig. 282. The force acting upon the unit will be numerically equal to the magnetic intensity R. The work done in taking the unit pole once around the circuit may be indicated by W; then,

$$W = 2\pi r \cdot \frac{2i}{r} = 4\pi i. \qquad \text{Eq. 144}$$

This is the general form (although here demonstrated only for a special case) of the second great principle mentioned above, namely, **the electromagnetic measure of any current is the work done by carrying a unit magnet pole once about the entire current; and the amount of work done is 4 π times the strength of the current inclosed by the path of the pole.**

Standard Current

378. The international ampere — that is, the ordinary unit of current used in commerce — has been specified by the International Electrical Congress at Chicago in 1893 to be a current which when passed through a solution of silver nitrate will deposit silver on the cathode at the rate of 0.001118 of a gram per second. This description of the standard current will be clear to the student after he has mastered Faraday's Laws of Electrolysis. See § 394.

Ohm's Law

379. Assuming that the student is now more or less familiar with the manner in which currents and electromotive forces are measured, let us consider a particular case where a definite and measurable electromotive force E is producing a definite and measurable current i in a circuit such, say, as that represented in Fig. 284. The following query now arises, **How does the strength of the current i in any circuit depend upon the electromotive force in that circuit?**

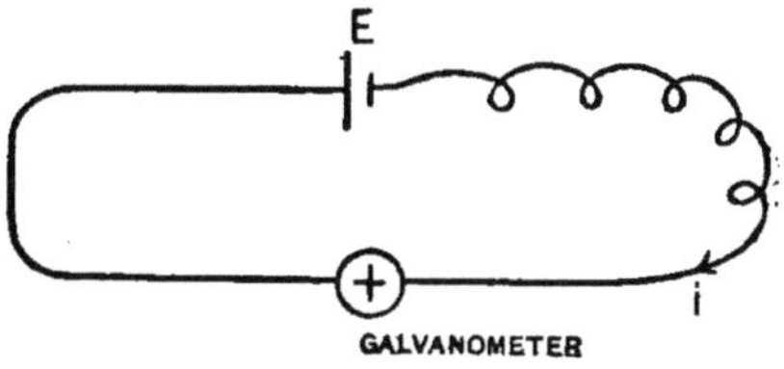

Fig. 284.

The answer to this question is contained in the discovery of the German physicist Ohm, in 1827, namely, **for any given**

circuit the current varies directly as the total E.M.F. in the circuit. In other words, the ratio of the E.M.F. to the current is a constant for any given circuit. This general fact is known as Ohm's Law. **The proportionality constant of the circuit which represents the ratio of the E.M.F. to the current strength is called the "resistance" of the circuit.**

Let us denote this resistance by R, then we may obtain an algebraic expression for Ohm's law by writing

$$\frac{E}{i} = R. \qquad \text{Ohm's Law, defining equation for resistance.} \qquad \text{Eq. 145}$$

But if, instead of considering the entire circuit, we consider only a portion of it, such as that between the points a and b (Fig. 285), then it is found that Ohm's Law holds in the same way for this portion of the circuit as for the whole circuit, *always provided there is between a and b nothing of the nature of a battery or dynamo,* i.e. *no source of electromotive force.*

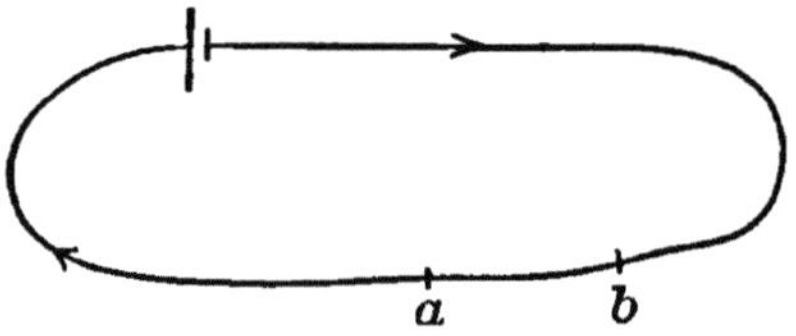

FIG. 285. — Ohm's Law applied to a portion of a current.

The Unit of Resistance. The Ohm

380. To obtain the unit of resistance one employs the same method which is used in the other departments of physics, namely, takes the defining equation and gives the other variables such values as will make the desired quantity unity. Thus in Eq. 145, if we make $E = i = 1$, then $R = 1$. **A circuit is therefore said to contain one unit of resistance when one unit of E.M.F. will produce in it one unit of current.**

The electromagnetic C.G.S. unit of resistance thus defined is entirely too small for laboratory use and hence a practical unit — called an "ohm" — has been defined as one which is a thousand million times the C.G.S. unit: —

1 Ohm $= 10^9$ electromagnetic C.G.S. units of resistance.

Standard of Resistance

381. By international convention agreed upon in Chicago in 1893, the ohm is represented sufficiently well by the resistance of a column of mercury of uniform cross section, at 0° C., when its length is 106.3 centimeters and its mass is 14.4521 grams. When a wire is said to have a resistance of three ohms it is meant that a given E.M.F. applied to the ends of this wire will produce exactly one third as much current as when the same E.M.F. is applied to the above described mercury column.

Resistivity

382. Having discovered that the ratio between E.M.F. and current is a constant for any given circuit, we now proceed to inquire what physical features of the circuit determine the value of this proportionality constant to which the name resistance has been given. **Upon what physical properties of the circuit does its resistance depend?** The most direct experimental answer is obtained, perhaps, by taking a circuit such as that shown in Fig. 286.

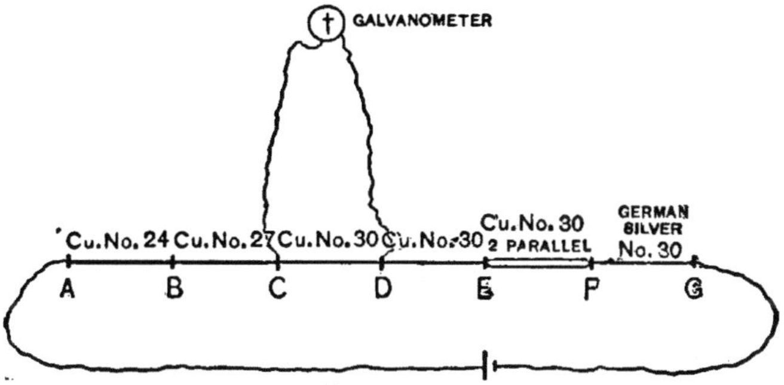

Fig. 286. — Illustrating Ohm's Law.

Pass through this circuit a small steady current from any source. Now since i is constant and since $E = iR$ it is clear that we may study the variations of R by observing the variations of E. To measure the E.M.F. between any two points such as C and D on this circuit, one has only to attach to these points the terminals of a sensitive galvanometer whose resistance is high compared with that portion of the battery circuit which lies between C and D. A branch circuit such as that from C to D *via* the galvanometer is generally called a "shunt."

To each pair of points on the main circuit will correspond a certain deflection of the galvanometer and this deflection will measure (in arbitrary scale divisions) the E.M.F. between that pair of points.

The length of each section in Fig. 286 is the same; that is, the linear distance between each point and its next door neighbor is constant.

Now it is observed that the E.M.F. between *B* and *C* is just twice that between *A* and *B*, No. 24 wire having just twice as large a cross section as No. 27. This illustrates the general fact that **the resistance between any two points on a conductor varies inversely as the area of the cross section of the conductor.**

This rule may again be verified by measuring the E.M.F. between the points *C* and *D*, which are connected by No. 30 wire, which has just half the cross section of No. 27. Here also it is found that the deflection corresponding to the interval *CD* is just twice that due to *BC*.

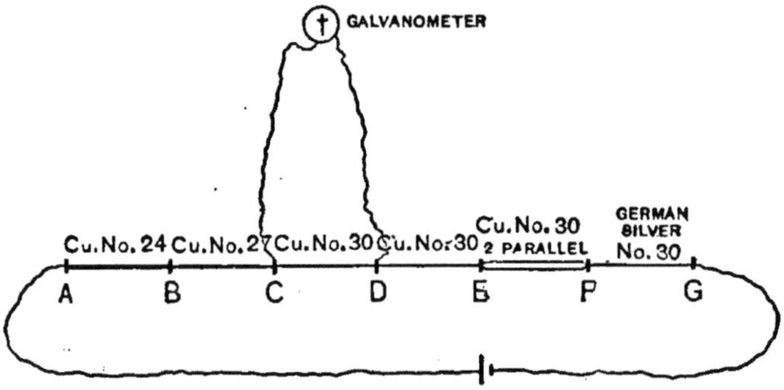

FIG. 286.—Illustrating Ohm's Law.

Next measure the potential difference between *C* and *E*; it will be found just twice that between *C* and *D*, which illustrates the general fact that **for any uniform conductor the resistance varies directly as the length.**

The last section of the wire, that between *F* and *G*, is the same in length and cross section and is carrying the same current as the section *CD*. But the material is different, the chemical composition of the wire is not the same. Comparing the deflection between *F* and *G* with that between *C* and *D*, the former is found to be many times greater; from which one may infer that, other things being equal, the resistance of German silver wire is many times greater than that of copper; and in general, that **the resistance of any conductor depends upon the chemical composition of the substance from which it is made.**

This property of a conductor which determines how its resistance depends upon its chemical composition is called its "*resistivity*," a quantity which we may denote by ρ and may measure as follows: —

Let the cross section of a wire be s, and its length l. We may then summarize the preceding results by writing

$$R \propto \frac{1}{s},$$

$$R \propto l,$$

$$R \propto \rho.$$

Hence $R \propto \frac{\rho l}{s}$ and $R = k\frac{\rho l}{s}$, where k is a proportionality constant. But it has been agreed to call the resistivity of a substance unity when its length, cross section, and resistance are each unity, from which it follows that k is always unity, and hence

$$R = \frac{\rho l}{s}. \quad \text{Defining equation for resistivity.} \quad \text{Eq. 146}$$

The resistivity — or, as it is frequently called, specific resistance — **of any substance is therefore numerically equal to the resistance in ohms between two opposite faces of a centimeter cube of that substance.**

The conductivity of any substance is defined as the reciprocal of its resistivity. In like manner the conductance of a body is defined as the reciprocal of its resistance.

383. Indeed, it has been adopted as a general principle in electrical nomenclature that specific properties, properties of substances as distinguished from bodies, shall be denoted by nouns ending in *-ity;* while properties which belong only to particular bodies shall be described by nouns ending in *-ance.* The advantage of this is that terms such as "impedance" and "inductance," which may be unfamiliar to the reader, carry on their face a part of their meaning.

384. As a matter of fact the resistance of any conductor depends slightly upon its temperature. This fact is, indeed, the basis of one of the best methods of measuring moderately high temperatures, namely, the electrical resistance thermometer, suggested by Siemens, perfected by Callendar. See § 249, above.

Practically all metals increase in resistivity as their temperature rises. But this is a question which must be left for the more advanced student. So also the slight variation of resistance which a conductor suffers when placed in a magnetic field. The student who pursues the subject farther will find this latter effect the basis of one of the most convenient modes of measuring the intensity of the magnetic field in a dynamo.

Table of Resistivities

Substance	Resistance at 0° C. of a wire 1 cm. long 1 cm.² in section	Resistance at 0° C. of a wire 1 foot long and $\frac{1}{1000}$ inch in diameter
Aluminum annealed	2.906×10^{-6} ohms	17.48 ohms
Antimony pressed	35.42×10^{-6} ohms	213.1 ohms
Bismuth pressed	130.9×10^{-6} ohms	787.5 ohms
German silver	20.89×10^{-6} ohms	125.7 ohms
Gold annealed	2.088×10^{-6} ohms	12.56 ohms
Copper annealed	1.584×10^{-6} ohms	9.529 ohms
Copper hard drawn	1.619×10^{-6} ohms	9.841 ohms
Iron annealed	9.693×10^{-6} ohms	58.31 ohms
Platinum annealed	9.035×10^{-6} ohms	54.35 ohms
Silver	1.6×10^{-6} ohms	9.7 ohms

From the *Smithsonian Tables*, p. 254.

Definition of the Electromagnetic Unit of Quantity

385. Quantities of electricity are measured just as we should measure a quantity of water delivered by a stream of known strength in a given time. Hence

$$Q = it. \qquad \text{Eq. 147}$$

becomes the defining equation for electrical quantity Q. If a Leyden jar is charged by a current i flowing for a time t, the charge Q on the jar is it.

The practical unit of quantity is called the coulomb, and is defined as the quantity of electricity transferred by a current of one ampere in one second.

The Electromagnetic Unit of Capacity

386. The capacity of a condenser having already been defined (§ 340) as the ratio of its charge Q to the potential difference E between its plates, one has

$$Q = CE. \quad \text{Defining equation for capacity.} \qquad \text{Eq. 148}$$

Hence the so-called "practical" unit of capacity is that of a condenser charged to a potential of one volt by one coulomb of electricity.

In honor of Faraday this unit of capacity is called the "farad"; but it is entirely too large for laboratory purposes, and hence standard condensers which are made for the purposes of measuring capacities are generally graduated in terms of a unit which is one millionth of a farad and is called a "microfarad." Sometimes a third of a microfarad is used because this represents approximately the capacity of a mile of Atlantic Cable.

SUMMARY

387. A bird's-eye view of the entire subject of electric measurements may be obtained by recalling that what we have done in the preceding discussion is to demonstrate five important laws of nature and to describe them in terms of the following five equations: —

$$E = -\frac{N}{t}. \qquad \text{Eq. 140}$$

$$W = 4\,\pi i. \qquad \text{Eq. 144}$$

$$i = \frac{E}{R}. \qquad \text{Eq. 145}$$

$$Q = it. \qquad \text{Eq. 147}$$

$$Q = CE. \qquad \text{Eq. 148}$$

In these equations appear five, and only five, unknown quantities, namely, the electromotive force E, the current i, the resistance R, the quantity Q, and the capacity C. These unknown quantities are expressed in terms of the three known quantities, namely, the time t, the number of magnetic lines N, and the work W. By elimination, then, one is enabled to express any of the five electrical quantities in terms of known or directly measurable quantities.

Problems

1. A current of 1 ampere and an E.M.F. of 50 volts are required to feed an incandescent lamp. What is the resistance of the lamp in ohms?

2. A Leclanché cell is used to ring a door bell. The resistance of the wire in the bell is 2 ohms, the resistance of the line is ½ an ohm, and the resistance of the cell is 1 ohm. The E.M.F. of the cell is 1.5 volts. What current will be produced when the circuit is closed? *Ans.* 0.429 ampere.

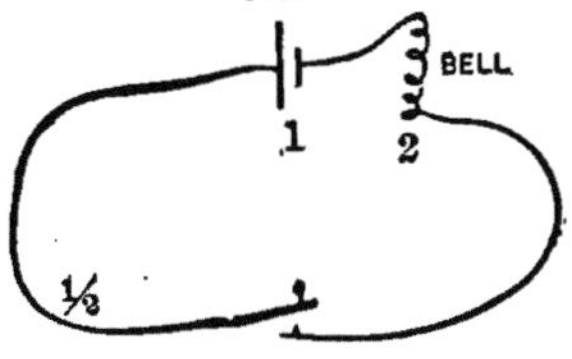

FIG. 286, *A*.

3. The E.M.F. of a single gravity cell is 1.1 volts; the resistance of each cell is 3 ohms. If three of these cells be joined up in series as indicated in

Fig. 286, B, and the poles of the battery be connected by a resistance of 1 ohm, what current will flow in the circuit? *Ans.* 0.33 ampere.

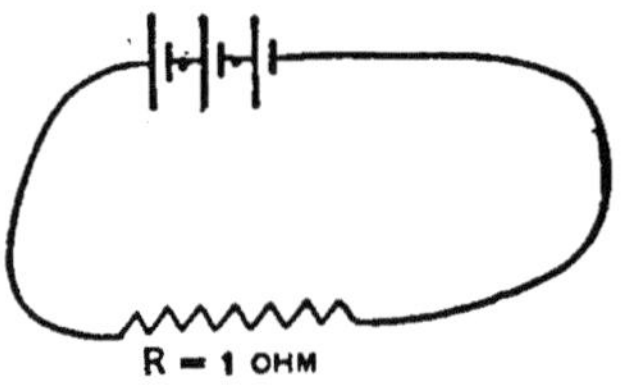

FIG. 286, *B*.

4. A telegraph sounder has a resistance of 70 ohms, and it requires a current of 0.2 ampere to work it. Will 20 gravity cells such as described in the preceding example be sufficient? Will 40 cells suffice? Be able to describe your reasons at the blackboard.

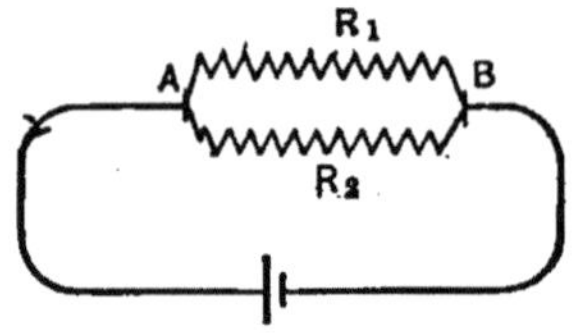

FIG. 286, *C*.

5. Two wires R_1 and R_2, alike in all respects, and having each a resistance of 20 ohms, are joined "in parallel" as shown in Fig. 286, C. What is the sum of their resistance, *i.e.* what is the resistance of that part of the circuit which lies between the points A and B? Use equation for resistance given in § 382, viz.,

$$R = \frac{\rho l}{s}.$$

One might imagine the two wires fused into one of equal length; the numerical value of s would then be doubled, but as an electrical conductor the large wire would be equivalent to the two smaller ones.

6. What is the total resistance of the two wires R_1 and R_2 when joined "in series" as indicated in Fig. 286, D?

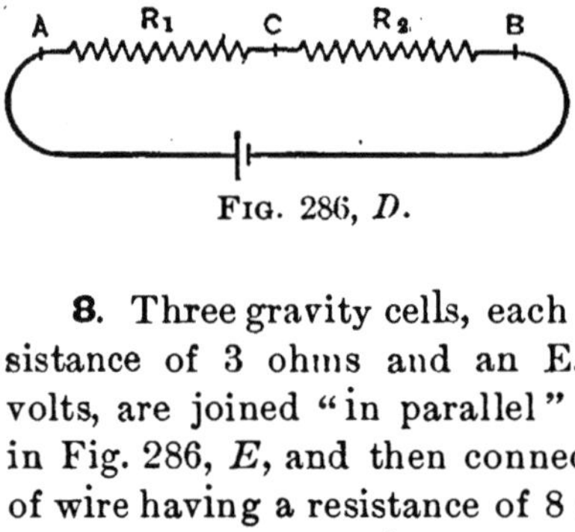

FIG. 286, *D*.

7. A gravity cell of 1.1 volts is joined up in opposition to a storage cell of 2.1 volts, *i.e.* their positive poles are connected. What is the resultant E.M.F. of the two cells? *Ans.* 1 volt.

8. Three gravity cells, each having a resistance of 3 ohms and an E.M.F. of 1.1 volts, are joined "in parallel" as indicated in Fig. 286, E, and then connected to a coil of wire having a resistance of 8 ohms. Find the current in the coil. How much current passes through each cell?

Ans. 0.122 ampere; 0.041 ampere.

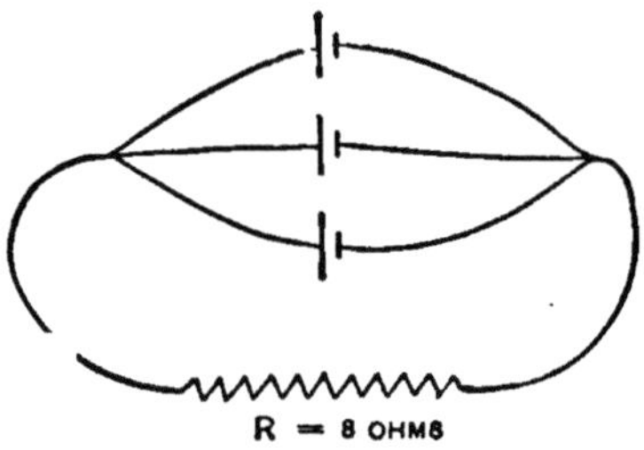

FIG. 286, *E*.

9. Siemens proposed a unit of resistance equivalent to that of a column of mercury 1 m. in length and 1 mm.² in cross section. Find the value of this unit in terms of the ohm.

10. What magnetic field will be produced at the center of a circle of wire conveying a current of 2 amperes, when the radius of the circle is 9 cm.?

11. A small circular wire inclosing an area of 3 cm.² is brought into a magnetic field in such a way that 10^5 lines of force are uniformly added to the circuit in 2 sec. Find the average E.M.F. generated. If the

resistance of the circuit is $\frac{1}{100}$ ohm, what current will be produced during the 2 sec.?

12. Which will affect a magnetic needle at the center more strongly, a current 8 amperes in a circle of 8 cm. or a current of 100 amperes in a circle of 1 m. radius?

13. The armature of a dynamo cuts lines of force at the rate of 1.2×10^{10} per second. What is its total voltage? Suppose the resistance of the armature between the brushes is $\frac{1}{50}$ ohm, what will be the voltage between the brushes when carrying a current of 40 amperes?

14. A room is illuminated by 40 incandescent lamps joined up in parallel. Each lamp has a resistance of 200 ohms when hot and requires a voltage of 110. Find the current necessary to light the room.

15. Taking the resistivity of copper as 1.6×10^{-6}, find the resistance of a pair of lead wires whose diameter is 3 mm. and whose joint length is 40 m.

16. A storage cell whose resistance is $\frac{1}{10}$ ohm and whose electromotive force is 2 volts sends a current through a wire of 3 ohms joined in series with a wire of $\frac{1}{2}$ an ohm. Find the potential difference between the terminals of each of the two wires.

17. The E.M.F. of a gravity cell on open circuit is 1.12 volts. But when the circuit is closed with an external resistance of 20 ohms, the voltage between the terminals of the cell drops to 1.00. Find the internal resistance of the cell.

18. Ten cells are arranged so as to have five in series and two of these series in parallel. The E.M.F. of each cell is 1.5 volts; the resistance 1 ohm. Find the E.M.F. and resistance of the combination.

19. The terminals of a cell are joined by three parallel wires whose resistances are r_1, r_2, r_3, respectively. Prove that the current passing through the cell will be distributed among these three conductors inversely as their resistances.

20. Explain by means of diagrams the difference between an ammeter and a voltmeter.

III. Effects of an Electric Current

A. Chemical Effects. Electrolysis

388. Every one has heard of plating spoons with silver, and bicycles and skates with nickel, even if he has never seen the process. All of these are instances of the general fact, largely established by Sir Humphry Davy, that when a current is passed through a solution, that solution is, in general, broken up into two other substances.

In the first year of the nineteenth century, the same year in which Volta described the Voltaic cell, it was found by Carlisle and Nicholson that a current passing through water breaks the water up into hydrogen and oxygen. Hydrogen is given off at the cathode and oxygen at the anode.

389. *Experiment* 1. In order to pass a current through water it is necessary to dissolve some salt or acid in it to make the liquid a good conductor. The most convenient thing is to add 10 or 15 per cent of strong sulphuric acid.

Take a Wolff's bottle half full of the dilute acid; through two of the holes in the bottle insert two strips of sheet platinum soldered to copper wires, as indicated in Fig. 287; connect these to a source of electromotive force which will send five to ten amperes through the acidulated water. Gas will be given off from each electrode, and the mixture may be collected for examination by attaching one end of a rubber tube to the remaining neck of the bottle, and immersing the other end of the tube in a soap solution. The escaping gas will thus blow a series of soap-bubbles which may be exploded by a lighted taper; for it has been shown that the gases are evolved in just the right proportion to explode, without leaving behind any oxygen or any hydrogen, *i.e.* the proportions of hydrogen and oxygen present are chemically equivalent. It is safer, of course, to remove the glass nozzle from the soap solution before applying the taper.

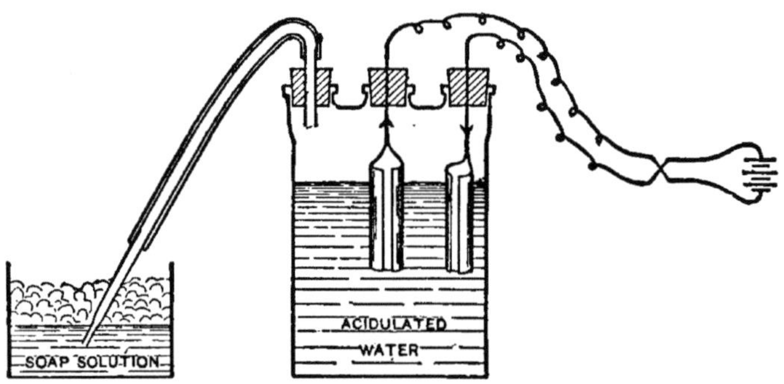

Fig. 287.—The decomposition of water by an electric current.

390. *Experiment* 2.—By using a U-tube with electrodes, one sealed into each branch, the gases can be collected before they have a chance to mix. If the arms of the tube are graduated, one can measure directly the volume of hydrogen set free, and also the volume of oxygen. Accurate measurements of this kind show that when the gases are collected under equal pressures, the volume of hydrogen is exactly double that of oxygen.

But chemists tell us that these are exactly the volumes of hydrogen and oxygen that unite to form water, and that the density of oxygen is sixteen times as great as that of hydrogen. What then will be the *mass* of oxygen set free as compared with the mass of hydrogen?

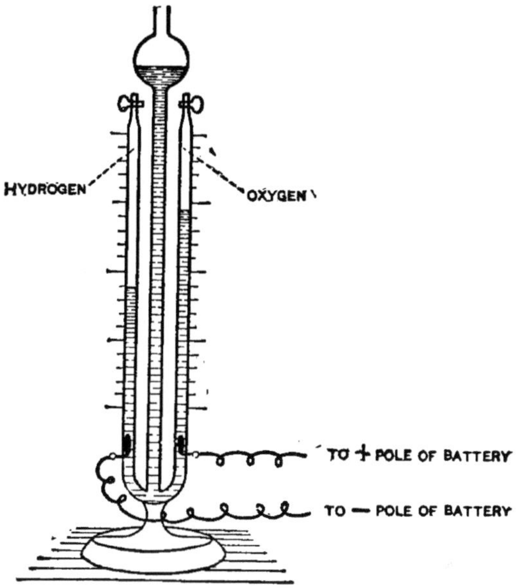

FIG. 288. — Electrolysis of water, in which the hydrogen and oxygen are collected in separate vessels.

391. *Experiment* 3. — Instead of acidulated water, we may use a solution of lead acetate as an electrolyte. The phenomenon is most conveniently seen or projected when the solution is placed in a flat vessel such as is represented in Fig. 289.

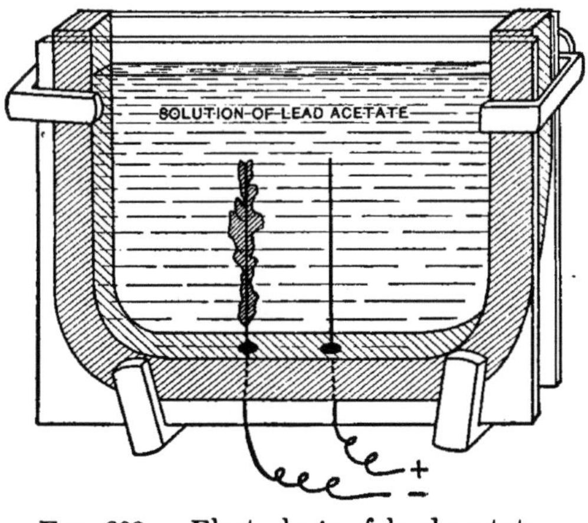

FIG. 289. — Electrolysis of lead acetate.

If we use a pair of platinum wires as electrodes, it will be seen that the lead acetate, which is a clear transparent solution, is broken up so as to yield dark metallic lead at the negative pole of the battery, *i.e.* at the same pole where the hydrogen appeared in the preceding experiment. In general, hydrogen and all the metals behave as if they rode through the electrolyte on the current, *i.e.* they collect at the same pole at which the current leaves the electrolyte.

392. *Experiment* 4. — Moisten a strip of filter paper with a strong solution of phenol phthaleïn * and sodium chloride, table

* This is a substance which chemists very commonly use to detect the presence of an alkali. The phenol phthaleïn is colorless; but a drop of it in an alkaline solution will yield a red color.

salt. If the positive pole of a battery be connected with one end of the strip, the negative pole will, when drawn over the strip, cause an electric current to pass along the strip, for the moistened paper is a conductor. This current will deposit sodium at the negative electrode, the sodium will unite with the water in the strip and produce caustic soda, *i.e.* a strong alkali, which, in the presence of the phenol phthalein, turns a brilliant red. The cousequence is that a brilliant red line may be traced when the negative electrode is moved over the strip of filter paper.

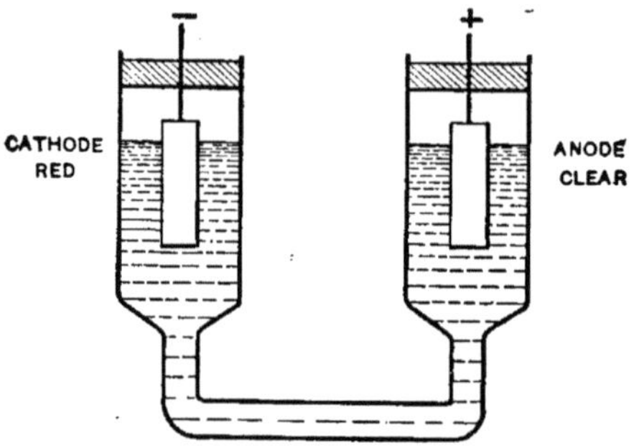

FIG. 290. — Illustrating the chemical reactions which sometimes occur at the electrodes.

A more easily visible method of demonstrating this phenomenon is to take a U-tube as shown in Fig. 290 and provide it with a pair of platinum electrodes. If now the salt solution (NaCl) be placed in this tube and a few drops of phenol phthalein be added, it will be found that as soon as the current begins to pass, the solution in the immediate neighborhood of the cathode turns red. This indicates the presence of an alkali in that region and in no other; in other words, the metallic sodium which the current deposits upon the cathode unites at once with the water to form sodium hydroxide. The chlorine which is set free at the anode dissolves in the water and forms a bleaching solution. The result of this is that when the current is reversed there is sufficient chlorine water about what is now the cathode to immediately bleach out the red due to the phenol phthaleïn. Accordingly it is necessary for the current to flow some little while after reversal before the red color remains about the cathode.

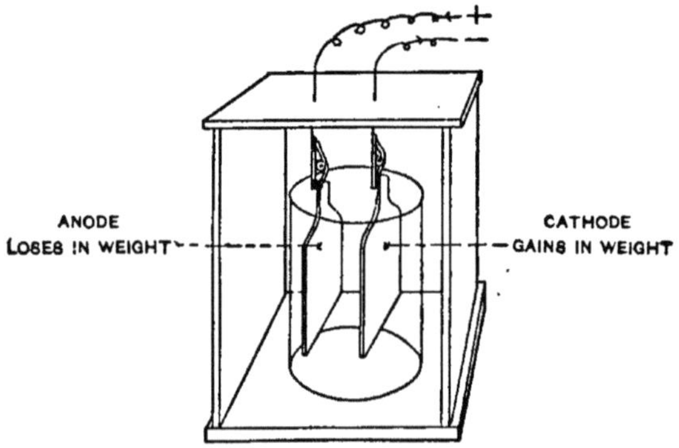

FIG. 291. — The copper voltameter.

393. *Experiment* 5. — If we immerse two copper plates in a solution of copper sulphate and pass a current from one plate to the other, the mass of the plate by which the current enters will diminish, while the mass of the plate by which the current leaves will grow larger; but the copper solution remains practically unchanged. By accurately weighing the plates before and after the passage of the current, and noting the duration of the current, one can determine just how much copper any given current will deposit in one second of time. **A cell fitted with electrodes and electrolyte in this manner is called a voltameter.** The copper thus deposited is very pure, and this process is now largely employed for refining copper.

394. *Experiment* 6. — When several voltameters are connected in series, after the manner shown in Fig. 292, the same current will pass through each of them. But on weighing the electrodes it is found that the amount of element deposited upon any particular electrode, by a given current flowing for a given time, varies directly as the atomic weight of the element and inversely as the valency of the element. Thus silver is univalent and has an atomic weight of 108, while copper is bivalent and has an atomic weight of 63. Accordingly in the arrangement described in Fig. 292 the amount of silver deposited will always be $108/\frac{63}{2}$, or a little over three times as great as that of copper. The ratio of the atomic weight of an element to its valency is generally known as its "chemical equivalent."

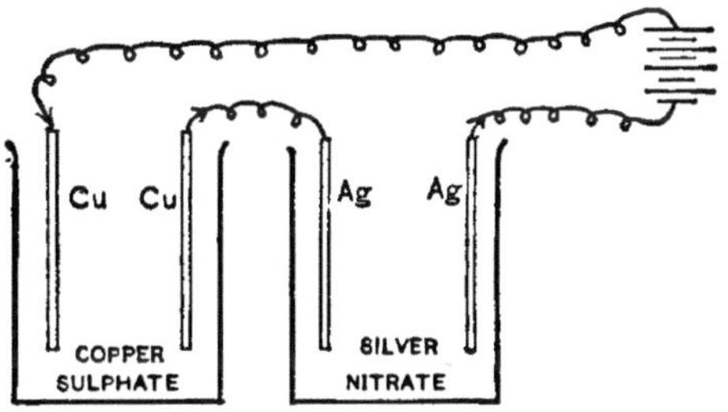

FIG. 292. — The current produces equivalent decomposition in cells containing different electrolytes joined up in series.

Faraday made a very careful study of many phenomena such as we have just been considering, and he summarized his results as follows: —

Faraday's Laws of Electrolysis

First Law

The amount by weight (i.e. the mass) of an electrolyte which is decomposed by an electric current is proportional to the current flowing and to the time during which it flows.

In order to express this law in terms of algebra let us denote by i the current in amperes, by t the time in seconds, and by w the increase in mass of cathode measured in grams. Then

$$w = \epsilon i t, \qquad \text{Eq. 149}$$

where ϵ is a proportionality constant which is called the "electrochemical equivalent" of the element deposited.

Second Law

When an electrolyte, or a series of electrolytes, is decomposed by an electric current, the components into which it is separated are always chemically equivalent, i.e. they are set free in just such amounts as may recombine and form a chemical compound without any left-over material.

For purposes of computation it is often convenient to put this law in terms of symbols as follows: Let subscripts "1" and "2" refer to two different electrolytic cells. Then if atomic weights be denoted by m, valencies by v, and electrochemical equivalents by ϵ, we have

$$\frac{\epsilon_1}{\epsilon_2} = \frac{m_1 v_2}{m_2 v_1}. \qquad \text{Eq. 150}$$

395. These two laws of Faraday would seem to indicate that there is associated with each atom of matter a certain definite charge of electricity which is the same for all atoms of the same valency and is quite independent of the "atomic weight." The charge associated with each unit of valency would then appear to be a natural unit of electricity.

The atom with its associated charge is called an "ion." But the charge alone, this natural electrical unit, is called an "electron," a name which we owe to the Irish physicist Dr. Johnstone Stoney. If we assume that each unit valency carries with it one electron, then the passage of the current through the electrolyte would be accomplished by the positively charged ions delivering up their positive charges to the cathode and the negatively charged ions giving theirs to the anode at the same time. This view affords us not only a simple explanation of both of Faraday's laws, but also gives us a helpful working hypothesis for electric currents as consisting of this double procession of charges dragging the atoms through the electrolyte.

The electrochemical equivalent ϵ is numerically equal to the mass of any element deposited by the passage of one coulomb of electricity through the electrolyte, while the reciprocal of ϵ is the amount of electricity associated with one gram of the element in question.

TABLE OF ELECTROCHEMICAL EQUIVALENTS

SUBSTANCE	RELATIVE ATOMIC WEIGHT	VALENCY	ELECTROCHEMICAL EQUIVALENT IN GRAMMES PER COULOMB
Aluminum . . .	26.9	3	0.0000936
Chlorine .	35.18	1	0.0003672
Copper . .	63.1	1	0.0006588
Copper . .	63.1	2	0.0003290
Hydrogen . .	1.000	1	0.0000104
Iron	55.5	2	0.0002895
Iron	55.5	3	0.0001930
Oxygen	15.88	2	0.0000829
Silver	107.12	1	0.001118
Zinc	64.9	2	0.0003387

From the *Smithsonian Tables*, Third Edition, p. 272.

When the C.G.S. unit of current — 10 amperes — is employed, it is evident that each value in the last column of the above table must be divided by 10.

B. HEATING EFFECTS OF ELECTRIC CURRENTS

396. In these days every one is familiar with the fact that when an electric current is passed through a carbon filament, the temperature of the filament may be increased to such a point that the carbon becomes white-hot. Trolley cars are heated by electric currents passing through coils of wire. Iron chains are now made by welding together the two halves of each link by means of an electric current. Even the rails in railway tracks are now welded together by the same process.

Between the years 1840 and 1843 the English physicist Dr. Joule made a very careful study of the heat produced by an electric current. He found that *for any one piece of wire* the amount of heat produced by any current depends upon two things only: —

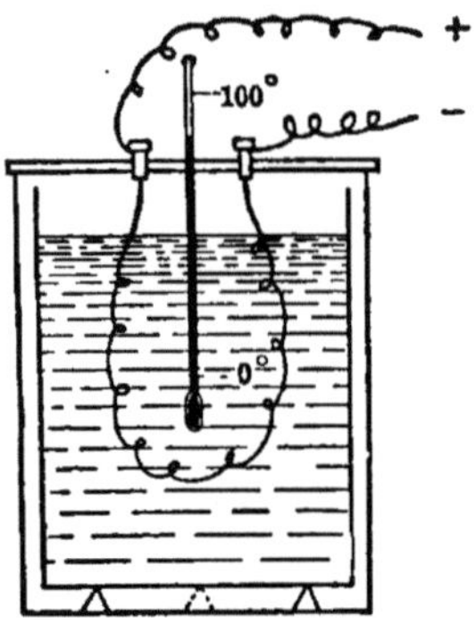

FIG. 293. — Calorimeter for measuring the rate at which heat is produced by an electric current.

1. Upon the amount of current flowing, *i.e.* upon the number of amperes, and

2. Upon the time during which it flows.

By use of a calorimeter containing a coil of wire, it was shown that when the current in the coil of wire was *doubled*, the heat was developed *four* times as rapidly as before; and when the current was *trebled*, the heat was imparted to the water *nine* times as fast as before.

But if any particular current was allowed to flow first for five minutes, then for ten minutes, it was found that the amounts of heat (calories) developed in these two intervals were in the ratio of 1 : 2. And Joule found that in general he could describe the results of his experiments as follows: —

Joule's Laws of Heating

1. **In any given conductor, the heat developed by an electric current in any given time varies directly as the square of the current.**

2. **In any given conductor, the heat developed by any given current is directly proportional to the time during which the current flows.**

397. Hitherto we have been considering the effect of a current upon a single conductor. Suppose, however, we take a series of different conductors and join them end to end, and pass one current through the entire chain. Will the heating effect be the same on each link of the chain? An experimental answer to this question may be obtained as follows: —

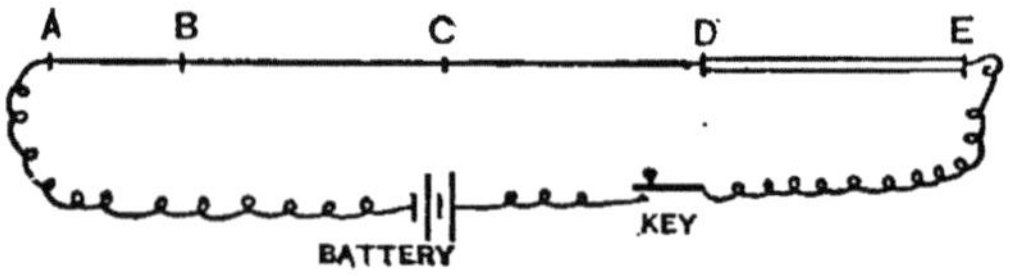

FIG. 294. — Amount of heat produced by an electric current depends upon length, diameter, and material of wire.

Join together four wires as represented in Fig. 294, where

AB is 10 cm. of iron wire;	0.25 mm. diam.
BC is 20 cm. of iron wire;	0.25 mm. diam.
CD is 20 cm. of copper wire;	0.25 mm. diam.
DE is 20 cm. of iron wire;	1.00 mm. diam.

Through this series passes current sufficient to heat the part *AB* red-hot. This will require about four amperes. Observe, then, —

(1) That the part *BC* (which is exactly similar to *AB* except in length) is just as hot as *AB*, and consequently is giving off twice as much heat as *AB*;

(2) That the copper wire *CD* (which is exactly like *BC* except in chemical composition) is very much cooler than *BC*;

(3) That *DE* (which is exactly like *BC* except that its area of cross section is larger) is also much cooler than *BC*.

From experiments of this kind, in which, however, all the quantities involved were accurately measured, Joule concluded that the amount of heat which any given current develops in any conductor in a given time depends upon the following three factors: —

(1) *The length of the conductor*, the amount of heat being directly proportional to the length;

(2) *The chemical composition, i.e.* the kind of material of which the conductor is made; thus we have seen that the iron wire becomes much hotter than copper wire of the same size; in like manner a carbon filament becomes much hotter than an iron wire of the same size, other things being equal. In a chain made of alternate links of silver and platinum wire each of the same size, a current will heat the platinum red-hot while the silver remains comparatively cool.

(3) *The area of cross section*, the amount of heat being inversely as this area. For this reason the insurance companies require copper wire of not less than a certain diameter to be used in leading a current to a given number of incandescent lamps. If the wire be too small, there is danger of its getting hot and setting fire to the building.

These results we may summarize as follows: —

Let H = heat developed by any current;

i = strength of current in C.G.S. units;

t = duration of current in seconds;

l = length of the conductor in centimeters;

s = area of cross section of conductor in square centimeters;

ρ = a constant depending upon the material of which the conductor is made.

Then the most careful experiments by Joule and his successors show that

$$H = \rho \frac{l}{s} i^2 t \text{ ergs,} \qquad \text{Eq. 151}$$

provided the constant ρ for each different substance has the same numerical value which we have already found for the resistivity of that substance. This being so, the factor $\rho \frac{l}{s}$ is simply the resistance of the conductor. See Eq. 146. Accordingly Joule's results assume the form,

$$H = Ri^2 t \text{ ergs,} \quad \text{Joule's Law.} \qquad \text{Eq. 152}$$

where the resistance R as well as the current i and the time t is measured in C.G.S. units. We may, therefore, write as the third and last of Joule's laws, —

3. **In any conductor, the heat developed by a given current flowing for a given time is directly proportional to the resistance of the conductor.**

If, however, the current be measured in amperes, the resistance in ohms, the time in seconds, and the heat in calories, then we have the following laboratory equation: —

$$H = \left(\frac{1}{4.2}\right) Ri^2 t \text{ calories} \qquad \text{Eq. 153}$$

In the laboratory it is practicable to measure any three of the four quantities in this equation, so that we can employ the equation to determine the remaining quantity. By measuring H, C, and t we may thus determine R.

The whole story of the heating effect of an electric current is contained in the above equation.

The Incandescent Lamp

398. We may increase the heat either by increasing the current or by increasing the resistance. Heat is developed wherever resistance of any kind is located. It is for this reason that long slender carbon filaments of high resistance are used

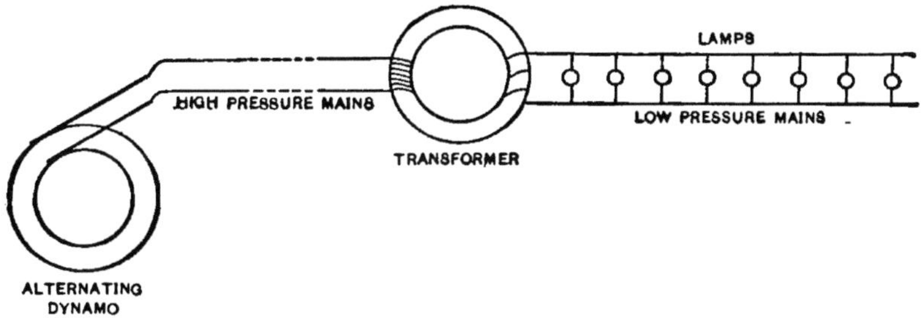

Fig. 295. — The alternating current system.

in incandescent lamps. The heat is located almost entirely in the filament, scarcely at all in the wires which lead the current to the filament.

Note also that it is the heat, not the electric current, which makes the filament luminous. If the carbon filament were raised to the same temperature by any other means, it would be just as bright.

Figs. 295 and 296 indicate the manner in which incandescent

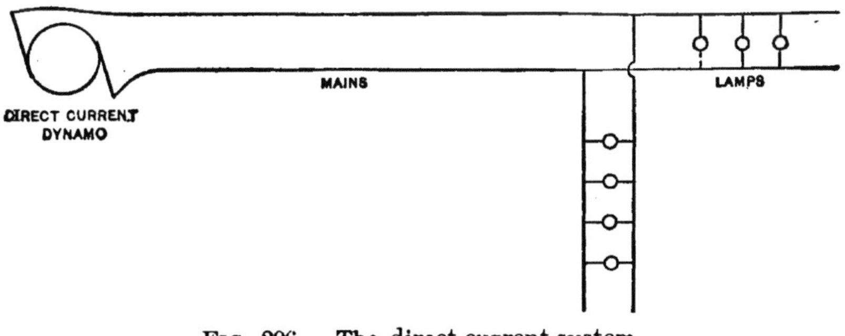

Fig. 296. — The direct current system.

lamps are usually connected in the alternating and direct current systems respectively.

The Theory of Electric Heating

399. Before leaving this brief outline of the heating effect it is well for the student to see that Joule's Law follows as an immediate consequence of Ohm's Law and the definition of potential difference.

Consider any electric circuit, say that represented in Fig. 294. Between any two points such as B and C which include between them no source of E.M.F. there is a difference of potential E which is given by Ohm's Law, namely, $E = iR$, where R is the resistance between B and C. If the current i flows for a time t, a qnantity of electricity it will pass from B to C. By definition of potential difference, E (or iR) is the amount of work done in carrying one unit of electricity from B to C. The amount of work done in carrying it units will therefore be i^2Rt. If we assume that all this work which is done in transferring electricity from B to C is spent in heating the conductor, then it follows that

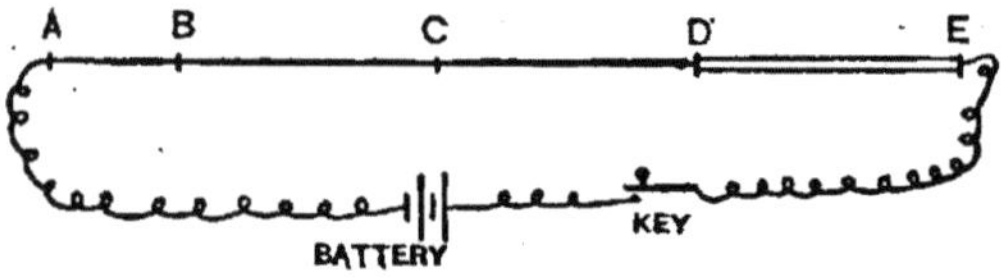

FIG. 294. — Amount of heat produced by an electric current depends upon length, diameter, and material of wire.

$$H = i^2Rt,$$

which is Joule's law. Now since experiment shows that this equation is true, we have good reason to think that all the work of the battery between the points B and C is transformed into heat.

If an electrolytic cell, a storage battery, or an electric motor be inserted into the circuit between B and C, Ohm's Law no longer holds for this part of the circuit where the work of the current is partly spent in overcoming the counter E.M.F. of the cell, the battery, or the motor.

Joule's equation thus shows us that with any given current the heat will be distributed throughout the circuit in exactly the same ratio as the resistance. It is for this reason that in incandescent electric lamps the resistance of the circuit is concentrated in the carbon filament of the lamp. An ordinary 16-candle power 110-volt lamp has about 200 ohms resistance. For this same reason fine wires for protecting electric circuits are made of a material which has high resistivity as well as a low melting point.

400. Joule's equation also shows us that, since the amount of heat developed in any given conductor depends upon the square of the current, heating effect is independent of the direction of the current; the energy transformed into heat is the same for a current $+i$ as for $-i$. Hence the heating effect offers an excellent method for measuring alternating currents. An exquisite illustration of this is to be seen in the thermogalvanometer recently devised by the English physicist Duddell for the measurement of small alternating currents. A minute closed circuit composed of one piece of bismuth and one of antimony is suspended between the poles of a permanent magnet in such a position as to include as few lines of force as possible.

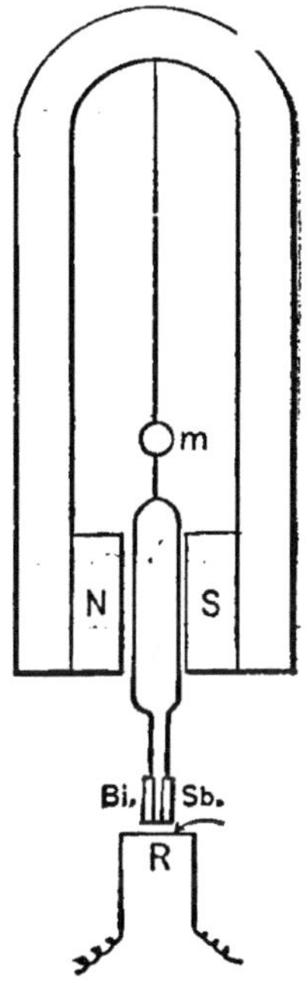

FIG. 297. — Duddell's thermogalvanometer.

Just below the lower junction of the bismuth and antimony is placed a short but exceedingly fine wire, one which has therefore a high resistance. The alternating current to be measured is passed through this wire R, Fig. 297. A certain fraction of the heat generated in R is communicated to the lower junction of the suspended circuit and thus produces a thermoelectric current, and causes the suspended circuit to rotate through a certain angle in the magnetic field. This deflection is read off by means of a mirror m attached to the suspension fiber. The scale is calibrated by passing through the instrument a direct current of known intensity.

Other Forms of Joule's Law

401. In virtue of Ohm's Law it is clear that we may write Eq. 152 in either of the three following forms: —

$$H = i^2Rt = iEt = \frac{E^2}{R}t \text{ ergs,} \qquad \text{Eq. 152}$$

where E is the E.M.F. between the extremities of the resistance R. The first form shows us that when i is a constant, $H \propto R$; the last form shows that when E is a constant, $H \propto \frac{1}{R}$. The importance of clearly distinguishing the conditions under which an experiment is performed and of taking account of them in our equations will be seen from the following experiment: —

Join in series two pieces of wire, about No. 36, having each approximately the same length and diameter, but one made of copper, the other of iron. If now a current be passed through these two wires, the value of i will be the same for each.

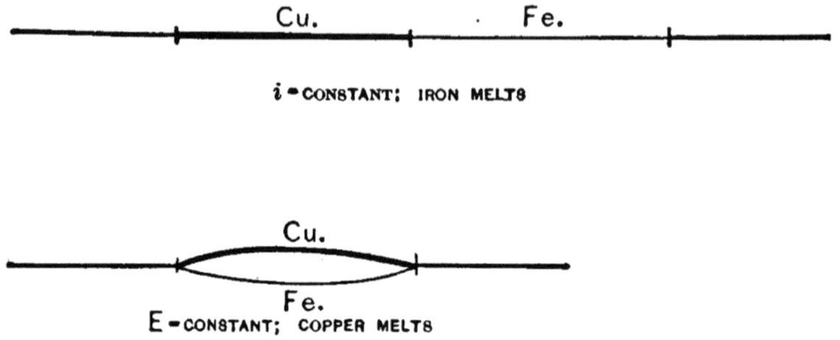

FIG. 298. — Interpretation of Joule's Law under different experimental conditions.

Hence $H \propto R$, and since the resistance of the iron is much greater than that of copper, the heat is localized mainly in the iron, so that as the current is increased the iron is the first to become red-hot and fuse.

When, however, these two wires are placed in parallel, the current is no longer the same in each; it is now the E.M.F. that is constant as we pass from one wire to the other. Hence $H \propto \frac{1}{R}$, and we may now expect the heat to be concentrated mainly in the good conductor, that is, in the copper. As the current is increased this expectation is justified, for we now observe that the copper wire is the first to turn red and melt.

In the laboratory it is well to remember this principle; for when a set of resistance coils is joined to a source of constant E.M.F., such as a storage battery, the low resistances are apt to be burned out if put in circuit alone. But if the coils are joined to a source of constant current, it is the high resistances which are in danger.

C. MAGNETIC EFFECTS OF AN ELECTRIC CURRENT

402. The student who has employed a galvanometer to detect electric currents already knows that a wire conveying an electric current is surrounded by a magnetic field. But it is not sufficient to know merely that the region about a current is a magnetic field, we want to know what kind of a field it is, what its direction is at any point, and where the field is most intense.

Perhaps as good a way as any to obtain a view which is at once sound and general is to follow some half dozen experimental illustrations of this magnetic effect, such as the following: —

403. *Experiment* 1 To illustrate the field about a straight portion of an electric circuit pass a current through a bare copper wire of small diameter. The copper wire should be just large enough not to melt. On dipping this wire into a box of fine iron filings, we observe that the filings, which, by induction, become temporary magnets, arrange themselves in circles about the wire, and as the wire is lifted out these rings of filings remain clinging about it, as indicated in Fig. 299. But when the current is interrupted they fall off, showing that these circular lines of force disappear with the current.

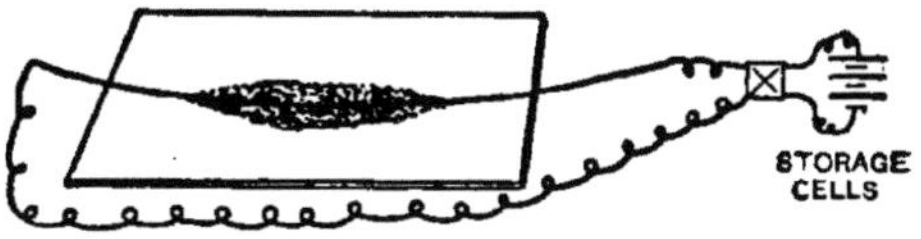

FIG. 299. — Iron filings arrange themselves in rings about an electric current.

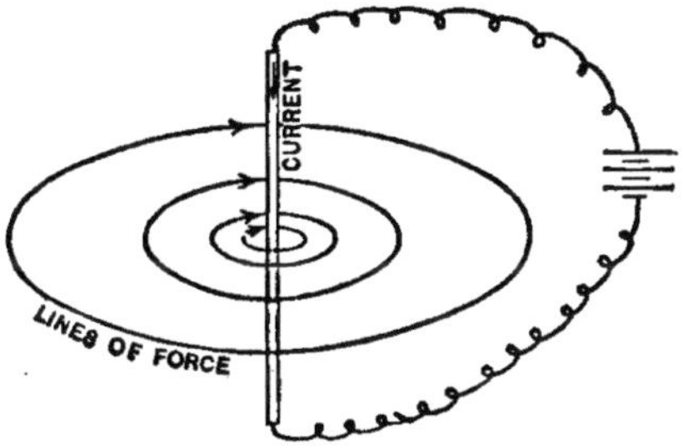

FIG. 300. — Showing right-handed screw relation between current and lines of force due to that current.

The magnetic field about a straight wire conveying a current is then represented by a series of circles drawn about the wire as center and in planes perpendicular to the wire, as illustrated in Fig. 300. As an experimental fact it is observed that the direction of these lines of force is related to the direction of the current, as indicated in Fig. 300, that is, in the right-handed screw relation. But this is exactly the state of affairs described by Biot and Savart's Law, Eq. 141, which takes the following form when the C.G.S. unit of current is employed: —

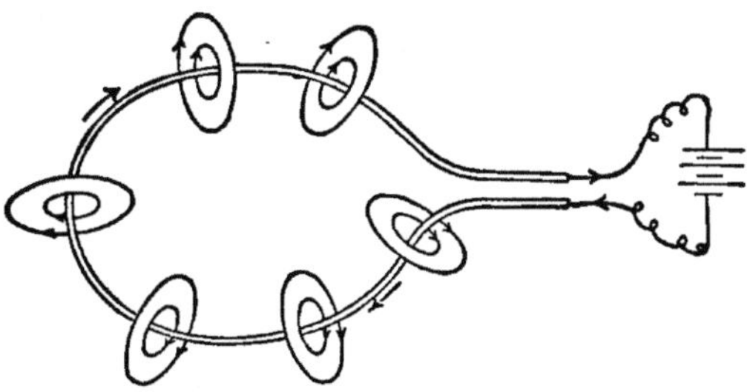

FIG. 301. — Lines of force about a single loop conveying a current.

$$H = \frac{2i}{\ }. \qquad \text{Eq. 154}$$

If the wire be bent into a loop, as in Fig. 301, lines of force will pass through the loop in a direction perpendicular to the plane of the loop. These are the lines of force which we have already (§ 365) used for the induction of currents.

404. *Experiment* 2. If a current be passed through a coil of wire (*i.e.* through a pile of loops), the lines of force distribute themselves about the coil very much as indicated in Fig. 302. Such a coil is called a "helix." And by use of a compass needle we easily satisfy ourselves that such a helix behaves as a magnet — one end repelling the north pole, the other end attracting it. Remembering that the arrow on a line of force indicates the direction in which a north pole would travel if left free at that point in the field, it follows that **the north pole of a helix is that in which the current appears to circulate in a counter-clockwise direction when viewed from outside the helix.**

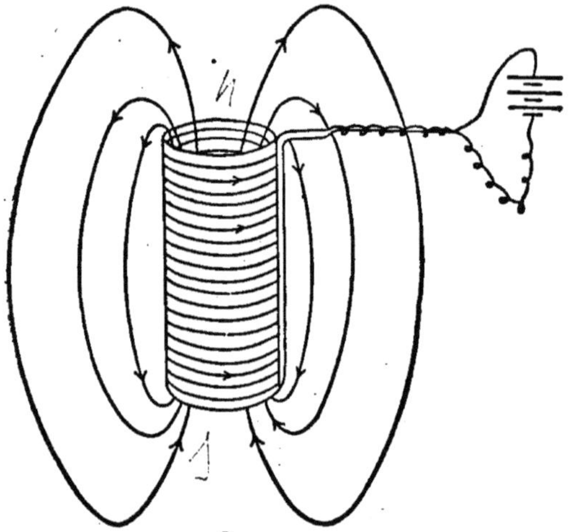

FIG. 302. — Lines of force about a helix.

Observe that here the magnetic lines of force in the helix are straight, while the path of the current is circular, the two being connected by the right-handed screw relation. **No exception, indeed, has ever been discovered to the general principle that about every electric current, whatever its path, there are linked lines of force in the right-handed screw relation.**

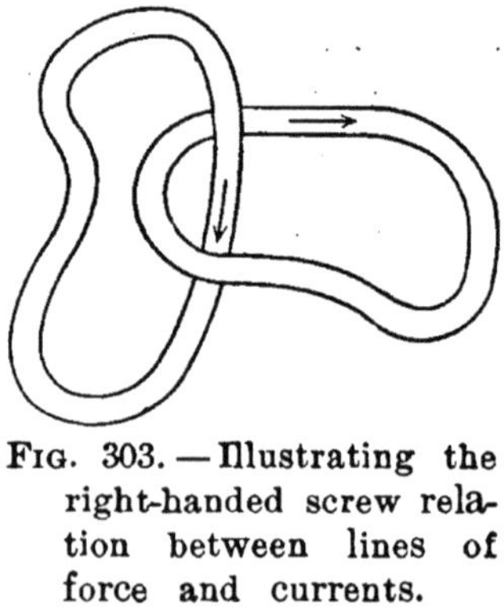

FIG. 303. — Illustrating the right-handed screw relation between lines of force and currents.

The Electromagnet

405. *Experiment* 3. A helix which is wound upon a rod of iron is called an "electromagnet." Such a device enormously increases the number of lines of force passing through the helix. The behavior of the electromagnet is well illustrated in the ordinary telegraph sounder. Here the operator at a distant station closes the circuit; the electromagnet furnishes a strong magnetic field, attracting to its poles a small piece of iron *A*, called an "armature." This armature is pivoted at *B*, and hence can rotate slightly about an axis through *B*. When the circuit is closed, the armature hits a stop which gives the sharp metallic click familiar to every one who has ever been in a telegraph office. When the circuit is broken, the armature is pulled back against another stop by means of a spiral spring *S*. The details of the "sounder" are shown in Fig. 305.

FIG. 304.—Representing the key at the sender's end and the sounder at the receiver's end of the telegraph line.

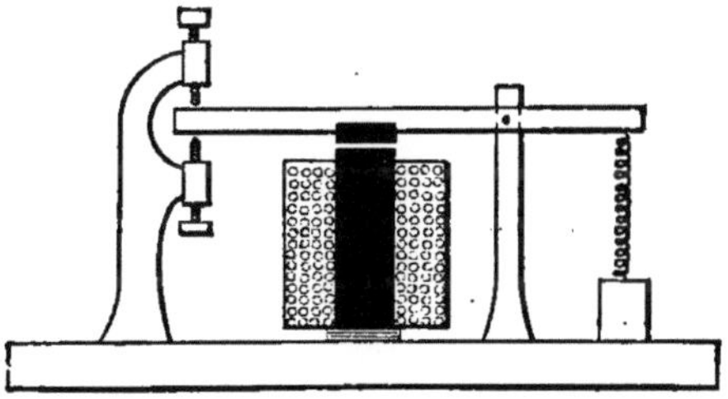
FIG. 305.

The electric door bell, which makes and breaks the circuit automatically, is very much like the telegraph sounder, only here when the circuit is first closed, by pushing the button at *K* (Fig. 306), the current runs through the armature; but the iron core of the magnet immediately becomes magnetized; the armature is pulled away from the metallic stop on which it rested, and the current is thus interrupted; but as soon as the current is broken the electromagnet ceases to attract, the armature falls back by means of a spring into its initial position, and the circuit is again closed.

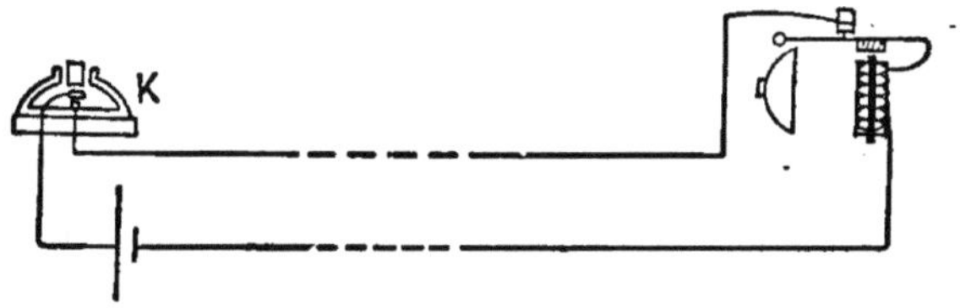

FIG. 306.—The electric door bell.

In this manner the armature is kept in continual vibration, and may, therefore, be employed as a bell clapper, the bell continuing to ring so long as the push button is held down at *K*.

406. *Experiment* 4. **The induction coil is merely an electromagnet provided with an automatic circuit breaker, and wound with a second and entirely independent coil of wire.** The wire which contains the automatic interrupter and which is usually thick and short so as to carry large currents is called "the primary." The other circuit which is usually very long and made of comparatively fine wire is known as "the secondary." This secondary is so long that it is wrapped many times about the lines of force which pass through the primary. The result is that when the primary is quickly broken, an enormous number of lines of force N are suddenly removed from the secondary. And since the electromotive force induced in the secondary is given by

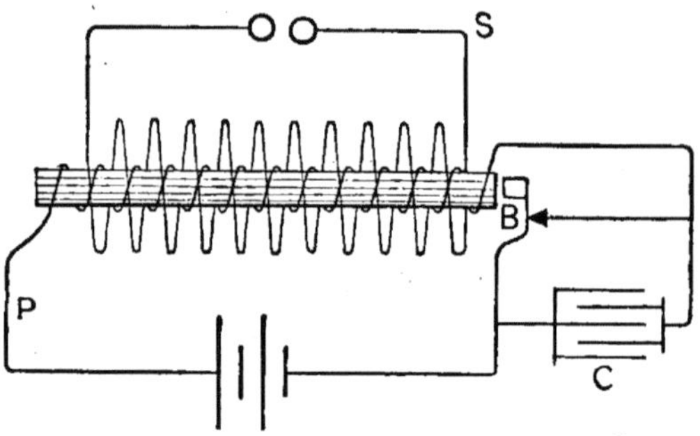

FIG. 307. — Diagram of induction coil.

$$\text{E.M.F.} = -\frac{N}{t}, \qquad \text{Eq. 140}$$

it is clear that the voltage between the terminals of the secondary may become enormous, so large, indeed, as to discharge through a very considerable gap of air or other insulator. It is for this purpose, namely, to produce a high electromotive force starting with a source of low electromotive force that in-

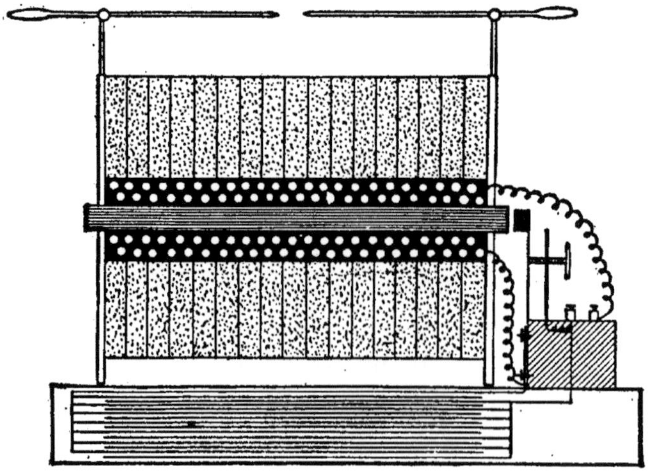
FIG. 308. — Longitudinal section of induction coil.

duction coils are mainly used. As will be seen from Neumann's equation (140), it is quite as important to make t, the duration of break, small as it is to make N large.

Accordingly, for the purpose of reducing t, it is customary to join one terminal of a condenser to each side of the break in the primary as indicated in Fig. 307. The effect of this condenser is to hasten the break very much, by receiving the charge which would otherwise pile up at each terminal of the primary gap and thus prolong the primary current.

Behavior of a Movable Circuit in a Magnetic Field

407. *Experiment* 5. Up to this point we have been studying principally the field produced by a given electric current; but we are now ready to determine by experiment how an electric current behaves when placed in another magnetic field, *i.e.* in a field other than its own. To answer this question we shall need a circuit which is free to move while a current is flowing in it. The classical method of accomplishing this is to float a small Voltaic cell in a large dish of water. A small coil of copper wire, consisting of half a dozen turns, with zinc and copper plates soldered to its respective terminals, can be fitted, through a cork, into a small and rather deep glass beaker, as shown in Fig. 310.

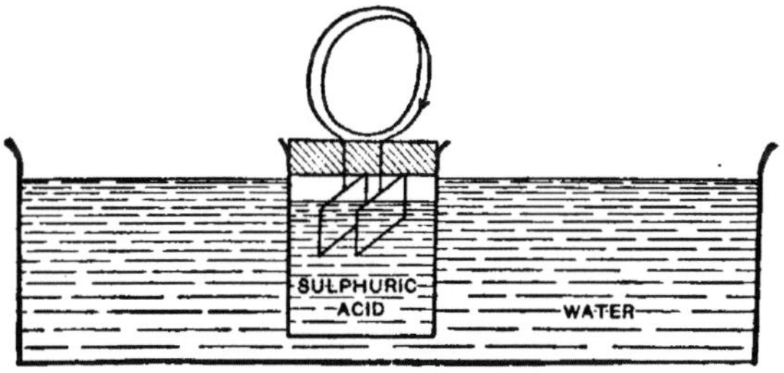

FIG. 309. — An easily movable electric circuit.

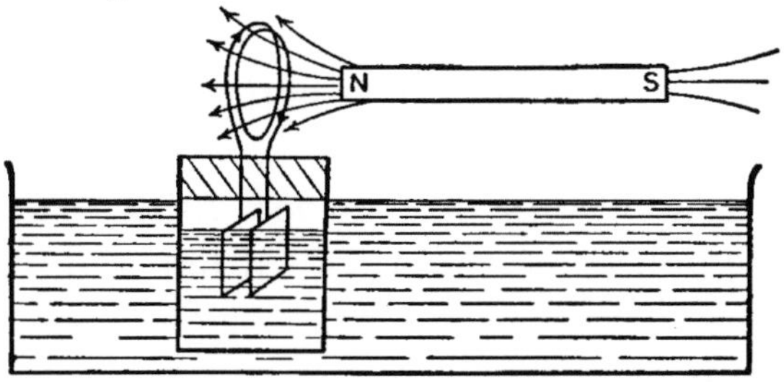

FIG. 310. — Movable coil *rotates* so as to include line passing through in the right-handed direction.

As soon as the glass is half filled with dilute sulphuric acid and floated on water we have at our disposal a current in a circuit which is free to move. If the floating cell is inclined to be unstable, a small piece of sheet lead laid in the bottom will remedy the difficulty.

Two phenomena are observed when a strong bar magnet is brought near this floating coil, viz.: —

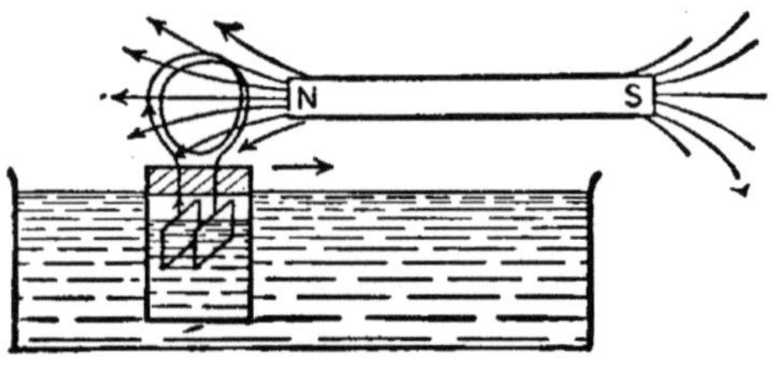

FIG. 311. — Movable circuit is *translated* so as to include as many lines of force as possible.

(1) If the *N* pole of the magnet be extended toward the coil, the circuit will always *rotate* so as to present that face in which the current is flowing in a clockwise sense, *i.e.* the south face of the coil is attracted by the north pole of the magnet.

(2) But this motion of rotation is not the whole of the phenomenon. For if the magnet be held in a fixed position, end on *in front of the coil*, it is observed that the coil is also *translated* as a whole, moving so as to include more lines of force.

And, in general, it has been proved that an electric current always sets itself so as to include as many lines of force as possible in the right-handed screw direction. A slack wire of gold tinsel carrying a current shows the same tendency to wrap itself about a straight magnet, that lines of force have to wrap themselves about a straight wire conveying a current.

The advanced student will discover that this general electrodynamical principle is merely a special case of the general dynamical principle that any body tends to move from rest into such a position as to make its potential energy a minimum.

The D'Arsonval Galvanometer

408. *Experiment* 6. A still simpler and more elegant illustration of this behavior of a current in a magnetic field is the following: —

Mount a Weston magnet upon a block of wood, as indicated in Fig. 312. Between the poles suspend a coil of half a dozen turns of No. 20 copper wire, using for the suspension a spiral of, say, No. 30 wire. Connect the coil by means of another spiral below to a single cell and key. If the plane of the copper coil be set parallel to the magnetic lines of force, as indicated, the coil will rotate whenever the circuit is closed; because by rotating it can set itself so as to include a much larger number of lines of force. This experiment (which can be prepared in ten minutes) should be thoroughly examined, repeated, and mastered by every student.

This is essentially the instrument which Lord Kelvin devised in 1867 for receiving messages over the Atlantic Cable. The suspended coil is connected with the copper wire inside the cable, and whenever a current is sent over the line, the suspended coil turns itself so as to include more lines of force. When the current ceases, the coil returns to its position of equilibrium. The angular deflection of the coil is indicated by the motion of a spot of light; and the operator reads the message by the motion of this spot. In later instruments the coil is also provided with a fountain pen so that it records its own deflections on a moving strip of paper.

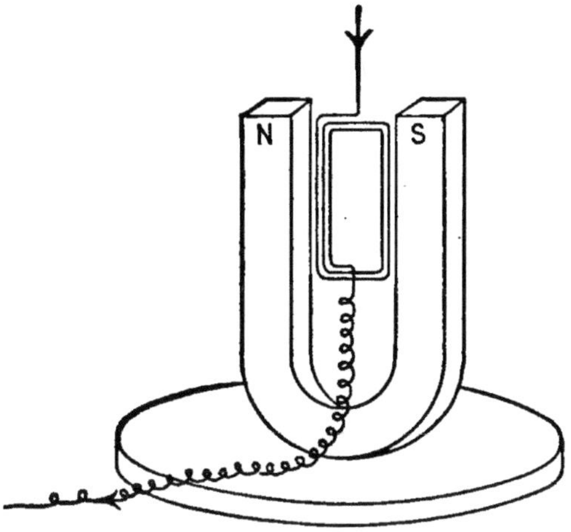

FIG. 312. — A galvanoscope in which a part of the wire conveying the current is movable, while the magnet is fixed.

Nearly all the voltmeters and ammeters which are seen on the switchboards of electric light stations and power houses are constructed on this same principle, only here the movable coil is wound on a rigid frame which is pivoted between a pair of jeweled bearings. The ammeter is merely a galvanometer whose resistance is very small compared with the rest of the circuit in which it is placed; while a voltmeter is a galvanometer whose resistance is very high compared with the resistance lying between the two points whose potential difference is desired.

The Electric Motor

409. *Experiment* 7. The electric motor is also a suspended coil of this kind. Here the coil is generally wrapped about an iron core on a horizontal steel shaft which is suspended in bearings so that the coil rotates with very little friction. This coil is called the **armature** and is placed between the poles of an electromagnet so as to be in a strong magnetic field. If a current is now passed through any one coil of wire on the armature coil, the coil will tend to rotate so as to include more lines of force; but when it has once set itself in a position where it includes as many lines as possible, the commutator (simply a split ring fastened on the shaft and insulated from it) changes the direction of the current in the coil, so that now the wire tends to rotate 180° farther so as to include more lines of force in the right-handed direction. Each turn of wire on the armature behaves in the same way, so that there is a continual moment of force acting upon the armature. By means of a belt and pulley a large part of the energy furnished to the motor may be transmitted to other machines and there employed to do work.

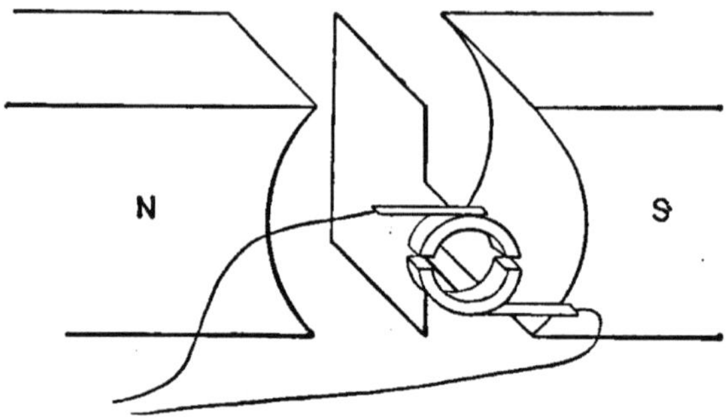

FIG. 313. — Electric motor, showing commutator.

It should be clearly grasped that the electric motor is not in itself a source of energy, but merely transforms into mechanical energy the electrical energy which is given to it. At the same time, by allowing us to utilize the energy of distant lakes and mountain streams where this source of power would not otherwise be available, the electric motor has had the practical effect of adding largely to our available resources. So influential have been the dynamo and the motor in modern civilization and so much have they done for the advancement of electrical science, that we here append a brief outline of

THE THEORY OF THE DYNAMO AND MOTOR

410. The cardinal discovery of the dynamo was made by Faraday in 1831. His machine consisted merely of a copper disk mounted on an axle so as to rotate between the two poles of a horseshoe magnet as shown in Fig. 314. The brushes make contact, one at the periphery, the other at the axle, of the disk. The radius *CP* is continually cutting lines of force in such a direction as to produce a current in the direction indicated by the arrow on the lead wire.

A *motor* operating upon the inverse of this principle had

some years previously (1823) been invented by Barlow. But the general principle that an efficient dynamo will make an efficient motor when supplied with current — the principle of reversibility — was enunciated by the Italian, Pacinotti, as late as 1864.

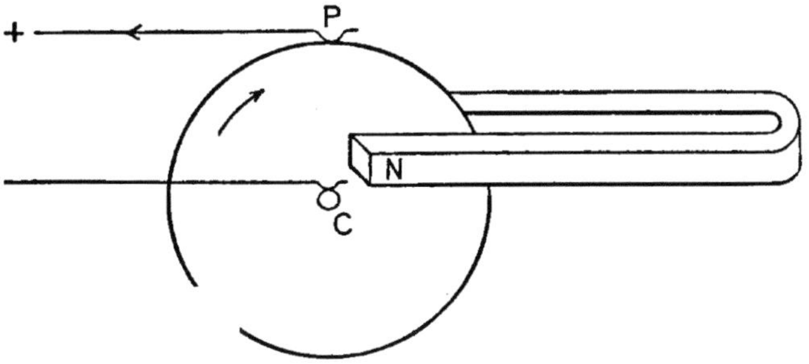

Fig. 314. — Faraday's disk machine, the earliest type of electromagnetic generator.

Between this date and 1876 Gramme, Wilde, Siemens, Wheatstone, and others showed how a part or all of the current generated by the machine might be led around the arms of the magnet and thus convert it into a powerful eloctromagnet, which in turn would enormously increase the output of current, and which gave to the machine the advantage of being able to start itself from the residual magnetism retained by the iron.

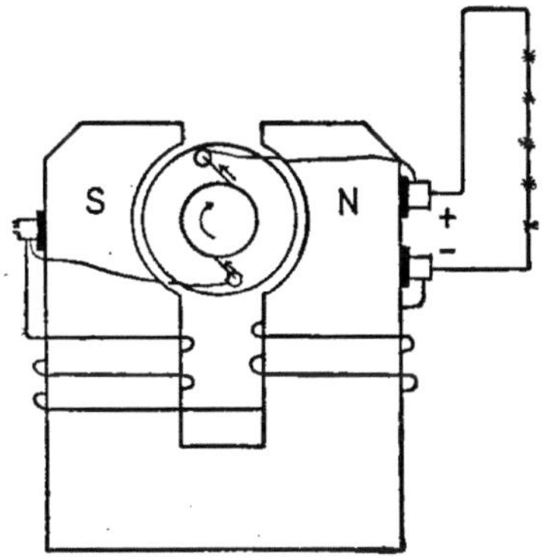

Fig. 315.

In Fig. 315 is seen a dynamo whose magnetic field is produced by passing the entire current about the iron. Such machines are said to be "series wound." The machine shown in Fig. 316 is the type used for lighting incandescent lamps. Here only a part of the current from the armature is diverted to excite the field. These dynamos are said to be "shunt wound." When both methods are combined on one machine, the winding is said to be "compound."

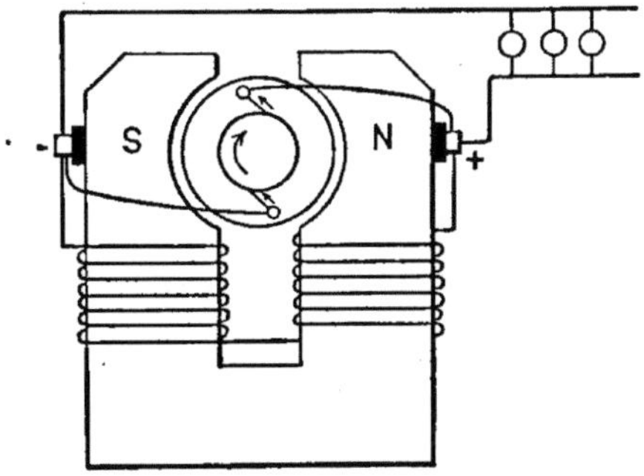

Fig. 316.

411. When it comes to a quantitative consideration of the dynamo, the engineering equations upon which the machine is designed and operated are nothing else than the two fundamental equations (139 and 144) upon which are based nearly the whole of electrical science — the first of these describes Faraday's great discovery; the second, Oersted's; the first is Neumann's equation; the second, Lenz's Law.

Consider any single straight copper wire on the armature of a dynamo. Suppose it to lie parallel to the axis of rotation so that its motion as well as its direction is nearly everywhere perpendicular to the lines of force — magnetic flux, as it is called — which run from one pole piece to the other through the armature. Let us call this single straight wire an "inductor"; then if

L = length of the inductor in centimeters;

V = linear velocity of inductor in centimeters per second, and

H = intensity of magnetic field — lines per square centimeter — in the air gap, between pole piece and armature,

the rate at which the inductor cuts lines of force will be HLV, and the electromotive force E generated in this inductor will be

$$E = HLV \times 10^{-8} \text{ volts.} \qquad \text{Eq. 155}$$

The various inductors are laid on the armature in such a way that all their electromotive forces are added together; and it is the sum of these E.M.F.'s which make up the voltage of any particular machine.

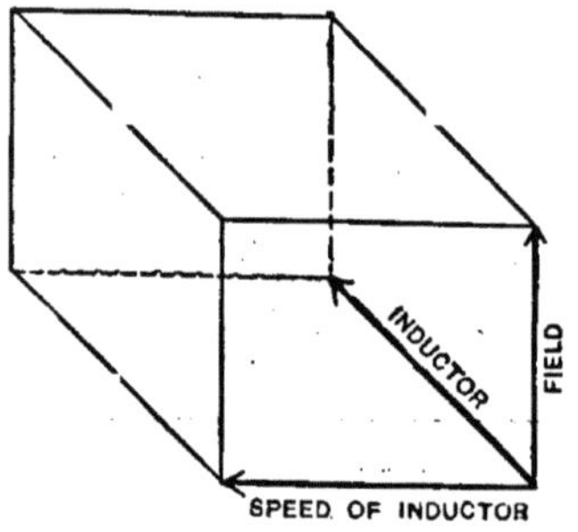

FIG. 317. — Three factors which determine E.M.F. of dynamo.

The above equation gives us the E.M.F. of the dynamo in terms of its speed.

We next proceed to derive the second equation, which will give us, in terms of the current, the force required to drive the inductor through the field.

From Joule's Law and from the definition of E.M.F. we have already learned that the rate at which a current does work between any two points is the current strength i, multiplied by the E.M.F. between those two points. Neglecting friction, the rate of work — the power — which must be exerted upon the inductor is iE. But we also know from mechanics that the power used in moving any body is the product of the force F exerted upon the body, and the velocity V with which the body is moved.

Hence

$$FV = iE,$$

or, substituting for E its value from Eq. 155, we have

$$F = HLi \text{ dynes}$$

as the mechanical force required to drive the inductor through the field. If i be measured in amperes instead of C.G.S. units, then

$$\mathrm{F} = HLi \times 10^{-1} \text{ dynes,} \qquad \text{Eq. 156}$$

which is the second fundamental equation of the dynamo.

The student of electric engineering will find that in practice these equations call for certain corrections and modifications depending upon whether the inductors are wound upon a ring or upon a drum; but he will also find that they, together with Ohm's Law, cover essentially the entire electrical theory of the dynamo and the motor, for the motor is merely a reversed dynamo. The magnetic behavior of iron is also a vital factor in dynamo design.

412. *Experiment* 8. An exceedingly valuable illustration of electromagnetic induction is found in the telephone, which was invented by Graham Bell and Elisha Gray in 1876.

The instrument consists merely of a thin iron diaphragm, underneath which is supported a permanent magnet, carrying on the end, near the diaphragm, a helix of fine wire. See Fig. 318. The lines of force which spread out from the upper end of the magnet tend to pass along the iron disk and keep within it as much as possible. Connect the two ends of the helix to a moderately sensitive D'Arsonval galvanometer. It is observed that when one pushes the disk with his finger a little nearer to the end of the magnet, a current shows itself on the galvanometer; and when the diaphragm is released and springs back into its position of equilibrium, a current in the opposite

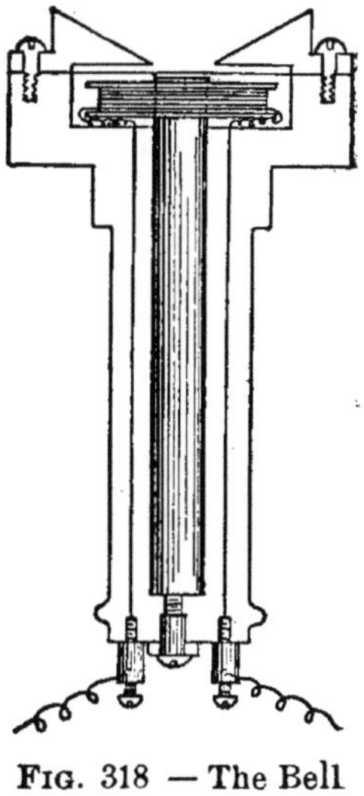

FIG. 318 — The Bell telephone.

direction is seen. Evidently the distribution of the lines of force which pass through the helix is, in each case, changed in such a way that the total flux through the helix is altered, thus producing an E.M.F. in the helix. When for the push of a finger one substitutes the vibrations of the air caused by the human voice he has the "Bell transmitter." When the two ends of the helix are connected to a pair of line wires having a similar telephone at the distant end, the currents induced in the transmitter will pass through the helix at the distant end and will alternately magnetize and demagnetize — to some extent — the magnet at the receiving end. The effect of these variations in the magnetism of the receiving telephone is to set into vibration the diaphragm of the receiver and reproduce the voice which is talking into the transmitter. Such was the early telephone, the transmitter being practically an alternating current dynamo, the receiver being a motor.

In 1877 it was discovered by Berliner that the efficiency of the transmitter could be vastly improved by allowing the vibrations of the diaphragm merely to vary the resistance of a battery circuit. The battery thus furnishes the energy for the induced current and the voice merely controls its distribution. A form of this instrument, which is due to Blake and which has been widely used in the United States, is shown in Fig. 319. The diaphragm *D* presses a small platinum point against a carbon disk *C*; this diminishes the resistance and hence increases the current in a local battery circuit; the variations of the local current, acting through a small transformer coil, send out over the line the induced currents, which reproduce at the receiver each vibration of the transmitter disk.

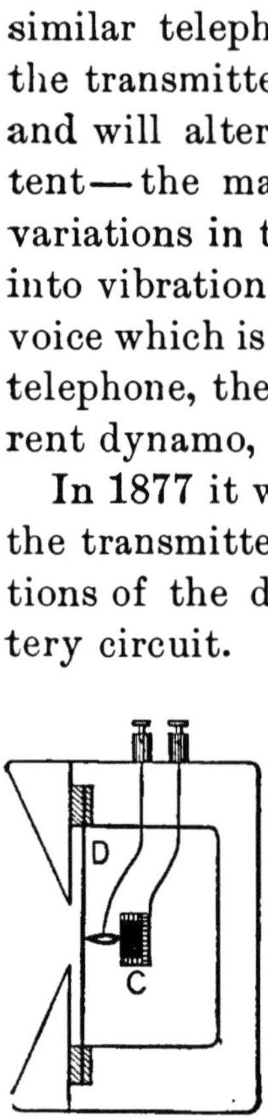

FIG. 319. — The Blake transmitter.

The essential principle which the world owes to Graham Bell and to Elisha Gray is that the successful reproduction of human speech requires a continuously varying, instead of an intermittent, current.

413. With this brief outline, we must leave the vast domain of electrical science, only reminding the student that before him lie many beautiful fields of inquiry into which we have not yet taken even a hasty glance. Among these is the problem of alternating currents, of which direct currents are, as it were, a mere special case. Telephone engineering — perhaps the most difficult of all the applied sciences — is based upon a study of alternating currents. We have passed over the entire question of electrical waves, an exquisite science, created by Maxwell and Hertz, utilized by Marconi and others for the purpose of wireless telegraphy. To one side also we have left the discharge of electricity in gases, a most attractive study, which bids fair to give us a unitary view of the entire subject in terms of the electron theory.

Problems

1. How long will it take a current of one ampere to deposit a gram of pure silver? *Ans.* 14 m. 55 sec.

2. An ampere is 10^{-1} C.G.S. units; a volt is 10^8 C.G.S. units. Find the heat in ergs developed by a current of one ampere flowing for one second between two points whose potential difference is one volt.

3. How much power in watts (see § 107) will be required to keep up a current of 40 amperes charging a storage battery at 55 volts?

4. The resistance of a 50-volt lamp is 50 ohms. How much heat in joules will be developed in the filament in 10 seconds? What power in watts is required to feed such a lamp? If the lamp is 16-candle power, how many watts per candle power are required?

5. How many amperes will be required to deposit 0.75 gram of copper in 2 hr.?

6. Suppose the current in the preceding problem to be furnished by a battery of gravity cells. How much zinc will be used up in the battery while the 0.75 gram of copper is being deposited?

7. A chain is made of alternate links of platinum and silver. If the links are identical in all other respects, find the ratio of the heat developed in two adjacent links.

8. What weight of water will be decomposed by a current of 10 amperes flowing for 1 hr.?

9. Suppose the earth is used as the return wire on a direct current lighting circuit. If the ground connection is made by use of two iron plates, and if 100 lights each using $\frac{1}{2}$ ampere are employed for 4 hours each day, find the loss of weight at the iron anode in the course of one year.

10. A dynamo supplies a circuit of 600 incandescent lamps. Each lamp requires 100 volts and $\frac{1}{2}$ ampere. What must be the resistance of the leads and the armature in order that the energy loss in them shall amount to 5 % of the total electrical output of the machine?

11. Explain what is meant by a "watt," a "kilowatt," and a "kilowatt-hour."

12. A waterfall is situated 10 mi. from a certain town which is lighted by electrical energy derived from this fall. The power required to light the town is 150 kilowatts. Will it be more economical to transmit this power, using 1500 amperes and 100 volts, or by employing a voltage of 10,000 to transmit 15 amperes? Suppose the resistance of the 10 miles of line wire to be 40 ohms; compute the loss on the line in these two cases.

13. Taking the atomic weight of silver as 108 and the electrochemical equivalent of hydrogen as 0.0000104, find the electrochemical equivalent of silver.

14. A copper wire of given length is replaced by one of aluminum. What cross section must the aluminum wire have in order that a given current will produce no more heat in it than in the copper?

15. Two cells are joined in series with a given resistance and produce a current of 1.5 amperes. One of the cells is now reversed, its poles being interchanged with their previous position. The current is now 2 amperes. Find the ratio of the E.M.F.'s of the two cells.

CHAPTER XI

LIGHT

414. A very large proportion of all that we know concerning the external world is obtained from an examination and comparison of our sensations of sight.

As under the head of Sound we considered those phenomena which produce the sensation of hearing, so now, under the head of Light, we proceed to consider those phenomena which are recognized by our visual sense. And just as in the case of sound we did not consider either the physiological or psychological side of the question, so here **the discussion is limited to light as an external physical process giving us the sensation of sight.** There may be in a room objects to be seen, also an eye to see them; but until light from some source is supplied there is no seeing.

415. Following the method used in previous chapters of this volume, we shall begin and end with a study of some of the fundamental phenomena of light, accompanying the discussion with certain definitions and applications, hoping thus to learn something of modern views concerning the "nature of light."

Of these fundamental phenomena the following seven are selected as being perhaps the more important for our purpose:

I. Rectilinear Propagation.
II. Finite Speed.
III. Diffraction and Interference.
(Light a wave motion.)
IV. Polarization.
V. Reflection.
VI. Refraction and Dispersion.
VII. Interference and Diffraction.

Neither in thought nor in practice are these phenomena separated by water-tight bulkheads. On the contrary, they are, indeed, intimately connected, from which it follows that any grouping of facts such as that indicated above is purely a matter of convenience. For the sake of orienting ourselves and starting together, we shall now take a hasty and rather qualitative glance at the first four of these fundamental facts, and shall then proceed to take up the other three in a more quantitative way.

Some Preliminary Definitions

416. When we are looking at the various objects in a room, it is not difficult, in general, to select those which we might still be able to see in case no light were admitted to the room from the outside. If the room were completely shut off from any outside light, the chairs and tables in it would become quite invisible, while a burning match, a red-hot coal in the fireplace, a lighted gas jet, or an incandescent lamp would each illuminate the room, and at the same time be seen by its own light. **Bodies of this latter class which shine by their own light are said to be self-luminous; while bodies of the former class, being seen only by reflected or diffused light, are described as non-luminous.** Luminous bodies are generally, but not always, hot. A phosphorescent clock face is easily seen at night, but can hardly be called a hot body. The student who recalls his daily duties will find that most of his work is performed by the aid of borrowed, or diffused, light. For instance, you are now reading this page by light which is reflected from the white paper to the eye. Some substances, such as water, air, thin layers of gelatine, many kinds of glass, and some crystals, permit us to see with distinctness other objects through them. But there are other substances, such as iron, wood, ink, through which bodies cannot ordinarily be seen.

Bodies through which one can see other objects distinctly are said to be "transparent"; bodies through which vision is impossible are said to be "opaque." But every one will recall certain other substances, such as milk in thin layers, thin tissue paper, lightly ground glass, even ink in thin layers, thin shavings of wood, through each of which objects are faintly or indistinctly seen. These bodies form a class intermediate between transparent and opaque bodies. **Substances which transmit some light, but not enough for distinct vision, are said to be "translucent."**

If bodies are taken in layers of proper thickness, it may be easily shown that there is no sharp line between the most opaque and the most transparent substances. For metals, such as gold and silver, which are very dense and ordinarily very opaque, become translucent, and even transparent, when taken in sufficiently thin layers, as may be easily shown by depositing a thin film of silver on glass; while, on the contrary, what we call "perfectly clear" water, when taken in sufficiently thick layers, permits almost no light to pass through it. Thus geographers who have studied the bottom of the sea have found that those parts of the ocean which lie at a depth of several kilometers are in almost complete darkness. No substance is either perfectly transparent or perfectly opaque.

I. Light moves in Straight Lines

417. An object behind an opaque screen is invisible to a person in front of the screen. As every one knows from experience, the reason of this is that the light proceeding from any part of, say, a man behind the tree cannot pass through the opaque tree, cannot take a curved path around the tree, and, therefore, cannot reach the eye of the observer. Evidence of this kind, such as we have all been collecting since early childhood, shows us that if light does not travel in straight lines, it travels in lines which are very nearly straight.

One of the simplest illustrations of the rectilinear propagation of light is obtained by placing a naked arc light in front of a white screen. Intercept the light which falls on the screen by means of a sheet of lead foil, placed rather close in front of the arc. If now this sheet of foil be punctured with a pin, an inverted image of the arc will appear upon the screen. Each new puncture will create a new image. The image thus produced is, as regards both size and position, what we should expect if it had been traced by a straight line, always passing through the pin hole, one end tracing out the arc while the other end traces out the image on the screen.

In order to test the matter in another way in the laboratory one may proceed as follows: —

As a source of light, use a lamp, either gas or oil, provided with a tin chimney in which has been drilled, at a point on a level with the flame, a hole from one to two millimeters in diameter. This round hole, near the flame, may be considered as a luminous point, and not only for this, but for many other purposes in the study of optics, is a most excellent source of light. At some distance, say one meter, in front of this point-source clamp a sheet of cardboard with a pin hole in it. The pin hole *P* (Fig. 320) and the source of light *S* are two points which serve to define a straight line. We might think of this line as represented by a thread stretched tightly between these two points. Now place your eye behind the pin hole, and look toward the hole in the opaque lamp chimney. You will find that any opaque object placed on this straight line, *PS*, will render the luminous point invisible. This means that the light from the luminous point can reach the pinhole in the card *only along the straight line* joining these two points.

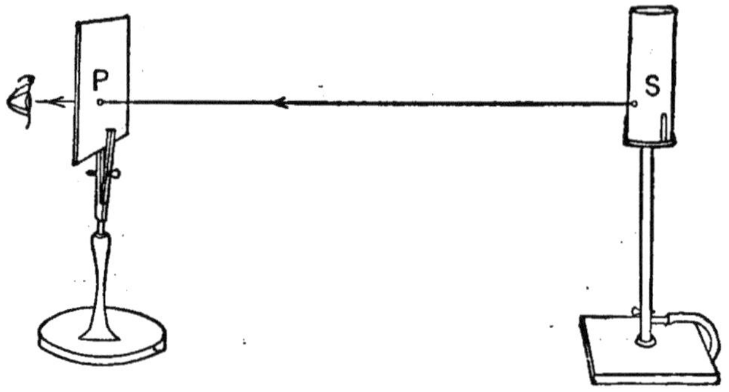

Fig. 320. — Light travels in straight lines.

418. But from the following it will be seen that this statement is not quite true as it stands. Suppose that we place in this straight line an ordinary spectacle lens, as represented in Fig. 321, so that an *image* of the luminous point is produced upon the pin hole on the cardboard. To do this we have placed the spectacle lens so that the straight line *SP* passes approximately through the center of the lens. Let us now interpose a coin somewhat smaller than the lens immediately in front of the lens so that the straight line passes through the center of the coin. We observe that this coin does not prevent all the light from the luminous point reaching the pin hole. The

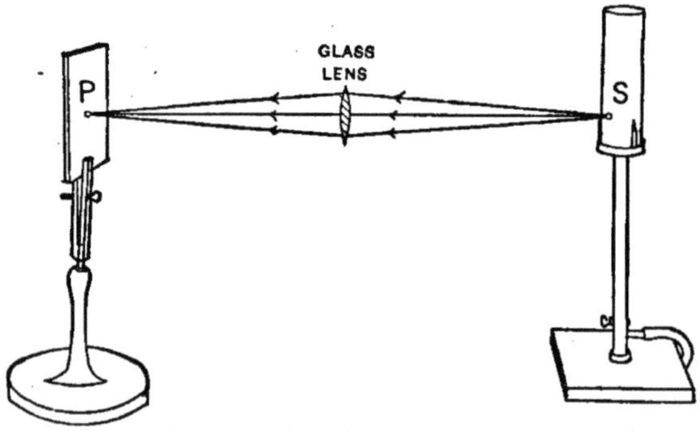

Fig. 321. — If a medium is not homogeneous, light does not travel through it in straight lines.

image still remains at the pin hole, the only difference being that it is now not quite so bright as before. The path of the light from the source to the pin hole is no longer a straight line. All the light which forms the image now passes around the edge of the coin. But the reason of this is very evident, for we have now interposed a piece of glass in the path of the light, and, indeed, it has been found that, in general, when light has to pass through materials of different kinds, it does not travel in the same straight line through them all.

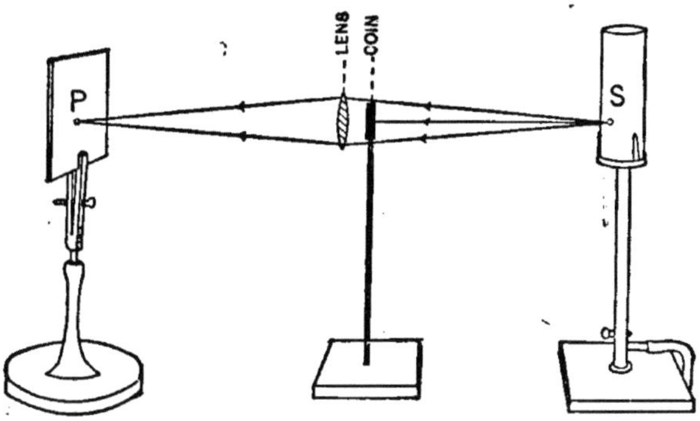

FIG. 322.

If now we remove the spectacle lens and allow the penny to remain, in Fig. 322, we shall find that the coin prevents any light from the source reaching the hole in the cardboard. The entire path of the light is here through air alone.

Definition. A body such as the air in a room, or a piece of pure glass, every part of which is like every other part, is said to be homogeneous.

Accordingly we may say that when the medium is not homogeneous light does not travel through it in straight lines.

419. But there are certain circumstances in which light does not move in straight lines, even when the medium is perfectly homogeneous. If a luminous object be viewed through a very narrow aperture, it will in general be found that the image is distorted. This experiment is most easily realized perhaps by making a narrow slit in the tin lamp chimney represented in Fig. 320. Viewed by the naked eye the slit appears to have a certain width. But now examine it through a narrow slit — a single cut made with a pocket knife — in a visiting card.

The card should be held in the hand very close in front of the eye. The slit *S* appears now very much diminished in brightness, but very much wider than to the naked eye. This phenomenon is still better seen if we substitute for the visiting card a single straight cut made on the opaque film of an ordinary photographic plate. A small piece of a lantern-slide is best; for it is difficult to cut a slit in the film of a "fast" plate without making the edges ragged. The effect of a small hole on a bundle of rays passing through it is then to spread the rays out, so that the slit S appears to have a width CD. For the source S would not appear to have the width CD unless the rays were spread out. The ray PH, prolonged backward, gives us one edge of the apparent slit; the ray QK, prolonged backward, gives us the other. For it is a general fact that we appear to see any object, not in the direction in which the rays leave the object, but in the direction in which the rays enter the eye. The light from the gas flame does not, therefore, pass in straight lines through the small opening HK.

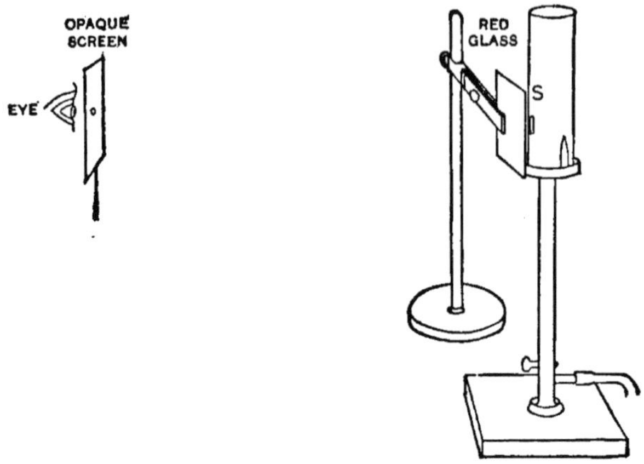

Fig. 323. — A luminous line appears widened when seen through a small aperture.

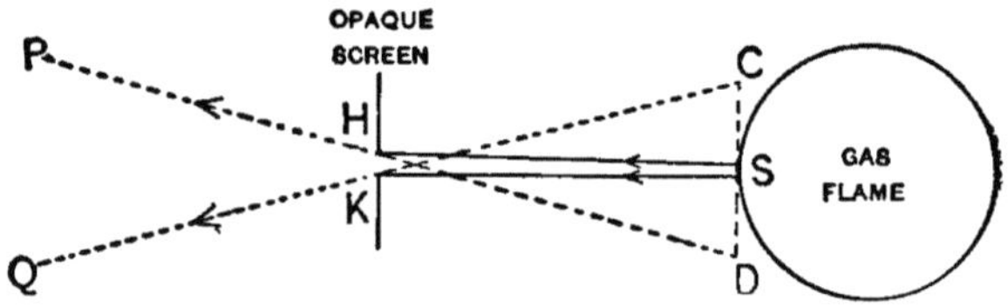

Fig. 324. — Horizontal section through preceding figure.

From the preceding experiments, it is evident that light travels in straight lines: —

(i) **When the medium is homogeneous**; and

(ii) **When the rays are not compelled to pass through any very small openings.**

The most accurate evidence for this general conclusion is, however, derived from the computation of lenses, mirrors, and other optical instruments. Predictions which rest upon the assumption that light travels in straight lines through homogeneous media and large openings have always been verified by the experimental result.

II. Light travels with Finite Speed

420. Passing now to the second fundamental fact of optics, it is not surprising that the ancients should have considered the velocity of light as infinite. For every part of a landscape illuminated by a flash of light is seen apparently at the same instant as the flash. It is evident that if light did not travel at an enormous speed, this would not be the case. If, for instance, light traveled with the same speed as sound, it would be a full second after the flash of lightning before we could see points as far away as 1200 feet; and four seconds later we could see points a mile away. It was such evidence as this, perhaps, that led the ancients to think that light traveled with an infinite speed.

421. But Galileo suspected that the velocity of light was not infinite, and proposed to *measure* it as follows: Two observers, each provided with dark lanterns, were to be stationed on two distant hilltops at night. When the first observer, at *A* (Fig. 325), sent a flash of light from his lantern, the second observer, at *B*, was to make a similar signal as soon as the first flash was perceived. The interval between sending the first signal and receiving the answer might be supposed to be the time taken by the light in making one round trip between the hills. This experiment was tried by the Florentine Academy.

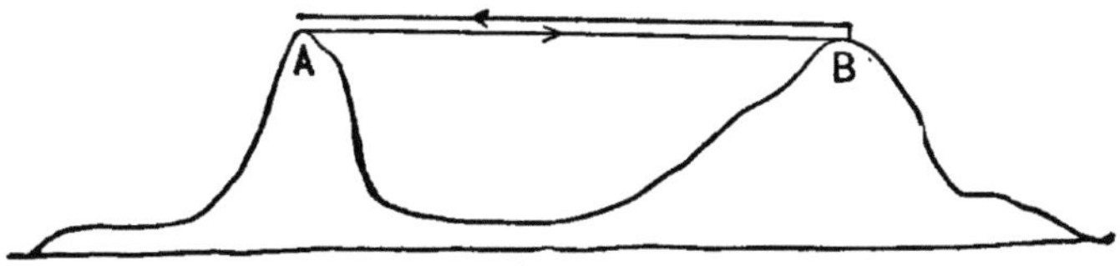

Fig. 325. — Galileo's plan for measuring the speed of light.

The conclusion was that the time required was not appreciable. We now know, however, that the time lost by the second ob server in perceiving the first flash and deciding to send a return signal would occupy more time than is required for light to make a trip around the world.

422. But the following most ingenious method was discovered and tried by the Danish astronomer, Ole Roemer, during the years 1675–1676 at the Paris Observatory.

Jupiter has seven moons, one of which is larger and brighter than any of the others, and is called the "first satellite."

Every time this moon M revolves about Jupiter J it is eclipsed to an observer on the earth, *i.e.* it passes into the shadow of Jupiter, as indicated in Fig. 326. Roemer measured the period of this first moon very carefully, *i.e.* determined very exactly the *average* interval of time between two of its successive eclipses. Knowing this, he could of course predict the time of future eclipses. This he did; but on comparing the predicted times with the observed times, he found that whenever the earth was in that half of its orbit next to Jupiter ABC (Fig. 326), these eclipses occurred each a little earlier than the predicted time. But when the earth was in that half of its orbit away from Jupiter, *i.e.* in the half marked ADC, the eclipses were each a little tardy in their appearance. And in particular when the earth was at the point B, nearest to Jupiter, the eclipse occurred about eight minutes earlier than

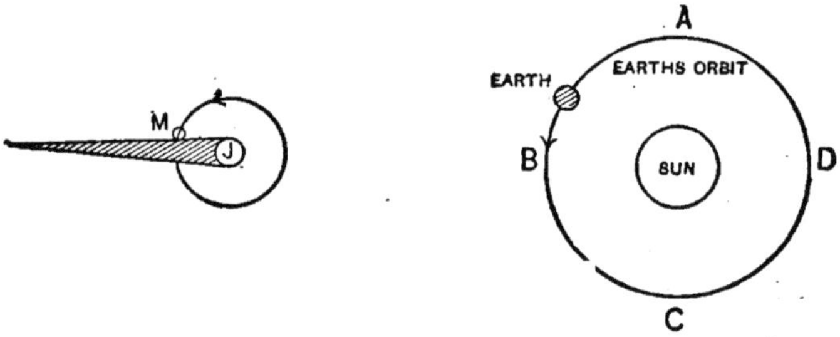

FIG. 326. — Roemer's method for measuring the speed of light.

the predicted time; while at D, a point most remote from Jupiter, the eclipse took place about eight minutes late. From this Roemer concluded that eight minutes is the time required for light to travel from the point B, or the point D, to the center of the earth's orbit, *i.e.* from the sun to the earth. This distance is 154 million kilometers. Hence, for the speed of light S he obtained approximately —

$$S = \frac{154 \times 10^9 \text{ meters}}{498 \text{ seconds}} = 309 \text{ million meters per second.}$$

This, of course, is the rate at which light travels in a vacuum; for the region between the earth and sun is probably a very perfect vacuum. But experience shows that the speed in air is only about one part in four thousand less than that in a vacuum. Such a speed as this, which would carry a point seven times around the earth's equator in one second, is practically beyond the grasp of the imagination.

423. Roemer's measurements remained unconfirmed for half a century, until, in 1729, Bradley, then Professor of Astronomy at Oxford, discovered that each of the "fixed" stars during the course of a year appears to describe a small ellipse in the sky. This he explained by assuming that the apparent position of a star depends not only upon its actual direction from the earth, but also upon the velocity of light and upon the velocity of the observer. The matter will be clear to one who considers how he would have to hold a straight tube in order that raindrops might fall axially through it without touching the sides.

Let us assume that the rain is falling vertically; then if the observer is at rest, the tube must be held vertical. But if the observer is traveling with a speed v in any direction, the bottom of the tube must be held back toward the observer by an amount just equal to the distance which the observer would travel while the raindrop is passing through the tube.

If V be the speed of the raindrop and v that of the observer, it will be clear from Fig. 327 that the angle θ at which the tube must be inclined is given by

$$\tan \theta = \frac{v}{V}. \qquad \text{Eq. 157}$$

The velocity of the earth in its orbit about the sun is approximately nineteen miles a second. This stupendous speed, which would carry one across the American continent in less than three minutes, is the speed of every one who observes a star from the earth. Accordingly, if we imagine the path of the raindrop to be replaced by a ray of light from a fixed star, and the tube to be replaced by a telescope, provided with cross hairs in the eyepiece, we shall have all the conditions for observing the aberration of light. Bradley found that in looking at a star whose rays fall perpendicularly upon the plane of the earth's orbit, as indicated in Fig. 327,

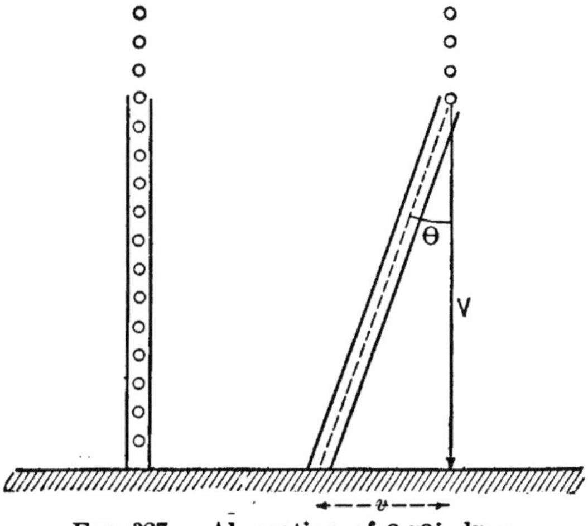

FIG. 327.—Aberration of a raindrop.

he had to incline the telescope at an angle of 20″ with the mean direction of the star.

Knowing θ and v, it became a simple matter to substitute these values in Eq. 157, and thus determine the value of V, that is, the speed with which light travels down the tube of the telescope. The numbers which he obtained were 308,300,000 meters per second. The velocity of light thus measured by Bradley coincided so nearly with the value obtained by Roemer as to leave little doubt that each of these men had correctly explained the phenomena which they observed.

424. More than a century elapsed before any one succeeded in measuring the speed of light along a path confined to the surface of the earth. This was first done by the French physicist Fizeau, in 1848. He employed essentially the method suggested by Galileo, only, instead of placing a *second* observer on a distant hill, he placed there a mirror so arranged as to send back to the *first* observer any flashes of light which he directed upon it.

These flashes of light were obtained by rapidly rotating a toothed wheel at a known rate, and in such a way that each successive tooth intercepted a beam of light directed steadily towards the mirror. Since the wheel was rotated at a measured rate, the time required for any one tooth to pass from its present position to the position occupied by its predecessor was a known quantity. Call this T. When the wheel was rotated at a certain critical speed, it was observed that while the beam was traveling from the source to the mirror and back again an open space between two teeth through which the beam had passed out to the mirror had been replaced by its successor, so that an observer looking along the path of the beam would see the reflected beam. At other critical speeds the reflected beam was eclipsed by an opaque tooth, so that the observer looking in the direction of the reflected beam saw nothing.

If we call the distance from the wheel to the mirror L, and denote by T the interval between two successive flashes, so timed that the light which passes out *just before* any particular tooth is reflected back *just after* that tooth has passed, then we may write, for the speed of light —

$$v = \frac{2L}{T}.$$

In this manner Fizeau obtained a value of 315,000,000 meters per second. But his method has been greatly improved, especially by Cornu and by Michelson. These improvements, together with the advantages and disadvantages of various methods, are clearly set forth in Edser's *Light for Students*.

The best of these modern measurements make it practically certain that the velocity of light does not differ from the round number, 300 million meters per second, by as much as one part in three thousand.

III. Light is a Wave Motion

425. We have already seen (§ 195) that energy is, in general, transmitted in one of two ways, viz. either by currents or by waves. From the time of Aristotle until the time of Huygens, and even later, men believed that light consisted in the motion of small particles—corpuscles—emitted either from the eye or from the luminous body, with great speed. In other words, they believed light to be propagated by currents rather than waves.

This hypothesis is, indeed, recommended by the ease with which it explains the rectilinear propagation of light through a vacuum; but it involves other difficulties which, as we shall presently see, are insuperable. Newton made a masterly attempt to avoid these difficulties; but the hypotheses which he was compelled to introduce led to consequences which contradicted the experimental facts.

426. The first decisive evidence which led men to consider light a wave motion was given by Dr. Thomas Young (1773–1829), a London physician. He argued that if light is a wave motion, we ought to be able to add two trains of light waves together and produce darkness, just as (§ 193) we add together two trains of waves in strings and thereby produce rest. If, however, light consists of currents of matter emitted by the luminous body, as Newton had suggested, one cannot obtain darkness by adding two currents together. On the contrary, the two currents would produce *more* disturbance than either one alone, *i.e.* more light than either one alone.

The manner in which Dr. Young added two rays of light is very simple and clear to any one who has examined the manner in which a beam of light is flared out—diffracted—on passing through a narrow aperture. This phenomenon has already been described in § 419. We there found that a beam of light on passing through a small slit was spread out into a wedge of light.

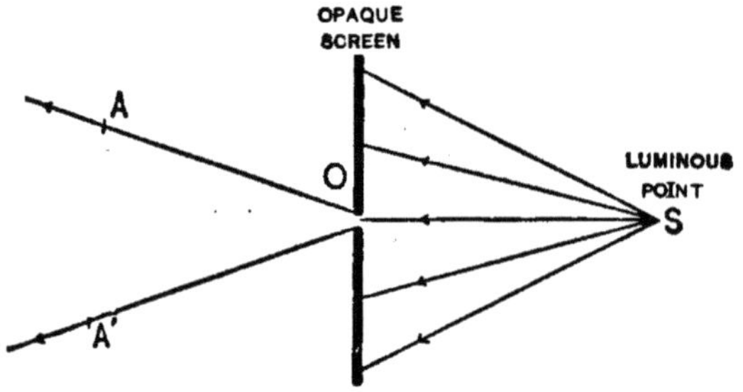

FIG. 328.—Diffraction of light through a single opening.

This spreading apart which a bundle of rays experiences on passing through a narrow aperture is called diffraction.

An eye placed anywhere between the points A and A' (Fig. 328) will receive upon its retina, as upon a screen, one single series of disturbances through the opening at O. The small cone of rays which leaves the luminous point S and passes through the opening O behaves very much as a series of water waves striking against an opening in a breakwater.

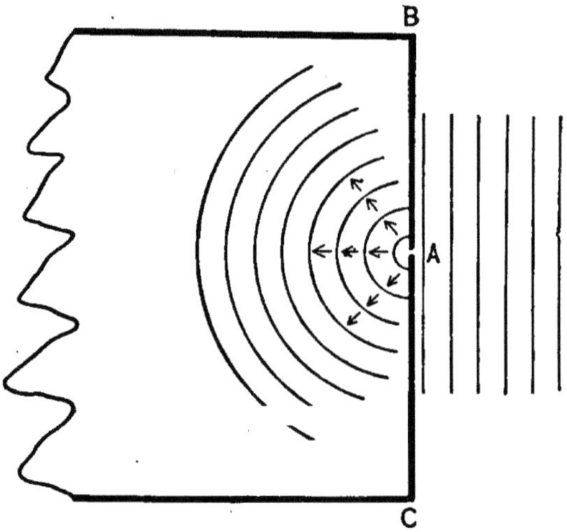

FIG. 329.—The effect produced upon water waves by a small opening in a breakwater.

Let BAC (Fig. 329) be a wall separating the quiet water from the open sea. The crests of the waves, before they enter the opening at A, are straight; when they strike the wall BC only a small portion of the wave enters at A. And it is observed that this small portion of the wave front does not proceed in one straight line, but spreads out so that the successive crests are nearly circular lines, as shown in Fig. 329. The line which is normal to the crest or front of a wave *at any point* indicates the direction in which that wave is traveling *at that point.* This normal is analogous to a "ray" of light, which is a line indicating the direction in which the light travels at any point. This spreading apart of the wave normals in water is also called diffraction.

427. If, in the case of the breakwater, the opening at *A* be large, the wave crests proceed through it and remain straight, as shown in Fig. 330; so also it is found, if we use a large opening in the visiting card (Fig. 323), the light rays proceed straight through without change of direction, *i.e.* the apparent size of the opening in the lamp chimney is not appreciably changed by viewing it through a large hole in the visiting card.

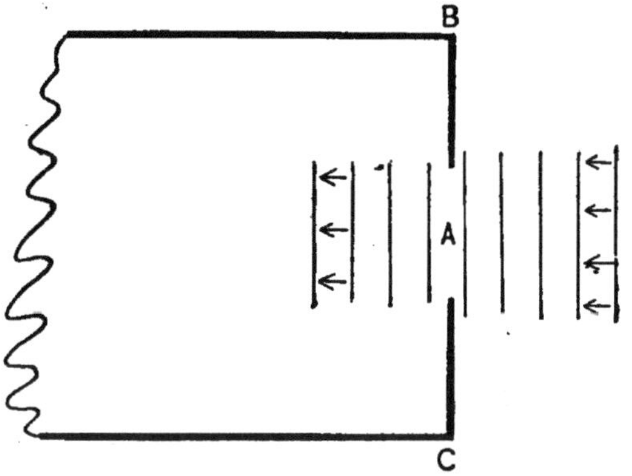

Fig. 330.—*A* is a large opening in a breakwater. Waves do *not* pass a small opening in this manner.

"Some few years ago a powder hulk exploded on the river Mersey. Just opposite the spot there is an opening of some size in the high ground which forms the watershed between the Mersey and the Dee. The noise of the explosion was heard through this opening for many miles, and great damage was done. Places quite close to the hulk, but behind the low hills through which the opening passes, were completely protected, the noise was hardly heard, and no damage to glass and such like happened. *The opening was large compared with the wave length of the sound.*"—Glazebrook, *Physical Optics*, p. 149.

428. What is really seen under the circumstances of Fig. 328 is the image of *S* produced by the small opening at *O*, and by the lens in the eye. Remember, the whole of the region included in the angle AOA' is filled with light.

But if *two* pin holes, instead of a single slit, be made side by side in the opaque screen, each will furnish a cone of light, as represented in Fig. 331. These two cones, AOA' and $BO'B'$, will overlap in the region between

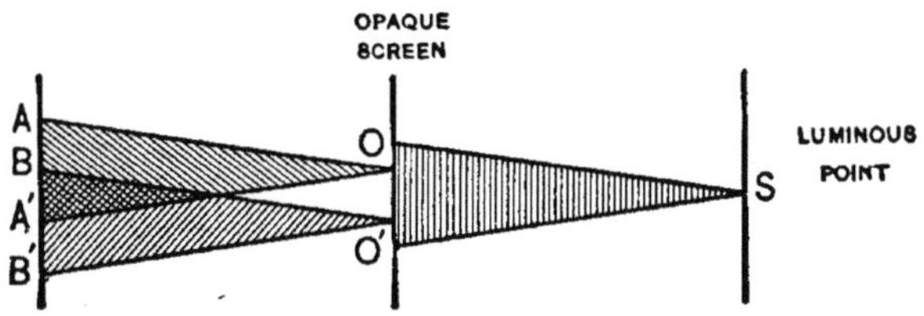

Fig. 331. — Showing how diffraction allows two rays to interfere.

B and *A'*. An eye placed anywhere in this region will receive upon its retina two different series of disturbances, one from *O*, and another from *O'*.

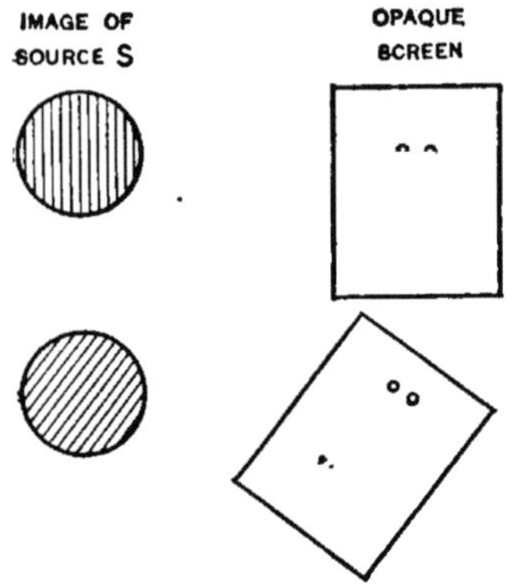

FIG. 332. — Showing that the dark bands depend upon the two pin holes.

Under these circumstances one sees in the region between *B* and *A'* an image of the opening at *S*; and the image is filled with an alternation of bright and dark bands. These bands are symmetrically placed on each side of a bright central band.

It will be observed also that as the opaque screen of cardboard is rotated in its own plane, these bright and dark bands also rotate, remaining always at right angles to the line joining the two pin holes, as indicated in Fig. 332.

429. Accordingly (Dr. Young argued), if light consists in a wave motion, we may expect to find that at any point *P* (Fig. 333) where **the respective distances of the** two openings *PO* **and** *PO'* **differ by one half a wave length, these two** trains of waves **annul one another and** produce **darkness**; but at any other point *Q*, where the two paths *QO* and *QO'* differ by one whole wave length, or any whole number of wave lengths, there the brightness will be greater than that due to either train alone. And this is exactly what he found to be the fact. **Light from a** bright point **shining** through **two** small openings **in a** card **behaves like** water waves from **the sea** rushing **through** two small **openings in a breakwater.** Two trains of waves which combine with each other to produce alternately bright and dark bands in this manner are said to interfere; and the phenomenon which is called interference of light appears to be strictly analogous to the interference of waves in strings and of waves in air which we have already observed and studied.

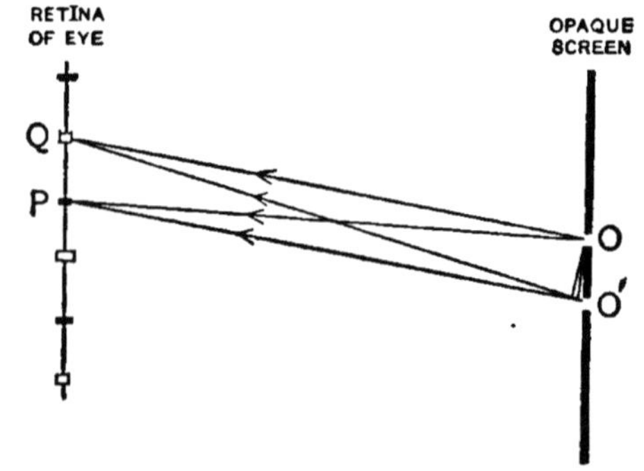

FIG. 333. — Showing the conditions which determine whether an interference band is bright or dark.

These and other similar experiments seem to furnish irrefutable evidence for thinking light to be a wave motion.

IV. Light Waves are Transverse

430. We have already learned (§ 191) that there are in general two fundamental types of wave motion, namely, *transverse* and *longitudinal.* If we accept the evidence which has just been adduced for thinking light a wave motion, the next pertinent inquiry is, are these light waves transverse or longitudinal? Are they similar to the waves which run along a stretched clothesline when struck with a stick, or do they resemble sound waves?

To answer this question it is necessary to first establish a criterion which will distinguish between these two wave types. Such a test is found in the symmetry which accompanies longitudinal waves and the asymmetry which is characteristic of transverse waves. When a spherical raindrop is viewed *along the direction of its path*, its appearance is precisely the same however the eye or the raindrop may be turned about its path as an axis; so also, in hearing a sound the same sensation is produced however the ear or the sounding body be rotated about the direction of the sound as an axis. In general, a longitudinal wave is symmetrical about its direction of propagation.

With transverse waves such as those in a stretched string, the matter is quite otherwise. Let two boys stand holding a rope stretched between them, each boy holding one end. If one boy shakes his end of the rope transversely in a single direction it is an easy matter for the other boy to tell what this direction is merely from the wave pulses which strike his hand. In other words, the waves in the string are not symmetrical with respect to the direction of their propagation, *i.e.* with respect to the direction of the string. If a third boy were to stand between the other two and hold a slotted board so as to allow the string to pass through the slot, the vibrations may or may not be stopped. If the direction of the slot is parallel to the motion of the particles of the rope, the vibrations will pass through the slot undisturbed; but if the slot is held perpendicular to the motion of the particles of the rope, the vibrations will be at once stopped in that part of the rope behind the slot; no waves can pass through the slot.

431. In 1669 it was discovered by **Bartholinus**, a Danish philosopher, that a single ray of light on passing through a crystal of calcium carbonate (calcite) is split up into two separate rays traveling in different directions. So that if one lays a piece of transparent calcite over an ink spot on a piece of white paper, he will in general see two images of the spot. This phenomenon, which is called double refraction, now concerns us only because a few years later Huygens (1629–1695) discovered that each of these two rays refracted by the calcite exhibits precisely the phenomenon which we have seen in the stretched string. In other words he found that, on looking through one piece of calcite at a beam of light which had just passed through another piece of calcite, the intensity of the light which reaches the eye depends very much upon the angular position of the crystal about the ray of light as an axis. The ray of light is not therefore symmetrical about its direction of propagation

432. An experiment due to the French physicist Biot illustrates this fact in an even simpler way. Biot discovered that tourmaline, unlike most other crystals, gives only a single refracted ray, the second ray being absorbed by the substance of the crystal. When light is passed through one tourmaline and then examined through a second tourmaline, it is found that the light may be completely extinguished by rotating either tourmaline, with reference to the other, about the ray as an axis (see Fig. 334). Here again light waves exhibit asymmetry with reference to their direction of propagation.

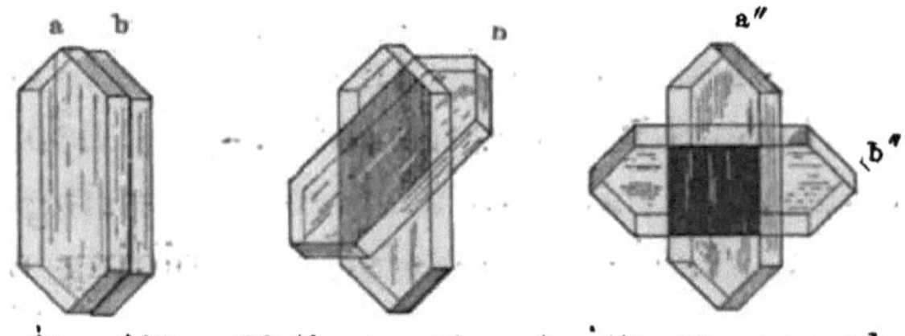

FIG. 334.—Biot's experiment with the crossed tourmalines.

433. Another illustration of this asymmetry was discovered in 1808 by **Malus**, a French army officer, who found that light which has been reflected from a surface of water or from glass or from any polished surface, when viewed through a tourmaline, has its intensity changed in a marked manner as the tourmaline is rotated in front of the eye. He showed also that for each substance there is a certain angle at which, if the light be reflected, it may be completely extinguished by the tourmaline. In other words he discovered that *reflected* light presents this same lack of symmetry which had already been observed by Huygens and Biot in the case of transmitted light.

Malus in 1811 introduced the word "polarized" to describe light which exhibits different properties in different directions at right angles to the line of propagation. This peculiar asymmetry he called polarization. Common light, such for instance as that by which you are reading this page, does not possess this peculiarity. The explanation of this must be taken up later.

As might have been predicted from Malus's discovery, polarization can be detected, as well as produced, by reflection. This is best shown perhaps by a pair of glass mirrors, such as those drawn in Fig. 335. The incident ray is so directed as to make an angle of 57° with the normal to the mirror. In the *left-hand* figure the mirrors are parallel, the vertical beam is polarized by reflection at the lower mirror; the upper mirror reflects it along the direction *BC*. In the right-hand figure the mirrors are crossed, *i.e.* at right angles to each other, the vertical ray is polarized precisely as before, but the upper mirror in this position is unable to reflect any ray.

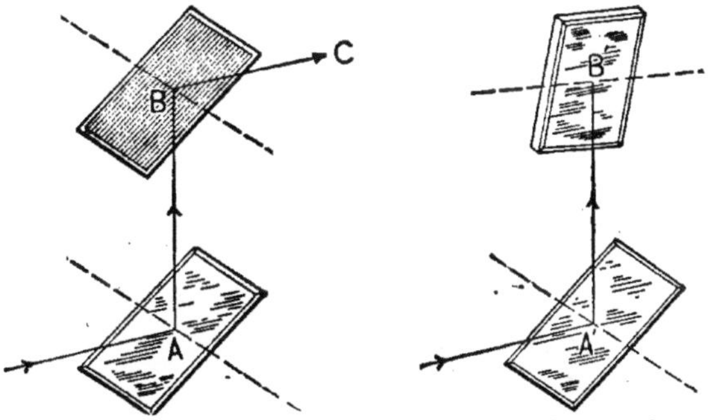

FIG. 335. — Polarization produced and detected by reflection.

434. From evidence such as the foregoing a brilliant young French engineer, Fresnel, was led to the conclusion that light waves are transverse. For more than a century following the publication of Newton's *Optics*, in 1704, his corpuscular theory had held undisputed sway. But between the years 1812 and 1826 Fresnel produced and coördinated such a mass of cogent evidence for thinking that light consists of transverse waves that there has since been almost as little doubt concerning this principle as concerning any of the principles of ordinary mechanics.

Summarizing, then, we may consider it a fact, well established by such evidence as the four fundamental phenomena just studied, that light is a transverse wave motion propagated through what we sometimes call a "vacuum" — sometimes the "free ether" — with a velocity closely approximating 300 million meters per second.

Upon this assumption, if you prefer, and with this general view of the entire subject, we proceed to examine some of the other fundamental facts of optics. As will be seen later, many of these facts might have been deduced as mere inferences from the four which precede.

V. The Reflection of Light

435. We have just been considering the behavior of light when it has a free path in which to travel. Suppose, however, a beam of light strikes upon some body lying in its path.

If the body has a very smooth surface, it will generally happen that a large part of the light which falls upon it from any direction is sent off in some one other direction. This is called **reflection**. If the body is a rough one, such as ground glass or a piece of newspaper, the light which falls upon it is scattered in all directions. Such light is said to be **diffused**; and the process is called **diffuse reflection**.

But some of the light which is incident upon a body will penetrate the body; this we know because objects can be seen through all substances if only the body is made thin enough. The rays which pass through a body are said to be **transmitted**.

If, however, the body is so thick as to be opaque, then those rays which enter the body never leave it. Such rays are said to be **absorbed**; and the process is called **absorption**.

Reflection

436. The simplest and most elegant manner of studying the reflection of light is perhaps the following. A cross is made of two pieces of pine wood mortised together, as indicated in Fig. 336. One piece is slotted so as to receive a strip of plate glass *G*.

If a pin or an incandescent lamp be placed at *P*, this strip of glass will act as a mirror. To an eye placed somewhere in front of the glass, say at *E*, the image of the pin or lamp will appear at P' We can now easily place a *second* pin or lamp at the point P', so that it coincides with the *image of the first* pin. The second pin at P' should be viewed over the top of the glass strip. The glass strip may well be ground or greased on the rear surface so that the reflection from the first surface only is seen.

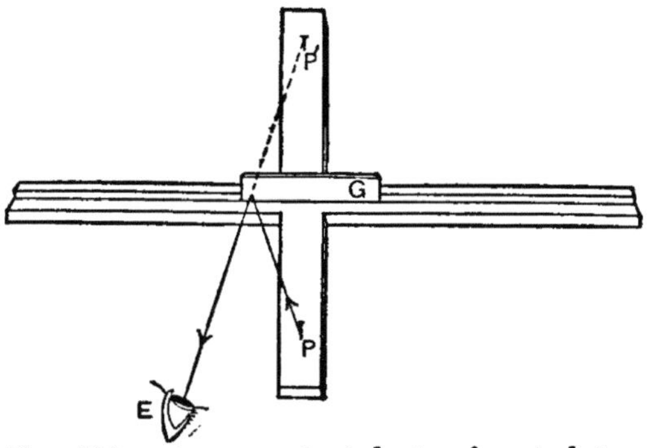

FIG. 336.—A convenient device for studying the laws of reflection.

The first pin now locates the position of an object; while the second pin locates the position of its image, produced by a plane mirror. Concerning this image it is important to note the four following facts:—

1. On observing this image from various points of view, it is seen that the image does not change position as the eye changes position. The image has a fixed position in space quite independent of the observer.

2. If, however, either the object or the reflecting surface be moved, the image also moves.

3. If now we draw a straight line from P to P', we find by use of a square that the reflecting surface is at right angles to this line.

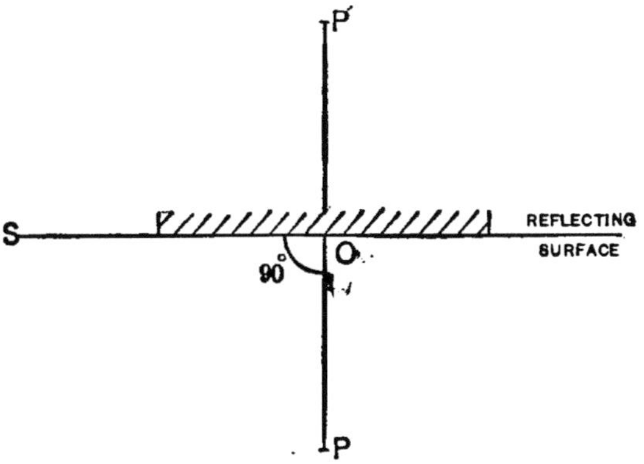

FIG. 337.—$P'O = PO$.

4. Not only so, but on measuring the perpendicular distance $P'O$ from the image to the reflecting surface, and also the perpendicular distance PO from the object to the reflecting surface, it is found that these distances are equal.

These facts, which are familiar to nearly every one, may be summarized by saying that *the image of a point in a plane mirror lies on the perpendicular let fall from the point to the mirror, and lies as far behind the mirror as the point lies in front of the mirror.*

The Law of Plane Reflection

437. The general fact may be more simply described as follows: —

The point P is sending out rays of light in all directions. Let O' (Fig. 338) be a point at which one of these rays strikes the mirror. Then the two right-angled triangles $P'O'O$ and $PO'O$ are equal, for they have one common side, OO', and the sides OP' and OP are equal. Hence the angles $O'PO$ and $O'P'O$ are equal.

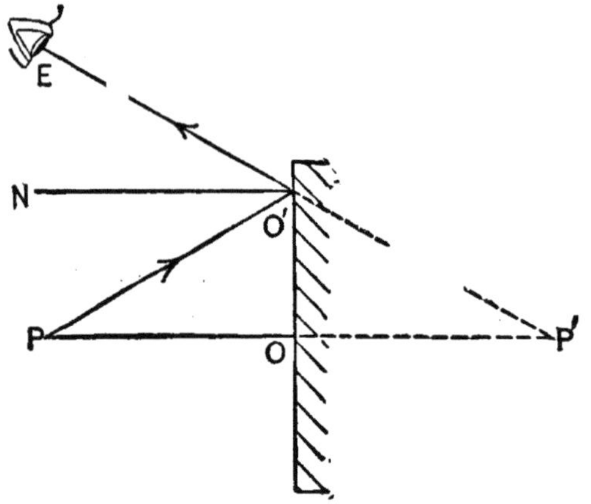

Fig. 338. — Illustrating the law of reflection at plane surfaces.

Draw the line NO' perpendicular to mirror at O'. Then the angles $NO'E$ and $NO'P$ are equal, since they are respectively equal to $O'P'O$ and $O'PO$.

Definitions

The angle between the normal to the reflecting surface and the incident ray $NO'P$ is called the angle of **incidence.** The angle between the normal and the reflected ray $NO'E$ is called the **angle** of **reflection.**

But these are the angles which we have proved to be equal, in the case of the ray PO'; the same evidently is true of any other ray. Not only so, but this experiment shows that the reflected ray lies in the plane of the normal NO' and the incident ray PO'.

Accordingly, if we denote the angle of incidence by i and the angle of reflection by r, we may state the laws of reflection as follows: —

(i) **At each** point of **the** reflecting **surface the** angle of **reflection** is **equal** to the **angle** of **incidence.**

(ii) **The reflected** ray **lies in the plane defined by the normal and the incident** ray.

Or, in terms of algebra

$$i = r. \qquad \text{Eq. 158}$$

But this is precisely the manner in which we have found (§ 183) water waves behaving. When a wave coming from any direction strikes a solid pier at a definite angle of incidence, the reflected wave leaves the pier at the same angle on the other side of the normal. If we employ Professor Hastings'

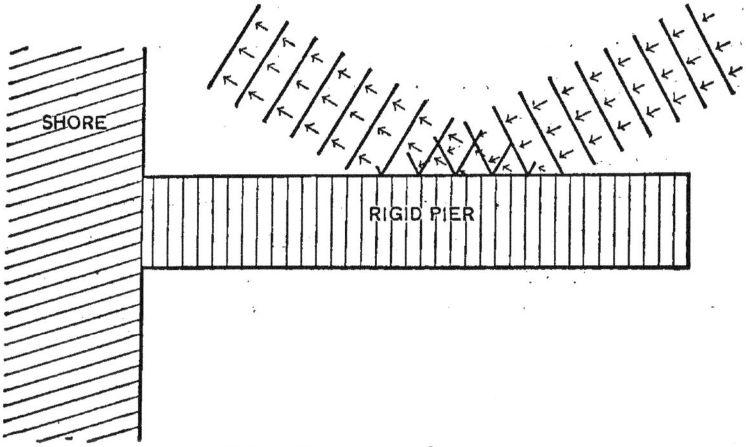

FIG. 339. — Reflection of water waves.

excellent definition of a mirror as "a barrier to the further progress of the waves which does not destroy their motion," it is clear that the rigid pier is a mirror for water waves quite as truly as plate glass for light waves.

Huygens' Explanation of the Laws of Reflection

438. The first man who was able to reduce the reflection of light to anything simpler and more general was Huygens, who, in his *Treatise on Light*, published in 1690, enunciated a general principle which has proved to be of the utmost fertility in almost every branch of optics.

The principle is derived from a consideration of what must happen according to the principles of ordinary dynamics if a wave motion started by any luminous particle is propagated through any material medium, including the free ether. Let *A* denote

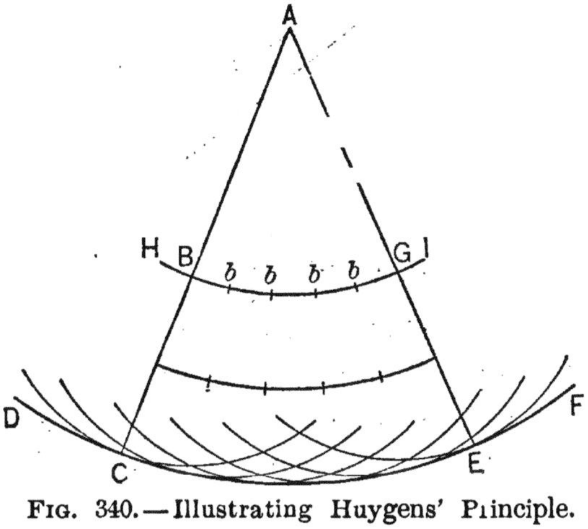

FIG. 340. — Illustrating Huygens' Principle.

a luminous particle, and *DCEF* the position of the front of a spherical wave at any instant *t*. Let *b*, *b*, *b*, *b* denote particles of the medium through which this wave has passed on its way from *A* to its present position. Then *at some instant earlier than t*, each of these particles *b* must be considered as a source of disturbance sending out spherical wavelets of its own. This being true, we must consider the disturbance at any point on the wave front *CE* as the resultant of a large number of disturbances arising from all the particles of the medium over which the wave has passed.

Huygens' Principle consists, then, in the statement that the wave front in a train of light waves is a surface of disturbance which results from and envelops (i.e. is tangent to) the secondary waves sent out by each particle lying in the wave front at an earlier instant.

439. Let us now apply this principle to determine the image of a luminous point in a plane mirror, the case which we have already solved in § 436. If *P* be the luminous point, and *AOB* a plane mirror, a series of spherical waves will be emitted by

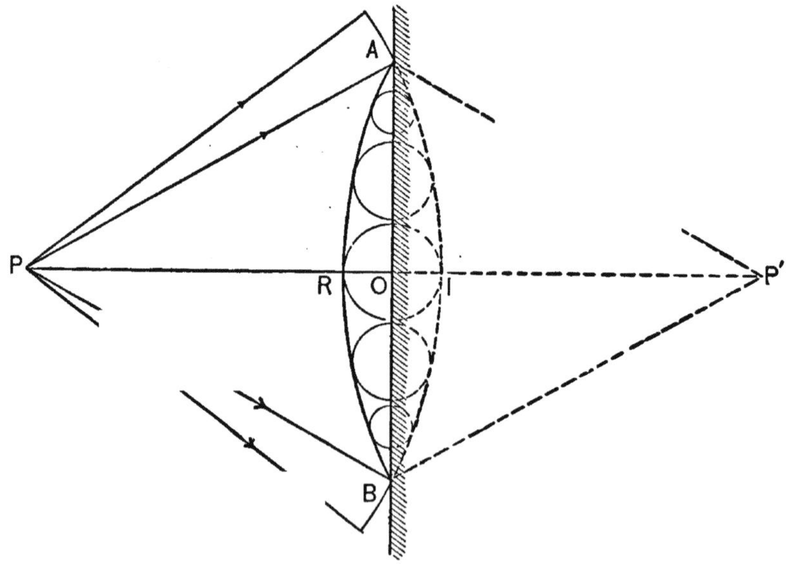

Fig. 341. — Reflection of a spherical wave in terms of Huygens' Principle.

the point *P* and will proceed toward the mirror with the speed of light. The first point on the mirror which will be struck by this advancing wave is the point which is nearest *P*. Call this point *O*. Then *O* becomes, according to Huygens' Principle, a source of light at the instant when the wave first touches the mirror. At immediately succeeding instants, the successive points between O and A on the mirror will be reached by the incident wave front and they will in turn become sources of secondary waves. If there had been no mirror at AB, the incident wave would have gone on and occupied the position AIB. As it is, the spherical wavelets emitted by each successive point on the mirror have for their envelope the spherical surface whose trace is $A'RB$. This envelope is therefore the reflected wave surface. By symmetry the center of curvature P' of ARB is at the same distance behind the mirror that the source P is in front of the mirror. To an observer in front of the glass the reflected light appears to come from the point P'; this point is therefore the image of P.

In this manner Huygens derives from his principle the laws of reflections at plane surfaces. The student will find it profitable to draw a diagram, corresponding to Fig. 341, but showing the particular case in which the luminous point P is at an infinite distance, thus giving rise to a plane wave which is reflected at a plane surface.

The Reflection of Spherical Waves at Spherical Surfaces

440. This is the most general case of reflection met with in ordinary optics, inasmuch as spheres are the only surfaces which can be accurately ground with ordinary machinery. The position of the image can here be easily determined by means of Huygens' Principle; but the method which depends upon the use of rays is more generally useful and practical, for which reason we here adopt it.

In Fig. 342 let SP indicate a spherical mirror whose center of curvature is at C. Let L denote a source of light — a lumi-

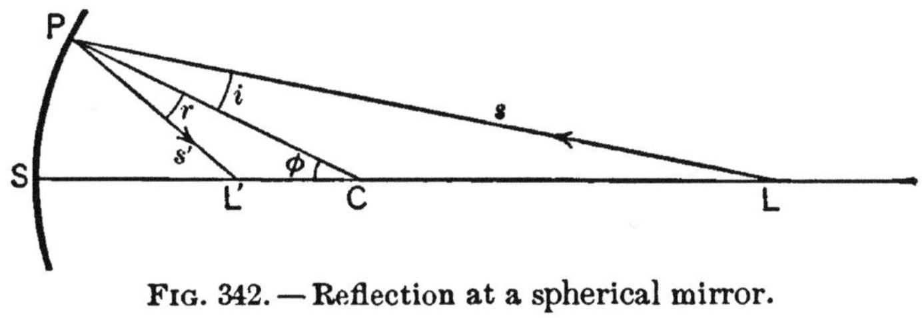

FIG. 342. — Reflection at a spherical mirror.

nous point — and L' its image in the mirror. In general we shall denote points by capital letters and distances by small letters. The line joining the apex of the surface S and the center of curvature C is call the *axis of the mirror.* We shall consider only that case in which the luminous point is situated somewhere on the axis of the mirror.

The center (or apex) of the mirror S is chosen as the origin; and, as in ordinary geometry, distances measured to the right will be called positive, and those measured to the left, negative.

Let us now consider any ray of light LP leaving the source L and incident upon the mirror at P. The mirror surface at P is normal to the radius CP. The reflected ray will therefore cross the axis of the mirror at some point L', such that the angle of reflection $L'PC$ is equal to the angle of incidence LPC.

Let $LP = s$, $L'P = s'$, $CP = r$, and the angle $PCS = \phi$; then if ϕ be small, $CL = s - r$ and $CL' = r - s'$, approximately. By geometry, $\frac{s}{s-r} = \frac{\sin \phi}{\sin i}$ and $\frac{s'}{r-s'} = \frac{\sin \phi}{\sin r}$. Since, however, $\sin i = \sin r$, Eq. 157, it follows that

$$\frac{s}{s'} = \frac{s-r}{r-s'}, \qquad \text{Eq. 159}$$

an equation which gives us s', the distance of the image from the mirror, as soon as we know the radius of the mirror and the position of the source. Clear of fractions and then divide each term in the equation by $ss'r$; you obtain

$$\frac{1}{s} + \frac{1}{s'} = \frac{2}{r}. \qquad \text{Eq. 160}$$

441. Thus if one be asked to find the image of a luminous point placed 50 cm. in front of a convex mirror having a 20 cm. radius of curvature, he has merely to substitute the following values in Eq. 160 : —

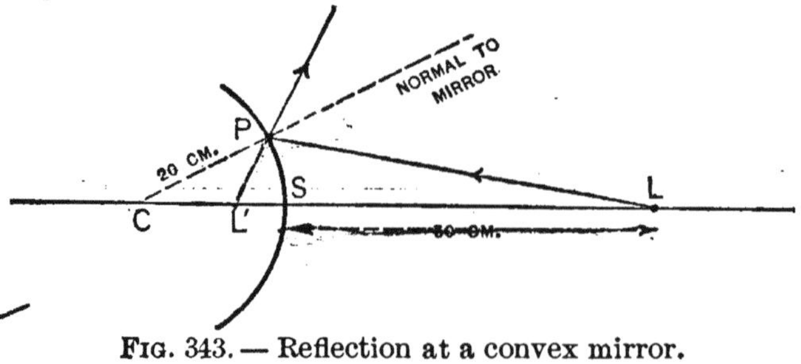

Fig. 343. — Reflection at a convex mirror.

$$s = +\ 50 \text{ cm.},$$
$$r = -\ 20 \text{ cm.}$$

From which it follows that $s' = -8\frac{1}{3}$ cm.; that is, the image lies $8\frac{1}{3}$ cm. to the left of the mirror; in other words, the reflected rays appear to come from an image $8\frac{1}{3}$ cm. behind the mirror.

An image of this kind from which rays merely appear to proceed, but do not actually proceed, is called a virtual image. But when a mirror so changes the direction of a group of rays coming from one point so as to make them intersect at another point and actually proceed from that point, this second point is said to be a real image of the first.

442. The case of reflection at a plane mirror is also included in Eq. 160; for a plane mirror is described by writing $r = \infty$, whence

$$s = -s'. \qquad \text{Eq. 161}$$

In other words, the image is just as far behind the mirror as the object is in front of it, a conclusion which we have already (§ 436) derived from experiment. The two positions L and L', defined by s and s' in the general equation, are sometimes spoken of as "conjugate points."

Focal Length of Mirror

443. If the source L be at an infinite distance, *i.e.* if the incident rays be all parallel to the axis of the mirror, this state of affairs may be described by writing

$$s = \infty$$

from which it follows that

$$s' = \frac{r}{2} = +f. \qquad \text{Eq. 162}$$

The length f, defined by this equation, the distance from the mirror to the image when the incident rays are parallel, is called the focal length of the mirror. The advanced student will find that this definition of focal length is a very special case of a much more general and better definition. See Drude, *Theory of Optics*, p. 21. Introducing this definition into the general equation (160), we may write,

$$\frac{1}{s} + \frac{1}{s'} = \frac{1}{f}, \qquad \text{Eq. 163}$$

from which we may always obtain the position of the image in terms of the focal length of the mirror. From the symmetry with which the two variables s and s' enter this general equation, it is evident that, when the object is shifted so as to occupy the position of its former image, the image will shift to the former position of the object. Values of s and s' being interchangeable, the positions of object and image are interchangeable.

Size of Image produced by Mirror

444. So far we have considered only the position of the image of a luminous point situated on the axis. We now pass to a body of finite size situated on or near the axis. The position of the image of each point of this object will be found exactly as before. But in order to determine the *size* of the image as compared with that of the object, it will be necessary to first observe two very convenient facts, namely:

(i) **An incident ray which passes through the center of curvature of the mirror has its direction unchanged by reflection.**

(ii) An incident ray which is parallel to the axis of the mirror passes, on reflection, through the principal focus of the mirror; and *vice versa*, an incident ray which passes through the principal focus will be reflected in a direction parallel to the axis of the mirror.

With these two rules in mind, we can see at once, from Fig. 344, that the object and the image each subtend the same angle at the center of curvature of the mirror, from which it follows that their sizes are proportional to their respective distances from the center. For let LP_1 be an incident ray from the point of the arrow parallel to the axis; the reflected ray must pass through the principal focus F; and hence the image of the arrow point must lie somewhere on the line P_1F.

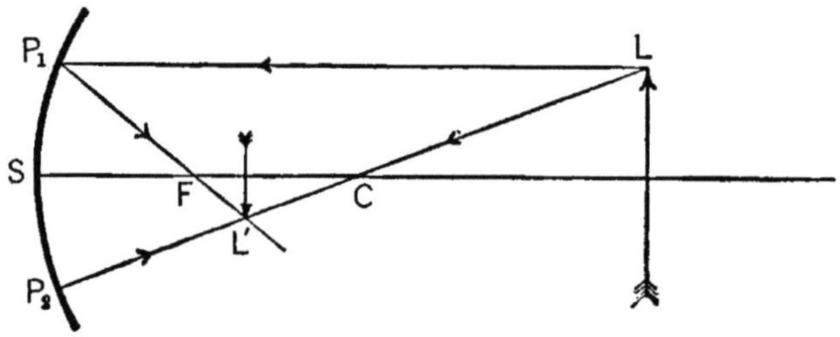

Fig. 344. — Showing how to determine size of image.

Let LP_2 represent an incident ray passing through the center, then the image of the point L must lie also on this line

P_2C. And since it lies on both P_1F and P_2C, it must lie at their intersection, L'. If the angle SCP_2 is not very large, we may equate LC to $s - r$, and $L'C$ to $r - s'$, which enables us to describe the size of the image, as follows: —

$$\frac{I}{O} = \frac{\text{Size of Image}}{\text{Size of Object}} = \frac{r - s'}{s - r}. \qquad \text{Eq. 164}$$

Or, in virtue of Eq. 159,

$$\frac{I}{O} = \frac{s'}{s}.$$

The student will find it an interesting problem to show that this construction holds as well for convex mirrors as for concave; only in the case of a convex mirror it is the *prolongations backward* of the two incident rays which pass through the center and focus respectively.

Problems on Reflection

1. An incandescent lamp is placed 40 ft. in front of a concave mirror whose radius of curvature is 8 ft. Find the position of the image.

2. A candle placed 7 cm. in front of a concave mirror gives an image at 21 cm. behind the mirror. Prove that the radius of curvature of the mirror is 21 cm.

3. Find the focal length of a concave mirror required to produce at a distance of 20 cm. from the mirror an image of an object which is located 40 cm. in front of the mirror.

4. Where must an object be placed in front of a concave mirror in order to give an image which shall be erect and of the same size as the object?

5. Where must an object be placed in front of a concave mirror in order that the image may be inverted and of the same size as the object?

6. An object 6 in. high is placed 12 ft. in front of a concave mirror whose focal length is 2 ft. Find the size of the image.

7. Prove that when an object lies between a concave mirror and its principal focus, the image is always erect and virtual.

8. An object lying 16 cm. in front of a convex mirror gives an image at a distance of 3 cm. behind the mirror. Find the radius of the mirror and then the size of the image.

9. Why do convex surfaces such as polished door knobs and the rounded wooden knobs on furniture give such "high lights" when photographing "interiors"?

10. If you were required to project an arc light upon a screen and to magnify it three times, what kind of a mirror would you use and where would you hold it?

11. Show that, in the case of a plane mirror, the image will move away from the object twice as fast as the mirror moves away from the object. Show also that when the mirror is rotated the reflected beam will rotate twice as fast as the mirror.

12. Two plane mirrors are hinged together and placed so that their polished surfaces include an angle of 90°. Prove that a candle placed anywhere within this right angle will give three images in the two mirrors.

13. Interpret Eq. 159 to prove that, while the size of the image bears to the size of the object the same ratio as their distances from the *center*, the ratio of their sizes is also the same as the ratio of their distances from the *mirror*.

14. Prove that when an object is placed midway between a mirror and its principal focus the image is twice as large as the object.

15. Assume the sun to be 1 million mi. in diameter and 100 million mi. distant. Find the size of the sun's image produced by a concave mirror having a focal length of 52.8 ft.

16. Explain how it is that a man standing in front of a plane mirror can see his complete image by use of a portion of the mirror which is only half as high as he is.

17. Two plane mirrors are placed with their polished surfaces parallel and facing each other. An incandescent lamp placed between them is 6 inches from one mirror and 8 inches from the other. Locate the positions of the first two images in each of the mirrors.

18. An incandescent lamp is moved from a position which is 120 cm. to one which is 80 cm. in front of a convex mirror of radius 10 cm. Find through what distance the image is shifted by this motion of the object.

19. Show that when an arc is shining obliquely on a mirror an object in front of the mirror may cast two shadows instead of one.

20. Two concave mirrors, each of 12 cm. radius, stand facing one another. The distance from the apex of one mirror to the apex of the other is 36 cm. A luminous object is placed 12 cm. in front of the first mirror. Prove that its image in the second mirror after one reflection occupies the same position as the image in the second mirror after two reflections, *i.e.* after being reflected once in the first mirror and again in the second.

VI. The Refraction of Light

445. The first man who succeeded in measuring the speed of light in any medium other than air or "vacuum" was the French physicist Foucault, who in 1850 proved beyond all doubt that light travels at a much slower rate in water than in air or vacuum. Since then Michelson has measured the ratio of the velocity in air to that in water and finds it 1.33. The ratio of the velocity in carbon bisulphide to that in air is 1.77. Each of these results are for white light.

All experiments show that light travels more slowly in any kind of ordinary matter than it does in a vacuum. Kundt, Drude, and others have proved that there are exceptions to this statement under certain highly special conditions. See § 469. We are now prepared to trace some of the consequences of this general result.

Indians in spearing fish show themselves very familiar with the fact that the fish is not exactly where he appears to be in the water, but is always a little lower down. Boys in wading a creek generally learn that they have to roll their trousers higher than at first they thought necessary; for the creek is always deeper than it appears to be. A lead pencil placed in a tumbler of water and looked down upon from one side appears to be sharply bent at the point where the pencil enters the surface of the water.

446. One explanation of these three, and of all similar, phenomena is easily obtained by an experiment which is described by Lord Bacon, but which probably belongs to a century still earlier than his. He places a small piece of metal, say a coin, in the bottom of an opaque bowl; and then gives his eye such a position *E* that he just *cannot* see any part of the coin *P* (Fig. 345). If now water be carefully poured into the bowl while neither the position of the eye nor that of the coin is changed, it will be seen that the coin is lifted into view, as indicated in Fig. 346. The change in direction of the ray *PS*, when it leaves the water and passes into air, is called refraction.

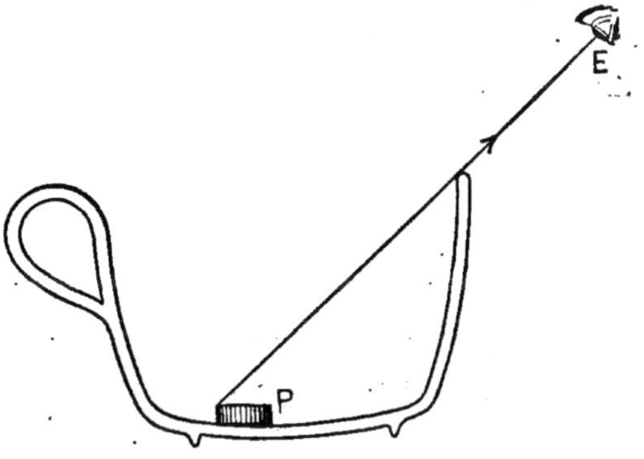

FIG. 345.

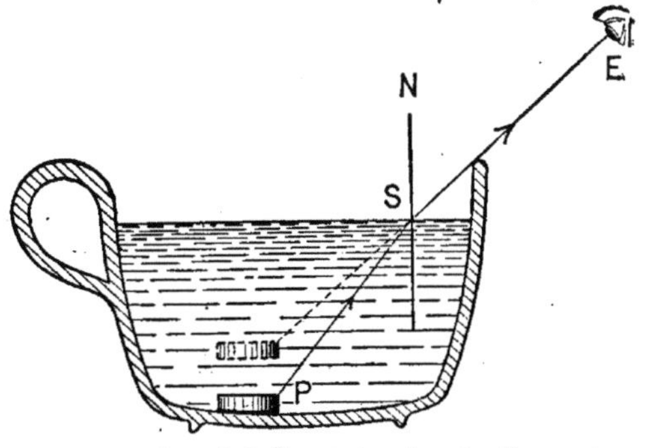

FIG. 346. — Coin is lifted into view by the water.

In general, it is found that the ray emerging from water to air is bent away from the normal *NS*, while a ray which enters the water from the air is always bent toward the normal. The same phenomenon happens when a ray enters the air from glass, as may be easily seen by making a straight scratch *BC* on a small piece of plate glass. If this glass be laid on a sheet of white paper in such a way that the scratch lies in the prolongation of a straight pencil mark *AB*, the observer who views the scratch through the glass will see it appear to shift position to the right or left of the pencil mark according as the scratch is viewed from the left or from the right, *i.e.* the emergent ray is bent away from the normal. And what happens in the case of water and glass happens in general when a ray passes from any one medium to any other medium.

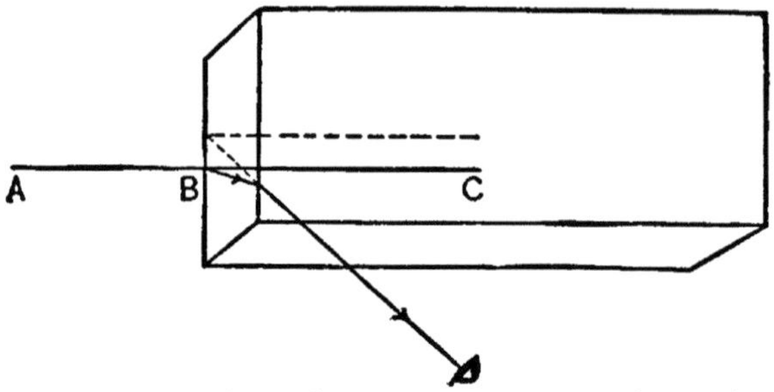

Fig. 347.—Refracted ray is bent away from the normal on emerging from glass into air.

447. The man who first described all these phenomena in a simple manner was Willebrord Snell (1591–1626), a Dutch mathematician. His description is as follows:—

Let *AS* (Fig. 348) be any bounding surface separating two different media, say water and air. Take any point *S* on this surface and describe a circle about it as a center. Let *QS* be any ray incident at *S*; the problem is to find the direction of the refracted ray *SP*. Snell discovered that this could always be done as follows; viz. from *Q* let fall a perpendicular *QL* on the line *NS*, which is drawn normal to the surface at *S*. Then, if we define *PM* as the per-

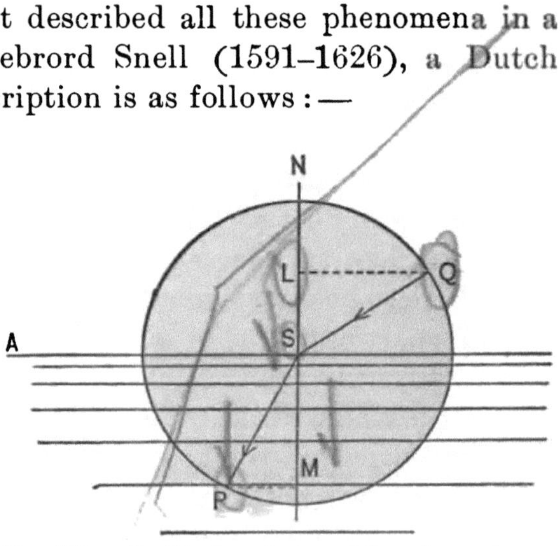

Fig. 348.—Snell's description of the refracted ray.

pendicular distance of P from this same normal, the ratio of QL to PM is constant for any two media. This constant is usually indicated by n, and is called the **index** of **refraction**.

$$QL \,/\, PM = \text{constant} = n = \text{index of refraction.}$$

For water and air this constant is about 1.33; for ordinary glass and air its value is about 1.52. Experiment shows also that the refracted ray SP lies in the same plane with the normal NS and the incident ray QS.

Definitions. The angle QSL, between the incident ray and the normal, we have called the **angle** of **incidence**; it is generally denoted by i. The angle PSM, between the refracted ray and the normal, is called the **angle** of **refraction**; it is generally denoted by r.

The Two Laws of Refraction

Accordingly we may summarize these two results as follows:—

(i) **The angle** of refraction r **bears** to **the angle** of **incidence** i **the** relation

$$\frac{QL}{PM} = \frac{\sin i}{\sin r} = n = \text{constant.} \qquad \text{Eq. 165}$$

(ii) **The** refracted ray lies **in the same plane** with **the incident** ray **and the normal to the refracting** surface.

Explanation of Refraction

448. Assuming a fair familiarity with the phenomena of refraction, we proceed to inquire whether it is like any other phenomenon which we have already studied; whether there is any simpler class of facts to which we can refer it as a special case; whether, in short, we can "explain" the phenomenon of the bending of the rays.

An Illustration of Refraction

Let us suppose that a regiment of soldiers is marching over a country interspersed with plowed fields; and suppose also that the regiment moves with a speed of 3 miles an hour over the smooth ground and 2 miles an hour over the plowed ground. Let AB (Fig. 349) represent the line which separates the smooth ground from the plowed. The arrows indicate the line of march, *i.e.* the arrows are at every point perpendicular to the front of the ranks.

Consider any rank, say No. 8, which is just entering upon the rough ground. Evidently the right-hand end will be slowed up, for the left end is now traveling 1 mile an hour faster than the right — the front will now form a slightly bent line, until the whole rank has crossed into the plowed region; and then the rank will again be straight, **but it will be headed in a direction slightly different from that which it previously had.** It will, in fact, be "refracted." Not only so, but the line of march will be refracted toward the normal to the line *BA*.

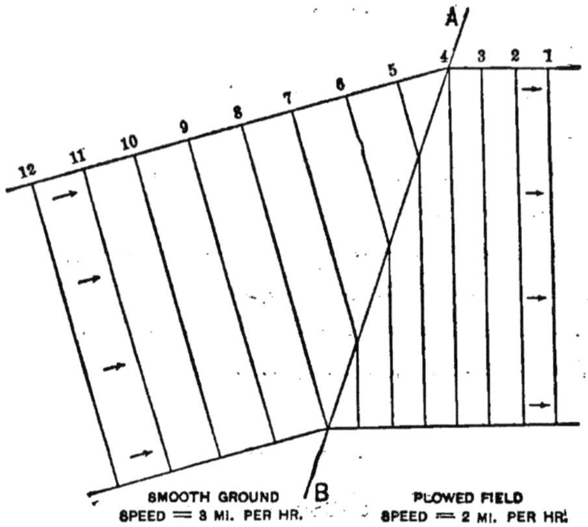

FIG. 349. — Refraction due to change of speed.

We have already seen (§ 445) that light travels more slowly in water than in air. The refraction of rays of light is therefore strictly analogous to the refraction of the line in march, each depending upon a change of speed in passing from one medium to another.

Another Instance of Refraction

449. We have already found (§ 182) by experiment that the speed of water waves depends upon the depth of the water in which they travel — the deeper the water the faster the wave travels. If now we consider a series of waves, whose crests are represented by the straight lines in Fig. 350, it is evident that when these waves cross the boundary line *AB* into more shallow water, they will be deflected just as the soldiers are, and just as the rays

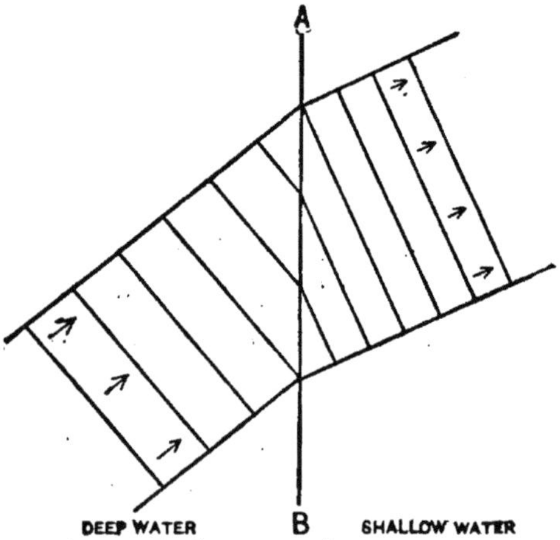

FIG. 350. — Refraction of water waves due to change of speed.

of light are. This analogy between waves and light is essentially due to Huygens.

From what precedes it will be evident that the amount of bending (refraction) which a ray undergoes in passing from one medium to another depends upon the ratio of the speeds of light in these two media. If the speed is the same in each, there is no change in the direction of the ray.

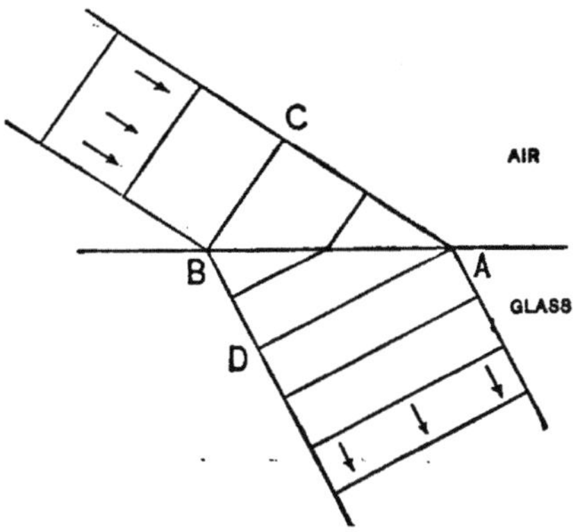

FIG. 351.—The physical meaning of refractive index.

From Fig. 351 it is evident that, if CA represents the distance which light travels *in air*, during the time required for it to travel from B to D *in glass*, we have

$$\frac{\text{Speed in air}}{\text{Speed in glass}} = \frac{CA}{BD} = \frac{\sin i}{\sin r} = n, \qquad \text{Eq. 165}$$

as defined in § 447. For all transparent liquids and solids n is greater than unity.

450. Since n for any particular medium varies inversely as the speed of light in that medium, we may generalize the above equation by writing the refractive index for a ray passing from, say, water to glass, thus,

$$\frac{\text{Speed in water}}{\text{Speed in glass}} = \frac{\sin i}{\sin i'} = \frac{\frac{1}{n}}{\frac{1}{n'}},$$

where n is the refractive index for air and water and n' the index for air and glass; i is the angle of incidence and i' the angle of refraction.

A more significant way of writing this equation is the following: —

$$n \sin i = n' \sin i' = \textit{optical invariant}. \qquad \text{Eq. 166}$$

Total Reflection

451. Up to this point we have been considering the passage of a ray of light from a medium where it travels with a certain speed to one where it travels less rapidly. Let us now reverse the sense in which the light moves so that the incident ray travels along the path of what was formerly the refracted ray. The angle of refraction will in this case, by Huygens' construction, Fig. 352, always be greater than the angle of incidence,

$$r > i.$$

But the maximum value of r is $\frac{\pi}{2}$. Hence the maximum value of i is given by

$$\frac{\sin i}{\sin \frac{\pi}{2}} = \sin i = \frac{1}{n}, \qquad \text{Eq. 167}$$

where n is the refractive index of the first medium with respect to air or vacuum.

We now proceed to consider what happens when the angle of incidence is greater than the maximum value indicated above. The phenomena which occur under these circumstances are familiar to every one who has examined minutely a freshly fallen snowflake and found it to be made up of small but transparent crystals of ice. How does transparent ice acquire the brilliant whiteness of snow? The same phenomenon has been observed by every one who has ever broken up a piece of clear ice with a hammer. The moment the ice is broken into small pieces it appears white. The same is true of finely broken glass. To explain this phenomenon, let us consider any point in the interior of a solid piece of glass. Suppose a ray of light to leave the point P_1 and pass on through the surface AS. The angle of incidence is P_1SM_1. The incident ray has the direction P_1S, but is refracted in the direction SQ_1. Imagine another ray to start from the point P_2; the angle of incidence now becomes P_2SM_1, the angle of refraction increases to, say, Q_2SL_1. As the angle of incidence increases still farther, the refracted ray finally takes the direction SQ_3, that is, the refracted ray just grazes the refracting surface. The perpendicular from Q_1 let fall upon the vertical line NM_1 has now its largest possible value. What will hap-

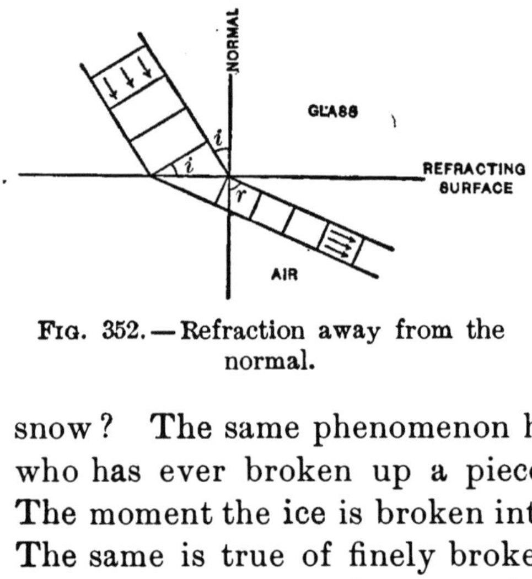

FIG. 352.—Refraction away from the normal.

pen, then, if we increase the angle of incidence to P_4SM_1? If refraction now occurs, the ratio $\frac{QL}{PM}$ can no longer remain constant. It is observed that, **under these circumstances, refraction does not occur.** But if the angle of incidence ever exceeds the

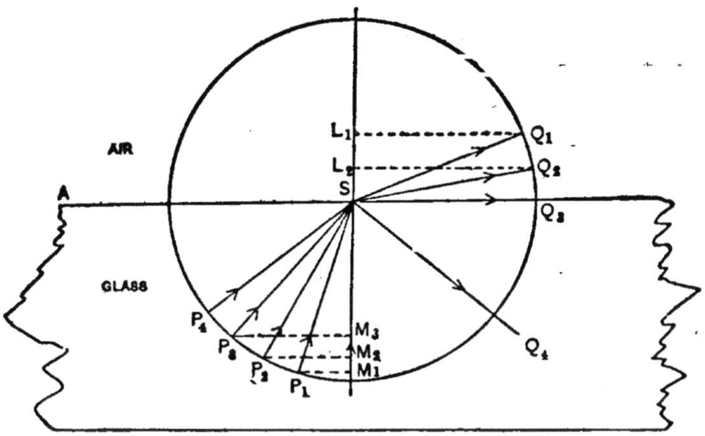

FIG. 353. — The phenomenon of total reflection.

limit P_3SM, the ray is reflected back at S into the glass. This phenomenon is called **total reflection. There is then a certain limiting angle of incidence — called the critical angle — for any ray in glass, defined by the fact that if this angle of incidence be exceeded, the ray will not escape from the glass but will be sent back into it.** This is the angle defined by Eq. 167.

The following table gives the critical angles for a few of the more important substances: —

TABLE OF CRITICAL ANGLES

Diamond	24°
Ruby	36°
Rock Salt	41°
Crown Glass	42°
Turpentine	43°
Water	48°

452. A test tube containing only air and held in a tumbler of water shows total reflection very nicely. Much of the light attempting to pass from the water through the test tube fails, because the angle of incidence in the glass is greater than 42°.

A tumbler of water held a little above the level of one's eyes, as in Fig. 354, readily shows total reflection from the lower surface of water.

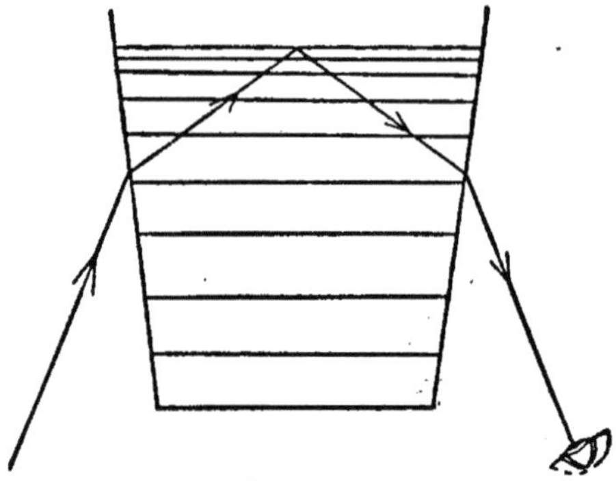

FIG. 354.—Total reflection in a tumbler of water.

We now see how snow appears so white. Any stray ray of light which enters a body of snow will, in general, penetrate only a few of the transparent crystals before it will strike some refracting surface at an angle greater than the critical angle. It will then be totally reflected, and will finally find its way out into the air again. A large proportion (not all) of the rays of light which fall on snow will therefore be sent back. But this is all that we mean by a white body, viz. one which reflects a large portion of all the light which falls upon it. From the preceding table it is clear also why rubies and diamonds are more brilliant than glass bodies of the same size, shape, and polish.

CASE I. *Refraction at a Single Spherical Surface. Axial Rays*

453. Up to the present we have been considering the general law of refraction, which we now proceed to apply to some particular cases which are highly important in the arts.

In Fig. 355 let ***PS*** denote a spherical surface separating two media whose refractive indices are n and n' respectively. We

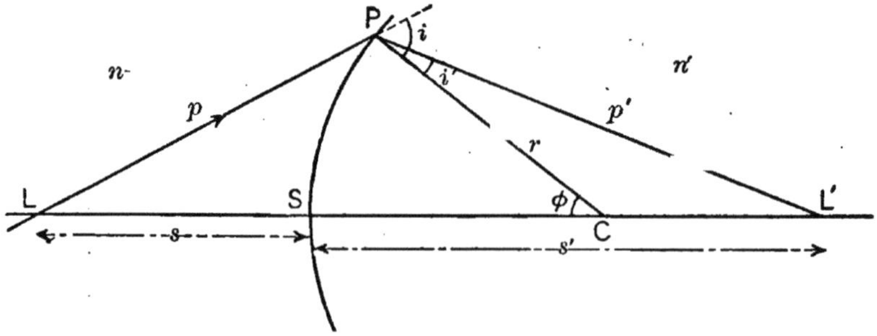

FIG. 355.—Case of refraction at a single spherical surface.

proceed to determine the position of the image produced by refraction at this surface, for which purpose we adopt the following nomenclature:—

Let r = radius of curvature of refracting surface — positive when convex side of surface is toward incident ray,
C = center of curvature,
L = luminous point, source on axis,
LC = axis of refracting surface, an unlimited straight line,
L' = point at which refracted ray crosses the axis, image of L,
S = apex of refracting surface,
s = distance from apex of surface to luminous point; negative to left, positive to right,
s' = distance from apex to image L'; negative to left, positive to right,
$c = s - r$,
$c' = s' - r$,
ϕ = angle subtended at C by the arc SP,
i = angle of incidence,
i' = angle of refraction.

Starting now with the optical invariant, Eq. 166,

$$n \sin i = n' \sin i',$$

divide each side by $\sin \phi$, whence

$$\frac{n \sin i}{\sin \phi} = \frac{n' \sin i'}{\sin \phi}$$

But for axial rays, that is, for those which make a small angle with the axis, we have, by geometry,

$$\left.\begin{aligned} &\frac{\sin i}{\sin \phi} = \frac{s-r}{p} = \frac{s-r}{s} \\ \text{and} \quad &\frac{\sin i'}{\sin \phi} = \frac{s'-r}{p'} = \frac{s'-r}{s'} \end{aligned}\right\} \text{very approximately,}$$

where the meanings of p and p' are evident from Fig. 355.

$$\text{Hence} \quad n\left(\frac{s-r}{s}\right) = n'\left(\frac{s'-r}{s'}\right) = \text{optical invariant.}$$

Dividing through by r, we have

$$n\left(\frac{1}{r} - \frac{1}{s}\right) = n'\left(\frac{1}{r} - \frac{1}{s'}\right) \quad \text{Laboratory equation for spherical surface.} \qquad \text{Eq. 168}$$

As soon as s, the position of the source, is given, this equation enables one to determine s', the position of the image.

Special Case II. *First Medium, Air; Second Medium, Glass, Water, etc.*

454. Here $n = 1$ and $n' = \mu$, whence Eq. 168 becomes

$$\frac{\mu}{s'} - \frac{1}{s} = \frac{\mu - 1}{r}. \qquad \text{Eq. 169}$$

Illustration. Let it be required to find the image of a luminous point, placed 10 cm. to the left of a block of glass ($\mu = 1.5$) bounded by a convex surface of 12 cm. radius as shown in the figure.

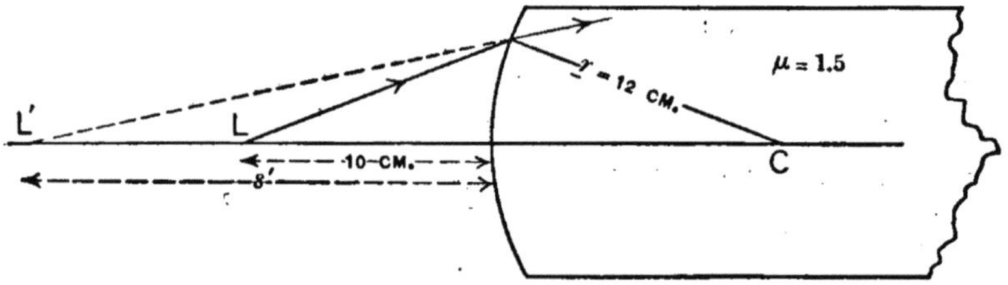

Fig. 356.

Here $\mu = 1.5$

$r = +12$ cm.

$s = -10$ cm. Find s'

Hence $$\frac{1.5}{s'} + \frac{1}{10} = \frac{0.5}{12}.$$

Or $s' = -25.7$ cm.

Accordingly the image is virtual, and lies 25.7 cm. to the left of the surface of the glass.

Special Case III. *Incident or Refracted Rays Parallel. Focal Lengths*

455. Let $s = \infty$, then the incident rays are parallel to each other, and our general equation (169) becomes

$$s' = SF' = \frac{\mu r}{\mu - 1} = f' = \text{second focal length.}$$

Or if, instead of making $s = \infty$, we make $s' = \infty$,

$$s = SF = -\frac{r}{\mu - 1} = f = \text{first focal length.}$$

Observe that the second focal length, *i.e.* the one in the glass, is μ times greater than the first focal length. Note also that these focal lengths have opposite signs, and that the sign of each depends upon the sign of r.

The advanced student will learn a more practical and general definition of focus, introduced by Gauss. In a qualitative way, Gauss' idea of a focus is that of a point such that when the object passes through it, a discontinuity occurs in the position of the image, *i.e.* the image shifts from a position infinitely distant on the right, say, to another position infinitely distant on the left, without passing through the intermediate positions.

Employing the first focal length f, we may write, Eq. 169,

$$\frac{\mu}{s'} - \frac{1}{s} = -\frac{1}{f}. \qquad \text{Eq. 170}$$

SPECIAL CASE IV. *Refraction at a Plane Surface*

456. To describe a plane refracting surface in algebraic terms we have merely to write $r = \infty$. On introducing this condition into our general equation (169) we obtain,

$$s' = \mu s. \qquad \text{Eq. 171}$$

From this it follows that the image in the glass or in the water is μ times as far from the surface as the object is. A fish swimming in the water and looking at a fishing rod above the surface sees it 1.3 times as far away as it really is.

Conversely, since $s = \frac{s'}{\mu}$, an Indian spearing a fish sees it only $\frac{1}{\mu}$, or $\frac{3}{4}$ as far below the surface as it really is. In *Harper's Weekly* for November 16, 1907, will be found an entertaining account, by Professor R. W. Wood, of how that portion of the world above the surface of the water appears to a fish. The photographs were obtained by means of a camera and plate totally immersed in water.

SPECIAL CASE V. *Reflection*

457. By placing $\mu = -1$, and again measuring r from the origin according to the rules of ordinary geometry, we may obtain the entire theory of *reflection* at plane and spherical mirrors; for then our general equation (169) becomes,

$$\frac{1}{s'} + \frac{1}{s} = \frac{2}{r}. \qquad \text{Equation of spherical mirror.} \qquad \text{Eq. 172}$$

If, in addition, we put $r = \infty$, we have the description of a plane mirror,

$$s = -s' \qquad \text{Eq. 173}$$

Special Case II in Terms of the Wave Theory

458. In the preceding demonstration we have assumed only that light travels in straight lines in a homogeneous medium and that $n \sin i$ is an optical invariant from one medium to another. We shall now assume that light consists in waves which travel in each medium with a speed which is inversely as the refractive index μ for that medium, and we shall show that *this assumption leads to results the same as those obtained above.* As before let L (Fig. 357) be the luminous point whose image is sought. Let AIP be the position which any spherical wave front starting from L would have assumed at an instant t, if the glass had not been there. Owing to the presence of

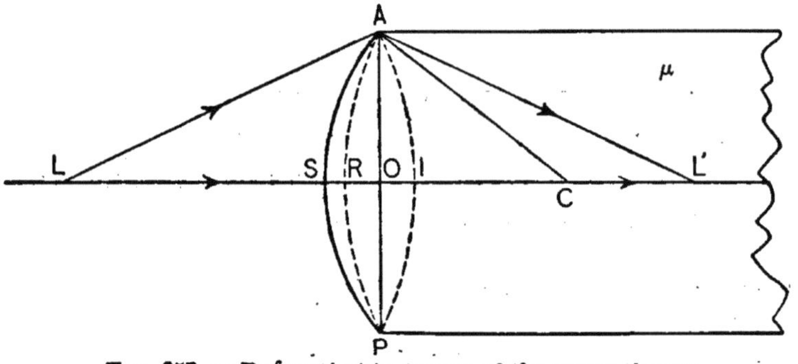

FIG. 357.—Refraction in terms of the wave theory.

the glass, however, the speed of the wave has been very much slowed up. That portion which has been *most* retarded is the part which has been longest traveling in the glass, *i.e.* the center of the wave which was incident at S.

Now on the wave theory the focus which is conjugate to L, that is, the image of L, is simply that point where the disturbances due to each element of the wave front meet in the same phase. In other words the image L' of L is a point such that the light occupies the same time in traveling from L to L', whatever be the path. Evidently this point is identical with the center of curvature of the wave surface in the second medium.

Let ARP denote this wave surface in the glass. Draw the chord AOP perpendicular to the axis LS. Then if a denote half this chord, we may, by simple geometry, write for SO, which is called the "sagitta" of the refracting surface,

$$SO = \frac{a^2}{2r}, \text{ for refracting surface.}$$

And in like manner,

$$RO = \frac{a^2}{2s'}, \text{ for the refracted wave front;}$$

$$IO = \frac{a^2}{2s}, \text{ for the incident wave front.}$$

With these preliminaries, let us now equate the time occupied by the outer edge of the wave in traveling over the path LAL' to the time consumed by the central portion of the wave in going over the route $LSOL'$. If we call the speed of light in air unity, this gives us

$$LA + \mu AL' = LS + \mu SL' \qquad \text{Eq. 174}$$

or

$$LA + \mu AL' = LI - (SO + OI) + \mu(SR + RL').$$

But since $LA = LI = -s$

and $AL' = RL' = +s'$

we have, on substituting and canceling in the above equation,

$$0 = -SO - OI + \mu SR$$

$$= -SO - OI + \mu(SO - RO).$$

Using the sagitta,

$$0 = -\frac{a_2}{2r} + \frac{a^2}{2s} + \mu\left(\frac{a^2}{2r} - \frac{a^2}{2s'}\right).$$

Dividing by $\frac{a^2}{2r}$, and transposing, one obtains

$$\frac{\mu}{s'} - \frac{1}{s} = \frac{\mu - 1}{r}, \text{ which is} \qquad \text{Eq. 169}$$

This equation which we have just obtained by two different methods is perfectly general and holds equally well for concave, convex, and plane refracting surfaces.

Problems

1. Find the refractive index when the angles of incidence and refraction at a plane surface are 45° and 30°, respectively.

2. The critical angle for a certain medium is 45°. Find its refractive index.

3. A pebble lies at the bottom of a pool of water which is four feet deep. How far below the surface will this pebble appear to a man looking vertically down through the water? *Ans.* 3 ft.

4. An air bubble imprisoned in a piece of glass appears to lie $\frac{4}{19}$ cm. below the surface of the glass, which has a convex surface of 20 cm. radius. If the refractive index of the glass is 1.5, what is the real distance of the air bubble below the surface? *Ans.* 0.3 cm.

5. In what direction must an incident ray strike a spherical glass surface in order that this direction may not be changed by refraction?

6. A luminous point is placed $6\frac{2}{3}$ cm. to the left of a surface of glass of which the refractive index is 1.6. The image which is virtual lies 16 cm. to the left of the surface. Find the radius of curvature of the glass surface. *Ans.* + 12 cm.

7. Show from the general equation (169) that there are two positions on the axis at which a luminous point may be placed such that, after refraction at a single spherical surface, the image will coincide with the object.

8. What refractive index will be required in a glass ground with a concave surface of 40 cm. radius in order that a point situated in air 25 cm. to the left of the concave surface may be brought to focus at a distance of 90 cm. to the right of the surface? *Ans.* $\mu = 1.8$.

9. A microscope is focused upon a scratch on the upper surface of a flat-bottomed glass beaker. Liquid is now poured into the beaker to the depth of y cm., when it is found that the microscope must be lifted through a distance h in order to remain focused upon the same scratch. Prove that the refractive index of the liquid μ is given by the following equation,

$$\mu = \frac{y}{y - h}.$$

10. From the general equation (169) find the two principal foci for a refracting surface S, such as that indicated in Fig. 358.

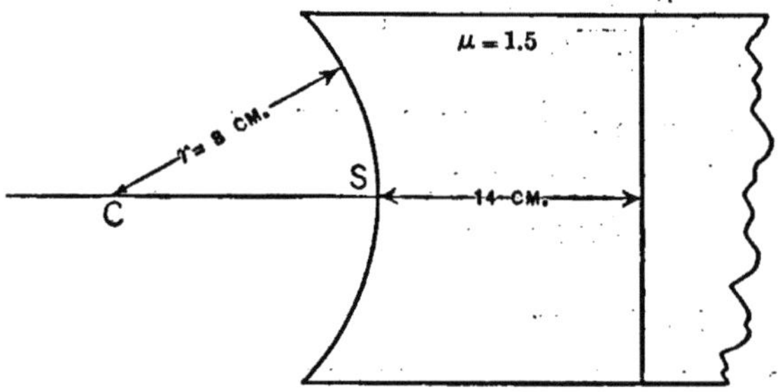

FIG. 358.

11. Referring to Fig. 358, the block of glass, instead of being left unfinished at the right hand, is ground to a plane perpendicular to the axis and is silvered. A small incandescent lamp is placed 48 cm. to the left of the surface S. Find the position of its image after refraction at S and reflection at the silvered surface.

Ans. 10 cm. to right of S.

CASE VI. *Refraction of a Spherical Wave Surface at Two Refracting Surfaces in Succession. Lenses*

459. DEFINITION. **A portion of a refracting medium which is limited by two curved surfaces is called a** lens.

The only surfaces which can be ground in lathes with accuracy and ease are spherical surfaces, hence all lenses, practically, are made with spherical surfaces, *i.e.* the shape of a lens may be considered as the shape of a figure bounded by two spheres.

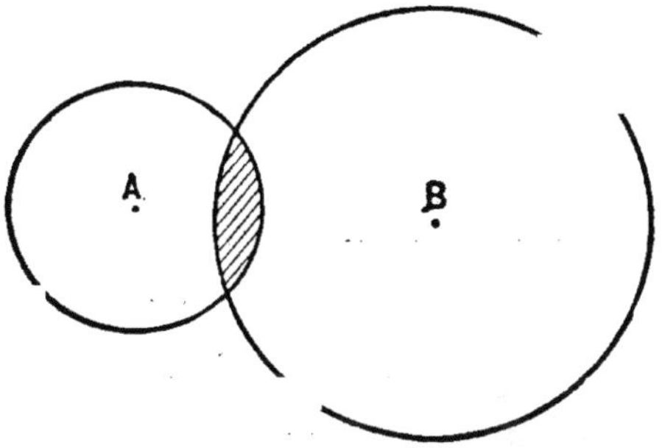

FIG. 359. — A converging lens.

DEFINITION. **The line joining the centers of curvature of the two refracting surfaces is called the axis of the lens.**

If one face of the lens is a plane, we may consider this plane as a portion of a sphere of infinite radius. The axis of such a lens is the line through B (Fig. 360) normal to the plane surface.

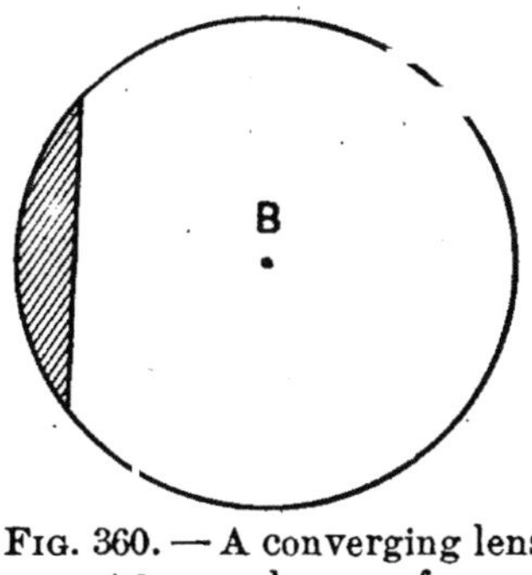

FIG. 360. — A converging lens with one plane surface.

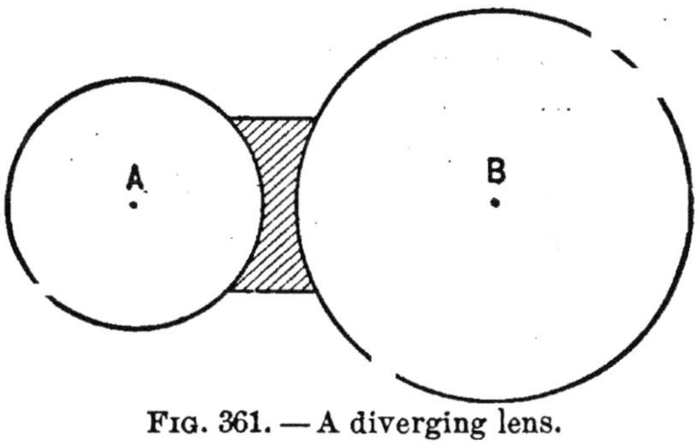

FIG. 361. — A diverging lens.

It will be observed that the axis of a lens is normal to each surface of the lens at the point where it passes through the surface.

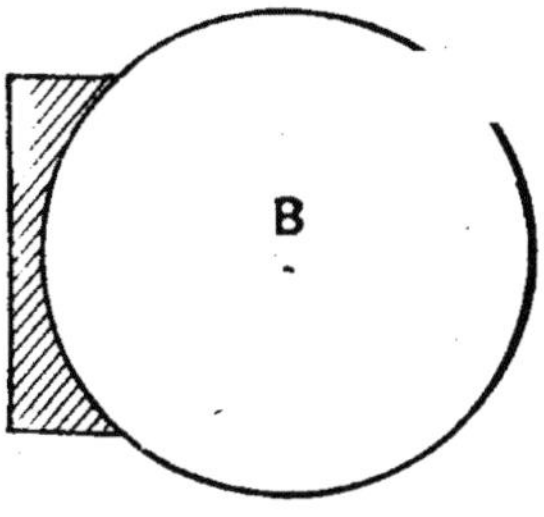

FIG. 362. — A diverging lens with one plane surface.

We proceed now to trace the history of a wave surface through a lens defined as above. And for the purpose of obtaining a general solution, we shall choose a lens limited by two spheres of radii r_1 and r_2 which are each positive, as indicated in Fig. 363. For, as before, we shall consider the radii positive, when the convex surface is turned toward the incident light. Here also we shall use the same nomenclature as before, following the conventions of ordinary geometry and allowing the subscript "1" to refer to the first surface, and the subscript "2" to the second surface.

The first problem is to obtain the position of the image L' due to a luminous source L; in other words to find s' in terms

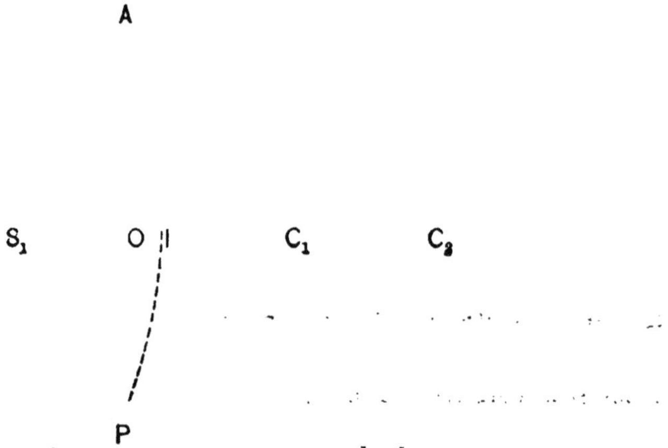

FIG. 363. — Illustrating refraction through a medium bounded by *two* spherical surfaces.

of s, r_1, r_2, and μ. This is most simply accomplished as in § 458, by equating the times occupied by two different parts of the wave surface in passing from L to L'. For this purpose the most convenient portions of the wave surface are the center and the extreme edge. Again calling the speed of light in air unity, we have

$$LA + AL' = LI - S_1I + \mu S_1S_2 + S_2L' \qquad \text{Eq. 175}$$
$$- LI - (S_1O + OI) + \mu(S_1O - S_2O)$$
$$+ S_2O - RO + RL'$$

But since $LA = LI = -s$
and $AL' = RL' = +s'$ } approximately for small pencils,

we have $0 = - S_1O - OI + \mu(S_1O - S_2O) + S_2O - RO.$

Introducing now the value of the sagitta, § 458,

$$0 = -\frac{a^2}{2r_1} - \frac{a^2}{2s} + \mu\left(\frac{a^2}{2r_1} - \frac{a^2}{2r_2}\right) + \frac{a^2}{2r_2} - \frac{a^2}{2s_1}.$$

Dividing by $\frac{a^2}{2}$ and transposing, we obtain

$$\frac{1}{s'} - \frac{1}{s} = (\mu - 1)\left(\frac{1}{r_1} - \frac{1}{r_2}\right), \qquad \text{Eq. 176}$$

which is a perfectly general equation *for pencils of small aperture coming from points near the axis and incident upon thin lenses.* If r_1 is positive and r_2 negative, we have a double convex lens; but if r_1 is negative while r_2 is positive, we have a double concave lens. Eq. 176 describes completely all the following special cases, which need therefore be merely indicated.

SPECIAL CASE VII. *Incident rays parallel. Focal length*

460. Let $s = \infty$.

Hence $s' = \frac{1}{\mu - 1}\left(\frac{r_1 r_2}{r_2 - r_1}\right) = f$, say = second focal length.

Let $s' = \infty$.

Then $s = \frac{-1}{\mu - 1}\left(\frac{r_1 r_2}{r_2 - r_1}\right) = -f$ = first focal length.

In terms of f, the general equation (176) becomes

$$\frac{1}{s'} - \frac{1}{s} = \frac{1}{f}. \qquad \text{Eq. 177}$$

The point at which the refracted ray cuts the axis when the incident ray is parallel to the axis is called the principal focus. The power of a lens, which we may denote by P, is then defined as **the reciprocal of the focal length.** For, the shorter the focal length of a lens, the greater the change it produces on the direction of rays passing through it; hence its power is said to vary inversely as its focal length.

$$P = \frac{1}{f}. \qquad \text{Eq. 178}$$

Defining equation for power of a lens.

A lens whose focal length is 1 meter is said to have a power of 1 dioptric. The dioptric is the unit of power for lenses. A lens whose focal length is ½ meter has a power of 2 dioptrics; and so on. The power of a lens is simply its ability to change the divergence or convergence of rays incident upon it.

SPECIAL CASE VIII. *One Refracting Surface Plane*

461. Let $r_1 = \infty$, and r_2 be negative, then we have a plano-convex lens, for which

$$\frac{1}{s'} - \frac{1}{s} = \frac{1 - \mu}{r_2}. \qquad \text{Eq. 179}$$

When, however, r_1 is positive and $r_2 = \infty$, we have a convexo-plane lens, for which

$$\frac{1}{s'} - \frac{1}{s} = \frac{\mu - 1}{r_1}. \qquad \text{Eq. 180}$$

Graphical Construction of Image

462. So far we have concerned ourselves only with the *position* of the image produced by a lens. Before we attempt to determine its size, it will be convenient to add one more term to our optical vocabulary.

Definition. A ray of light falling upon any point of a lens may have such an angle of incidence that after refraction in the lens it will emerge in a direction parallel to the direction of incidence. If we draw a tangent plane to the front surface of the lens at the point of incidence, and then draw another plane, tangent to the back surface and parallel to the first tangent plane, we may at once construct the ray in the glass; for its path will be the line joining these two points of tangency. Indeed, the lens will behave as a plate of plane parallel glass for a ray which is incident at A and emergent at B in Fig. 364.

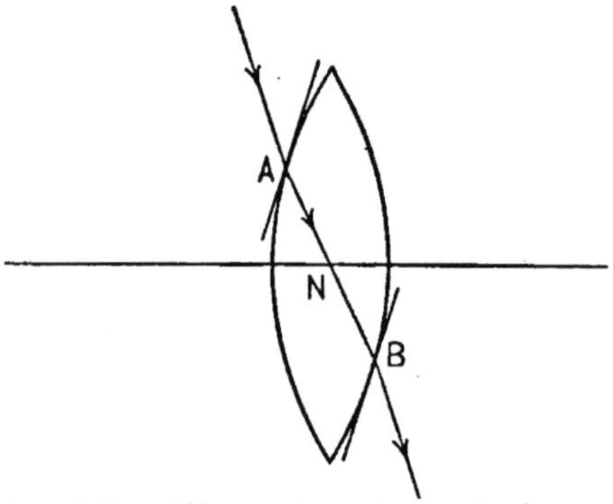

Fig. 364. — Illustrating the optical center of a lens.

That point where an undeviated ray in passing through the substance of a lens crosses the axis is called the optical center of the lens.

In any actual lens, that is in any lens of finite thickness, the emergent portion of an undeviated ray is more or less shifted at right angles to the direction of the incident ray while it remains parallel to the incident ray. But in thin lenses, such as we are now discussing, this shift is negligible. The advanced student will find this part of the subject beautifully simplified by the consideration of "principal points" and "nodal points," whose interesting properties would here lead us too far afield.

To construct the optical image of a point (or of any aggregation of points) we have therefore only to employ any two of the three following general principles: —

(i) An incident ray parallel to the axis passes, on emergence, through the principal focus.

(ii) Or, conversely, an incident ray which passes through the principal focus will emerge in a direction parallel to the axis.

(iii) An incident ray which passes through the optical center emerges without change of direction.

Just how these principles are applied will be clear from a careful consideration of Figs. 365 and 366. From Fig. 365

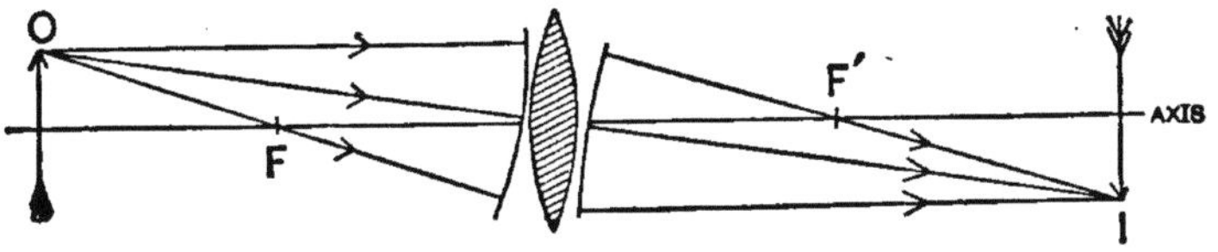

FIG. 365. — Illustrating the construction of an image in a double convex lens.

will be seen how the three typical incident rays are drawn when the lens is double convex and when the object is placed at a distance from the lens which is greater than the focal length. From Fig. 366 will be seen the behavior of a double concave lens.

In each case the top ray, being parallel to the axis, emerges in a direction which passes through the focus. The middle ray, being directed toward the optical center, emerges without change of direction. The lowest of the incident rays passes through the focus and hence emerges parallel to the axis. In the particular cases here cited, it will be observed that the double convex lens gives an image which is real but inverted; while the double concave gives an image which is virtual but erect.

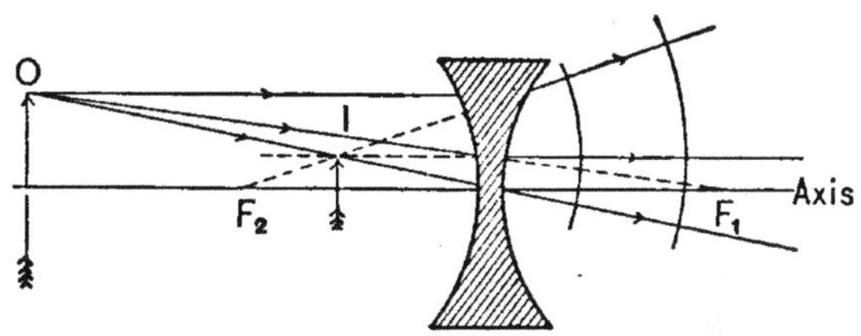

FIG. 366. — Illustrating the construction of an image in a double concave lens.

Magnification

463. *Definition.* The ratio between the linear magnitude of an image and the corresponding linear magnitude of the object, both measured at right angles to the axis of the lens, is called the linear magnification of the lens. As will be seen from the typical Fig. 367, this ratio $A'L'/AL$ is identical with the ratio s'/s, as will be seen by comparing the two similar triangles ANL and $A'NL'$.

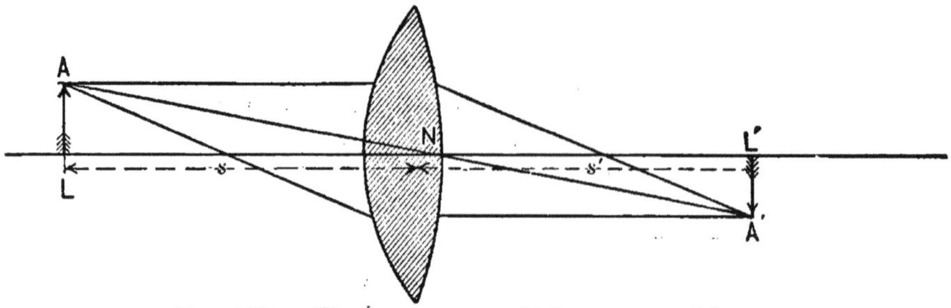

FIG. 367. — Illustrating magnifying power of lens.

Spherical and Chromatic Aberration

464. It would be interesting if time permitted to consider a more general case in which the incident ray makes a larger angle with the axis of the lens, and in which the luminous point is at some distance from the axis. The lens would then begin to show an imperfection known as "spherical aberration." The image of a point would no longer be a point.

Up to this point we have assumed that μ is a constant, but if we use light of more than one color we find that μ varies from color to color and that the image of a luminous point on the axis is drawn out into a string of images each of a different color. The table illustrates this variation of refractive index.

μ	RED LIGHT C	YELLOW LIGHT D	GREEN LIGHT F
Flint glass . .	1.630	1.635	1.648
Crown glass .	1.527	1.530	1.536

Indeed, it is at once evident from § 460 that the focal length of a lens is shorter in proportion as $\mu - 1$ is greater. This failure of a lens to bring all colors to the same focus is called "chromatic aberration." The manner in which these errors are corrected by combining several lenses of different refractive indices and different radii of curvature forms a chapter of great interest to the advanced student, who should consult Edser's *Light* and Czapski's already classical *Theory of Optical Instruments*. The general plan is, in a word, to combine lenses into a system in such a way that the error of one element is equal to the error of another element, but is of opposite sign.

CASE IX. *The Prism*

465. An interesting and important special case of refraction at two surfaces is that of the prism.

Definitions. (i) **The angle** of **the** prism is the dihedral angle between the inclined planes which bound the refracting medium. This angle α is represented by $B\widehat{A}C$ in Fig. 368. The line in which these inclined planes meet is called the **refracting edge** of the prism.

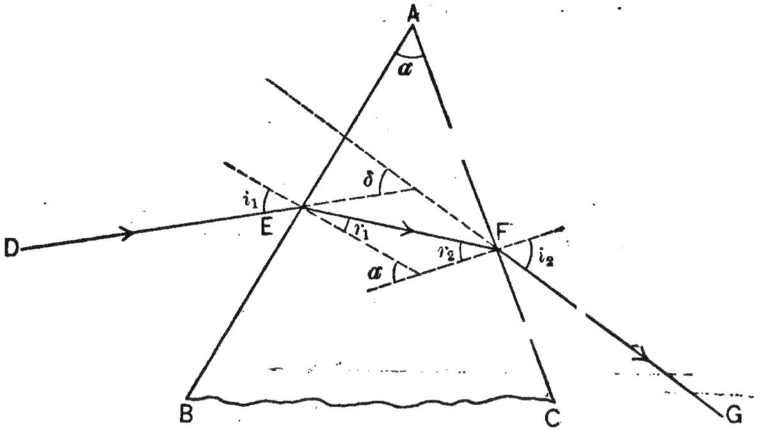

FIG. 368. — Refraction through a prism.

(ii) Consider an incident ray DE falling upon one surface of the prism and, after two refractions as indicated in Fig. 368, emerging from the second surface as the ray FG. The angle between the incident and emergent rays, which we here denote by δ, is called the **angle** of **deviation.**

If we denote angles of incidence by i and angles of refraction by r, and distinguish between the first and second surfaces by subscripts as before, it is evident from the figure that the

$$\text{angle of the prism} = \alpha = r_1 + r_2, \qquad \text{Eq. 181}$$

and that the

$$\text{angle of deviation} = \delta = (i_1 - r_1) + (i_2 - r_2). \qquad \text{Eq. 182}$$

To derive these two equations we have merely to recall that the exterior angle of any triangle is equal to the sum of the opposite interior angles. These two equations enable us to determine the deviation of a ray as soon as we know i_1, μ, and α.

Angle of Minimum Deviation

466. As has already been pointed out, it is a consequence of the law of refraction that when any emergent ray is replaced by an incident ray which has the same direction and the same position, but an opposite sense, the path of the reversed ray through the prism is identical with that of the direct ray. In other words, if we consider, in Fig. 368, *GF* as the incident ray, then *ED* will be the emergent ray.

By applying this principle of reversibility to a prism it may be shown that **the angle of deviation is a minimum when the path of the ray inside the prism is symmetrical with respect to the two refracting surfaces.**

For since the ray is reversible, it follows that corresponding to any one angle of deviation there are two possible angles of incidence, namely i_1 and i_2 (see Fig. 368). But it is an experimental fact that there is only one angle of incidence at which the deviation is a minimum; and the only way in which this single value for the angle of incidence can be obtained is by making $i_1 = i_2$; from which it follows that $r_1 = r_2$ and hence that the ray inside the prism is symmetrical with respect to the two refracting surfaces.

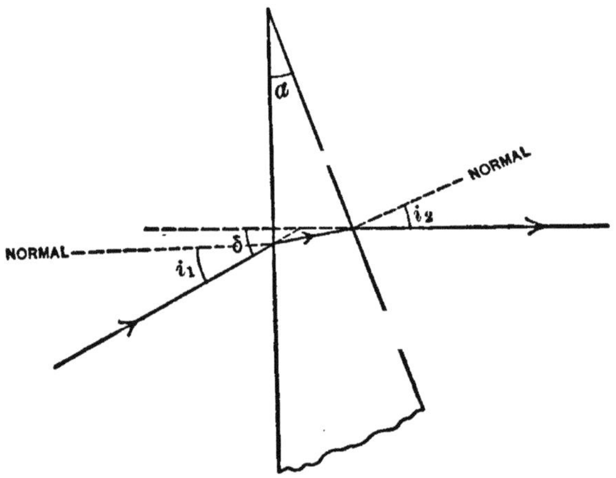

FIG. 369.—Refraction through a prism of small angle.

The position of minimum deviation is one of great usefulness in work with the spectroscope, as well as in the accurate measurement of refractive indices.

SPECIAL CASE X. *Prism of Small Refracting Angle*

467. We have already seen (§ 465) that in general

$$\delta = (i_1 - r_1) + (i_2 - r_2)$$

and

$$\alpha = r_1 + r_2.$$

By definition,

$$\mu = \frac{\sin i_1}{\sin r_1} = \frac{\sin i_2}{\sin r_2}.$$

But when i_1 and i_2 are small, as indicated in Fig. 369, we may, approximately, write

$$\mu = \frac{i_1}{r_1} = \frac{i_2}{r_2}. \qquad \text{Eq. 183}$$

Eliminating i_1 and i_2 from these equations, we have

$$\delta = (\mu - 1)(r_1 + r_2) = (\mu - 1)\alpha, \qquad \text{Eq. 184}$$

a very convenient expression for computing the effect of thin prisms such as are used in spectacles and stereoscopes.

Determination of Refractive Indices

468. When any transparent material can be obtained in the form of a prism, the principle of minimum deviation enables one to measure the refractive index of the material with considerable ease. For since the path of the ray inside the prism is symmetrical, we need consider only one half of it, as shown in Fig. 370. The conditions described are as follows:—

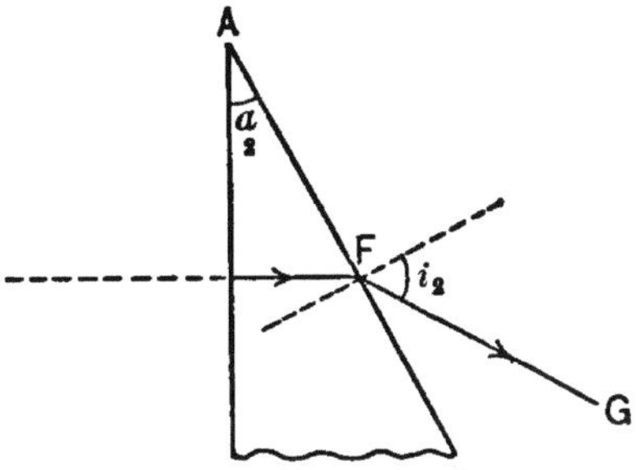

FIG. 370.—The measurement of refractive indices.

$$i_1 = i_2 \text{ and } r_1 = r_2.$$

Hence Eqs. 181 and 182 become

$$\alpha = 2r$$

and

$$\delta = 2\,(i - r).$$

Hence

$$= \frac{\alpha}{2}$$

and

$$i = \frac{\alpha + \delta}{2}.$$

Accordingly,

$$\mu = \frac{\sin i}{\sin r} = \frac{\sin\left(\frac{\alpha + \delta}{2}\right)}{\sin\frac{\alpha}{2}}, \qquad \text{Eq. 185}$$

which is the laboratory equation for the measurement of μ. The problem then reduces itself to the measurement of the two angles α and δ. These measures can be made by use of a protractor and a few pins; but they can be carried out with a very high degree of accuracy by means of a divided circle and a pair of telescopes.

Dispersion. Color

469. In tracing wave surfaces through various prisms and lenses we have hitherto tacitly assumed that the ratio of the speed of light in air to the speed in, say, glass was constant;

but experiment shows that this assumption is not allowable except for a single color. Michelson has proved this directly; for in measuring the speed of *red* light he found it 1.4 per cent greater in water and 2.5 per cent greater in carbon bisulphide than that of the *blue.* In a vacuum such as that which probably exists between us and the moon it appears to be rigidly true that rays of all colors travel with the same speed; for when a white star is occulted by the moon, all the various colors in the star's light disappear *at once;* and when the star emerges from behind the moon, it flashes out instantly and perfectly white. If in the medium between us and the moon there were as much difference between the speed of red and blue light as there is in flint glass (see § 464), the red rays from the emerging star would reach us $\frac{1}{70}$ second earlier than the blue rays. This interval of time might be too small to detect. Accordingly the eclipse of the bright component of the star Algol has been adduced as much stronger evidence for the view that lights of all colors travel with the same velocity in free space. The star Algol is a rapid variable star—supposed to be double, but with one component dark. Every three days the bright component is eclipsed. But on reappearance it does not show the various colors in succession, as it certainly would at that enormous distance, if the speed of light varied from one color to another.

The advanced student will discover, indeed, that the velocity of light in vacuo—300 million meters per second—is one of the fundamental constants of optics and of electromagnetism as well.

In all other media the variously colored rays travel at rates which differ not only from that in vacuo but also from each other. Indeed, the wave theory of light is so well established that the speed of any colored ray in any medium may be considered as determined when the refractive index of that medium has been measured. Thus in the following table are given refractive indices for certain metals which are ordinarily opaque. But Kundt prepared prisms of these substances which were so thin as to be transparent, and thus discovered that there are certain substances for which the refractive index is less than unity, which is interpreted to mean that the speed of light in each of these bodies is greater than in vacuo.

TABLE OF REFRACTIVE INDICES FOR DIFFERENT COLORS

Substance	Index of Refraction for		
	Red (C)	Yellow (D)	Blue (F)
Flint glass	1.5826	1.5867	1.5967
Crown glass	1.5311	1.5339	1.5404
Gold	0.38	——	1.00
Platinum	1.76	——	1.44
Iron	1.81	——	1.52
Fluor spar	1.4325	1.4340	1.4372
Water	1.3318	1.3336	1.3377
Carbon disulphide	1.6336	1.6433	1.6688
Alcohol	1.3596	1.3614	1.3657
Diamond	——	2.4699	——

Smithsonian Tables, Third Edition.

Observe that the color which is produced by a prism is due, not to the refraction, but to the difference of the refraction for various rays. **This transformation of a beam of parallel light into a divergent pencil is called dispersion.** The measure of dispersion for any substance is generally defined as the ratio of the difference of deviation for red and blue rays to the mean deviation for these two rays.

This leads, through Eq. 184, to the following definition: —

$$\text{Dispersive power} = \frac{(\mu_r - 1)\alpha - (\mu_b - 1)\alpha}{(\mu - 1)\alpha} = \frac{\mu_r - \mu_b}{\mu - 1}, \qquad \text{Eq. 186}$$

where μ_r, μ_b, μ, are the refractive indices for the red, blue, and mean rays, respectively.

Composition of White Light

470. The fact that different rays travel at different rates enabled Newton to separate white light by means of a prism into its various constituents. For on transmitting a small beam of sunlight through a prism, he observed that the deflected light did not form a white spot on the wall beyond but was drawn out into a brilliant strip, red at one end and violet at the other. Between the red and the violet are distributed an infinite variety of colors which are usually grouped under the following four names: orange, yellow, green, blue. But Newton's capital discovery in this domain is that any one of these colors after dispersion in one prism undergoes no further change of this kind on passage through a second prism.

Newton succeeded also in compounding or adding together these various colors which he had found in sunlight, thus obtaining again white light; in this manner he demonstrated in a synthetic way the composition of ordinary daylight, a fact which he had already proved by analysis.

The majestic phenomenon of the rainbow he showed to be merely a case of refraction in raindrops.

Another Illustration of Refraction

471. The laws of refraction may be illustrated by means of the "optical disk," one form of which is shown in Fig. 371. Half of a circular disk of glass *A* is mounted on a wooden base *B*. A horizontal beam of light from the sun or from a projecting lantern falls on the mirror *M* and is reflected thence to a second mirror *N*. After leaving *N*, the beam is incident on the upper plane face of the glass disk *A*.

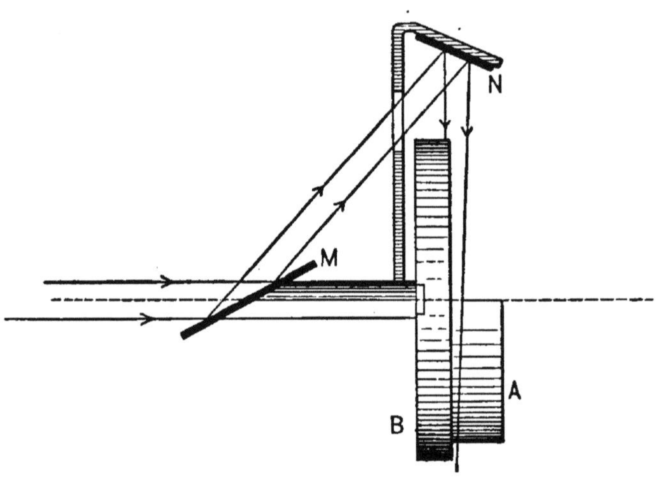

FIG. 371. Tatnall's form of optical disk.

The surface of the mirror *N*, with the exception of a narrow strip, is covered with black paper, as shown in Fig. 372. By means of this diaphragm, the light incident on the refracting surface is restricted to a narrow beam. This beam falls upon the whitened face of the plate *B* at a small angle, so that the course of the incident ray in air and the refracted ray in glass may be plainly

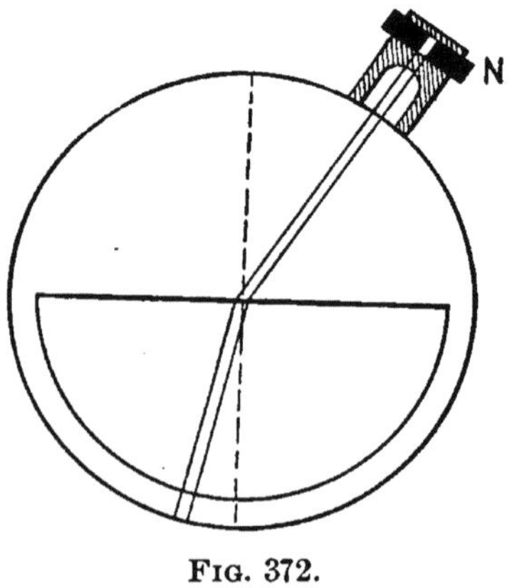

FIG. 372.

seen. The frame which carries the two mirrors may be revolved around the pivot by which it is attached to the base *B*, and in this way the angle of incidence may be varied at will.

If the mirror *N* be placed in the position shown in Fig. 373, the refraction will take place on passing from glass into air. The incident beam *GD*, the refracted beam *DE*, and the reflected beam *DF* are now all seen at the same time. The refracted beam is bent away from the normal, while the incident and reflected beams make equal angles with the normal.

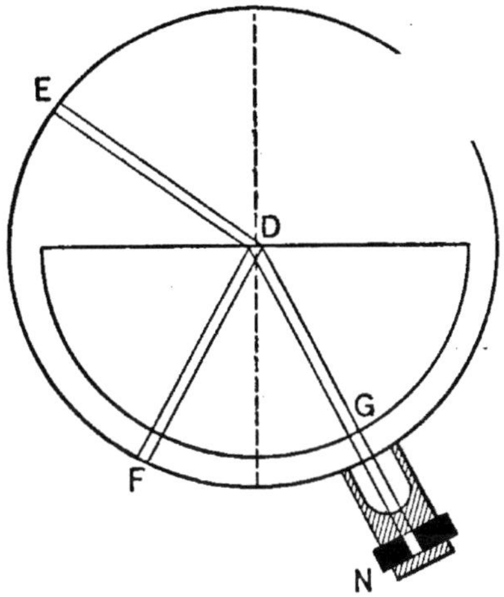

FIG. 373.

By moving the mirror *N* toward the right, in Fig. 373, the angle of incidence is increased; but the angle of refraction increases more rapidly than the angle of incidence. When the critical angle of incidence is reached, the angle of refraction has become a right angle, and the beam *DE* just grazes the refracting surface. On increasing the angle of incidence beyond this critical value, it will be found that there is no refracted beam, the light being now totally reflected.

Problems

1. Find the focal length of a thin double convex lens made of glass whose mean refractive index is 1.50, the radius of curvature of the first surface being + 25 cm. and that of the second being − 30 cm.

2. An object is placed on the axis of a double convex lens whose focal length is + 20 in. The object is placed 50 in. to the left of the lens. Find the position and size of the image.

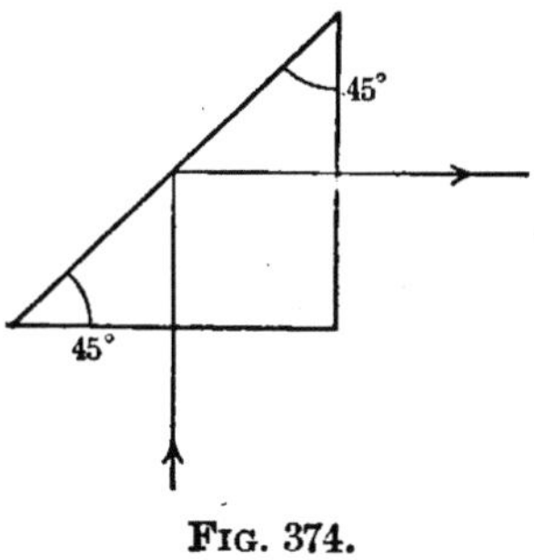

FIG. 374.

3. The projection lens of a lantern (Fig. 392) has a focal length of 1 ft. and is used to throw a picture upon a screen which is 24 ft. away. How far back of the lens must one place the glass slide?

4. A 45-degree prism is used as a totally reflecting mirror in the manner indicated in Fig. 374. What is the least value of refractive index which the glass can have?

5. A gas flame placed in front of a thick glass mirror silvered on the back gives a series of images. Explain the relative intensity and position of these.

6. How far in front of a double convex lens must an object be placed in order to give an inverted image of the same size as the object?

7. A simple pocket magnifying glass has a focal length of 2 in. Find the size and position of the image when the object is placed between the lens and the focus at a distance of $\frac{1}{2}$ in. from the focus.

8. Locate and describe the image produced by two thin lenses situated as shown in Fig. 375,

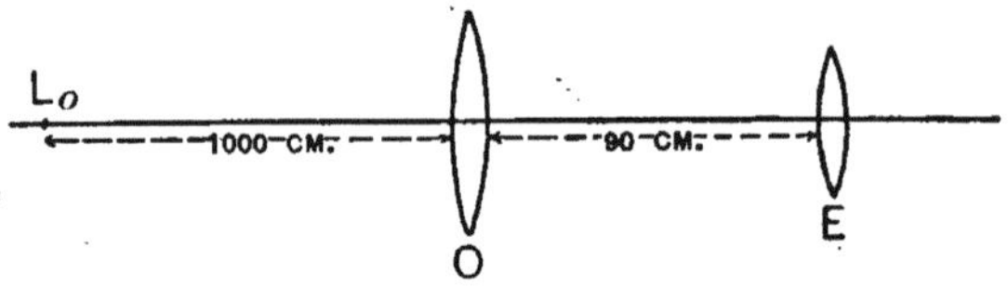

FIG. 375.

$$f_O = +80 \text{ cm.},$$
$$f_E = +4 \text{ cm.},$$
$$s_O = -1000 \text{ cm.},$$

where the subscripts indicate the lenses referred to.

9. Locate and describe the image produced by the combination of thin lenses shown in Fig. 376.

$$f_O = +6 \text{ in.}$$
$$f_E = -2 \text{ in.}$$
$$s_O = -2 \text{ mi.}$$

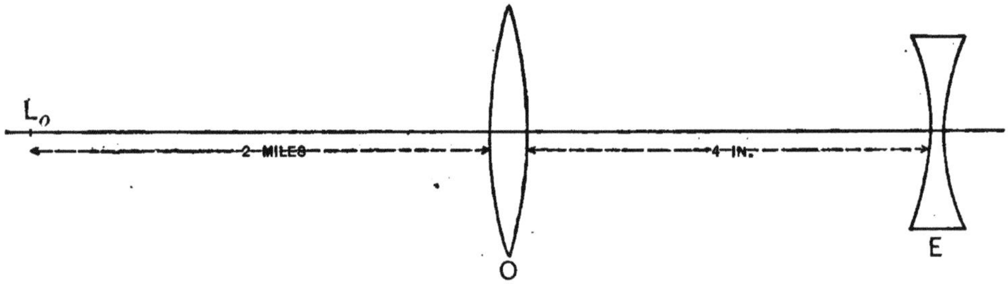

FIG. 376.

10. What must be the refracting angle of a prism of ordinary glass ($\mu = 1.52$) in order that its minimum deviation should be 15°?

11. Prove that if $\mu = 1.333$ for water and 1.5 for glass, the focal length of a glass is increased four times by immersing it in water.

12. Prove that in the case of the ordinary pocket lens—double convex magnifying glass—the magnification is D/f, where D is the distance of distinct vision.

13. Two parallel walls are 20 ft. apart. It is desired to project upon one wall, with a magnification of 6, an object situated on the other wall. Where must the lens be placed and what must be its focal length?

VII. INTERFERENCE AND DIFFRACTION

472. Up to the present we have been considering the consequences of only two fundamental facts, namely (i) the rectilinear propagation, and (ii) the finite, but variable, speed of light.

But there is a large group of important phenomena which find their explanation in the fact that the light waves producing them are compelled to pass through apertures which are small when measured in terms of a wave length of light. These beautiful phenomena were first cleared up by the brilliant young French physicist Fresnel (1788–1827), whose particular achievements in optics are —

(i) The introduction of the idea that light consists in transverse vibrations, and

(ii) The combination of the principle of Huygens with that of interference; in other words, the union of the principles of Huygens and Young. Just how this second idea is employed to explain what happens when light passes through small apertures will be clear from the following illustrations.

CASE I. *Passage of Plane Waves through a Narrow Slit: Effect upon a Distant Screen*

473. When the glower of a Nernst lamp or any other nearly linear source of light is viewed through a narrow slit, held parallel to the source and close to the eye, a series of colored bands separated by dark bands something like those shown in Fig. 377 are seen. If monochromatic light is used, the spaces in the figure represented as white are colored. The central space

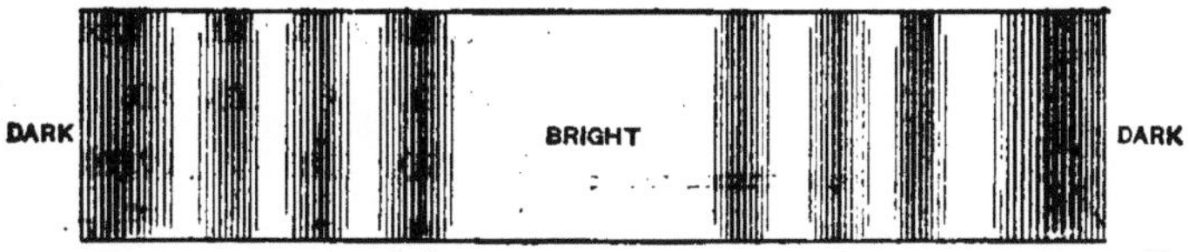

FIG. 377. — Image of a linear source produced by a narrow slit.

is always broader and brighter than any of those on either side. For this experiment a single ruling having a breadth of from $\frac{1}{20}$ to $\frac{1}{10}$ millimeter, made upon the back of a piece of silvered mirror, is well adapted.

It is often convenient also to observe this phenomenon by viewing a small brilliant source through the narrow opening between two fingers of one's hand. These bands were called by Fraunhofer, who was one of the earliest to study them, "spectra of the first class."

Fresnel's explanation is as follows: Let *AB* (Fig. 378) be a horizontal section across a vertical slit. Let *w* indicate the successive incident wave fronts, and *MM'* the screen illuminated by the light passing through the slit.

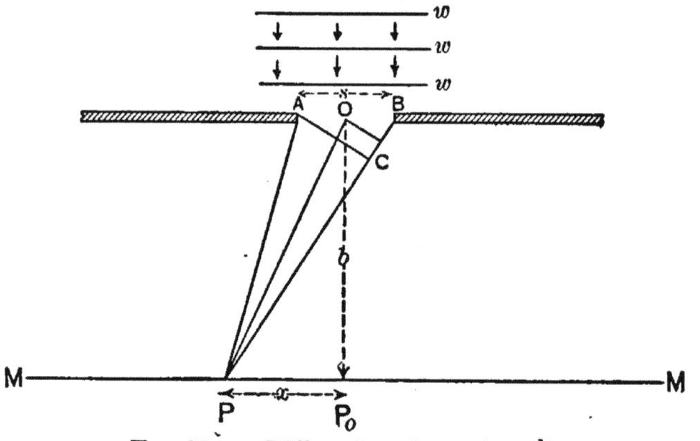

FIG. 378. — Diffraction through a slit.

Two queries now demand answer: (i) what happens at various points along the screen, the width of the aperture *s* remaining constant? and (ii) what happens at any one point on the screen, say *P*, while the width of the aperture varies?

474. (i) Taking these questions in order, let us consider any wave front passing through the slit as divided up into narrow vertical strips such that when viewed from any given direction, say *AD* (Fig. 379), the edges of each strip will differ by half a wave length in their distance from the line *AC*, which is drawn perpendicular to this given direction. These strips differing by half a wave length in their distance from *AC* are called, after their inventor, "Fresnel zones."

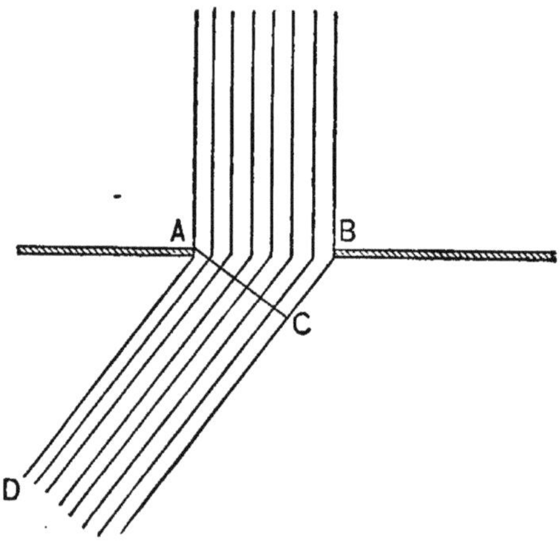

FIG. 379. — Fresnel zones.

If the screen *MM'* (Fig. 378) be at a sufficient distance from the slit, the arc of a circle drawn through *A* and *C*, with *P* as a center, will differ inappreciably from the straight line *AC*. Under these circumstances the different Fresnel zones in the aperture will each, on Huygens' Principle, send out light to every point of the screen and will each produce a luminous disturbance (displacement of the ether) at *P*, with a phase which is exactly 180° different from that produced by its next door neighbor. The rays of light which reach the point P from the various zones will differ in phase at P exactly as they differ in phase at the line AC.

When the incident wave (w in Fig. 378) is plane, the phase of each zone on reaching P increases in arithmetic progression from A to B. It is then an easy matter to compute the effect of all the Fresnel zones at any point P, provided we assume what is here practically true, namely, that the effect of every zone at P is the same *except in phase:* in other words, we shall assume that so far as regards distance, direction, etc., the zones are similarly situated.

Let P_0 represent that point on the screen which is nearest the central point of the incident wave, and let b denote the distance of this point P_0 from the center of the slit O. Consider now the total illumination at any point of the screen P distant x cm. from P_0. Let n denote the number of Fresnel zones in the slit with respect to the point P. Note that "number of Fresnel zones" has no meaning except with reference to some one point. Since the triangles ABC and OPP_0 are similar, it follows that

$$\frac{x}{b} = \frac{BC}{AC} = \frac{BC}{s}, \text{ approximately,}$$

where s is the width of the slit; and since there are n zones in the slit, it is evident from the definition of the zones that

$$BC = n\frac{\lambda}{2},$$

where λ is the wave length of the incident light. Hence

$$x = \frac{bn}{s} \cdot \frac{\lambda}{2}. \qquad \text{Eq. 187}$$

This equation tells us just how n varies as the point P moves along the screen, while b, s, and λ remain constant.

For any given position of P, either one of two things may happen, according as n is an odd or even number.

(a) *When n is even*, it is clear that all the zones of the slit (1, 2, 3, 4, ⋯ n, Fig. 380) will just annul each other in pairs. Suppose $n = 2$; then $BC = \lambda$; and the aperture may accordingly be divided into two parts, any point of one of which will be in exactly opposite phase to the corresponding point in the other. The effect at P will therefore be *nil.* If P move further away from the center of the screen, so that $n = 4$, then $BC = 2\lambda$. Now each half of the aperture may again be broken up into two parts, each of which will interfere with the other, since they differ in phase by $\frac{\lambda}{2}$. Hence they produce no displacement of the ether — no light — at P. **The condition for darkness (black band) at any point P then is that the slit shall comprise an even number of Fresnel zones with respect to P; or, in terms of algebra, n shall be even in Eq. 187.**

(*b*) *When n is odd*, the zones of the slit will interfere in pairs as before; but there will always be one left over which will produce a disturbance in the ether at P. **And hence the condition that P shall lie on a bright band is that the slit width be an odd number of Fresnel zones, i.e. n in Eq. 187 shall be odd.**

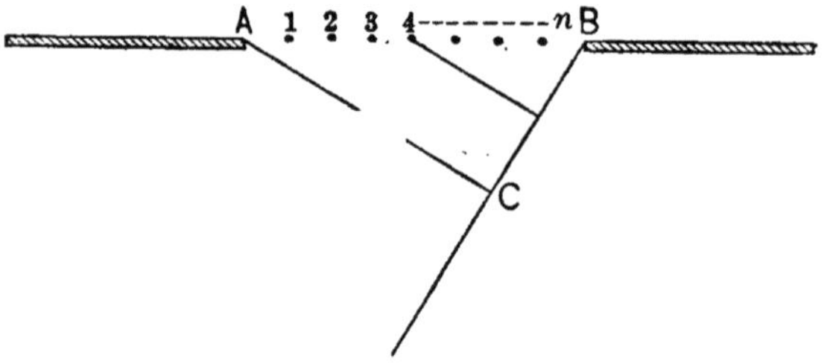

FIG. 380.

Thus, at the center of the screen P_0, the aperture will contain at most one Fresnel zone; $n = 1$; the light from every part of the slit reaches P_0 in nearly the same phase. Hence the center of the screen is nearly always bright. If $n = 3$, $BC = \frac{3}{2}\lambda$. The aperture may therefore be divided into three zones, two of which will exactly annul one another, while the third will remain to illuminate the screen at P. This condition gives rise to the first bright band on either side of the center. For the second bright band $n = \frac{5}{2}\lambda$; for the third, $n = \frac{7}{2}\lambda$, etc.

Under the above conditions which we have assumed in this case, the plane wave, narrow slit, and distant screen, it is clear that with respect to P_0, which is the middle of the geometrical image, the slit will contain in general less than one Fresnel zone. There is, therefore, no chance for interference between different zones, and the center of the screen is always bright. But as one recedes from the center to either side, the zones increase in number, and as n passes from odd to even and back again, the illumination passes from bright to dark and back again in the manner shown in Fig. 377 and described in Eq. 187.

475. (ii) Let us now take up the second query asked above. What happens at any one point on the screen when the slit width is varied, other conditions remaining constant? As one jaw of the slit is made to approach the other it must cut off one Fresnel zone after another. As the zones diminish in number, n in Eq. 187 will become alternately odd and even; which, in turn, means that the fixed point on the screen becomes alternately bright and dark. This prediction is easily verified by viewing the filament of an incandescent lamp through a cut in a visiting card which can be bent so as to make the slit wider or narrower. In this case the eye acts as a telescope, receiving the parallel rays which come through the slit and bringing them to focus on the retina.

We pass now to another case of diffraction through a narrow slit, namely,—

CASE II. *Incident Wave Plane: Screen near Aperture.*

476. Let the aperture be again divided into Fresnel zones such that the center of any one is half a wave length further away from P than its next door neighbor. The zones will now differ in size, direction, and distance, making the problem too difficult for any treatment more elementary than that of the integral calculus. But if we confine our attention to the center of the screen, *i.e.* to the point P_0, the solution becomes simple again. For we have now only to consider each half of the slit: divide these up into Fresnel zones with respect to P_0. If either half contains an *even* number of zones, they will very nearly annul each other so far as the effect at P_0 is concerned, and *darkness* will result. But if each half contains an *odd* number of zones, they may interfere in pairs, but there will always be one zone uncompensated, and this will illuminate P_0,

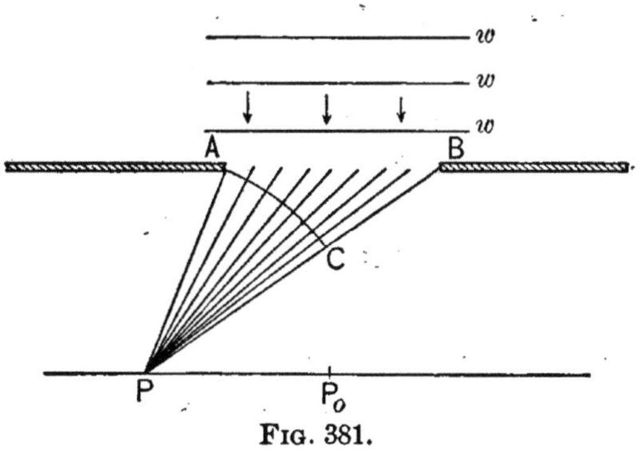

FIG. 381.

making it bright. As P_0 approaches or recedes from the center of the slit, the number of zones in the half slit will pass alternately from even to odd, and the illumination will change alternately from darkness to brightness. This effect is easily observed by placing a narrow aperture, say $\frac{1}{20}$ millimeter, at a distance of, say, 1 meter from a linear source. If then a compound microscope be focused upon a region lying within the first millimeter from the slit, the alternations of bright and dark centers will be easily seen.

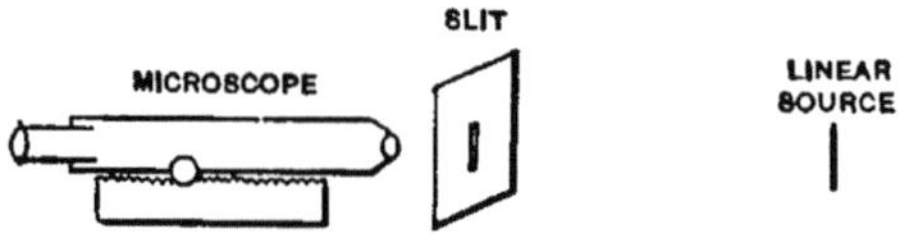

FIG. 382. — Fresnel's diffraction phenomenon.

Intensity of Diffracted Light

477. Heretofore we have been concerned mainly with the positions of the maxima and minima of intensity upon the illuminated screen. But the late Professor Cornu (1841–1902), a distinguished French physicist, has given us a method of great simplicity and elegance for graphically measuring the intensity at any point on the screen so soon as we know the phase and displacement due to each of the various elements of the wave surface illuminating the slit.

To obtain the resultant effect of any number of disturbances at the point P, we have merely to lay off on any diagram (Fig. 383), beginning with O as origin, the various displacements, each with a length which shall measure the amount of the displacement and each at an angle to the axis of X which shall measure its phase. We know that the resultant of any number of vector quantities laid off in this manner is represented by the side which closes the polygon, in this case by the dotted line OP. Any element in this spiral represents the amplitude of the vibration transmitted by the corresponding element of the wave surface passing through the slit.

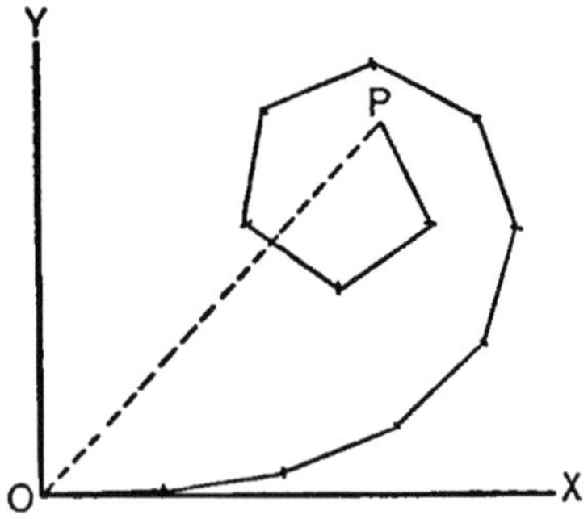

FIG. 383. — Illustrating the construction of Cornu's spiral.

If we consider the illumination at the center of the screen, we may divide the wave front into two symmetrical halves, each of which will give rise to a spiral like that in Fig. 383. The resultant amplitude then, in a case of this kind, will be given by joining the two extremities of the spiral as indicated by the dotted line in Fig. 384. If we consider a point on the screen which is at one side of the center, then the number of elements in one branch of the spiral will, of course, be larger than in the

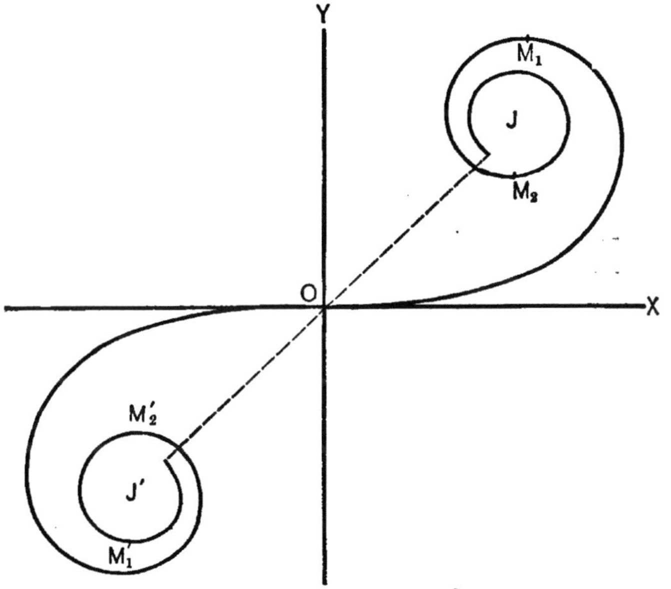

FIG. 384. — Cornu's spiral.

other, and the resultant amplitude will vary accordingly. To obtain the effect of the first Fresnel zone it will be necessary to lay off elements of the wave surface, starting from any given point, until one reaches a point where the total phase difference from the starting point O is 180° or $\frac{\lambda}{2}$. This point is indicated in Fig. 384 by M_1, at which point the tangent is parallel to OX, but is drawn in the opposite sense. At M_2 the phase difference is 360° or $2\left(\frac{\lambda}{2}\right)$. Hence the vector $\overline{OM_2}$ represents the amplitude resultant from the first two Fresnel zones; and so on.

Since the intensity of light at any point is measured by the luminous energy which falls upon unit area in unit time at that point, we must next find how to derive the intensity from the amplitude. The energy of a particle m moving with a simple harmonic motion is $\frac{1}{2}mv^2$. And since the velocity at the center varies as the amplitude, the energy of any simple harmonic disturbance will vary as the square of the amplitude. To obtain the intensity at any point of the screen we have therefore only to square the amplitude as given by Cornu's spiral.

CASE III. *Diffraction of a Plane Wave at a Pair of Narrow Slits. Spectra of the Second Class*

478. Imagine two narrow slits, such as the one which we have just been studying, ruled side by side about $\frac{1}{5}$ millimeter apart. On viewing, through this pair of apertures, any linear source such as a Nernst filament or even an ordinary gas jet turned low, a series of alternately bright and dark bands are seen to occupy the center of the field, *i.e.* the region covered by the broad central bright band *due to either slit alone.* We assume that the source is sufficiently distant from the apertures to make the incident waves practically plane, and that the screen, which may be the retina of the eye, is sufficiently distant to permit us to consider the arc AC, described about P as a center, as a straight line. To each element of the incident wave surface in the aperture A will correspond an element in the aperture B. The effect at any point P on the screen will be obtained when we add up the effects of all the elements in these two slits. Use red glass so as to make the incident light monochromatic. It is at once seen by the method of Case I

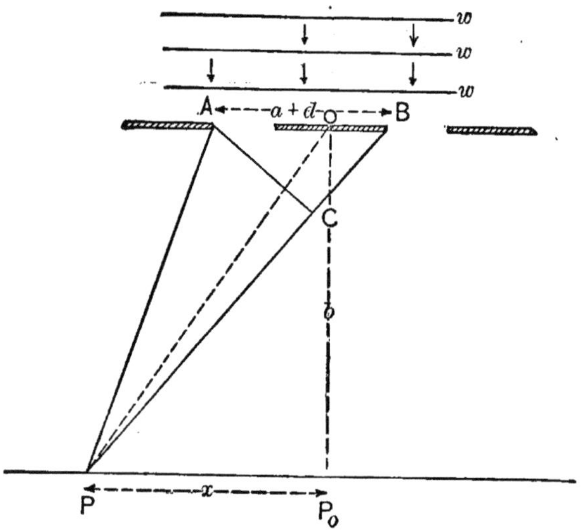

FIG. 385. — Diffraction through a pair of slits.

that, if P is so chosen as to make $BC = \frac{\lambda}{2}$, we shall have darkness at P; for *each* element in the aperture A will reach P with a phase which is just 180° ahead of the corresponding element in B; the resultant amplitude at P will therefore be zero. The same thing will happen if P be moved along the screen to any position, say x, where BC is any odd number of half wave lengths, *i.e.* when $BC = (2n+1)\frac{\lambda}{2}$, n being the series of natural numbers, beginning with zero. The triangles ABC and OPP_0 are similar; therefore if we denote the distance between the slits by $a+d$, and the distance of the screen by b, we have

$$BC = (a+d)\frac{x}{OP} = (a+d)\sin\theta,$$

where θ is the angle of diffraction P_0OP. The condition that P shall lie on a *dark* band is therefore

$$(a+d)\sin\theta = (2n+1)\frac{\lambda}{2} \qquad \text{Condition for dark band.} \qquad \text{Eq. 188}$$

The condition for a *bright* band is obtained in the same way, only now the difference of path between two corresponding elements — one from each slit — must be an *even* number of half wave lengths. For two rays of light which are any whole number of wave lengths apart will reënforce each other and, if their amplitudes are equal will, produce an intensity four times as great as that of either one alone. The position x of a bright band on the illuminated screen is accordingly expressed as follows in terms of algebra: —

$$(a+d)\sin\theta = 2n\frac{\lambda}{2}. \qquad \text{Condition for bright band.} \qquad \text{Eq. 189}$$

Let us now remove the red glass and view these bands in white light. It is then seen that the bright bands are colored, bluish green on the side toward the center of the field and red on the side away from the center of the field. Each of these bands is in fact a spectrum of the source. Fraunhofer called these "spectra of the second class" to distinguish them from those which are seen through a *single* slit.

The advanced student will find the distribution of intensity among these various spectra an enticing problem. In a case such as the present, where both the incident and diffracted waves are plane, it may be easily shown that Cornu's spiral becomes a circle.

CASE IV. *Diffraction through a Large Number of Equidistant Parallel Slits. The Plane Diffraction Grating*

479. Here again we shall consider the incident and diffracted wave fronts as plane, since this is the case principally used in spectroscopy.

Let AE (Fig. 386) represent an opaque screen in which have been cut a large number of equidistant parallel apertures. Let a be the width of each aperture and d the width of the opaque portion between any two consecutive apertures. Then $a + d$ will be what is called "the grating space," that is, the distance between any two corresponding elements of the wave surface in two consecutive apertures.

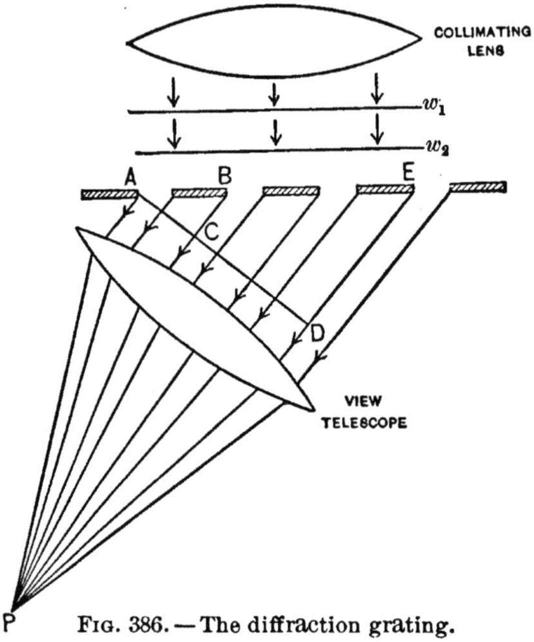

FIG. 386. — The diffraction grating.

If the source of light is placed in the principal focus of the collimator lens, the emergent wave fronts w_1, w_2, etc., will be plane. In like manner, the diffracted wave AD will reach the principal focus of the view telescope without any change of phase in its various parts; for the "optical distance" to the principal focus P is the same for all parts of AD — it being supposed that the axis of the telescope is perpendicular to AD. Hence, the phase difference with which rays from the various apertures reach the plane AD is precisely the phase difference with which they will reach the focus P. We need, therefore, to consider only the phase in which the light reaches the plane AD.

480. Two principal questions now arise: (i) what angle must the axis of the view telescope make with the normal to the grating in order that, for any one wave length λ, the point P may be a maximum of intensity? and (ii) what is the intensity of light at any point P, in terms of the intensity of the light incident upon any one of the n apertures? The first only of these queries will be here considered; the answer to the second will be found in Eq. 196.

It is evident from previous cases that when the telescope is directed normally upon the grating, the light from each aperture will reach P in the same phase and we shall have the greatest possible intensity. This bright band is called the "central image." As P moves to the left, say, the field will remain more or less bright until the distance DE (which is the retardation of the ray at one end of the grating over that at the other end) becomes equal to one wave length. When this occurs, the light from the aperture A will be just 180° out of phase with the light from the middle of the grating; in like manner the light from B will be just 180° out of phase with that from the ruling next to the middle, and so on, — the effect being that the light from one half the grating is annulled by the light from the other half. When $DE = 2\lambda$, the point P will again be dark; for then the grating will interfere by quarters, the light from each quarter being 180° out of phase with the light from the adjoining quarter, giving absolute darkness at P. At points between these there will be slight illumination; for instance, at the point where $DE = 3\frac{\lambda}{2}$; for now the first third of the grating will interfere destructively with the second third, leaving the last third to illuminate the point P. A slight maximum, due to light from one fifth of the grating, will also occur when $DE = 5\frac{\lambda}{2}$; and so on.

The student who is interested will do well to read Professor Wood's simple, clear, and clever discussion of these maxima (called secondary maxima, or "spectra of third class," by Fraunhofer) in the *Philosophical Magazine* for October, 1907.

But the first intense illumination on either side of the central image will occur when DE is equal to as many wave lengths as there are spaces on the grating; for then CB will be equal to one wave length, and the light from each aperture will reach the surface AD in precisely the same phase. The light which is thus concentrated at P is called the "spectrum of the first order." If the difference of phase between corresponding elements in two successive apertures is $n\lambda$, the light which then reaches P in maximum intensity is said to be the "spectrum of the nth order." So long, therefore, as the incident wave front is parallel to the grating, the condition that P shall be bright is that CB shall be equal to a whole number of wave lengths, or, in terms of algebra,

$$(a + d) \sin \theta = 2n\frac{\lambda}{2}, \qquad \text{Eq. 190}$$

where θ is the angle of diffraction, DAE.

If, however, the angle of *incidence* i is not zero (as in Fig. 386) but is an angle HAE such as indicated in Fig. 387, then the retardation of the extreme ray, between the two wave fronts AH and AD, is $HE + ED$. Hence, the general condition that P shall be bright becomes

$$(a + d)(\sin \theta + \sin i) = 2n\frac{\lambda}{2}. \qquad \text{Eq. 191}$$

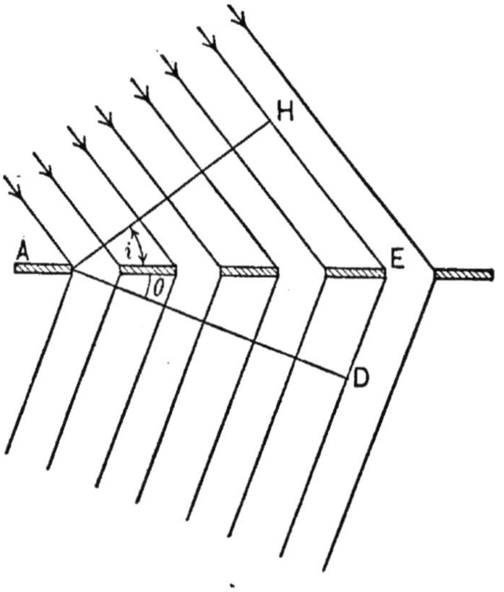

Fig. 387.

A grating of the kind just described, in which the diffracted light passes through parallel apertures, is called a **transmission grating**.

Reflection Gratings

481. The opaque regions of a transmission grating are generally produced by ruling glass surfaces with a diamond point. But the diamond point deteriorates very rapidly during this work and hence it is customary to employ highly polished speculum metal for the ruled surface. The speculum metal is softer than the glass and will permit the diamond point to cut many more lines upon it than upon glass. A grating ruled upon a polished opaque surface behaves exactly as a transmission grating in which the incident wave is *the image of the actual wave* falling upon the opaque grating. Thus, in Fig. 388, where the grating is supposed to be one of speculum metal, the effect is precisely the same as if the incident wave AH were replaced by its image AH' and the grating were at the same time made transparent. In either case the retardation of the extreme ray is $HE + DE$, and the position of the spectrum of the nth order is given by

$$(a + d)(\sin i + \sin \theta) = 2n\frac{\lambda}{2}, \qquad \text{Eq. 191}$$

where, as above, i = angle EAH, and θ = angle EAD. Ruled surfaces at which the reflected wave front is diffracted are called **reflection gratings**.

The best gratings of the world are ruled on Rowland's engine at Baltimore, and on Michelson's at Chicago. In most cases they contain somewhere between five and twenty thousand lines to the inch; so that in many cases the polished surface between two adjacent rulings becomes vanishingly small. Notwithstanding this fineness Wallace, Thorpe, and Ives have shown that it is possible, by means of transparent celluloid, to make a "cast" of a reflection grating and to mount this cast upon a piece of plane glass in such a way as to give a very perfect spectrum. A transmission grating produced in this way, by using an opaque grating as a mould, is called a **replica**.

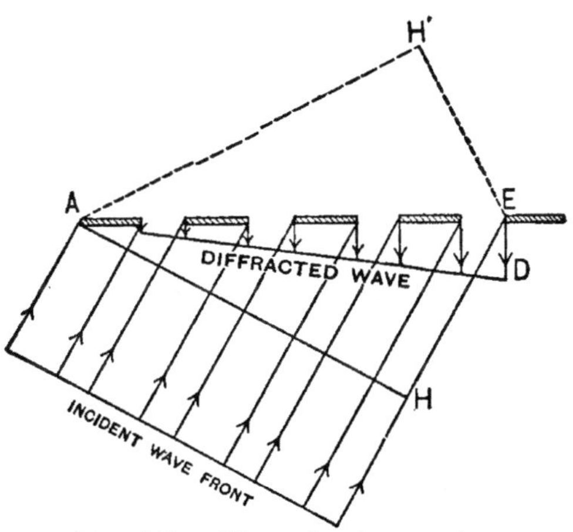

FIG. 388. — The reflection grating.

The Angstrom Unit

482. The numerical value of a wave length of light is usually expressed, not in terms of centimeters, but in terms of a much smaller unit introduced by the Swedish physicist Ångstrom, and called the "Ångstrom unit." Its value is the hundred-millionth of a centimeter. But by action of the International Solar Union at Paris in 1907 this unit is now defined as $\frac{1}{6438.4696}$ of the wave length of the red cadmium ray in air at normal pressure and 15° C.; and is called the "angstrom."

Resolving Power of a Grating

483. A general answer to the question raised above (§ 480) as to intensity which a grating will yield in any one direction must be postponed for advanced study. But it is not difficult to show that the sharpness of the image of any linear source of light, such as the illuminated slit of a spectroscope (§ 500), viewed through a grating, varies directly as the number of lines on the grating. The same source presents very different appearances according to the number of apertures through which it is viewed. If ordinates represent intensities and abscissas wave lengths, (*a*) in Fig. 389 will illustrate the broad diffuse appearance of line when seen through a very short grating: (*b*) and (*c*) show the same line through gratings which contain more lines. The reason for this is as follows: suppose the spectrum of the *n*th order to be seen in any direction θ; then if the incident ray be normal to the grating, we have

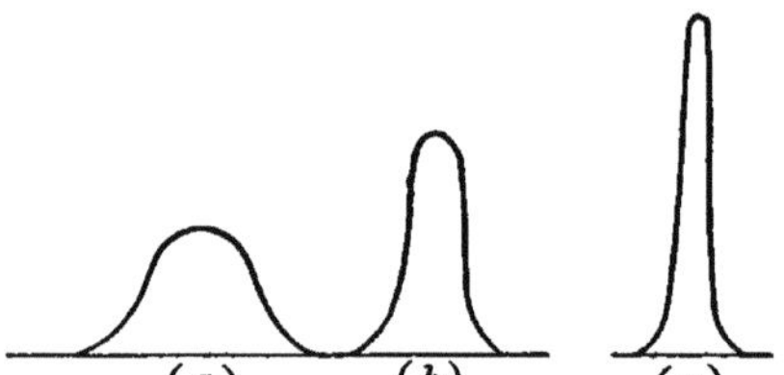

FIG. 389. — Illustrating the effect of small and large resolving power.

$$(a + d) \sin \theta = n\lambda. \qquad \text{Eq. 190}$$

If N be the total number of lines in the grating, we have $N(a + d) \sin \theta = Nn\lambda =$ retardation of last ray over the first ray. This retardation would amount to ED in Fig. 386. Suppose now that the direction of view be changed from θ to θ', a very small amount, so that

$$N(a + d) \sin \theta' = Nn\lambda + \lambda = Nn\lambda',$$

in other words, so that the retardation of the last ray becomes one wave length greater than before. We shall then have destructive interference between the light from the two halves of the grating. If it contain one thousand lines, the light from the first aperture will be just half a wave length ahead of that from the 501st aperture: in like manner the second aperture interferes with the 502d; and so on.

From the two preceding equations, it follows that

$$(a + d)(\sin \theta' - \sin \theta) = n(\lambda' - \lambda) = \frac{\lambda}{N}, \qquad \text{Eq. 192}$$

which means that for any given wave length the angular distance from a point of maximum intensity θ to the next minimum θ' is inversely proportional to the total number of lines N on the grating and to the spacing of the lines.

The ability of a grating to separate the images of a single source, emitting two nearly equal wave lengths λ and λ', is called its *resolving power*. The numerical value of the resolving power is measured by the following expression: —

$$\text{Resolving power} = \frac{\lambda}{\lambda - \lambda'} = nN, \qquad \text{Eq. 193}$$

where $\lambda - \lambda'$ is the difference of wave length between the closest images which the grating will show as double. As will be seen from Eq. 192, this **resolving power depends** (*if the grating be perfect in other respects, which it seldom is*) **upon the product of the total number of rulings on the grating N by the order of the spectrum n.**

The phenomena which are seen when a source of light emitting more than one wave length is viewed through a grating will be taken up under the head of the spectroscope.

Case V. *Rectilinear Propagation of Light*

484. When, on a clear night, one observes a star with the naked eye, why is it that the plane wave front which strikes him in the face, so to speak, gives the impression of light coming from a single direction? How is it possible to screen off all of this wave front except a small portion immediately bounding the direction of the star without appreciably affecting the light which reaches the eye from the star? This is the great problem of the rectilinear propagation of light which even one hundred years ago was still demanding explanation.

The simple and direct answer which Fresnel gave, and which was perfected by Stokes near the middle of the nineteenth century, is based upon the principles of diffraction and interference which we have just been studying. The time at our disposal will not permit of the reproduction of their argument here; but the interested student will find himself easily able to follow the discussion of this subject in some of the larger compendia of physics. The same remark might be made concerning diffraction in the case of light passing a straight edge, a circular aperture, or a narrow wire.

Problems

1. A linear aperture whose width is $\frac{1}{500}$ cm. is placed parallel to and 60 cm. in front of a screen. If this aperture be illuminated with a beam of parallel rays whose angle of incidence is zero, and whose wave length is

$\frac{1}{20000}$ cm., what will be the distance between the central image and the first dark band of the diffraction image on the screen? *Ans.* 1.5+ cm.

2. A glass grating is ruled with 4250 lines to the centimeter. When a plane wave of yellow light strikes this grating at perpendicular incidence, it is observed that the spectrum of the second order is deviated (diffracted) through an angle of 30°. What is the wave length of the yellow light? *Ans.* $\lambda = 0.00005880$ cm.

3. A source of light emits two colors. When examined through a grating, it is found that the spectrum of the 4th order for one color is seen in the same direction as that of the 5th order for the other color. What is the ratio of the wave lengths for these two colors?

4. A certain spectrum of the 2d order is produced by a grating which is ruled with 400 lines to the millimeter. Another spectrum of the same source but of the 4th order is produced by a grating which contains 300 lines to the inch. Compare the lengths of these two spectra.

5. Two point sources of light, each in the same phase, are situated at a distance of $\frac{1}{20}$ mm. apart. If the wave length of the light which they emit is 6500 A, how far apart will be the spectra of the second class on a screen one meter away from the sources and parallel to the line joining the sources?

6. How much will the phase of a ray of light be retarded by inserting in its path a cover glass one millimeter thick having a refractive index of 1.6? Assume the light to be that of the yellow sodium flame for which $\lambda = 5893$ *A*.

7. Prove that, if the opaque spaces are exactly equal to transparent spaces in a transmission grating, spectra of the fourth order will be missing.

CHAPTER XII

OPTICAL INSTRUMENTS

I. *The Photographic Camera*

485. A photographic camera, which is perhaps the simplest and best known of all optical instruments, is merely a box fitted at one end with a converging lens and at the other end with a device for holding a plate which is sensitive to light. Etymology of *camera?* The sides of this box are usually made of folded leather so that the distance of the plate *P* (Fig. 390) from the lens *L* may be varied at will.

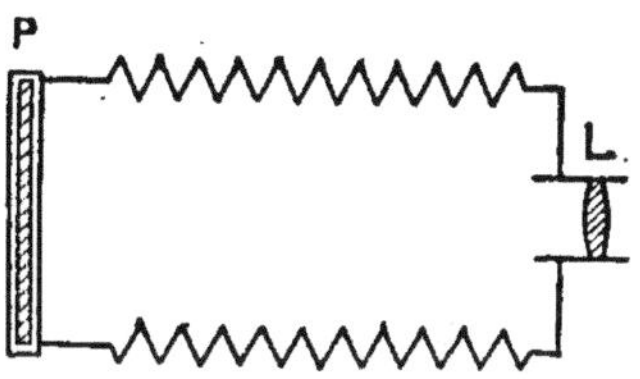

FIG. 390. — Section of photographic camera.

The rear end of the camera is furnished also with a ground glass which may be placed in the same position that the sensitive plate is to occupy later.

The *first* step in taking a photograph consists in placing this ground glass in such a position that the lens will produce upon it a sharp image of the object whose photograph is desired. This is called "focusing."

The *second* step is to cover the lens so that no light can enter the camera; then put the sensitive plate in the position formerly occupied by the ground glass and "expose" the plate by again uncovering the lens for the proper length of time.

The *third* step is to "develop" and "fix" the plate. The lights and shadows are now just reversed from what they are in the subject. The plate in its present condition is therefore called a "negative."

The *fourth* step, called "printing," again reverses the lights and shadows and gives the finished picture, which is called the "positive."

A good photographic lens is one: —

(*a*) Which has the same focal length for yellow as for violet light; yellow rays being those which are principally used in focusing on the ground glass, violet rays being those which most affect the sensitive plate. In other words, the lens must be corrected for chromatic aberration, § 464.

(*b*) Which has the same focal length for the edges as for the center of the lens. When this is so, the whole of the lens can be used, and the photograph taken with a short exposure. Such lenses are said to be "quick"; they are corrected for spherical aberration.

(*c*) Which produces an image distinct, not only at the center, but clear out to the edges of the sensitive plate for which it is intended. Such a lens is said to have a "flat field."

(*d*) Which gives the same magnifying power at the center and at the edge of the photographic plate. If the lens is not carefully corrected so as to give this result, buildings and geometrical figures will be distorted. A lens so corrected is said to be "rectilinear."

For special purposes, such as portrait or landscape photography, it is obviously unnecessary that the lens should possess all of these four qualities.

II. *The Human Eye*

486. The eye, from an optical point of view, is a camera in which the sensitive photographic plate is replaced by a sensitive membrane, called the **retina**. This retina is connected with a nerve called the optic **nerve**, which conveys the impression to the brain.

In the front part of the eye is placed a converging lens O· this is called the crystal**line lens**. This lens projects upon the retina an image of external objects. The distance between this lens and the retina remains practically constant. How, then, can the eye see objects which lie at different distances from the eye? For it is evident from Fig. 365 that the distance of the image from the lens ordinarily varies with the distance of the object from the lens. The answer to this question lies in the fact that the muscles of the eye are able to change the focal length of the crystalline lens at

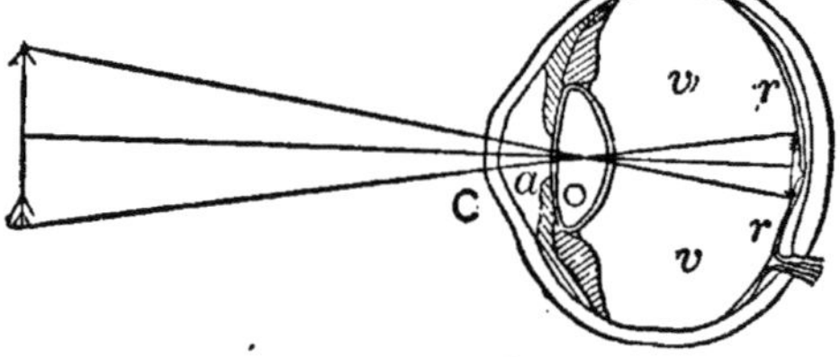

FIG. 391. — Longitudinal section of the human eye.

will. The power of this lens is thus varied automatically to suit the distance of the object. This process of changing the focal length of the eye to suit the distance of the object is called **accommodation.**

In front of the eye hangs a dark-colored diaphragm, in the center of which is an opening called the *pupil.* This aperture varies with the amount of light falling upon the eye. As the light grows brighter, the pupil contracts; as the light grows weaker, the pupil enlarges. Examine the pupil of your own eye in a mirror after remaining for a few minutes in a dark room; in a moment after coming into the light the pupil contracts again to its natural size.

Every student should examine carefully one of these little pasteboard models, called "skiascopic eyes"; nothing will make the structure of the eye so plain. The model is provided with a lens which enables one to easily examine its retina with the ophthalmoscope of Helmholtz, and to try the effect of various spectacle lenses.

III. *The Projection Lantern*

487. This instrument consists essentially of three parts, namely: —

(i) A brilliant source of light *A*, Fig. 392.

(ii) A pair of large short focus lenses *C*, of large aperture, which shall distribute this strong light uniformly over the object *S* which is to be projected, and

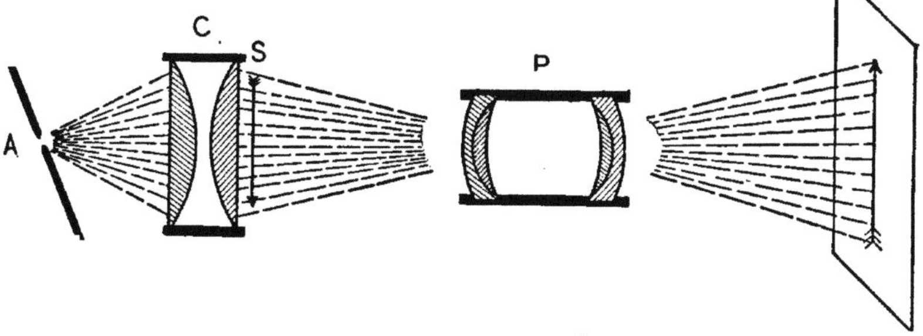

Fig. 392. — The projection lantern.

(iii) A fairly well corrected lens *P*, which throws upon the screen or wall *W* an enlarged image of the picture on the glass slide. This second lens is generally of the rectilinear type and is called the "projection" or "focusing" lens.

All that here needs special mention is the function of the condensing lens *C*, which has nothing to do with producing the picture on the screen, except to deflect towards the axis those rays which would otherwise pass through the edge of the slide and never enter the projection lens. If the condenser is removed, the effect is to diminish the area of the slide which is visible on the screen. Evidently the slide and the screen must be in conjugate foci of the projection lens. If the object to be projected is opaque, the light *A* and the lens *P* must of course both be on the same side of the object.

IV. *The Simple Microscope*

488. This instrument consists usually of a single converging lens of short focal length. The distinguishing feature of the simple microscope is that the lens is always placed at a distance from the object which is less than its focal length. Now, if the object *O* (Fig. 393) were placed exactly at the focus, the power

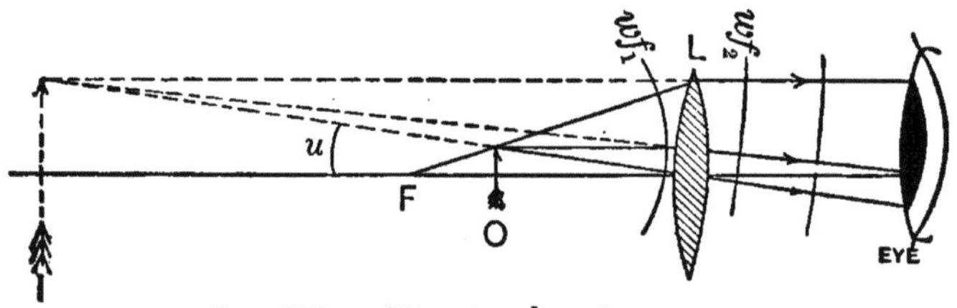

Fig. 393. — The simple microscope.

of the lens would be just sufficient to flatten out the curved wave front wf_1, so that it would emerge as a plane wave front. But if the object be placed between the focus and the lens, as in this case, wf_1 will be too convex for the lens to overcome its curvature entirely; the emergent wave front wf_2 will therefore still be convex on its advancing side. The image *I* will therefore lie on the same side of the lens as the object *O*. The image will, therefore (§ 441), be virtual and erect. Since the image and the object subtend the same angle θ at the center of the lens, it is evident that the image will be larger than the object in the following ratio: —

$$\frac{\text{Size of image}}{\text{Size of object}} = \frac{\text{Distance of image from lens}}{\text{Distance of object from lens}}.$$

But since the lens is always held very near the eye, the distance of the image is practically the distance of distinct vision, say 10 inches; and since the object is always held very near the principal focus of the lens, its distance from the lens is practically the focal length. Denote this focal length by f, and the distance of distinct vision by v; then the magnifying power of the simple microscope may be described in terms of algebra as follows: —

$$\frac{I}{O}=\frac{v}{f}, \quad \text{Equation of simple microscope.} \qquad \text{Eq. 194}$$

where O is a linear dimension of the object, and I the corresponding linear dimension of the image.

We might, of course, bring an object up close to the eye so that it would subtend an angle as large as that of the image seen through the microscope, but unfortunately the human eye cannot accommodate for these short distances. The necessity for the microscope arises then from our limited power of accommodation.

Distinguish carefully between the "power" of a lens (§ 460), which is the reciprocal of its focal length measuring its ability to change the curvature of a wave surface, and its "magnifying power," which is the ratio of image to object.

V. *The Compound Microscope*

489. The compound microscope is an instrument which is rendered necessary by the fact that the magnifying power of the simple microscope cannot be pushed very far. As indicated in Fig. 394, the essential features of this instrument are: —

(i) A very perfect, but highly complex, lens placed close to the object and giving a magnified real and inverted image of the object. From its proximity to the object, this lens is called the objective.

(ii) The other essential of the compound microscope is a simple microscope with which to examine the real image produced by the objective.

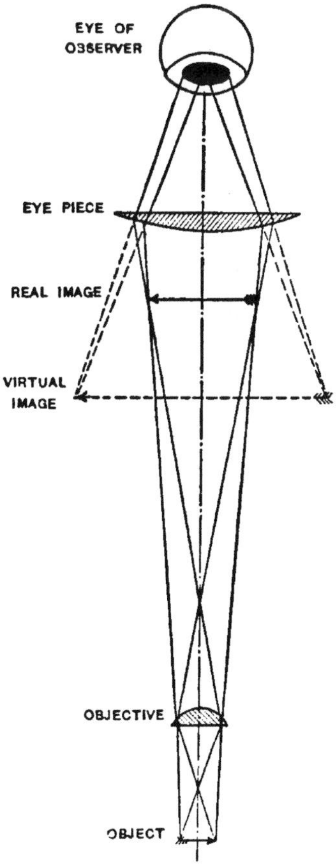

FIG. 394. — The compound microscope.

From its proximity to the eye, this lens is called the **eyepiece.**

If we denote by F the focal length of the objective and by T the length of the tube of the microscope, the magnification which the objective produces is practically T/F; for T and F are the respective distances of image and object from the objective. As we have seen above (§ 488), the magnification which the simple microscope — here the eyepiece — produces is the distance of distinct vision divided by its focal length v/f. Consequently the magnifying power of the compound microscope may be described as follows: —

$$\frac{I}{O} = \frac{Tv}{Ff}. \quad \text{Equation of compound microscope.} \qquad \text{Eq. 195}$$

The perfection of the modern compound microscope is due very largely to the late Professor Abbe of the University of Jena, working in conjunction with the optician Carl Zeiss. Its complete theory forms several of the most beautiful chapters in geometrical and physical optics.

VI. *The Astronomical Telescope*

490. The astronomical telescope is an instrument used for viewing very distant objects. The rays of light which fall upon the telescope are, therefore, generally parallel.

The purpose of the telescope is twofold: —

(1) To gather a large amount of light and condense it into a small bundle of rays, so that these can all enter the pupil of the eye and thus make any luminous *point* appear much brighter than it would be to the naked eye.

(2) To magnify the angle subtended by two luminous points, and thus make them appear farther apart than when seen by the naked eye.

These purposes are accomplished most simply by two converging lenses, L and E. The lens L is called the object lens, or simply the **objective**; the lens E is called the eye lens, or, more frequently, the **eyepiece.**

The wave front wf_1 sent out by a star is practically plane when it strikes the objective, as is indicated in Fig. 395. The effect of the objective is to turn this plane wave front into a spherical one wf_2. Let F denote the center of curvature of the wave front wf_2. This is called the principal focus of the telescope. The eye lens is so adjusted that its principal focus also lies at *F*. Therefore the rays which emerge from the eye lens are again parallel.

The effect, therefore, of a large telescope, such as that at the Lick Observatory or at the Yerkes Observatory, is to take a bundle of parallel rays forming a cylinder a yard or so in diameter and condense it into a small cylinder of parallel rays

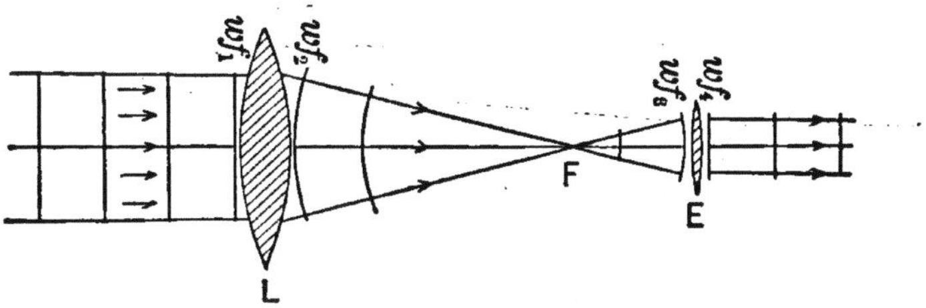

Fig. 395. — The astronomical telescope.

just large enough to enter the pupil of the eye. So that a star (which is practically a luminous point) appears immensely brighter when viewed through a telescope than it does to the naked eye.

491. If two stars close together, say a double star, be viewed through a telescope, *each* will send a beam of parallel rays through the instrument, as indicated in Fig. 396. But the two parallel beams which emerge from the eye lens make with each other a much larger angle than do the two parallel beams which enter the objective. The telescope thus magnifies the angular distance between two stars. It thus enables us to recognize certain stars as double when to the naked eye they appear single. For we can only distinguish two luminous points as separate when they subtend at the eye an angle which is greater than about two minutes of arc.

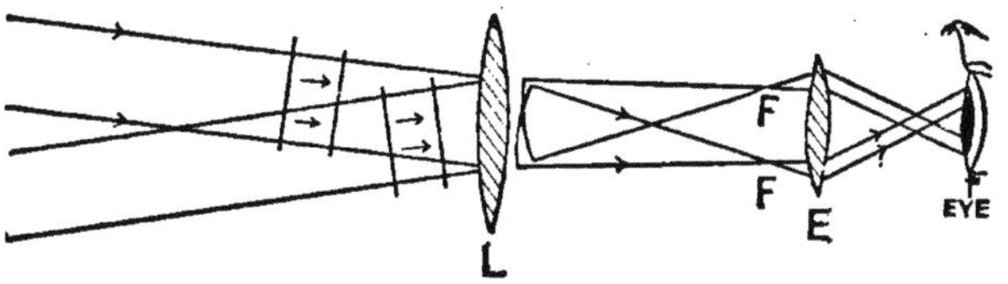

Fig. 396. — The telescope as used to magnify an angle.

VII. *The Opera Glass*

492. The earliest telescopes did not employ a converging eyepiece such as those represented in Figs. 395 and 396; but instead they used a double concave eye lens, the power of which was equal to and opposite in sign to that of the human eye.

The telescope with which Galileo discovered four satellites of Jupiter is shown in Fig. 397. On account of its compactness and because it gives an upright image, this same principle is employed in the modern opera glass.

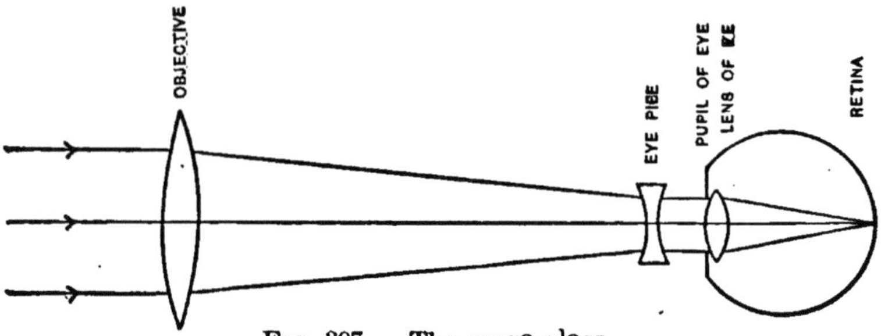

FIG. 397. — The opera glass.

If we assume what is nearly true, namely, that the eyepiece produces a curvature of the incident wave front, which is exactly equal and opposite that produced by the human eye, then by placing the objective of the opera glass at such a distance that its principal focus lies on the retina, we shall see the object distinctly; and since the image is much larger than that produced by the naked eye, we shall see the object magnified. In short, the eyepiece and the eye act together, as if they constituted a piece of plane glass. The ordinary opera glass here described has been immensely improved in recent years by the introduction of a pair of prisms between the eyepiece and the objective. Just how this serves to improve the definition, magnification, and uniformity of field forms a problem well worthy the attention of the advanced student.

VIII. *The Interferometer*

493. The interferometer is an instrument devised by Michelson (1881) for the study of aberration; it is now employed for a great variety of purposes, but its fundamental use is perhaps the accurate measurement of small distances. As the name would imply, the principle involved is that of the interference of light waves. But, as we shall see presently, the interference is not between two single point sources or two single linear sources, as in § 428, but between pairs of corresponding points which, taken together, constitute two extended wave surfaces.

494. Just what is meant by this will be clearer after the student has repeated the following experiment: Take two small strips of $\frac{1}{4}$-inch plate glass and clamp them together at one end, as shown in section in Fig. 398. At the other end *DB* place a single thickness of tissue paper between the plates. Now examine these plates in the light reflected from some monochromatic source, such as a sodium flame or mercury arc. The region between the two plates will be seen filled with a large number of alternately bright and dark lines. The explanation of these "interference bands," as they are called, is perhaps most simply given in terms of Fig. 399, which is taken from an article by Dr. Henry G. Gale in the *American Machinist*, July 11, 1901.

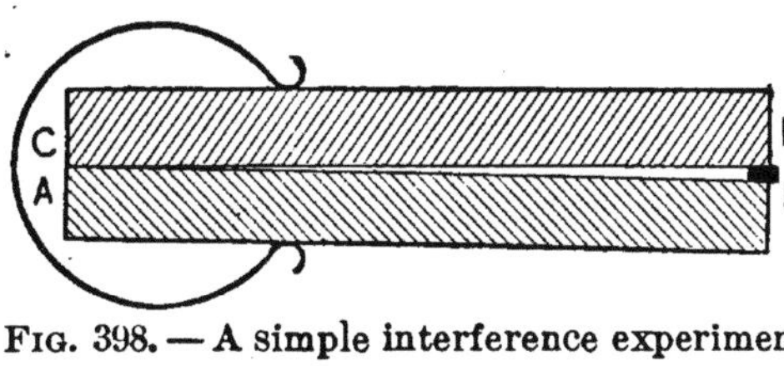

FIG. 398. — A simple interference experiment.

Imagine *AB* and *CD* to be the two glass surfaces inclosing the wedge-shaped air film in Fig. 398. The fine parallel lines in Fig. 399 are drawn $\frac{1}{2}$ wave length apart so that one can estimate the thickness of the film *at any point* to within a small fraction of a wave length, by merely looking at the diagram. Next imagine a ray of light coming from the right and incident upon the surface *AB* (the inner surface of the glass), at the point *e*; part of the light will be reflected from *AB* and part of it will be refracted into the air film, and there reflected from the surface *CD*. These two rays are now in a condition to interfere; for they each have their origin in the same luminous particle.

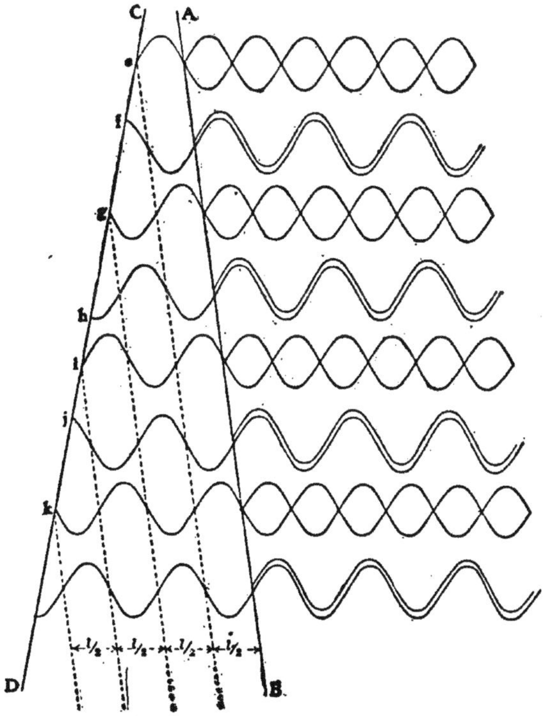

FIG. 399. — Illustrating interference between two large wave fronts.

Let us choose the point *e* such that the thickness of the film at that point is just one half wave length, and then ask what the retardation of the second train of waves is when compared with the first. It follows that the total retardation due to the air film is λ, where λ is the wave length of the light employed. The same thing happens, of course, at points such as *g*, *i*, *k*, etc., where the thickness of the air film is 2, 3, 4, etc., any whole number of half wave lengths. So far, therefore, as the difference in path between the two rays is concerned, they would be in the same phase, and hence reënforce each other, whenever the air was any whole number of half wave lengths in thickness. But consider what happens at *f*, where the air film has a thickness of $\frac{3}{4}\lambda$. Here the retardation due to one round trip across the film is $\frac{3}{2}\lambda$; in the same manner it will be seen that at *h* and *j* the difference in phase is $\frac{5}{2}\lambda$ and $\frac{7}{2}\lambda$, respectively; in other words, the two trains here differ by an odd number of wave lengths, and ought therefore to interfere destructively, giving rise to a dark band. We are thus led to expect exactly such a series of alternate bright and dark bands as those observed with the two pieces of plate glass in Fig. 398.

495. But oddly enough, experiment shows that the black band occurs at exactly the position where the above view would lead us to expect a bright band. The one consideration which is here needed to complete the theory, is very clearly set forth by Professor Michelson in terms of Kelvin's wave model, as follows: "This discrepancy is due to the assumption that both reflections took place under like conditions, and that the phase of the two trains of waves would be equally affected by the act of reflection. This assumption is wrong, for the first reflection takes place from the *inner* surface of the first glass, while the second occurs at the *outer* surface of the second glass. The first reflection is from a rarer medium — the air; while the second is from a denser medium — the glass. A simple experiment with the Kelvin wave apparatus (§ 177) will illustrate the difference between the two kinds of reflection. The upper end of this apparatus is fixed while the lower end is free; the fixed end, therefore, represents the surface of a denser medium, the free end that of a rarer medium. If now a wave be started at the lower end, by twisting the lowest element to the *right*, the twist travels upward till it reaches the ceiling, whence it returns with a twist to the *left*, *i.e.* in the *opposite* phase. When, however, this *left* twist reaches the lowest element, it is reflected, and returns as a twist to the *left* — so that the reflection is in the *same* phase.

"There is thus a difference of phase of one half a period between the two reflections, and when this is taken into account, experiment and theory fully agree." — *Light Waves and Their Uses*, p. 16.

Accordingly we see that destructive interference occurs when the thickness of the film is any even number of quarter wave lengths; reënforcement occurs when the thickness of the film is any odd number of quarter wave lengths.

496. Now, instead of thinking of these two surfaces *AB* and *CD* (Fig. 399) as made of glass, let us imagine them produced as follows: A Bunsen flame containing sodium is placed in the principal focus of a converging lens *L*. A certain luminous particle *S* in this flame emits a spherical wave. The lens *L* transforms this spherical wave into a plane wave, as indicated in Fig. 400.

This plane wave strikes a fixed plane parallel plate of glass *R* at an angle of 45°. The rear surface of this plate is thinly silvered. Part of the light, indicated by solid lines, is therefore transmitted; part of it, indicated by dotted lines, is reflected. The *transmitted* wave front proceeds next through a plane parallel plate *C* and is reflected from a plane silvered mirror *F*. It now retraces its path to the silvered surface of the mirror *R*, where it is reflected to the eye as shown in Fig. 400. The reflected wave front proceeds to a movable mirror *M* which is carried on the nut of a screw, whose axis lies in the

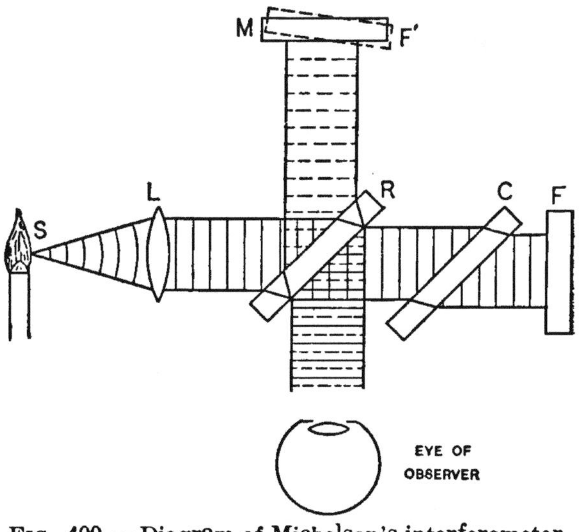

FIG. 400. — Diagram of Michelson's interferometer.

direction joining the mirrors *R* and *M*. From the mirror *M* the wave front is reflected normally and is transmitted by *R* so as to enter the eye from the same direction as the other wave front. In brief the incident wave front from the source *S* is split up by the mirror *R* into two wave fronts which enter the eye in a condition for interference. The wave front from the fixed mirror *F* behaves exactly as if it came from its image in the mirror *R*, *i.e.* from a position indicated by the dotted line *F′*. Any wave front leaving *M* and the corresponding wave front from *F* may therefore be regarded as a pair of wave fronts replacing the two which were reflected from *AB* and *CD* in Fig. 399; only these two surfaces are not fixed as were *AB* and *CD*. Here the mirror *M* can be moved with the screw so as to make these two surfaces coincide, or so as to put either one in advance of the other. Not only so, but one observes that as the mirror *M* advances the fringes (interference bands) move across the field at the rate of one fringe for each half wave length of advance in the position of the mirror.

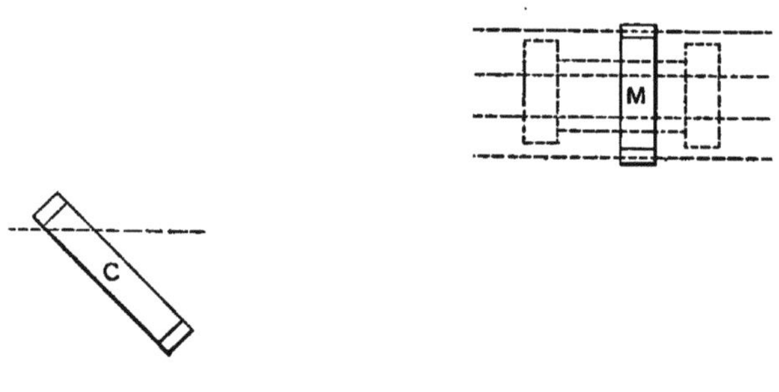

Fig. 401. Michelson's interferometer.

In this manner Michelson has succeeded in using the wave length of light as a new unit of length for the accurate measurement of very minute distances. In 1892 he, in conjunction with Bènoit, measured the standard meter at Paris in terms of the wave length of the red cadmium ray and found, with an error probably not greater than one part in ten million, that the meter is equal to 1553163.5 such wave lengths. The moral effect of this determination was to render the red cadmium wave length our standard of length.

$$\text{Wave length of red cadmium ray} = \frac{1}{1553163.5}\text{ meter}$$
$$= 6438.4722 \times 10^{-10}\text{ meter.}$$

IX. *The Spectroscope*

497. If one observes a Bunsen flame through an ordinary glass prism held close to the eye, he sees a succession of colored images of the flame. This series of colored images is called the spectrum of the flame.

An instrument such as the prism which will separate the various radiations emitted by any source of light is called a spectroscope. Accordingly we may define spectroscopy as that branch of physics which has for its object **the determination and description of the various radiations which bodies emit, reflect, and absorb.**

Sources of Radiation

498. In practice we are limited to two artificial sources upon which we can *experiment* — flames and electric currents. In addition to these there are a number of natural sources whose occurrence is so far beyond our control that we can only *observe* them. The following classification may be used to summarize these facts: —

- SOURCES
  - Natural
    - Lightning
    - Stars, including the Sun
    - Comets
    - Auroras
    - Meteors
    - Glowworms
  - Artificial
    - Flames
      - Candle
      - Alcohol
      - Bunsen
      - Oxyhydrogen
    - Electric Currents
      - Arc
      - Spark
      - Brush discharge

The **Bunsen flame**, devised by the German chemist of that name in 1856, has three marked advantages as a spectroscopic source, namely: —

(i) It is simple in construction and operation.

(ii) Its outer mantle is almost totally devoid of any characteristic spectrum of its own.

(iii) It possesses a high temperature, not less probably than 1788° C.

The **electric arc**, with a temperature lying probably between 3500° and 3900° C., is exceedingly efficient as a source because it easily volatilizes any substance placed in it.

The **electric** spark, obtained by the use of the induction coil, is the source of radiation which is most generally convenient and is most widely employed. In general, the electric spark produces two spectra, namely, the spectrum of the electrodes and the spectrum of the air or other gas between the two electrodes. But it was discovered by the German mathematician Plücker, in 1858, that the spectrum of the electrodes may be avoided by inclosing them in a partial vacuum. Accordingly, the universal method of studying the spectra of gases is now to seal platinum electrodes into a glass tube and then fill the tube under low pressure with the gas to be studied. The converse problem of getting the spectrum of the electrodes without that of the air between them has recently been solved by Schuster and Hemsalech, who have proved that the air ceases to become luminous when a sufficient amount of inductance has been placed in the circuit through which the spark passes.

Separation of Radiations

499. Let us suppose now that we have before us one or more of the above-mentioned sources which we wish to examine. The next step will be to separate the rays of various wave lengths so that we may examine them individually. For the student may here assume, what is the experimental fact, that each different chemical element introduced into the flame or arc gives a different spectrum, and one which is peculiar to itself.

The Prism Spectroscope

500. If the source of light be small or linear, and if it is necessary to make only a hasty visual examination, the most convenient plan is to view the source through a prism placed immediately in front of the eye, as indicated in Fig. 402. The chief difficulty in the naked prism is that the source, however small, is still so large that the successive colored images formed by the prism will overlap each other, leaving the separation incomplete. To avoid this difficulty a small astronomical telescope — called a **collimator** — is placed between the prism and

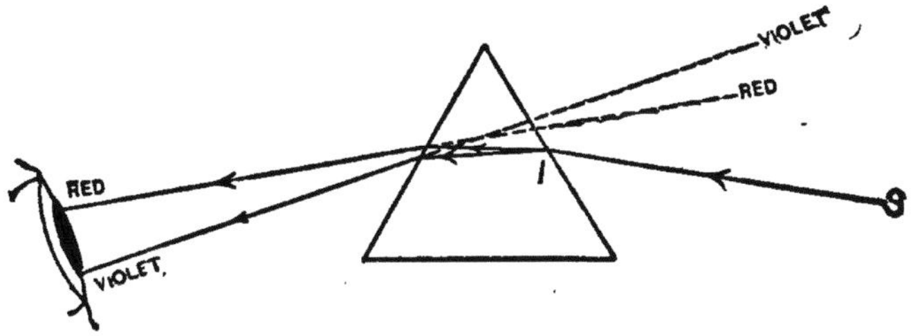

FIG. 402. — A simple form of spectroscope.

the source, as shown in Fig. 403. In the principal focus of its objective is placed a narrow straight **slit** with movable metal jaws. The source is now placed immediately back of this slit, or a real image of the source is focused upon the slit by means of a lens called the **image lens**. When the slit is properly placed and illuminated, each point of it will emit through the lens of the collimator a beam of nearly parallel light, whose cross

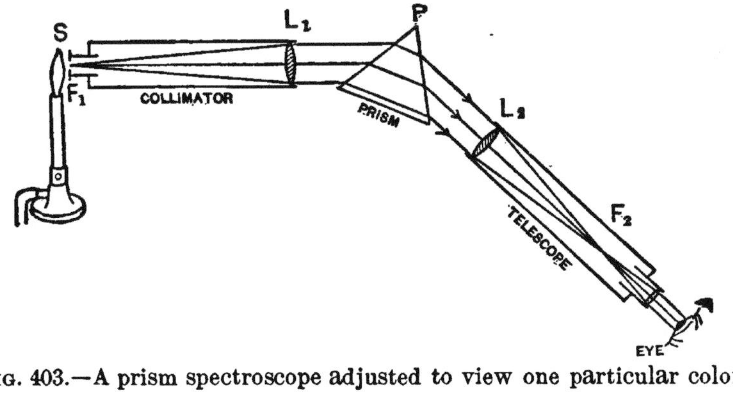

FIG. 403.—A prism spectroscope adjusted to view one particular color.

section is the **effective aperture** of the objective. If now this emergent beam be examined through a prism by the naked eye as before, a series of colored images of the slit will be seen, each appearing to be at an infinite distance from the observer; and if the slit is narrow, these images will be sharp and fine and easily separated. But they will not in general appear very bright, because the human eye is too small to admit more than a fraction of the emergent beam. Accordingly, these colored images, each at an infinite distance, are viewed through another small telescope called the **view telescope.** This serves to condense the entire beam which emerges from the prism and reduce it to another parallel beam just large enough to fill the aperture of the average human eye.

The Spectrograph

501. When it is desired to photograph a spectrum, the view telescope is removed and is replaced by a camera provided with a photographic objective. Such an instrument is known as a **spectrograph,** a convenient form of which is shown in Fig. 404.

FIG. 404. — Hilger's form of spectrograph.

The Spectrometer

502. When the typical spectroscope represented in Fig. 403 is provided with a graduated circle and so arranged that the view telescope and the prism can rotate about the axis of the circle, the instrument is known as a **spectrometer,** and can be employed for the measurement of prism angles (§ 468) and wave lengths (§ 481) as well as for the examination of spectra.

In the prism spectroscope it is nearly always necessary for the sake of good definition to place the prism in such a position that it will produce a minimum deviation for the pencil of rays under observation. For it may be shown by geometrical optics that, if the prism be placed in any other position, the image of a point source will not be a point, but a line.

The Grating Spectroscope

503. So far as auxiliary parts are concerned, this piece of apparatus is very like the prism spectroscope; but as regards its essential feature, the dispersing piece, the two instruments are radically different. For the dispersion which is produced by a prism rests ultimately upon the fact that waves of longer length travel through glass with a higher speed than those of shorter length; while the dispersing effect of a grating is due to the fact that when waves of different length are forced to pass through a narrow opening, they are all spread out like a fan, — diffracted, — the longer waves being more spread out, however, than the shorter ones.

If a gas flame, turned low, or the filament of an incandescent lamp be viewed through a single narrow slit in a visiting card, a number of spectra are seen on each side of the bright central image. These have been called by Fraunhofer "spectra of the first class." They are not used in spectroscopy. If, however, the same source be viewed through two or more fine parallel equidistant slits placed close together, say less than $\frac{1}{64}$ inch apart, another series of spectra will be observed. These are the combined effect of light passing through several openings; they are known as "spectra of the second class" and are practically the only ones used in the grating spectroscope.

These narrow equidistant parallel slits were first prepared by the German optician Fraunhofer, at Munich, in 1821. He made them by wrapping fine wire over the threads of two parallel screws, then soldering the wire to the screws, and afterwards cutting away one entire side of the little cage thus formed. Such a device he called a **grating**. The modern grating is made by ruling straight lines with a diamond point on a flat plate of highly polished speculum metal. Here the light is reflected from the unruled surface between two furrows made by the diamond; and diffraction occurs exactly as if the light had passed through a narrow aperture of the same width. Such a device is called a **reflection grating**, and when placed where the prism stands in an ordinary prism spectroscope, we have what is known as a **grating** spectroscope. The student should now review §§ 472 to 481, dealing with the passage of light through narrow apertures.

When a beam of parallel light emerges from the collimator and falls upon a grating placed with its rulings parallel to the slit of the collimator, part of the light is reflected according to the laws of ordinary reflection, and forms in the focal plane of the view telescope what is known as the **central image.** The remainder of the light, however, is spread out on either side of this central image, the blue light being deviated least and the red light (longer waves) being deviated most. Unlike the prism, the grating yields several spectra, distributed on each side of the central image.

If monochromatic light be viewed with such an instrument, a series of bright lines — sharp maxima — each an image of the slit, will be seen on each side of the central image. The image nearest the central, on either side, is said to be the "spectrum of the first order," the next is called the "second order spectrum," etc.

The properties of these images and of diffraction spectra in general are most briefly described by the following three equations: —

$$(a + d)(\sin i + \sin \theta) = n\lambda. \qquad \text{Eq. 190}$$

This equation which has been already derived in § 480, tells us in what direction θ to look for the spectrum of the nth order when the incident light makes an angle of incidence, i and has a wave-length λ, the distance between two consecutive rulings of the gratings being $a + d$.

$$R = \frac{\lambda}{\lambda - \lambda_1} = nN. \qquad \text{Eq. 193}$$

This equation, which has been derived in § 483, tells us the smallest difference of wave length, $\lambda - \lambda_1$, which two lines can have and yet be recognized as double, when N is the total number of lines on the grating employed and n is the order of the spectrum.

The third equation, which tells one how bright any particular spectrum will be, is here given, but its derivation must be left for the advanced student.

If I is the intensity, in any given direction, of the spectrum of a monochromatic source of wave length λ, produced by a grating which has N lines ruled upon it at the uniform distance of $a + d$ cm. apart, then

$$I = I' \frac{\sin^2 \dfrac{N\mu(a + d)}{2}}{\sin^2 \dfrac{\mu(a + d)}{2}}, \qquad \text{Eq. 196}$$

where I' is the intensity which a single aperture would give in the same direction. The "given direction" is here defined by μ, whose value is

$$\mu = \frac{2\pi}{\lambda}(\sin i + \sin \theta), \qquad \text{Eq. 197}$$

where the algebraic symbols have the same meaning as in Eq. 190.

All the best gratings of the world up to the present time (1907) have been ruled on Rowland's dividing engine at Johns Hopkins University. Most of these gratings carry from five to twenty thousand lines per inch. Within the current year, however, Michelson at the University of Chicago has produced gratings which appear to be superior in resolving power to the best of the Rowland gratings.

504. In 1883 Rowland invented a **concave grating spectroscope** the distinguishing feature of which is that it requires neither collimator nor view telescope, consisting as it does of a grating ruled upon a concave spherical mirror of speculum metal. If one has regard both to convenience and resolving power this instrument, for general spectroscopic purposes, is probably unexcelled.

The Echelon Spectroscope

505. Measured by resolving power the echelon spectroscope devised by Michelson in 1898 is a still more effective instrument. This is essentially a grating with only a few rulings, in which, however, the form of the ruling or groove is given a perfectly definite form. The result is obtained by using a pile of from ten to thirty plane parallel glass plates of equal thickness, the edge of each plate being slightly displaced over that of its neighbor. The high resolving power is here obtained by the use of spectra of high orders, even as high as twenty or thirty thousand.

PRINCIPLES OF SPECTROSCOPY

506. The science of spectroscopy is a large and independent subject which must be left for later study. But it may be here stated that the entire structure is built upon a foundation consisting of something like the following principles:

(i) *Spectra of Gases.* — The emission spectrum of a gas is practically always a spectrum of bright lines; while the emission spectra of solids and liquids are practically always continuous.

(ii) *Kirchhoff's Laws.* — (1) The ratio of the emission to the absorption of a body depends upon the temperature only; and (2) this ratio is numerically equal to the emission of an absolutely black body at the same temperature.

(iii) *Doppler's Principle.* — The effect of motion in the line of sight is to apparently shorten the wave length when source and observer approach each other; and to lengthen wave length when source and observer recede from eath other. See § 194. It is by use of this principle that the rate of approach of the "fixed" stars is measured.

(iv) *Principle of Humphreys and Mohler.* — Effect of increased pressure about the source is, in general, to slightly increase the wave length.

(v) *The Zeeman Effect.* — When a source of light is placed in a strong magnetic field, each individual line in its spectrum will, in general, become a triplet, or a still more complex line.

(vi) *Temperature Effects.* — (1) The maximum radiation of a solid body varies as the fifth power of its absolute temperature. (2) The wave length of the maximum radiation from a solid body varies inversely as the absolute temperature of that body. See § 261.

(vii) *Law of Spectral Series.* — It has been shown by Kayser and Runge and others that, in the case of many chemical elements, nearly all their spectral lines are arranged in series, and that their respective wave lengths, λ, may be computed from the following formula: —

$$\frac{1}{\lambda} = A + Bm^{-2} + Cm^{-4}, \qquad \text{Eq. 198}$$

in which A, B, and C are constants for each series and m denotes the natural numbers beginning with 3. **Each chemical element has its own different characteristic values for these three constants.** This fact is the foundation of spectrum analysis.

507. In conclusion the student is warned not to think of the preceding outline of optics as anything but a very incomplete sketch. For instance, the entire subject of color vision and color mixture has been omitted; likewise the subject of photometry, important in electrical engineering. The pressure which a ray of light exerts, measured with admirable precision by E. F. Nichols and G. F. Hull, has not even been mentioned. The phenomena of polarized light and astrophysics have scarcely been touched upon. These and many other interesting topics are discussed in the following books: —

References on Light

S. P. THOMPSON. — *Light, Visible and Invisible* (Macmillan Co.). Price $1.50. A course of six lectures addressed to beginners. The language is simple and clear; the important facts are emphasized. The book is one of great value to students and to teachers.

LOMMEL. — *The Nature of Light* (Int. Sci. Series). Price $2.00. Written by an eminent German scholar. Elementary, clear, reliable.

NEWTON. — *Opticks.* London, 1704. Price, approx., $5.00.

TAIT. — *Encyclopædia Britannica*, Art. "Light."

RAYLEIGH. — *Encyclopædia Britannica*, Art. "Optics."

CARPENTER. — *Encyclopædia Britannica*, Art. "Microscope."

EDSER. — *Light for Students* (Macmillan Co.).

MICHELSON. — *Light Waves and Their Uses* (Univ. Chicago Press, 1902).

HASTINGS. — *Light, a Consideration of the More Familiar Phenomena of Optics* (Scribners, 1901).

WOOD. — *Physical Optics* (Macmillan).

SCHUSTER. — *The Theory of Optics* (Arnold; London, 1904).

SCHEINER. — *Astronomical Spectroscopy.* Translated by Frost (Ginn).

LUMMER. — *Photographic Optics.* — Translated by Thompson (Macmillan).

DERR. — *Photography for Students of Physics and Chemistry* (Macmillan, 1906).

KAYSER. — *Handbuch der Spectroscopie* (Leipzig, 1900). Three volumes already (1907) published. The standard work on spectroscopy.

LIST OF TABLES

NUMBERS REFER TO PAGES

INDEX

NUMBERS REFER TO PAGES

Proper names are in italics

This book has been produced in line with the EU GPSR guidelines about the safety of products.

The General Product Safety Regulation is the European Union's updated framework for ensuring that all consumer products, including books, are safe for consumers.

This book has been printed by Libri Plureos GmbH. The printer has issued safety certificates for the materials - like ink, paper and glue - being used.

The product identifier is: 9786256004580

The author is responsible for the content of the book, is the publisher of the works and bears full responsibility for it.

The book has been published via Bookmundo. Bookmundo enables any author to share their stories with the rest of the world via printed books and ebooks and a broad distribution network.

Bookmundo will act as an intermediary in regard to questions about safety and will address them to the printer / author. Should there be any question in regard to the safety of the product, please contact us.

Bookmundo
Delftsestraat 33
3013AE Rotterdam
The Netherlands
info@bookmundo.com